CHINESE GLOBAL EXPLORATION
IN THE PRE-COLUMBIAN ERA

Evidence from an Ancient World Map

CHINESE GLOBAL EXPLORATION IN THE PRE-COLUMBIAN ERA

Evidence from an Ancient World Map

Sheng-Wei Wang
China-US Friendship Exchange, USA

World Scientific

NEW JERSEY • LONDON • SINGAPORE • BEIJING • SHANGHAI • HONG KONG • TAIPEI • CHENNAI • TOKYO

Published by

World Scientific Publishing Co. Pte. Ltd.
5 Toh Tuck Link, Singapore 596224
USA office: 27 Warren Street, Suite 401-402, Hackensack, NJ 07601
UK office: 57 Shelton Street, Covent Garden, London WC2H 9HE

Library of Congress Cataloging-in-Publication Data
Names: Wang, Sheng-Wei, author.
Title: Chinese global exploration in the pre-Columbian era : evidence from an ancient world map / Sheng-Wei Wang, China-US Friendship Exchange, USA.
Description: New Jersey : World Scientific, [2024] | Includes bibliographical references and index.
Identifiers: LCCN 2022061339 | ISBN 9789811271083 (hardcover) | ISBN 9789811271090 (ebook for institutions) | ISBN 9789811271106 (ebook for individuals)
Subjects: LCSH: Discoveries in geography--Chinese. | Cartography--China--History. | World maps--Early works to 1800. | Explorers--China.
Classification: LCC G322 .W36 2024 | DDC 910.951--dc23/eng/20230505
LC record available at https://lccn.loc.gov/2022061339

British Library Cataloguing-in-Publication Data
A catalogue record for this book is available from the British Library.

For any available supplementary material, please visit
https://www.worldscientific.com/worldscibooks/10.1142/13273#t=suppl

Desk Editors: Logeshwaran Arumugam/Jiang Yulin

Typeset by Stallion Press
Email: enquiries@stallionpress.com

About the Author

Sheng-Wei Wang is an independent scholar and writer, passionate about uncovering our shared history of maritime globalisation. She was born in Taiwan, lived in the U.S. and Germany, and currently resides in Hong Kong. She is Eastern and Western educated with a B.Sc. in Chemistry from the National Tsing Hua University in Taiwan and a Ph.D. in Theoretical Chemical Physics from the University of Southern California. Besides joining the research group at Munich University led by Professor Gerhard Ertl, the 2007 Nobel Laureate, she had also worked at top-ranking universities in the U.S. and the Lawrence Berkeley National Laboratory. She started research on Pre-Columbian Chinese world exploration in 2013, and this is the second English book published on related topics. In 2022, she founded WorldDiscovery.net to share her work and to promote an inclusive and open dialogue about Chinese maritime explorations since Antiquity. She was recently interviewed on her research findings by Azam Khan on the Talk The Walk Show of the Hong Kong International Business Channel (HKIBC). She also enjoys attending speaking engagements with academics and the general community to engage in the wider discussion on what it means to be a member of a globalised world and how it is shaped by our shared history as human beings; it is this excitement that she hopes to recreate on WorldDiscovery.net and in her books.

Contents

Introduction

In school, we all learn that Christopher Columbus discovered America and that Europeans ushered in the Age of Discovery. However, even the most rigorous scholars admit the existence of a very puzzling phenomenon on the ancient Western maps: many supposedly "unknown" areas were already mapped out before the Age of Discovery. To resolve this enigma requires considering China's role — then the world's strongest maritime power. In my book, I address two questions: Did the Ming Treasure Fleets,[1] led by Admiral Zheng He (郑和; 1371–1435),[2] explore a large fraction of the world before Columbus? And how early did the Chinese start to explore the world?

Since there can be no more than one truth, even if the process of seeking the truth may be tortuous, I have decided to thoroughly analyse an important ancient world map — Kunyu Wanguo Quantu 《坤輿萬國全圖》[3] or Complete Geographical Map of All the Kingdoms of the World, written with Chinese characters — abbreviated here as KWQ. The map was published in 1602 by Italian Jesuit priest Matteo Ricci.[4] For over 400 years, this map has been regarded as strictly of European origin and based on maps which Ricci brought with him to China in 1582. This may be explained as partly due to the *Hǎijìn* (海禁) or sea ban imposed by the Ming emperors: it abruptly stopped future voyages of the Treasure Fleets and destroyed the nautical maps and precious documents gathered by the Chinese mariners, resulting in the later Chinese losing geographical knowledge of the world. It may also be partly due to a coverup by Western authorities, navigators, and map makers who copied the Chinese maps or hid the true sources of their maps which guided the European explorers to "rediscover" many supposedly "unknown" areas.

However, Chapter 1 of this book, titled "Chinese Explored Cape Breton Island Long before the Europeans", compares KWQ with four major sixteenth century European maps and the latest archaeological findings, and shows the startling evidence that the eastern region of Canada on KWQ is of Chinese origin. The evidence also indicates different eras when Chinese settled on Cape Breton Island before the Europeans arrived, and Zheng He's mariners built or rebuilt an ancient canal across the village of St. Peter's, hence the canal separated Cape Breton Island into two islands as shown on KWQ. These findings have profound implications, including that the Chinese explored the Americas in the Pre-Columbian era.

Then what struck me was the expansive and contiguous area at the bottom portion of KWQ depicted as Mo Wa La Ni Jia (墨瓦蠟泥加; it is the transliteration of *Magallanica* or *Magellanica*) where on a modern map you would see, instead, mostly Oceania, Land of Fire and Antarctica. This huge area often appears on European maps between the fifteenth and eighteenth centuries as *Terra Australis Incognita*, a hypothetical continent first posited in Antiquity. Chapter 2, titled "Chinese Explored Australia, New Zealand, Land of Fire and Antarctica long before the Europeans", studies 墨瓦蠟泥加 on KWQ and the latest archaeological findings, and reveals that contrary to popular belief, Matteo Ricci and his Chinese collaborators did not base 墨瓦蠟泥加 or *Terra Australis Incognita* solely on European concepts. Instead, the chapter shows evidence that the ancient Chinese explored Australia, New Zealand, Land of Fire and Antarctica no later than the 1420s, long before the Europeans.

Even more surprising is the European part of KWQ. When we think of Sino-European relations, we think of the Silk Road or today's heated politics in Europe. But as today's world has shown us, reality is much more complicated. Chapter 3, titled "A Chinese-based World Map Depicts Europe between 1157 and 1166", presents the finding concerning the European part of KWQ and this map shows the complicated European political landscape, which is more than 400 years before Ricci's time, and overlaps with the Chinese Southern Song Dynasty (1127–1279). Moreover, British surveyor T.C. Bell had discovered that ancient Chinese had operated on the British Isles since Antiquity.

What about Africa? Africa is one of those places that people in the Global North either do not think about or only have a superficial knowledge about. However, like any other human-inhabited continent on Earth, Africa is made up of many countries and has a very long history. Chapter 4, titled "A Chinese-based World Map Depicts Africa in 1433", takes us on a journey to the early days of Sino-African exchanges. Despite the fact that KWQ has long been thought to be a copy of European maps, here also, the chapter argues that the geographical data on Africa depicted on KWQ were derived from Chinese sources in the year 1433, during the last of the seven Ming voyages.

Finally, we come to an essential question: Did Chinese explore the Americas before Columbus? This final chapter, Chapter 5, titled "Did Zheng He's Mariners Explore the Americas before Columbus? An Analysis Offers Essential New Insights", finds that, contrary to popular belief, the Americas as shown on KWQ are not the result of European information but from the Chinese sourced from Pre-Columbian times. Moreover, KWQ not only shows the political landscape of the Americas in the 1420s, but also provides evidence for much earlier Chinese migrations to the New World.

In summary, the book shows that (1) KWQ is a map of Chinese origin; (2) contrary to general perception, Zheng He's mariners reached far beyond the east coast of Africa; (3) Europe on KWQ reveals its political landscape over four centuries before Ricci's time; (4) the maps of Australia, New Zealand, Land of Fire and Antarctica and the Americas contain Chinese geographical data updated by Zheng He's mariners during their sixth voyage in the 1420s; (5) the date of the geographical data in the African part of the map is early 1433, which was the time of the Ming mariners' seventh (and last) voyage; and (6) Sino-European relationship goes back long before the Common Era, based on the latest archaeological findings.

My systematic and exhaustive research can serve as a reference basis for future China studies by students and experts in the field and offers new insights into the history of world maritime exploration. It also serves as an extensive database of many hundreds of carefully curated geographical and historical facts pertaining to large parts of the world in the Pre-Columbian era.

Endnotes

[1]Sheng-Wei Wang. *The Last Journey of the San Bao Eunuch, Admiral Zheng He*. Hong Kong, China: Proverse Hong Kong, 2019.
[2]*Ibid.*
[3]*Library of Congress*. "Kun yu wan guo quan tu". www.loc.gov, Library of Congress, 2022, loc.gov/item/2010585650.
[4]Ronnie Po-Chia Hsia. *Matteo Ricci and the Catholic Mission to China, 1583–1610: A Short History with Documents (Passages: Key Moments in History)*. Indianapolis, USA: Hackett Publishing Company, Inc., 2016.

Chapter 1

Chinese Explored Cape Breton Island Long Before the Europeans

Abstract

This chapter compares Cape Breton Island and Newfoundland on Matteo Ricci's 1602 world map — Kunyu Wanguo Quantu《坤舆万国全图》or Complete Geographical Map of All the Kingdoms of the World — with their counterparts on four other contemporary European maps. The four maps are: (1) the 1562 Map of America by Diego Guttierez; (2) the 1569 World Map by Gerardus Mercator; (3) the 1570 World Map by Abraham Ortelius; and (4) the 1594 World Map by Petrus Plancius. The comparisons give clear indications that Kunyu Wanguo Quantu is not a direct or adapted copy of these European maps. This chapter also makes an in-depth analysis of recent archaeological findings of ruins on Cape Breton Island, Nova Scotia, Canada. The analysis focuses on stonework on the island and the existence of an ancient canal across the village of St. Peter's. The canal separated Cape Breton Island into two islands as depicted on Kunyu Wanguo Quantu but not on these four other European maps. The results derived from the analysis offer compelling evidence showing that: (1) Chinese lived on Cape Breton Island before the arrival of the Vikings around 1000 A.D.; and (2) the mariners of Admiral Zheng He (郑和) explored the island and built (or rebuilt) the ancient canal during their sixth voyage to the "Western Ocean" in the 1420s, which was long before the arrival of the Europeans in the New World. Such findings render the Chinese origin of this portion of the Kunyu Wanguo Quantu undisputable.

Keywords: Cape Breton Island, Kunyu Wanguo Quantu, Matteo Ricci, Vikings, Zheng He

1. Introduction

For over four hundred years, Matteo Ricci's 1602 map — Kunyu Wanguo Quantu[1] (abbreviated as KWQ in this chapter) — written with Chinese characters and having latitudinal and longitudinal lines, has been widely regarded as a map drawn by Ricci and his Chinese collaborators based on the European maps which Ricci brought with him to China in 1582. At that time, it was assumed that the Chinese people did not know that the Earth is round, and hence, they would not be able to project geographical information from a round Earth onto a flat surface. This explains why it was believed that KWQ could only be a direct or an adapted copy of other European maps. However, this viewpoint has been challenged recently.[2] In this chapter, I shall show multiple lines of evidence to confirm that KWQ has a Chinese origin.

In the following sections, I shall: (1) examine two key geographical locations — Cape Breton Island and Newfoundland — on Ricci's 1602 map and compare them with their counterparts on four well-known contemporary European maps; (2) examine the latest archaeological findings on Cape Breton Island; and (3) make further map comparisons before drawing conclusions.

2. KWQ Shows Cape Breton Island and Newfoundland Differently from Other European Maps

Matteo Ricci arrived in Zhaoqing, China, in 1582 and started to make maps in 1584. Since he presented the map — KWQ — to the Ming Wanli Emperor (who reigned from 1572 to 1620) in 1602, I choose four other contemporary European maps made in 1562, 1569, 1570 and 1594 to compare with KWQ to find their differences in depicting Cape Breton Island and Newfoundland. These were the two earliest geographical locations in North America known to the European explorers. I include the 1594 map because the *Haijin* or sea ban policies in the Ming Dynasty had been lifted in 1567 for China to stop the closed-door policy, Ricci might have obtained the latest European map before he made the 1602 map.

Since Jacques Cartier[3] was the first European to enter the Gulf of St. Lawrence — in 1534 through the Strait of Belle Isle, finding that Newfoundland is an island — the European maps should depict this part of the world reasonably well, although none of Cartier's original maps have survived. This is also the reason why I believe that Matteo Ricci, as a European and a mapmaker, would have been aware of this event, and the 1602 KWQ reflects this important European exploration.

However, the 1562 Map of America by Diego Guttierez[4] in Fig. 1.1(a) shows Cape Breton Island as a part of the Canadian continent, not as a separate island, and Newfoundland is poorly represented. The 1569 World Map by Gerardus Mercator[5] in Fig. 1.1(b) shows that Cape Breton Island is at about the correct geographical location as a single island, while Newfoundland is also at about the right location but is broken up into several islands. The 1570 World Map by Abraham Ortelius[6] in Fig. 1.2(a) also shows Cape Breton Island as a single island and Newfoundland is broken up into several islands. The 1594 World Map by Petrus Plancius[7] in Fig. 1.2(b) shows Cape Breton Island at about its correct geographical location as a single island, and Newfoundland is also at about the right location as a single island; both islands are depicted closer to their actual shapes.

From the above qualitative comparisons, we can see that the European cartographers in the second half of the sixteenth century continued to improve their mapping of this part of the world. But since none of them ever explored or lived on Cape Breton Island, they were not aware that the island was not a single island during the period when their maps were drawn. This will become clear in Section 9.

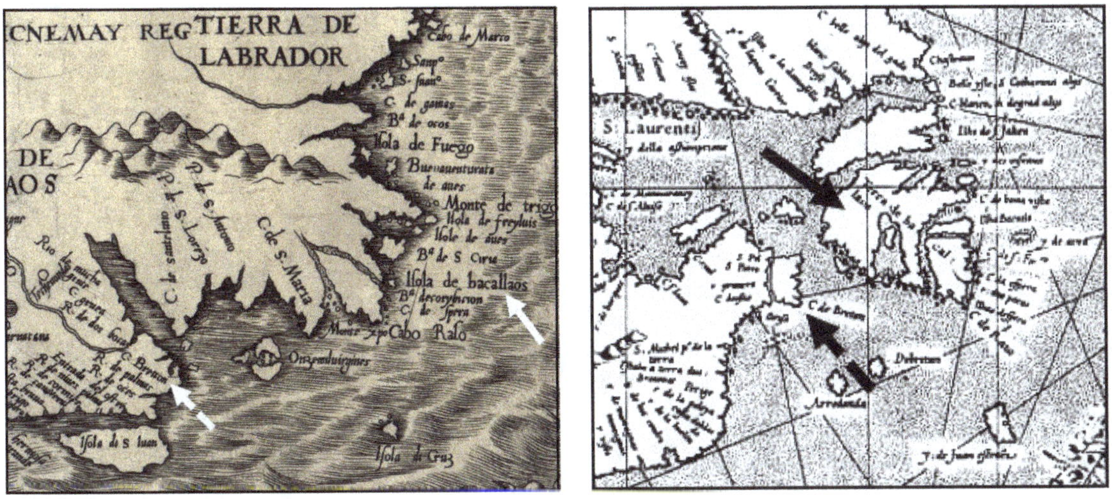

(a)

(b)

Fig. 1.1. (a) East coast of Canada extracted from the 1562 Map of America by Diego Guttierez (see endnote 4; public domain). (b) Also east coast of Canada, but extracted from the 1569 World Map by Gerardus Mercator (see endnote 5; public domain). Each figure uses a dotted-line arrow to point at the Cape Breton Island labelled as C. de Breton and a solid-line arrow to point at Newfoundland labelled as Terra de Baccalaos.

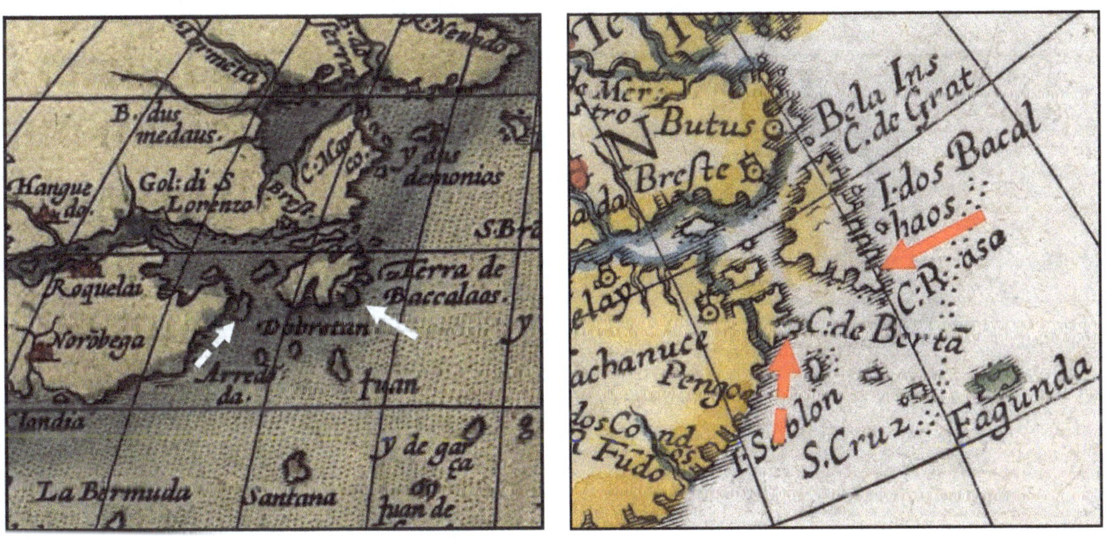

(a)

(b)

Fig. 1.2. (a) East coast of Canada extracted from the 1570 World Map by Abraham Ortelius (see endnote 6; public domain). (b) Also east coast of Canada but extracted from the 1594 World Map by Petrus Plancius (see endnote 7; public domain). Each figure uses a dotted-line arrow to point at the Cape Breton Island labelled as C. de Breton and a solid-line arrow to point at Newfoundland labelled as Terra de Baccalaos or Idos Bacalhaos.

Fig. 1.3. East coast of Canada extracted from Ricci's 1602 map of KWQ (see endnote 1c; public domain). This figure uses two dotted-line arrows to point at Cape Breton Island and two solid-line arrows to point at Newfoundland.

When I first examined KWQ in Fig. 1.3, I was startled to see that Cape Breton Island and Newfoundland are depicted so differently from what I saw in the above-mentioned European maps. Specifically, on KWQ: (1) a portion of Newfoundland becomes a peninsula labelled as Ke Er De Le Ya Er De (可尔得勒亚尔地) (to which the left solid-line arrow points; translated from "Terra Corterealis" which was the name shown in old Portuguese maps for the coast of the mainland opposite today's Newfoundland)[8]; (2) "Terra de Baccalaos" identified as the other portion of Newfoundland in the above-mentioned four European maps becomes, instead, Ba Ge Lao De (巴革老地) (to which the right solid-line arrow points; directly translated from the Portuguese "Terra de Baccalaos"; Spanish for cod; during the 1400s, when Portuguese sailors first fished the Grand Banks, they called the nearby island "Terra de Baccalaos" which means "the land of the dried codfish"; today the island is known as Newfoundland), also becomes a peninsula separated from Ke Er De Le Ya Er De (可尔得勒亚尔地) by a bay; and (3) Cape Breton Island becomes two islands: Gui Dao 鬼岛 or the Ghost Island (to which the left dashed-line arrow points; 鬼 (Gui), meaning ghost; 岛 (Dao), meaning island; 鬼岛 (Ghost Island), meaning notorious for its shipwrecks),[9] and Ru Li Han Island (如里漢岛) (to which the right dashed-line arrow points). They are correctly located on the map near the entrance of the Gulf of St. Lawrence and off the mainland of today's Nova Scotia Province.

The fact that Newfoundland is depicted so differently in KWQ from these four European maps tells me immediately that KWQ cannot be a direct or adapted copy from these maps. It may not be a European map dated after 1534, the year when Cartier entered the Gulf of St. Lawrence by the Strait of Belle Isle and discovered that Newfoundland is an island. In 1602, when Europeans already knew that Newfoundland is an island, why did Matteo Ricci depict it as two large peninsulas — 可尔得勒亚尔地 (Terra Cortereaiis) and 巴革老地 (Terra de Baccalaos) — separated by a huge bay?

Furthermore, why does Cape Breton Island become two islands on KWQ? This requires an in-depth investigation. But before that discussion, I would like to first address the question whether the Chinese people knew that the Earth is round and whether they could be cartographers who projected geographical information from a round Earth onto a flat surface, when Matteo Ricci showed KWQ to them. What was the truth?

3. Chinese Knowledge of a Round Earth, the Chinese Latitude, the Chinese Map Projection and the Dating of KWQ

The ancient Chinese had an explicit account of the shape of the Earth, reflected in the concept of Hun Tian Shuo "浑天说"[10] or the "Theory of Sphere-Heavens" or the "Geocentric Theory in Ancient Chinese Astronomy", which originated in the Warring States period (475–221 B.C.). The basic concept is as follows[11]:

。。。天地之体，状如鸟卵，天包地外，犹壳之果黄也；周旋无端，其形浑浑然，故曰浑天也
。。。半覆地上，半在地下。其二端谓之南极、北极。。。两极相去一百八十二度半强
。。。赤道带天之纮，去两极各九十一度少强。

My translation of the above passage is as follows:

... The shape of the celestial body is like a bird's egg. The celestial body has an outer space surrounding the Earth. The space is outside the Earth like the eggshell surrounds the egg yolk. The celestial body rotates like a wheel and does not stop. Its extent is uncertain, hence, it is called the fuzzy celestial body. ... Half of the mass is above the ground and the other half below the ground. The two poles at the two ends are called the South Pole and the North Pole. ... The two poles are a little more than 182.5° apart. The Equator is the celestial body's belt, being a little more than 91° from each pole.

The shape of the egg yolk inside an eggshell is spherical. In an article titled "Chinese knowledge of the spherical Earth centred on the Sun",[12] Gunnar Thompson points out that not only did the ancient Chinese know that the Earth is round, but some of the ancient Chinese astronomers also knew that the Earth and the moon revolve around the sun. Since they knew that it takes 365.25 days for the Earth to revolve around the sun, the Chinese set a complete circular rotation as 365.25°. Since the Earth is approximately round, in the above passage, the two poles of the Earth are therefore a little more than 182.5° apart (365.25°/2 = 182.625°). The Equator is the celestial body's belt, being a little more than 91° from each pole (182.625°/2 = 91.3125°).

Lam Yee Din (林贻典) observes that on KWQ, the North and South Poles lie a little beyond 90° from the Equator.[13] This can be seen more clearly on the seventeenth-century Japanese copy of the map shown in Fig. 1.4. Lam points out that the 1565 Farlani Map of the Known World[14] also has the 91° Chinese latitude, an indication that some early European world maps were based on Chinese source maps, since the Europeans used 90° for the poles.[15]

Liu Gang (刘钢; a Beijing lawyer, map collector and analyst) has written a paper[16] concerning Chinese cartographers' ability to project geographical information from a round Earth onto a flat surface. In this paper, he presents compelling evidence that a bi-hemispherical world map — Yu Di Tu《舆地图》or Terrestrial Map — was created in 1320 by a Chinese cartographer Zhu Siben (朱思本) in the Yuan Dynasty (1279–1368). This was long before the emergence of a map of the world (around 1527) by Franciscus Monachus showing two halves of the globe that he had constructed.[17] Liu's assertion is based on his interpretation of a key ancient Chinese passage written by Luo Hongxian (罗洪先), a famous cartographer who drew Guang Yu Tu《广舆图》or UNESCO Unfolded Terrestrial Atlas, based on Zhu Siben's Yu Di Tu《舆地图》or Terrestrial Map, during the Jiajing (嘉靖) period (1522–1566) of the Ming Dynasty. Liu Gang writes as follows[18]:

> Zhu Siben used the method of graticule to chart his map, which was in the format of a globe, and then from the globe centre and along a north–south direction, he divided the globe map equally into

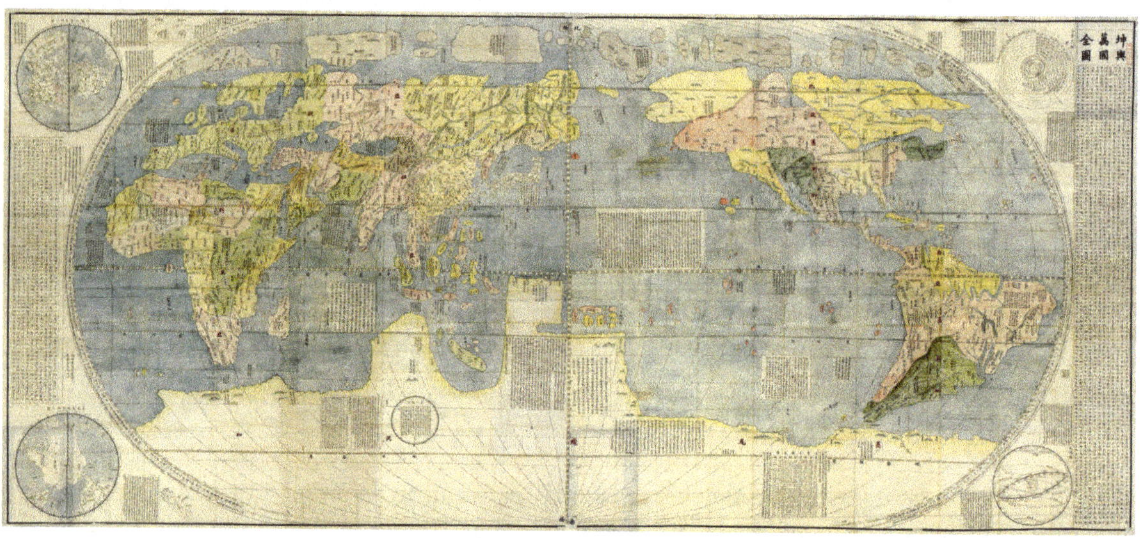

Fig. 1.4. Seventeenth-century Japanese copy of Ricci's 1602 map of KWQ (see endnote 1d; public domain; this map is in the Image Database of the Kano Collection, Tohoku University Library).

the eastern and western spheres and made them look alike and harmonious, by doing so the errors or confusion that might be caused by the two overlapping sides of the globe can be avoided.

The original copy of Zhu's map was still extant during the eighteenth century and Liu Gang points out that the cartographer of the 1411–1415 De Virga World Map copied the Chinese map and left some remarks unchanged on his copied map.[19] (In fact, De Virga revised his own map by incorporating Arab and the 1418 Chinese World Map/Zheng He Map and published his map with amendments in 1419; I shall discuss further the 1418 Chinese World Map/Zheng He Map in Section 9).

In Section 1, I already presented compelling evidence showing that KWQ is not a direct or adapted copy of European maps. Now I have another layer of evidence showing KWQ's Chinese origin, because the Chinese did in fact know how to project geographical information from a round Earth onto a flat surface using their own latitudinal system even before the Europeans knew how to do it.

If KWQ *indeed* is a Chinese map, then what is the political era revealed by this map? Is there a unique common era/date or not, for the different regions depicted on KWQ? This is an important issue to be investigated in this book, because *such information tells us when the geographical data were collected (when the places were explored). These dates were usually different from the date when the entire map was drawn for various reasons, one example is that in ancient days, it might take a year or longer for information to reach home for the cartographers to draw their maps.*

For determining the true date when KWQ was drawn, Siu-Leung Lee (李兆良) points out a Chinese annotation written on the map of KWQ near the location of Spain[20]:

此欧罗巴州有三十余国。。。去中国八万里, 自古不通, 今相通七十余载云。

My translation of the above passage is as follows (my comments are between square brackets)[21]:

This Europe has more than 30 countries. … The distance from China is 80,000 *li* [1 *li* = 0.5 km]. Since ancient times, the place had no interaction with China. Now they have had diplomatic relations [with China] for more than 70 years and less than 80 years.

Here, "now" can mean: (1) the year 1602 when Ricci published KWQ; or (2) the year when Ricci's European source map was made; or (3) the year when Ricci's Chinese source map was made. "七十余载" means for a length of time of more than 70 years but less than 80 years.

Case 1 would mean that China and the European countries established diplomatic relations between 1522 and 1532 (1602–80 = 1522; 1602–70 = 1532); this was unlikely, because the Ming emperor imposed the *Haijin* or sea ban, a closed-door policy, soon after Zheng He's seventh voyage ended around the mid-1430; it was unlikely to establish diplomatic relations with the European countries during this era. The sea ban was lifted starting from 1567.

Case 2 would mean that one of the four European maps dated 1562, 1569, 1570 and 1594 may be Ricci's source map, and this would imply that China and the Holy See established diplomatic relations between 1482 and 1492 (1562–80 = 1482; 1562–70 = 1492), or between 1489 and 1499 (1569–80 = 1489; 1569–70 = 1499), or between 1490 and 1500 (1570–80 = 1490; 1570–70 = 1500), or between 1514 and 1524 (1594–80 = 1514; 1594–70 = 1524), respectively. This was also unlikely, since all these years still fall in the period of Ming Dynasty's *Haijin*. This further disproves that KWQ is a direct or adapted copy of the European maps.

Finally, for Case 3, my viewpoint is this: China and the Papal State established formal diplomatic relations at the end of 1353 (the Pope's envoy Giovanni de' Marignolli returned to Europe to present a letter from the great khan to Pope Innocent VI for establishing a formal diplomatic relationship between the two countries). More than 70 years but less than 80 years from 1353 gives an era between 1423 and 1433 (1353 + 70 = 1423; 1353 + 80 = 1433). Siu-Leung Lee also points out that on the China portion of KWQ, the old name of Annan (安南) was Jiaozhi (交趾). The name change occurred in 1428.[22] On KWQ, the name Annan shows up, not Jiaozhi, implying that *the map must have been drawn between 1428 and 1433, a period after Zheng He's sixth voyage, but overlapping with the early part of his seventh voyage*. Given that, for the seventh voyage, the Ming Treasure Fleet embarked at Longwan near the Longjiang Shipyard on 19 January 1431, by then a map must have been completed to guide the new voyage. Hence, the true date when KWQ was drawn can be narrowed initially to the period between 1428 and 19 January 1431 (after the sixth voyage and right before the seventh voyage was launched).[23]

In Chapter 4 of this book, I show that this initial map was later updated after Zheng He's mariners explored Africa in 1433 during their seventh voyage. Hence, *the final date when KWQ was drawn must be 1433 in accordance with the interpretation of the above annotation*.

In Chapter 5 of this book, I offer essential insight to indicate that the Americas on KWQ were explored by Zheng He's mariners between the 1420s and 1428, a period that overlaps with the sixth voyage of Zheng He's Treasure Fleets but precedes their seventh (and last) voyage. Because Canada is part of the Americas, this means Canada was explored between the 1420s and 1428.

My analysis in Chapter 2 also shows that the geographical information contained in the region of Australia, New Zealand, Land of Fire and Antarctica was obtained by Zheng He's mariners in the 1420s during their sixth voyage. However, for Europe and Africa, my analyses in Chapters 3 and 4 suggest different eras for these regions being explored by Chinese mariners. Nevertheless, the final date when KWQ was drawn should still be before the end of 1433 based on Chapter 4's analysis. *Matteo Ricci merely made superficial editorial work to this map and published it in 1602.*

4. Discoveries Made on Cape Breton Island

The discussions in Sections 2 and 3 suggest that (1) KWQ has a Chinese origin, and it may have been drawn in 1433, but different regions on this map might have been explored by Chinese mariners during different political eras; and (2) the ancient Chinese cartographers knew that the Earth is round and how to project geographical information from a round Earth onto a flat surface as they did on KWQ. Now, we can try to find out: why did the Chinese cartographer depict Cape Breton Island as two islands? To answer this question, we must first examine recent archaeological discoveries made on Cape Breton Island.

In 2002, Paul Chiasson, a Canadian architect born on Cape Breton Island, discovered ruins with Chinese characteristics on the island while he walked on Cape Dauphin (see Fig. 1.5), the northern extension of Kellys Mountain. He writes in his 2006 book titled *The Island of Seven Cities: Where the Chinese Settled When They Discovered America*[24] that *no* European ever saw this city and it has not appeared on any European map to this day. Ten years later, in his second book titled *Written in Ruins: Cape Breton Island's Second Pre-Columbian Chinese Settlement*, he specifically writes that an ancient canal existed in the village of St. Peter's (see Fig. 1.5) *long before* the sixteenth century,[25] and he argues that the canal could have been built only by the Ming Chinese.

Why would someone build a canal across St. Peter's? The reasons may be explained as follows: St. Peter's is a small village on a narrow isthmus which separates the southern end of Bras d'Or Lake (an 80 km/49.71 mi tidal inland lake) to the north from St. Peter's Bay and the Atlantic Ocean to the south. Before the canal was built, if a sailor wanted to go from Bras d'Or Lake to the Gulf of St. Lawrence, he had two options: (1) exit Bras d'Or Lake from its north end (near Cape Dauphin), then sail southward along the east coast of Cape Breton Island to the southern end of the island. From there, he could take the Strait of Canso — a deep and narrow channel separating the Nova Scotian mainland from Cape Breton Island — to enter the relatively calm Gulf of

Fig. 1.5. Cape Breton Island on today's map; a solid-line arrow points to St. Peter's Bay on the Atlantic Ocean (drawn by Michel A. Van Hove).

St. Lawrence; however, this route requires sailing near the shallow rocky ledges along Cape Breton Island's eastern coast, famous for shipwrecks[26]; or (2) also exit Bras d'Or Lake from its north end, but then sail northward in the Atlantic Ocean, turning west into the Cabot Strait, known by mariners for its rogue waves, before entering the Gulf of St. Lawrence. Either route involves high risks. Besides, for inland exploration, sailors at this point would have to change their ocean-going ships to river-going boats, because the high risks of sailing smaller boats in the Atlantic Ocean's high winds and turbulent waves must be avoided.

This can be done if a canal is built in the village of St. Peter's to connect the southern end of Bras d'Or Lake with the nearby St. Peter's Bay, so that ships can sail from the quiet Bras d'Or Lake along the peaceful canal to enter St. Peter's Bay, the Strait of Canso, and the Gulf of St. Lawrence. Once there, the sailors can follow the St. Lawrence River, the Great Lakes, and major rivers to reach the heartland of America.

The presence of Cape Breton Island labelled as 鬼岛 and 如里漢岛 on KWQ (see Fig. 1.3), as I first point out in my 2019 book — *The Last Journey of the San Bao Eunuch, Admiral Zheng He*[27] — *does* seem to support Chiasson's theory of the existence of a waterway through the middle of the island, although it appears that the shapes and orientations of these two islands were not precisely surveyed. But did Chinese build the canal? It seems quite reasonable, because the map — KWQ — drawn by a Chinese cartographer shows the existence of such a waterway. However, is there other evidence besides the map? If the Chinese did build the canal, when did that happen? At this point of my research, there remain unanswered questions requiring the search of a missing chapter in Cape Breton Island's history to explain KWQ's depiction and Chiasson's argument.

5. Mysterious Stonework on Cape Breton Island

In 2005, T.C. Bell, a British surveyor specializing in Roman and Chinese engineering, explored Cape Breton Island extensively and reported his important discoveries there in remarkable photos,[28] which will be selectively shown in this chapter, with his kind permission. Parts of his findings have been cited and published in my 2019 book.[29] In this section, after giving a summary of his main discoveries, I will focus on mysterious stonework that was not examined fully in my 2019 book due to the book's limited scope on Cape Breton Island.

On Cape Dauphin of Cape Breton Island, T.C. Bell found many remains: extremely large, well-engineered, and well-defended settlements characteristic of a religious centre (this is the main site); an ore exploitation site, smelter ramps, ore crushers; a group of inhumations in Chinese burial styles; roads cut out of solid rocks (Fig. 1.6, left picture) and part of a defensive gateway (Fig. 1.6, right picture); canals, stonework alongside the site of a gateway (Fig. 1.7); cut stonework of a demolished gatehouse of a wall (Fig. 1.8, left picture) and a notched rock (Fig. 1.8, right picture); remains of stone gatehouses (Fig. 1.9); shore-side harbours suitable for treasure ships; quarried rocks (Fig. 1.10); the site of a stone gateway (Fig. 1.11, left picture) and a walled barrack block compound (Fig. 1.11, right picture); and two Viking Hogback stone grave markers[30] (Fig. 1.12; Hogbacks are stone-carved Anglo-Scandinavian sculptures, popular from tenth- to twelfth-century

Fig. 1.6. The left picture is a road cut out of solid rock, leading to the higher religious site on Cape Dauphin. The right picture is a defensive wall that ran from here down to the banks of the Bras d'Or Lake; note on the road the carefully cut stone face (circled), which was part of the defensive gateway. Photographed by T.C. Bell in May 2005.

Fig. 1.7. The left sketch shows the gate post structure. The right picture is part of a Roman fort's gateway in the Roman City of Ullswater, Cumbria, UK, part of which comprised an adjacent Roman granary. Note the fine scoring of the stonework to obtain a flat surface, which is identical to the stonework of the stone pad found alongside the site of a gateway, one of approximately 21 found, which accessed the religious and ore exploitation site on Cape Dauphin. Photographed by T.C. Bell in May 2005.

England and Scotland; Hogbacks gradually fell out of fashion by the beginning of the eleventh century) at the lower end of the main road to the main site from the Bras d'Or Lake. Bell's further visits to Canso, St. Peter's, and Louisbourg yielded evidence of yet more Chinese operations, but not all of them will be discussed, due to the limited scope of this chapter. Readers are

Fig. 1.8. The left picture is T.C. Bell standing on large cut stonework of a demolished gatehouse of a wall on Cape Dauphin. Photographed by Paddy Bell (T.C. Bell's wife) in May 2005. The right picture is a notched rock, part of a wall to the main road on Cape Dauphin. Photographed by T.C. Bell in May 2005.

Fig. 1.9. The left picture is remains of a stone gatehouse controlling access to the site on the upper plateau, on the road to Fairy Hole, at the north end of Cape Dauphin; note the cut stone (circled). The right picture has gatehouse remains north of English Town, on the west side of Cape Dauphin. Bell's friend stands alongside a recessed earthwork gatehouse which controlled access to the north and east up to the sites on the Cape Dauphin plateau and to the adjacent still visible stone quayside of Fader's Cove harbour. The road to the plateau is to the right of the gatehouse; the earthworks were constructed on stone foundations. Photographed by T.C. Bell in May 2005.

reminded to view the cut stones in these figures. These cut stones could not have been made by the Indigenous People of the region — the Mi'kmaq — because they today do not recognise the ruins on Cape Dauphin,[31] nor by any European, because no European ever saw this city and it has not appeared on any European map to this day.[32]

Fig. 1.10. Rocks have been quarried at Fairy Hole which is at the north end of Cape Dauphin. Photographed by T.C. Bell in May 2005.

Fig. 1.11. The left picture is the site of a stone gateway to the main site on Cape Breton Island, one of three to access the road; note the cut-out slot (left circle) and *carved* face (right circle). The right picture has ramparts of a walled barrack block compound (a double-line arrow is used to point towards it) at St. Peter's (a dashed-line arrow is used to point towards it); the canal to Bras d'Or Lake is adjacent (a solid-line arrow is used to point towards it). Photographed by T.C. Bell in May 2005.

Bell marvelled at the stonework in his description, but also expressed his sadness about the destruction he saw (my comment is between square brackets):

> One of my favourite photos of the ruin is of the walled road and alongside the carved pillar and carved pad stone of one of the many gatehouses on the route up to the temple [at the religious centre of the main site]. This proves the existence of a gatehouse, and the skill of the stone mason. My guess is that someone had removed all the stonework of the rest of the other gateways and had been interrupted, then forgot about it.

Fig. 1.12. The left picture is a carved face on rock in Viking style. The right picture is a Viking Hogback stone grave marker with bear head carving (the head points to the right) located nearby. Both were discovered at the lower end of the main road to the religious site on Cape Dauphin from the Bras d'Or Lake: they are the *first* and *only* Viking Hogback grave markers located to date in North America. Photographed by T.C. Bell in May 2005.

So, who quarried, cut, and carved these stones to build the walled roads, the pillars, and the gatehouses, if not the Mi'kmaq, or the Europeans? Bell wrote further, suggesting that the Chinese did it all (my comment is between square brackets):

> One does wonder at the experience and surveying expertise of the so-called archaeologists who claimed "no" visible evidence of Chinese on the Cape Breton site. Only a skilful man cuts stone into straight edges [but with no sharp edges], and carves faces and constructs rammed earth walls on stone foundations.
>
> How any intelligent person could claim that the road was made by a bulldozer to form a Fire Break is beyond me.
>
> I do not believe a bulldozer driver carved the gate post and the pad stone while making fire breaks. Nor the two Viking Hogback grave markers placed in the wall around 1000 A.D. (the Vikings themselves claimed that they reached the North American coast around 900 A.D.).

Perhaps, diligent research of the Chinese history of stone-cutting technology can help to resolve the mystery on Cape Breton Island to show that Bell is correct. That will be the goal of the next section of this chapter.

6. The Chinese Were Skilled Stone Craftsmen

The Chinese knew how to break rocks since ancient days,[33] when there were no explosives, mechanical levers, chisels or hammers. The imperial governor Li Bin (李冰), the builder of the Tungkjang dam in Sichuan (四川通江大坝), broke down a rock wall twenty meters high in

256 B.C., using the non-mechanical method of "fire and frost". He had the rock heated to make it brittle, then he had vinegar poured over it. On repeating the process several times, the rock gradually shattered.[34] This method of thermal cracking was used before the seventeenth century as an effective method to break rocks during many Chinese dynasties including the Ming Dynasty.[35] It was only abandoned in the second half of the nineteenth century after the rise of mechanical rock drilling.[36]

In another ancient civilisation, the Egyptians invented chisels around 8000 B.C., and later used chisels made of iron to cut granite: workers cut a series of holes in the granite with hammer and chisel, and then inserted wooden wedges; they soaked these wedges with water, which caused the wood to expand and the rock to split; the stone workers then used the chisel again to break the granite apart.[37]

We do not know when the ancient Chinese started to use chisels made of iron to drill holes in rocks before cutting stones. But we do know that, in 2009, two iron objects, some iron bars and rusted blocks were unearthed at the Mogou site in Lintan, Gansu (甘肃临潭磨沟遗址), dated between 1510 B.C. and 1310 B.C.[38] Later, during the Northern Wei Dynasty (386–557), a metallurgist named Qiwu Huaiwen (綦毋懐文) reportedly became the first person to create steel from the combination of wrought iron and cast iron. Chisels made of steel worked well for cutting stones.[39] During the Song Dynasty (960–1279), the iron industry grew enormously.[40] The ancient Chinese skilfully drilled holes in the surface of rocks, and with their wisdom and ability in geology and rock mechanics, they could hand drill, chisel, chip and split rocks into stones or slabs with the sizes and shapes they desired. The Huashan Grottoes (花山谜窟) in Huangshan City, Anhui Province, and the Longyou Grottoes (龙游石窟) in Quzhou City, Zhejiang Province, exhibit this kind of remarkable workmanship that the ancient Chinese practised over a thousand years ago.[41]

At some unknown point in time, the Chinese mastered the "drill and pour" technology using non-explosive demolition powders, for example, composed of lime, sulphur and alumina, to break the rocks by expanding them. It works like this: a series of holes are drilled in the rock, the powder is added to each hole, then water is added, and wooden plugs driven into the holes; the chemical creates a chemical reaction which expands and splits the rock. This new technique is non-explosive, convenient and dust-free. To this day, there are still well-known Chinese non-explosive (blastless) demolition powders on the market.

At this point of the present research, it is not clear whether the stones on Cape Dauphin were cut using the Chinese "drill and pour" method. When a rock breaks, most of the stone around the drilled holes (the evidence) also breaks off. Bell has spent days on the Cape Dauphin sites checking sections of stone walls and constructions, and going through his collection of photos, searching for evidence of blastless work, without success. He saw many corners of rocks that were broken off, but no clear evidence of drilled holes.[42] Furthermore, if the drilled area was hidden underneath the visible area by turning the stone upside down after blastless work, for the sake of appearance, ease of walking and safety (to avoid further breaking off from the exposed surface), then most drilled holes would be out of sight. The only convincing evidence is found in the construction of the canal in the neighbourhood of St. Peter's on Cape Breton Island, and this will be discussed thoroughly in the next section. However, this should not be an issue of concern now, because the following record will show that the trade of stone cutting, beyond doubt, flourished in the Ming Dynasty.

In the era of Ming Emperor Yongle (明永乐皇帝; who reigned during the period 1402–1424), the emperor ordered the Yangshan Quarry in Nanjing (南京洋山采石场) to cut a giant stele, for use in the Ming Xiaoling Mausoleum (明孝陵) of his deceased father.[43] Legend has it that workers there had to produce the daily quota of crushed rock to avoid being executed on the spot.[44] This was also the era when Emperor Yongle ordered the San Bao Eunuch — Admiral Zheng He — to lead Treasure Ships (giant ships) to the "Western Ocean" six times; Zheng He's seventh and last voyage was ordered later in 1430 by Emperor Xuande (宣德皇帝), the grandson of Emperor Yongle. Zheng He's crews also used a huge quantity of *ballast stones to stabilise treasure ships* in high winds and turbulent waves. It is quite apparent that the Chinese stone-cutting technique had matured for wide applications.

Since the Chinese history tells us that the ancient Chinese were very good at stone cutting, and since the Mi'kmaq and the Europeans did not even know the ruins on Cape Breton Island, it would then appear that the Chinese built the stone gateways, the gatehouses, the stone roads and the stone pillars, and carved stone faces on Cape Dauphin.

But did the Chinese also build the ancient canal across the village of St. Peter's, as Chiasson first suggested, thus splitting Cape Breton Island into two islands as depicted on KWQ? And can the cut stones used for building the ancient canal reveal what kind of technology the Chinese had applied?

7. Chinese Used Non-explosive Demolition Powder to Break Rocks to Build a Canal and a Fort in St. Peter's Prior to the Arrival of the Europeans

After reading my 2019 book in which I point out for the first time that Cape Breton Island is depicted as two islands on KWQ,[45] Bell emailed me his survey report on the village of St. Peter's. The report details his findings about the remains of an ancient canal across St. Peter's and an ancient fort nearby, from the time in 2005 when he and others were exploring the site. His main findings are as follows (my comments are between square brackets):

> When we surveyed St. Peter's, I told Paul Chiasson that there was an ancient fort on a platform at the St. Peter's end of the modern canal [the Canadian Government built the modern canal in 1869;[46] the ancient canal was still navigable in 1597[47] and for some years later] to the inland sea [the Bras d'Or Lake].
>
> Between the modern Canadian canal [built on an isthmus, the 800-m *St. Peter's Canal* joins the Atlantic Ocean to the Bras d'Or Lake][48] and the ancient fort's platform, there was a depression of about 50 m/164 ft in length and 12 m/39 ft in width, into which the fort's drains discharged. Originally this depression had been a harbour which could dock a vessel of size c. 47 m × 11 m/154 ft × 36 ft [standard size of a Chinese Treasure Ship in the Ming Dynasty]. Presumably, the crew stayed at the fort when in harbour.
>
> The large flagstones which had lined the original ancient canal [across the isthmus of St. Peter's] had been dumped around the mouth of the [original ancient] canal when it was rebuilt into a modern

> Canadian canal. The modern canal has a history of two periods of construction stages (the first modern canal was constructed from 1854 to 1869; the canal was rebuilt in 1985). My guess is that the first stage followed the ancient original canal by utilising its original stone block walls.
>
> One of these flagstones from the original ancient canal walls showed the unmistakable sign that it had been cut out of solid rock initially by Chinese non-explosive demolition powder [this will be shown soon]. This discarded stone (not used when the canal was rebuilt in 1985) could have originated from the harbour alongside the ancient fort, or the original ancient canal, we have no way of knowing. Furthermore, as found by a magnetic anomaly survey (a scan method), the original side field drains used to top up the original ancient canal are still in place; they are spaced approximately 18 m × 18 m/59 ft × 59 ft apart. The original quay for the Super Junks was observed in place but overlaid by the surplus flagstones from the original ancient canal.
>
> The depth of water alongside the quay would have taken a Super Junk. On a slightly raised platform above the quay side were the foundations of barracks capable of holding the entire Super Junk's crew (see the right-hand photo in Fig. 1.11).
>
> Alongside the mouth of the original ancient canal was an earth mound, which had been the base of a Lighthouse tower.
>
> We also located a harbour capable of holding four c. 47 m × 11 m/154 ft × 36 ft junks at the northern end of the Bras d'Or Lake. This harbour had just been exposed, together with a massive cut stone block, by bulldozer clearing of the area for new houses.

Bell writes that "One of these flagstones from the original ancient canal walls showed the unmistakable sign that it had been cut out of solid rock initially by Chinese non-explosive demolition powder"; this is vividly shown in Fig. 1.13. The left picture shows evidence of Chinese use of non-explosive material to split the stone slab; drilled holes are clearly visible. The right picture shows surfaces resulting from modern machine cutting with sharp edges and smooth faces, whereas surfaces smoothed manually with chipping hammer have rounded edges.

Neither the Vikings nor the local Mi'kmaqs left any record of using the same technology to crack rocks in those days.

In St. Peter's, the Mi'kmaq claim that two "raised mounds" were left by "white men before the French"[49] (the Chinese had a lighter skin colour than the Mi'kmaq). One mound may refer to the platform at the St. Peter's end of the modern canal, where the Chinese ancient fort stood; the other may refer to the earth mound which had been the base of a lighthouse tower. Both are recorded in Bell's passages above.

It was at least 1520 when the Portuguese, Spanish, Basque, French and English fishermen and whalers finally set up camps on Cape Breton Island.[50] By then, the Chinese settlers on Cape Breton Island must have all left, since the newly arrived European settlers have no record of seeing any Chinese or Chinese remains there. Only a Mi'kmaq legend tells the story of a great teacher who lived among them "before Europeans discovered the New World and left just before the first explorers arrived at the end of the fifteenth century".[51]

Cape Breton Island has a strategic location facing the Atlantic Ocean and right near the entrance of the Gulf of St. Lawrence. Ships coming close to Cape Breton Island from the Atlantic Ocean can be seen from the Chinese walled barracks on Cape Dauphin. On the northern access

Fig. 1.13. The left picture is a cut stone at St. Peter's on Cape Breton Island. These drillings (pointed at by the solid-line arrow) at the near end of this rock indicate that it has been split from its parent rock by the use of a Chinese "blastless demolition agent"; the rock's surface was smoothed with manual chisel and hammer; there is evidence of later second usage at the left corner in the form of a modern slot chipped out to receive a metallic metal strip (bedded in lead) to hold two stone blocks together (pointed at by the dashed-line arrow). The right picture has stone surfaces resulting from modern machine cutting with sharp edges and smooth faces (pointed at by the two dotted-line arrows), compared with a surface smoothed manually with chipping hammer to have rounded edges (pointed at by the solid-line arrow); are these the unusual rocks that the Vikings saw and recorded in the Vinland Sagas when they landed in the New World? Photographed by T.C. Bell in May 2005.

road to Cape Dauphin, Ciboux (Bird Island) is also visible from the Chinese walled barracks there, see Fig. 1.14.

Cape Breton has a temperate coastal climate. While the Chinese mariners went ashore to explore, they needed a base to dock their big ocean-going ships, and to serve as operational headquarters. The island would also appear to have been the Chinese North Atlantic base during Zheng He's mariners' sixth voyage as well as during their seventh voyage, when their fleet landed on the Labrador coast in the autumn of 1433.[52]

In Section 9, I will show that Zheng He's mariners built (or rebuilt) the ancient canal and the nearby fort in the neighbourhood of St. Peter's on Cape Breton Island, Nova Scotia, during their sixth voyage in the 1420s. This was long before the arrival of the European settlers.

An interesting question to ask next is: how many times have the Chinese explored Cape Breton Island?

8. Multiple Visits of Chinese to Cape Breton Island

In Section 5, I mentioned that T.C. Bell located two Viking Hogback stone grave markers, one on each side of the bottom of the Chinese road from the Bras d'Or Lake to their main settlement on Cape Dauphin. They are the *first* and *only* located to date in North America. This discovery proves

Fig. 1.14. View towards Bird Island from Chinese walled barracks, part of guard unit on the northern access road to Cape Dauphin. Photographed by T.C. Bell in May 2005.

that the Chinese road existed *prior to* the arrival of the Vikings in North America, contrary to what is claimed by the present government archaeologist and the people who extracted rocks from the area.[53]

Vikings were Norse people from their Northern European homelands in today's Norway, Denmark and Sweden. They used the Norwegian and Baltic Seas to engage with the world as looters, traders, colonists and mercenaries from the late eighth to the late eleventh centuries. In 1960, the discovery of the Norse site at L'Anse aux Meadows[54] at the northern tip of Newfoundland confirmed Leif Erikson's landfall in North America around the year 1000 A.D.,[55] long before Christopher Columbus set foot in the Bahamas in 1492. Once the Vikings reached Newfoundland, it would be natural for them to explore the nearby Cape Breton Island (the Norse ventured at least as far as the coast of New Brunswick),[56] hence they could have left their grave markers on the surface of the Chinese built road. Erikson and his crew did not stay long — only a few years — before returning to Greenland, perhaps due to hostile relations with the native North Americans.[57] This means that the earlier Chinese settled on Cape Breton Island before the Vikings' arrival around 1000 A.D. (during the Northern Song Dynasty).

The next wave of Chinese settlements on Cape Breton Island may be between the 1420s and 1428 and was related to the sixth voyage of the Ming Treasure Fleet. This era is derived from my analysis of the Americas on KWQ in Chapter 5. The fact that Cape Breton Island is depicted as two islands on KWQ suggests that Zheng He's mariners made landfall on Cape Breton Island and built or rebuilt (to be explained in Section 9) a canal near St. Peter's during their sixth voyage, thereby dividing the island into two parts.

As mentioned earlier in this chapter, it is very plausible that Zheng He's mariners also stayed at Cape Breton Island during their seventh and last voyage in the 1430s, based on the arguments and evidence presented in my 2019 book. With the experiences they gained from their previous voyage, Zheng He's mariners made landfall on the Labrador coast in the autumn of 1433. This time, they were at the door of Newfoundland. However, *no* map is known to record this last voyage to the Americas. But it is reasonable to state that the ancient Chinese made at least three visits to Cape Breton Island.

Like the earlier Chinese settlers, Zheng He's mariners also abandoned their temporary homes on Cape Breton Island and left (presumably returning to China). Their departure date must be before the arrival of the first European explorers at the end of the fifteenth century,[58] since there is no record of early European settlers seeing Chinese or Chinese remains when they arrived at Cape Breton Island.

Bell reported that harbours were discovered which were suitable for vessels of size c. 47 m × 11 m/154 ft × 36 ft. This is also consistent with the standard (medium) size of a Chinese Treasure Ship in the Ming Dynasty. It offers further circumstantial evidence of Cape Breton Island being an operational headquarters for Zheng He's mariners to explore the Canadian hinterlands and the American heartland during their sixth and seventh voyages.[59]

It would be nice to be able to carbon date the area of the Chinese settlement on Cape Dauphin. But one (with traditional bear's head) of the two discovered Viking Hogback grave markers shows signs of fire damage. This ties in with the known forest fires in the area. One Chinese smelter site which Bell dug into did show carbon deposits. But his only measured carbon date was approximately 100 years old, around the time of a forest fire.[60] Carbon dating in the area would be useless due to the known fires in the forest, unless carbon dating were applied to buried objects which are not affected by forest fires.

In summary, the analysis in this section shows that at least three visits of Chinese settlers to Cape Breton Island are likely to have occurred before the mid-1430s. The island would be a paradise for archaeologists for generations if they were encouraged to uncover its mystery!

9. Maps Uncover the Missing History of Cape Breton Island

The Vikings visited Cape Breton Island and left two stone grave markers on Cape Dauphin around 1000 A.D., but they did not use or make maps. The Vinland Map,[61] which is claimed to be a fifteenth-century *mappa mundi* with unique information about Norse exploration of North America, does not contain Cape Breton Island, and is now generally believed to be a forgery. The Cape Breton Island depicted in the four European maps shown in Section 1 was drawn by cartographers who had not personally explored or lived on that island. None of them seemed to know that there was an ancient canal across the village of St. Peter's, which separated the island into two islands in their times. The two earliest European maps of Cape Breton Island drawn by people who lived on the island are the work by Marc Lescarbot in 1604[62] and the work by Samuel de Champlain in 1613[63]; both lived in Acadia, a colony of New France in north-eastern North America. In both maps, Cape Breton Island has complicated island structures (fragmented) consisting of at least three segments. But these are not the two earliest European maps of Cape Breton Island which show details of the island.

The first European map of Cape Breton Island was drawn by Christopher Columbus in 1490.[64] He called it the Island of Seven Cities (the legendary name of Cape Breton Island), perhaps based on European legends or verbal accounts from sailors who had briefly visited the island.[65] His map was drawn seven years before John Cabot reached the island in 1497,[66] and over 100 years before Marc Lescarbot and Samuel de Champlain drew their maps. Unfortunately, no maps from John Cabot's 1497 voyage appear to have survived. All three maps share these common features: (1) a waterway exists from Bras d'Or Lake, through St. Peter's Bay, and into the Atlantic Ocean, because an ancient canal opened the isthmus in the village of St. Peters; and (2) Cape Breton Island is depicted as having at least three separate land masses,[67] whereas the correct depiction should be only *two* islands, as shown on KWQ in Fig. 1.5.

Lack of maintenance made the waterway non-navigable around the 1630s, and at that time, some of the European maps of Cape Breton Island started to show the opening of St. Peter's isthmus being closed. But when the French military arrived in 1713, the French records show that the canal had been sufficiently well made to still be useful in the early eighteenth century.[68] After the French left in the middle of the eighteenth century, the canal was gradually forgotten.[69] Hence, at least from the 1420s to the mid-eighteenth century, Cape Breton Island was basically two islands, not a single island.

It took the Canadian government fifteen years (1854–1869) to build the first of the two modern canals at the St. Peter's ancient canal site. The construction and administration work were mired in many difficulties, including resurveying, structural problems, poor planning, etc. "Whoever built the original canal (must) have been highly skilful and well organized",[70] Chiasson writes in his book. He specifically points out that "the stone of Mount Grenville that is just across the isthmus had to have been cut before the French arrived"[71] (if the stone had not been cut, the water of the old canal could not have flowed across it, since rivers do not flow uphill). According to the very earliest maps of the island, the cut was already there.[72]

Cut an opening in the stone ridge? This was an enormous job, even with modern technology in the mid-nineteenth century! But what about skilled Chinese craftsmen who cut stones and built the Huashan Grottoes and Longyou Grottoes over a thousand years ago, made giant steles for use in the Ming Xiaoling Mausoleum, cut b*allast stones to stabilise treasure* ships in high winds and turbulent waves, and used blastless material to split stones at St. Peter's, all around six hundred years ago? If Chinese did not cut the stone of Mount Grenville to make water flow through the isthmus before the French arrived, who else could have done that?

Cape Breton Island starts to show up in a 1763 copy of the 1418 Chinese World Map/Zheng He Map — Tian Xia Zhu Fan Shi Gong Tu《天下诸番识贡图》[73] or Map of the Barbarians from All under Heaven Who Offer Tribute to the Court — the copy is owned by Liu Gang. At present, the 1763 copy is still controversial: there is no certainty that it was forged by later generations, and there is also no certainty that it is authentic. I am going to discuss both cases separately:

Case 1: The copy is authentic. On this map, Cape Breton is depicted as a single island in Fig. 1.15 (top island pointed at by the white dashed-line arrow), indicating that there was *no* canal connecting Bras d'Or Lake with St. Peter's Bay in 1418. The second island below Cape Breton Island might be the Azores, although its position is too far west and too far south from what we know today. The Azores archipelago began to appear on portolan charts[74] during the fourteenth century, well before its official discovery date. Here, I correct a mistake made on pp. 289–290 in my 2019 book: there the third island below the Azores was mistaken as Bermuda. But examination of the high-resolution map in Fig. 1.15 shows that the third island-like depiction is the pair of Chinese characters "南洋" (Nan Yang, meaning Southern Ocean).

However, with KWQ showing Cape Breton Island as two islands — Ghost Island and Ru Li Han Island — and with the political era of the Americas on KWQ between the 1420s and 1428 as suggested in Chapter 5 of this book, it clearly signals that Zheng He's crews made landfall on Cape Breton Island and built or rebuilt a canal in the village of St. Peter's during their sixth voyage in the 1420s. I have explained clearly in Section 4 that the canal offers a lower-risk and direct waterway for ships to sail from Bras d'Or Lake to the St. Lawrence Bay. The chapter echoes strong support of Chiasson's viewpoint that an ancient canal existed in St. Peter's *long before* the sixteenth century.[75]

Case 2: The copy is a forgery. Since the original 1418 Chinese World Map/Zheng He Map is no longer in existence, and there is no earlier map of Cape Breton to compare with KWQ, we do not

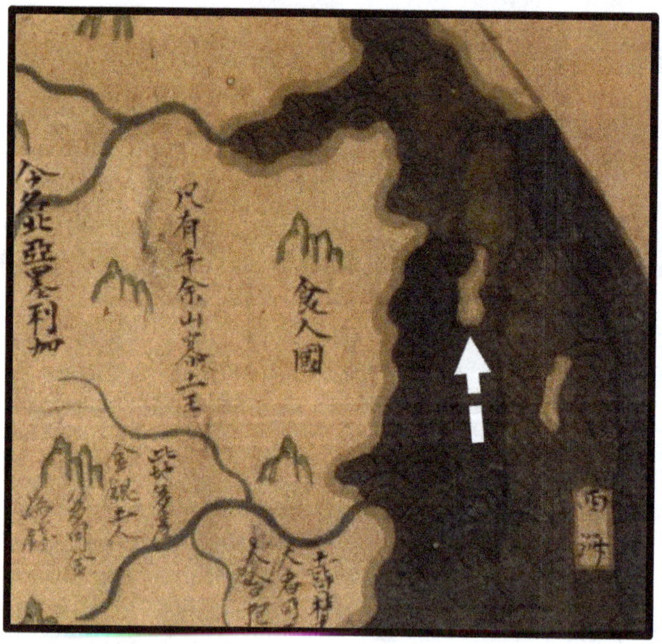

Fig. 1.15. East coast of North America extracted from the 1763 copy (see endnote 73; public domain) of the 1418 Chinese World Map/Zheng He Map; a dash-line arrow points to Cape Breton Island.

know whether the Chinese settlers had built the ancient canal at St. Peter's (although there were Chinese settlers on Cape Breton Island before the Vikings' arrival around 1000 A.D.). Hence, there are two possibilities: (1) the early Chinese settlers did not build the canal at St. Peter's; then that canal must have been built later by Zheng He's mariners in the 1420s when they explored Cape Breton Island, because KWQ shows that waterway; and (2) the early Chinese settlers did build the ancient canal at St. Peter's before they left the island. But when Zheng He's mariners arrived at St. Peter's, it was more than 400 years later. Due to lack of maintenance and usage, the canal was bound to have been rebuilt by Zheng He's crew to be navigable again.

Hence, no matter whether the 1763 copy of the 1418 Chinese World Map/Zheng He Map is genuine or forged, it cannot change the fact that Zheng He's mariners built or rebuilt in the 1420s the ancient canal at St. Peter's during their sixth voyage when they explored Cape Breton Island.

However, when I take another close look at Fig. 1.3, I must conclude that Zheng He's crew did not seem to discover or explore Newfoundland Island (Terra de Baccalaos on old Portuguese maps) during this same voyage in the 1420s. The island is missing from KWQ. But Ricci knew that Newfoundland Island had been discovered. He knew the island's name — Terra de Baccalaos — on some ancient Portuguese maps, and he knew that this island also appeared on other European maps (for example, the 1562, 1569, 1570 and 1594 maps discussed in this chapter), but he did not know where to place Terra de Baccalaos on a Chinese map — KWQ — on which no Newfoundland Island can be found. So, he ended up placing it as a peninsula named Ba Ge Lao De (巴革老地), directly translated from "Terra de Baccalaos", where Newfoundland Island is not supposed to be.

Zheng He's mariners explored Cape Breton Island and built up their knowledge about the North American coastal region during their sixth voyage to the West Ocean. Finally, during their seventh and last voyage, they landed on the coast of Labrador, very close to Newfoundland, in the autumn of 1433. This time, they were ready to explore the North American heartland. This is what I deduced in my 2019 book — *The Last Journey of the San bao Eunuch, Admiral Zheng He* — after an in-depth analysis of the 1597 epic work of Luo Maodeng (罗懋登), titled *San Bao Tai Jian Xi Yang Gi* 《三宝太监西洋记》[76] or *An Account of the Western World Voyage of the San Bao Eunuch*. From where they had landed, Zheng He's mariners could use Cape Breton Island as their headquarters and a launchpad — which they had prepared around ten years before — to enter the Gulf of St. Lawrence, the St. Lawrence River, the Great Lakes, the Mississippi and Illinois Rivers, and to plausibly arrive at Cahokia in the central Mississippi Valley.[77] This was more than 200 years before five French explorers "discovered" the Great Lakes from 1615 to 1669![78]

10. Conclusion

After a brief Introduction of this work, in Sections 2 and 3 of this chapter, I have compared Kunyu Wanguo Quantu (KWQ), published by Matteo Ricci in 1602, with four other contemporary European maps to show that the former has a Chinese origin and is not a direct or adapted copy of European maps. The Canadian east coast portion of KWQ reveals a political era between the 1420s and 1428 (the detailed proof is given in Chapter 5 of this book), a period after Zheng He's sixth voyage but before his seventh and last voyage. But the entire KWQ was drawn in 1433 based on the interpretation of an annotation on the map and the political era of Africa revealed on KWQ. The ancient Chinese cartographer of KWQ knew that the Earth is round and how to project geographical information from a round globe to a flat surface using the Chinese latitudinal system. Specifically, on KWQ, Cape Breton Island is depicted as two islands. Such a depiction is consistent with what we know today. My motivation was to make an in-depth analysis on multiple fronts to seek the truth of what had happened before the Europeans settled on Cape Breton, and the role played by Zheng He's mariners and the earlier Chinese settlers.

In Sections 4–9 of this chapter, I made further analysis and went through archaeological discoveries on Cape Breton Island made by Paul Chiasson in 2002 and T.C. Bell in 2005 to carry out an in-depth analysis of the island's history and the relevant Chinese history. My research enables me to draw the following conclusions:

(1) On Cape Dauphin, Cape Breton Island, Nova Scotia, there are remains of extremely large, well-engineered and well-defended settlements characteristic of a religious centre, a group of inhumations in Chinese burial styles, roads cut out of solid rocks, carved pillars, part of a defensive gateway, canals, stoneworks alongside the site of a gateway, stone gatehouse remains, smelter ramps, ore crushers, shore-side harbours suitable for treasure ships and so on. Especially, the stoneworks exhibit the skill of the stone mason, which is consistent with the long history of Chinese development of stone craftsmen starting from 256 B.C. The stonecutting industry flourished in the Ming Dynasty when Admiral Zheng He was ordered to lead giant ships seven times across the seas.
(2) Two Viking Hogback stone grave markers were discovered at the lower end of the main road to the religious site on Cape Dauphin from the Bras d'Or Lake. They are the *first* and *only* located to date in North America. The discovery proves that the road built by the early Chinese settlers was in place *prior to* the arrival of the Vikings around 1000 A.D.
(3) At the village of St. Peter's, the foundation of a barrack was discovered, which could hold an entire Super Junk's crew.
(4) Zheng He's mariners explored St. Peter's during their sixth voyage in the 1420s *long before* the arrival of the Europeans and built or rebuilt a canal across the village of St. Peter's to connect Bras d'Or Lake with St. Peter's Bay and the Atlantic Ocean. This explains why Cape Breton Island is depicted as two islands on KWQ.
(5) European cartographers of the sixteenth century, who had not explored or had not lived on Cape Breton Island, did not draw Cape Breton as two islands. However, Christopher Columbus in 1490, Marc Lescarbot and Samuel de Champlain (both had lived on Cape Breton Island) in the early seventeenth century, all drew a waterway from Bras d'Or Lake, through St. Peter's Bay, and into the Atlantic Ocean, hence splitting Cape Breton Island into separate islands. This ancient canal was navigable for a few hundred years. Even when the French military arrived in early eighteenth century, it was considered usable.
(6) KWQ is the earliest map known to show the existence of such a waterway, and the correct number of islands thus formed.
(7) Cape Dauphin is one of the world's fascinating historical sites and was home to an ancient Chinese temple and observatory. The harbours on Cape Breton Island enabled the Chinese

vessels to use the island as their North Atlantic base. It could become a paradise for archaeologists for generations if they were encouraged to explore the island's missing chapter of history!

The above conclusions render the Chinese origin of Kunyu Wanguo Quantu, and the Chinese exploration of Cape Breton Island in the 1420s, undisputable.

Acknowledgements

The author thanks T.C. Bell for sending his precious survey photos for their publication in this chapter. The author also thanks Paul Chiasson and T.C. Bell for sharing their insights and advice, and for approving the present chapter before its publication. The author is grateful for Professor Emeritus Michel A. Van Hove's technical assistance in preparing all the figures in published form in this chapter. Finally, special thanks go to M. Hayes-Shuptar (华铭叮) for the academic and editorial assistance with the endnotes.

Endnotes

[1]a. Library of Congress. "Kun yu wan guo quan tu". www.loc.gov, Library of Congress, 2022, loc.gov/item/2010585650; b. "File:Kunyu Wanguo Quantu by Matteo Ricci Plate 1-3.jpg". *Wikipedia Commons: The Free Encyclopedia*, Wikimedia Foundation, 23 Oct. 2020, commons.wikimedia.org/wiki/File:Kunyu_Wanguo_Quantu_by_Matteo_Ricci_Plate_1-3.jpg; c. "File:Kunyu Wanguo Quantu by Matteo Ricci Plate 4-6.jpg". *Wikipedia Commons*, 23 Oct. 2020, upload.wikimedia.org/wikipedia/commons/b/b3/Kunyu_Wanguo_Quantu_by_Matteo_Ricci_Plate_4-6.jpg; d. "File:Kunyu Wanguo Quantu (坤輿萬國全圖).jpg". *Wikipedia Commons*, 10 Aug. 2021, upload.wikimedia.org/wikipedia/commons/7/71/Kunyu_Wanguo_Quantu_(坤輿萬國全圖).jpg.

[2]Siu-Leung Lee (李兆良). *Kun Yu Wan Guo Quan Tu Jie Mi: Ming Dai Ce Hui Shi Je*《坤輿萬國全圖解密:明代測繪世界》or *Deciphering the Kunyu Wanguo Quantu, A Chinese World Map: Ming Chinese Mapped the World Before Columbus*. Taipei, Taiwan: Linking Publishing Company (聯經出版社), 2012; Sheng-Wei Wang. "Chapter nine: Secret revealed by ancient maps". *The Last Journey of the San Bao Eunuch, Admiral Zheng He*. Hong Kong, China: Proverse Hong Kong, 2019, pp. 290–301.

[3]Adam Woog. *Jacques Cartier (Great Explorers)*. New York, USA: Chelsea House Publications, 2009.

[4]"File:1562 Americæ Gutiérrez.JPG". *Wikipedia Commons*, 17 Sept. 2022, upload.wikimedia.org/wikipedia/commons/c/cf/1562_Americæ_Gutiérrez.JPG.

[5]"File:Mercator 1569 world map composite.jpg". *Wikimedia Commons*, 26 Nov. 2016, upload.wikimedia.org/wikipedia/commons/4/4b/Mercator_1569_world_map_composite.jpg.

[6]"File:OrteliusWorldMap1570.jpg". *Wikipedia Commons*, 12 July 2022, upload.wikimedia.org/wikipedia/commons/e/e2/OrteliusWorldMap1570.jpg.

[7]"File:1594 double hemisphere world map by Petrus Plancius.jpg". *Wikipedia Commons*, 2 Sept. 2022, upload.wikimedia.org/wikipedia/commons/0/0d/1594_double_hemisphere_world_map_by_Petrus_Plancius.jpg.

[8]J. Thomas Scharf. "Section 6 of Chapter I: The ocean path of discovery". *History of Saint Louis City and County; From the Earliest Periods to the Present Day: Including Biographical Sketches of Representative*

Men Vol. I. Illinois, USA: Northern Illinois University Digital Library, 1883, digital.lib.niu.edu/islandora/object/niu-twain:10944?msclkid=a9f9a031d0fa11ec83fb66f08e62b702.
[9]Sheng-Wei Wang. "Section 6 of Chapter I: The ocean path of discovery". *The Last Journey of the San Bao Eunuch, Admiral Zheng He, op. cit.*, p. 300.
[10]Zhang Heng (張衡; 78–139). *Hun Yi Zhu*《渾儀註》or *Commentary on the Armillary Sphere*; as found in the electronic version on the Internet at https://max.book118.com/html/2019/0724/5113014314002110.shtm.
[11]Sheng-Wei Wang. "Chapter nine: Secret revealed by ancient maps". *The Last Journey of the San Bao Eunuch, Admiral Zheng He, op. cit.*, pp. 277–278.
[12]Gunnar Thompson Ph.D. "Chinese Knowledge of the Spherical Earth Centered on the Sun". www.Gavinmenzies.net, Gavin Menzies, 18 Aug. 2011, gavinmenzies.net/Evidence/26-chinese-knowledge-of-the-spherical-earth-centered-on-the-sun-by-gunnar-thompson-ph-d.
[13]Charlotte Harris Rees. "Other maps that relate to the Tian Xia (Ch'onhado) atlases". *Chinese Sailed to America Before Columbus: More Secrets from the Dr. Hendon M. Harris, Jr. Map Collection*. Bloomington, USA: AuthorHouse, 2011, pp. 66–67.
[14]Vincint Virga and the Library of Congress. *Cartographia Mapping Civilizations*. New York, USA: Little Brown & Co., 2007, pp. 236–237.
[15]Charlotte Harris Rees. "Other maps that relate to the Tian Xia (Ch'onhado) atlases". *Chinese Sailed to America Before Columbus: More Secrets from the Dr. Hendon M. Harris, Jr. Map Collection, op. cit.*, p. 83.
[16]Liu Gang. "The Chinese inventor of bi-hemispherical world map". *e-Perimetron*, vol. 2, no. 3, summer 2007, pp. 185–193.
[17]Robert J. King. "Franciscus Monachus' *De Orbis Situ ac Descriptione*: A parallel translation". *The Globe*, no. 86, 2019, pp. 19–69.
[18]Liu Gang. "The Chinese inventor of bi-hemispherical world map". *Op. cit.*
[19]*Ibid.*
[20]Siu-Leung Lee (李兆良). "Dating Kunyu Wanguo Quantu (坤輿萬國全圖成圖斷代)". *Kun Yu Wan Guo Quan Tu Jie Mi: Ming Dai Ce Hui Shi Je*《坤輿萬國全圖解密: 明代測繪世界》or *Deciphering the Kunyu Wanguo Quantu, A Chinese World Map: Ming Chinese Mapped the World Before Columbus, op. cit.*, p. 51.
[21]Sheng-Wei Wang. "Chapter nine: Secret revealed by ancient maps". *The Last Journey of the San Bao Eunuch, Admiral Zheng He, op. cit.*, 2019, p. 291.
[22]Siu-Leung Lee (李兆良). "Dating Kunyu Wanguo Quantu (坤輿萬國全圖成圖斷代)". *Kun Yu Wan Guo Quan Tu Jie Mi: Ming Dai Ce Hui Shi Je*《坤輿萬國全圖解密: 明代測繪世界》or *Deciphering the Kunyu Wanguo Quantu, A Chinese World Map: Ming Chinese Mapped the World Before Columbus, op. cit.*, p. 54.
[23]Sheng-Wei Wang. "Chapter nine: Secret revealed by ancient maps". *The Last Journey of the San Bao Eunuch, Admiral Zheng He, op. cit.*, 2019, p. 292.
[24]Paul Chiasson. *The Island of Seven Cities: Where the Chinese Settled When They Discovered America.* Toronto, Canada: Random House Canada, 2006.
[25]Paul Chiasson. "Chapter four: The mark of an ancient canal". *Written in the Ruins: Cape Breton Island's Second Pre-Columbian Chinese Settlement*. Dundurn, Saskatchewan, Canada: Dundurn, 2016, pp. 62–68.
[26]Paul Chiasson. "Chapter thirteen: An unexpected discovery". *The Island of Seven Cities: Where the Chinese Settled When They Discovered America, op. cit.*, 2006, p. 213.
[27]Sheng-Wei Wang. "Chapter nine: Secret revealed by ancient maps". *The Last Journey of the San Bao Eunuch, Admiral Zheng He, op. cit.*, 2019, p. 300.
[28]T.C. Bell's booklet, "Cape Breton Island"; courtesy of T.C. Bell.

[29]Sheng-Wei Wang. "Chapter five: Travelling from the Gulf of St. Lawrence to the country of Fengdu". *The Last Journey of the San Bao Eunuch, Admiral Zheng He*, *op. cit*., 2019, pp. 163–166.
[30]Richard N. Bailey. *Viking Age Sculpture in Northern England.* Glasgow, UK: Collins, 1980.
[31]Private email communication with Paul Chiasson; courtesy of Paul Chiasson.
[32]Paul Chiasson. *The Island of Seven Cities: Where the Chinese Settled When They Discovered America*, *op. cit*., 2006.
[33]Xia Xiangrong (夏湘蓉), *et al*. *Zhong Guo Gu Dai Kuang Ye Kai Fa Shi* 《中国古代矿业开发史》 or *History of Mining Development in Ancient China*. Beijing, China: Geological Publishing House (地质出版社), 1980.
[34]Iulia Millesima. "A Very Ancient Method to Shatter Carbonate Rocks". www.labyrinthdesigners.org, Iulia Millesima & Hermolaos Parus, 2022, www.labyrinthdesigners.org/alchemy-science-history/a-very-ancient-method-to-shatter-carbonate-rocks/.
[35]Lu Rong (陆容). *Shu Yuan Za Ji* 《菽园杂记》 or *Miscellaneous Records from the Bean Garden*. Shanghai, China: Shanghai Ancient Books Publishing House (上海古籍出版社), 2012.
[36]*Ibid*.
[37]James A. Harrell and Per Storemyr. "Ancient Egyptian quarries — An illustrated overview". In N. Abu-Jaber, *et al*., eds. *QuarryScapes: Ancient Stone Quarry Landscapes in the Eastern Mediterranean, Geological Survey of Norway*. Geological Survey of Norway Special Publication, vol. 12, 2009, pp. 7–50. https://www.ngu.no/upload/Publikasjoner/Special%20publication/SP12_s7-50.pdf.
[38]Chen Jianli (陈建立), *et al*. "Gan Su Lin Tan Mo Gou Si Wa Wrn Hua Mu Zang Chu Tu Tie Qi Yu Zhong Guo Ye Tie Ji Shu Qi Yuan (甘肃临潭磨沟寺洼文化墓葬出土铁器与中国冶铁技术起源) or Iron artifacts of the Wa culture, unearthed from tombs in Mogou Temple, Lintan, Gansu and the origin of Chinese iron smelting technology". *Wen Wu Di 8 Juan* 《文物 第 8 卷》 or *Cultural Relics*, vol. 8, 2012.
[39]Li Baiyao (李百藥). "Biography No. 41 (列传第 41)". *Bei Qi Shu Di 9 Juan* 《北齊書第9卷》 or *Book of Northern Qi*, *Vol. 49*. Beijing, China: Zhonghua Book Company (中华书局), 1972.
[40]Asia for Educators. "The Song Economic Revolution". afe.easia.columbia.edu, Asia for Educators, Columbia University, 2020, afe.easia.columbia.edu/songdynasty-module/econ-rev-iron-steel.html.
[41]Li Lihui (李丽慧). "Da Xing Gu Di Xia Gong Cheng Ke Xue Wen Ti Ji Chang Qi Bao Hu Guo Ji Xue Shu Yan Tao Hui Zai Zhe Jiang Long You Xian Zhao Kai (大型古地下工程科学问题及长期保护国际学术研讨会在浙江龙游县召开) or International Symposium on Scientific Problems and Long-term Preservation of Large-scale Ancient Underground Engineering". www.igg.cas.cn, Chinese Academy of Science, 2015, www.igg.cas.cn/rssdy/201511/t20151102_4452136.html.
[42]Private email communication with T.C. Bell; courtesy of T.C. Bell.
[43]Barry Till and Paula Swart. *In Search of Old Nanking*. Hong Kong, China: Joint Publishing (Hong Kong) Company, Ltd., p. 131.
[44]Ann Paludan. *Chronicle of the Chinese Emperors: The Reign-by-Reign Record of the Rulers of Imperial China*. London, UK: Thames & Hudson, 2009.
[45]Sheng-Wei Wang. "Chapter nine: Secret revealed by ancient maps". *The Last Journey of the San Bao Eunuch, Admiral Zheng He*, *op. cit*., 2019, p. 300.
[46]Staff writer. "St. Peters Canal National Historic Site", www.pc.gc.ca, Government of Canada, 30 Aug. 2018, pc.gc.ca/en/lhn-nhs/ns/stpeters/decouvrir-discover.
[47]Paul Chiasson. "Chapter four: The mark of an ancient canal". *Written in the Ruins: Cape Breton's Second Pre-Columbian Chinese Settlement*, *op. cit*., 2016, pp. 61–63.
[48]Staff writer. "St. Peters Canal National Historic Site". www.pc.gc.ca, Parks Canada, 2020, https://www.pc.gc.ca/en/lhn-nhs/ns/stpeters.

[49]Private email communication with Paul Chiasson; courtesy of Paul Chiasson.
[50]*Ibid.*
[51]Paul Chiasson. "Chapter ten: Mysteries of the Mi'kmaq". *The Island of Seven Cities: Where the Chinese Settled When They Discovered North America, op. cit.*, 2006, p. 166.
[52]Sheng-Wei Wang. "Chapter five: Travelling from the Gulf of St. Lawrence to the country of Fengdu". *The Last Journey of the San Bao Eunuch, Admiral Zheng He, op. cit.*, 2019, pp. 157–166.
[53]Private email communication with T.C. Bell; courtesy of T.C. Bell.
[54]Staff writer. "L'Anse aux Meadows National Historic Site". whc.unesco.org, UNESCO World Heritage Convention, 2022, https://whc.unesco.org/en/list/4/.
[55]Leif Erikson was a Norse explorer from Iceland. He was the first known European to have discovered continental North America (excluding Greenland) before Christopher Columbus; Peter Archer. *The Book of Viking Myths: From the Voyages of Leif Erikson to the Deeds of Odin, the Storied History and Folklore of the Vikings*. New York, USA: Adams Media, 2017.
[56]*Ibid.*
[57]*Ibid.*
[58]Paul Chiasson. "Chapter ten: Mysteries of the Mi'kmaq". *The Island of Seven Cities: Where the Chinese Settled When They Discovered America, op. cit.*, 2006, p. 166.
[59]Sheng-Wei Wang. "Chapter five: Travelling from the Gulf of St. Lawrence to the country of Gengdu". *The last journey of the San Bao Eunuch, Admiral Zheng He, op. cit.*, 2019, p. 165.
[60]Private email communication with T.C. Bell; courtesy of T.C. Bell.
[61]Kirsten A. Seaver. *Maps, Myths, and Men: The Story of the Vinland Map*. Stanford, CA, USA: Stanford University Press, 2004.
[62]Paul Chiasson. "Chapter four: The mark of an ancient canal". *Written in the Ruins: Cape Breton's Second Pre-Columbian Chinese Settlement, op. cit.*, 2016, p. 64; the map is a collection item under the title of "Figure de la Terre-Neuve, Grande Rivière de Canada et Côtes de l'Océan en la Nouvelle France" by Marc Lescarbot, at Musée de la civilisation, fonds d'archives du Séminaire de Québec, T-4.
[63]Paul Chiasson. "Chapter four: The mark of an ancient canal". *Written in the Ruins: Cape Breton's Second Pre-Columbian Chinese Settlement, op. cit.*, 2016, p. 65; the map is a collection item under the title of "Carte geographique de la Nouvelle France" by Samuel de Champlain at Library and Archives Canada/Les voyages du sieur de Champlain, Xaintongeois, capitaine ordinaire pour le Roy en la marine collection/e010764734.
[64]Paul Chiasson. "Chapter four: The mark of an ancient canal". *Written in the Ruins: Cape Breton's Second Pre-Columbian Chinese Settlement, op. cit.*, 2016, p. 66; the map is a collection item under the title of "Columbus World Map" by Christopher Columbus at Bibliothèque nationale de France, Paris.
[65]Paul Chiasson. "Chapter four: The mark of an ancient canal". *Written in the Ruins: Cape Breton's Second Pre-Columbian Chinese Settlement, op. cit.*, 2016, p. 65.
[66]P. L. Firstbrook. *Voyage of the Matthew: John Cabot and the Discovery of North America*. San Francisco, USA: KQED Bookes, 1997.
[67]Paul Chiasson. "Chapter four: The mark of an ancient canal". *Written in the Ruins: Cape Breton's Second Pre-Columbian Chinese Settlement, op. cit.*, 2016, pp. 64–66.
[68]*Op. cit.*, p. 72.
[69]*Ibid.*
[70]Paul Chiasson. "Chapter four: The mark of an ancient canal". *Written in the Ruins: Cape Breton's Second Pre-Columbian Chinese Settlement, op. cit.*, 2016, p. 74.
[71]*Op. cit.*, p. 75.

[72]*Ibid.*
[73]"File:Zhenghemap.jpg". *Wikipedia Commons*, 23 Oct. 2021, upload.wikimedia.org/wikipedia/commons/b/bc/Zhenghemap.jpg.
[74]Portolan charts are ancient nautical charts known for their high cartographic accuracy mainly in the Mediterranean basin; Tony Campbell. *The Earliest Printed Maps 1472–1500*. Berkeley, CA, USA: University of California Press, 1987.
[75]Paul Chiasson. "Chapter four: The mark of an ancient canal". *Written in the Ruins: Cape Breton Island's Second Pre-Columbian Chinese Settlement*, *op. cit*., 2016, pp. 62–68.
[76]As found in the electronic version on the internet at https://www.zhonghuadiancang.com/wenxueyishu/sanbaotaijianxiyangji/.
[77]Sheng-Wei Wang. "Chapter five: Travelling from the Gulf of St. Lawrence to the country of Gengdu". *The Last Journey of the San Bao Eunuch, Admiral Zheng He*, *op. cit*., 2019, pp. 157–176.
[78]Jerry Dennis. *The Living Great Lakes: Searching for the Heart of the Inland Seas*. New York, USA: St. Martin's Griffin, 2004.

Chapter 2

Chinese Explored Australia, New Zealand, Land of Fire and Antarctica Long Before the Europeans

Abstract

It is found in this chapter that a world map — Kunyu Wanguo Quantu《坤舆万国全图》or Complete Geographical Map of All the Kingdoms of the World — published by Matteo Ricci in 1602 in China includes many geographical facts about Australia, New Zealand, Land of Fire and Antarctica. These facts are contained inside the huge and contiguous land mass — Mo Wa La Ni Jia (墨瓦蠟泥加) or *Magallanica* or *Magellanica* — at the southern edge of the map. The contour of that land looks strikingly like its counterparts on the three major European world maps of the sixteenth century, labelled there as *Terra Australis Incognita* (Unknown South Land), a hypothetical continent. Why does Kunyu Wanguo Quantu — said to be of Chinese origin in Chapter 1 of this book — contain this sixteenth century European map feature which was never present in the pre-1602 Chinese maps?

To answer this question, I analyse Kunyu Wanguo Quantu and the three European maps in great depth, while supporting my findings with those of the modern archaeological surveys made in New Zealand by British surveyor T.C. Bell. My conclusions are: (1) The Chinese explored Australia, New Zealand, Land of Fire and Antarctica long before the Europeans; (2) the European cartographers must have taken the Chinese geographical information and drawn the contour of a contiguous and extended Australia (but not New Zealand), and possibly Land of the Fire in South America and Antarctica as part of their *Terra Australis Incognita*; (3) the detailed geographical information revealed by the map of Mo Wa La Ni Jia (墨瓦蠟泥加) was due to the discoveries in the 1420s made by Zheng He's mariners during their sixth voyage to the Western Ocean, and most of that information is absent from the three European maps; (4) a *meteorite* hypothesised to have crashed into the sea near New Zealand in the fifteenth century may have done so around the mid-1430s; this is supported by T.C. Bell's discoveries that many carbonised Chinese shipwrecks were in their harbours and these likely include ships of Zheng He's Seventh Fleet; and (5) Australia, New Zealand, Land of Fire (*Tierra del Fuego*) and Antarctica are joined together as parts of the contiguous and expansive land area of Mo Wa La Ni Jia (墨瓦蠟泥加) on Kunyu Wanguo Quantu, implying that a compromise was likely made by Matteo Ricci and his Chinese collaborators between truthfully presenting the Chinese source map(s) and retaining the European representation of *Terra Australis Incognita*, to avoid offending either community.

Keywords: Antarctica, Australia, Kunyu Wanguo Quantu, Land of Fire (*Tierra del Fuego*), Matteo Ricci, meteorite, New Zealand, *Terra Australis Incognita*, Zheng He (郑和)

1. Introduction

For over four hundred years, the 1602 world map — Kunyu Wanguo Quantu[1] (abbreviated as KWQ in this chapter) — written with Chinese characters and having latitudinal and longitudinal lines, has been widely regarded as a map drawn by Matteo Ricci (an Italian Jesuit priest and one of the founding figures of the Jesuit China missions) and his Chinese collaborators, based on the European maps which Ricci brought with him to China in 1582. Recently, researchers have given arguments and provided evidence[2] to show that KWQ is of Chinese origin, and it is not a direct or adapted copy of major contemporary European maps.

In Chapter 1, titled "Chinese Explored Cape Breton Island long before the Europeans",[3] I compare KWQ with four major contemporary European maps: (1) the 1562 Map of America by Diego Guttierez; (2) the 1569 World Map by Gerardus Mercator; (3) the 1570 World Map by Abraham Ortelius; and (4) the 1594 World Map by Petrus Plancius. In that chapter, I also show that the southeastern Canada including Cape Breton Island on KWQ is of Chinese origin from the era between the 1420s and 1428.[4]

In this chapter, the 1562 Map of America by Diego Guttierez will not be included, because it is not a world map and does not contain the area I am trying to analyse.

Now when I take a close look at KWQ, what strikes me is the huge area near the bottom of the map, named Mo Wa La Ni Jia (墨瓦蠟泥加; it is the transliteration of *Magallanica* or *Magellanica*, see Fig. 2.1), which takes almost one-third of the map space. On the above-mentioned three European maps (Figs. 2.2–2.4), that huge area is labelled *Terra Australis Incognita* ("the unknown land of the south"; in Latin, *Terra incognita*, meaning unknown land, is a term used in European cartography for regions that have not been explored or documented, or regions the European cartographers might have heard about or seen on non-Europeans maps; *Australis* here means "of the south") or *Magallanica/Magellanica* (in Latin; the name of the Portuguese explorer Ferdinand Magellan was attached to the hypothetical continent because he supposedly skirted it in 1520).[5] The region of Mo Wa La Ni Jia (墨瓦蠟泥加) or *Terra Australis Incognita/Magallanica/Magellanica* looks distinctly different from modern maps but exhibits intriguing similarities in the KWQ and the three European maps, which prompted me to analyse them further.

This expansive and contiguous *Terra Australis Incognita* lived vividly in the Europeans' imagination. The concept was first proposed by Greek philosopher and polymath Aristotle Stagiritis (384–322 B.C.),[6] and was later expanded by Roman mathematician, astronomer, geographer and astrologer Claudius Ptolemy (c. 100–c. 170 A.D.) in his *Geography*[7] (c. 150 A.D.). They believed that the land in the northern hemisphere should be balanced by corresponding land in the south. The concept was reintroduced in the fifteenth century after the rediscovery of Ptolemy's work during the Age of Discovery. Major cartographers inherited that concept and started to draw a huge and contiguous *Terra Australis Incognita* into their maps during the sixteenth century.

A cartographic technique frequently used at the time was: w*hen new lands were discovered, they were often considered to be part of hypothetical continents*.[8] However, this was often incorrect; we of course now know that the so-called southern land mass is not in one piece, but includes separately Australia, New Zealand, Antarctica and many small islands, which is obviously different from the depictions on the above-mentioned three European maps as well as on KWQ.

In the following sections, I shall: (1) compare Mo Wa La Ni Jia (墨瓦蠟泥加) on KWQ with *Terra Australis Incognita/Magallanica/Magellanica* on three major contemporary European world maps to expose their similarities and differences; (2) extract detailed geographical information (evidence) from the map of Mo Wa La Ni Jia (墨瓦蠟泥加) to show Chinese exploration of Australia, New Zealand, Land of Fire and Antarctica long before the Europeans; (3) find out why the three sixteenth century European maps and KWQ drew their southern land the way they drew it; and (4) analyse the latest survey evidence on New Zealand collected by T.C. Bell to show Chinese presence in New Zealand since the pre-B.C. era, hence, uncover the true history embedded in the expansive and contiguous Mo Wa La Ni Jia (墨瓦蠟泥加) on KWQ.

2. The General Land Shape of Mo Wa La Ni Jia (墨瓦蠟泥加) on KWQ Shows Remarkable Similarity to Its Counterparts on the Three European Maps

In this section, I shall compare Mo Wa La Ni Jia (墨瓦蠟泥加) depicted on the KWQ (Fig. 2.1) with its counterparts on three other European maps (Figs. 2.2–2.4). The three maps are: (1) the

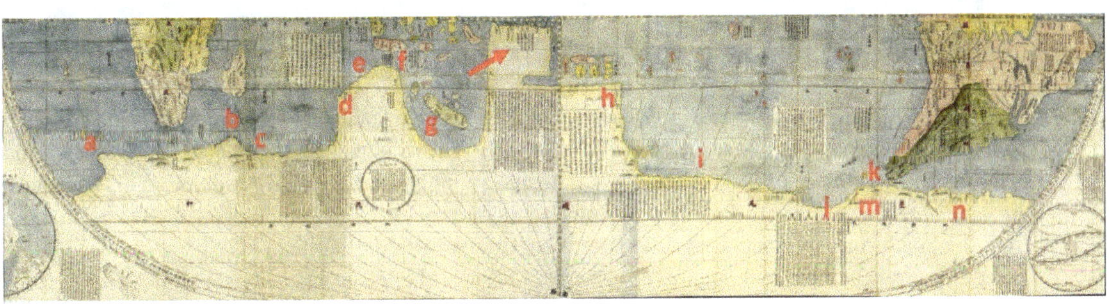

Fig. 2.1. Mo Wa La Ni Jia (墨瓦蠟泥加) extracted from the seventeenth century Japanese copy of the 1602 KWQ (see endnote 12; public domain). Labels from position a to position n ("l" is skipped to avoid confusion with "I") mark corresponding coastal features (capes, bays, inlets, islands, etc.) near those locations in this and the following figures (these labels help comparing locations among maps using different projections that produce variable distortions). The red arrow points to New Guinea, shown as part of Mo Wa La Ni Jia (墨瓦蠟泥加) (the huge beige-colour region bordered in yellow near the bottom of the map). The detailed geography of the map shown in this region will be discussed later in the text.

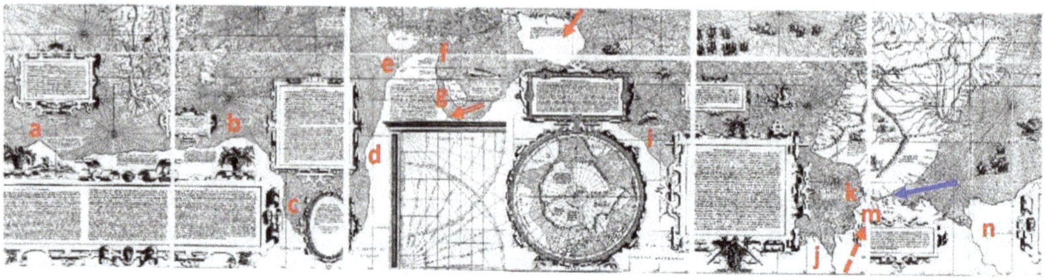

Fig. 2.2. *Terra Australis Incognita* extracted from the 1569 World Map by Gerardus Mercator (see endnote 9; public domain), showing the world south of the Equator (I have rearranged this map to place Australia near the centre of the display, for direct comparison with KWQ in Fig. 2.1). The solid red arrow at the top points to New Guinea, shown correctly as an island. The dashed red arrow near position m points to *Tierra del Fuego*,[13] Spanish for "Land of Fire", which should be an archipelago off the southernmost tip of the South American mainland, across the Strait of Magellan. But for Mercator, *Tierra del Fuego* was part of the huge land of *Terra Australis Incognita* across the Strait of Magellan (near position k, pointed out by a solid blue arrow) and separated from the South American mainland. Besides, he did not know what was further south of *Tierra del Fuego*. There is an unnamed big river (which may be the Antarctic Sound) near position n and a second unnamed smaller river to its left near position j. The river indicated by a double red arrow near the centre of this figure may be identified as today's Roper River in the Katherine region of the Northern Territory of Australia, although Mercator did not give it a name on his map. Whenever Mercator was not sure about what to fill into this empty land area, he covered it with a box containing some annotation (mostly explaining his cartographic methodology). Also, no mountains are shown on this map in *Terra Australis Incognita*.

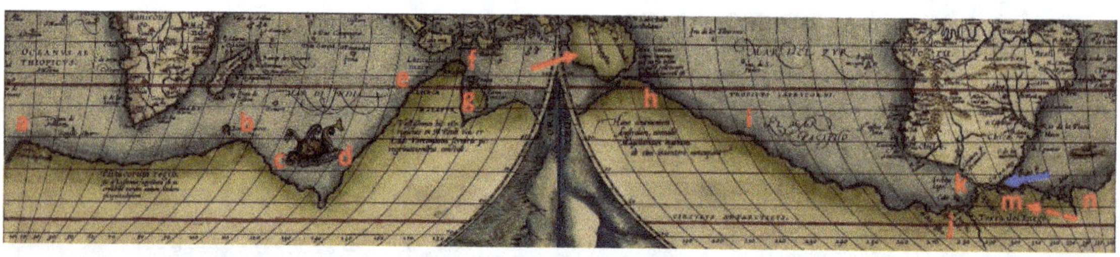

Fig. 2.3. *Terra Australis Incognita* extracted from the 1570 World Map by Abraham Ortelius (see endnote 10; public domain), rearranged to place Australia near the centre of the display. The solid red arrow points to New Guinea which is correctly shown as an island. The dashed red arrow near position m points to *Tierra del Fuego*, which is included as part of *Terra Australis Incognita* across the Strait of Magellan, pointed out by a solid blue arrow near location k, and separated from the South American mainland. Like Mercator, Ortelius also did not know what was further south of *Tierra del Fuego*. There is an unnamed large river (near position j) to the left of his label *Tierra del Fuego* and an unnamed shorter river near position n (which may be the Antarctic Sound). No mountains or other geographical information are given for *Terra Australis Incognita*.

1569 World Map by Gerardus Mercator[9]; (2) the 1570 World Map by Abraham Ortelius[10]; and (3) the 1594 World Map by Petrus Plancius.[11]

Clearly, the general land shape of Mo Wa La Ni Jia (墨瓦蠟泥加) on the Japanese copy of KWQ (Fig. 2.1) seems to be correlated to that of *Terra Australis Incognita* shown in Figs. 2.2–2.4:

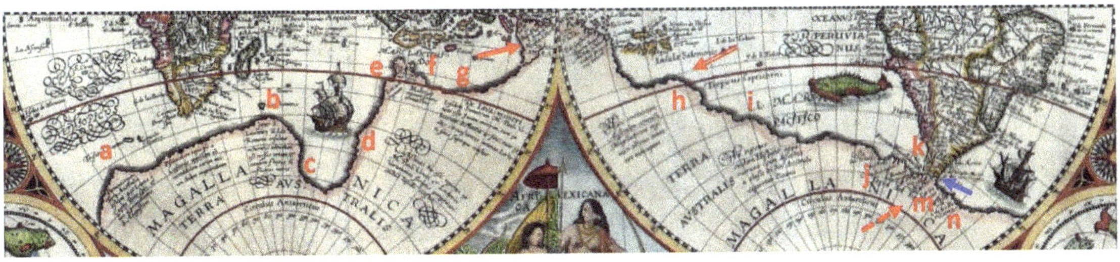

Fig. 2.4. *Terra Australis Incognita* extracted from the 1594 World Map by Petrus Plancius (see endnote 11; public domain), rearranged to place Australia near the centre of the display. The solid red arrow points to New Guinea; the fact that it is shown as part of *Terra Australis* marks a *retreat* in European cartographers' knowledge. A hump marked by the double red arrow in the middle of the map may indicate the shape of the northeast coast of Australia. A dashed red arrow near position m points to *Tierra del Fuego*, which is part of *Terra Australis Incognita* across the Strait of Magellan near position k pointed out by a solid blue arrow. Like Mercator and Ortelius, Plancius also did not know what was further south of *Tierra del Fuego* in the area he named *Magallanica*. No mountain and no river are depicted in *Terra Australis*.

the labels a–n point out corresponding coastal features that are remarkably similar in relative location and shape, in some cases even in detail. It is hard to imagine that chance alone would produce such similarity, except that these four maps may be copies of each other. But there is a major exception: both the 1569 World Map by Gerardus Mercator and the 1570 World Map by Abraham Ortelius show New Guinea as an island, whereas the 1594 World Map by Petrus Plancius and the 1602 KWQ show New Guinea as part of *Terra Australis Incognita* and Mo Wa La Ni Jia (墨瓦蠟泥加), respectively (both are incorrect). However, in Fig. 2.1, there is the following annotation inside an approximately square area (pointed out by a solid red arrow):

> 此地名為新入匿因其勢貌利未亞入匿相同歐羅巴人近方至此故未審其或為一片相連地方或為一島

My translation and comments (between square brackets) are as follows:

> This place is called New Guinea [新入匿] because the environment looks like Africa [利未亞入匿]. The Europeans have recently just got there, so the place has not been examined to see whether it is a connected land or an island.

Matteo Ricci's annotation on KWQ says that "The Europeans have recently just got there". But, a Portuguese explorer was the first European to explore the principal island of Papua New Guinea around 1526–1527,[14] which is more than 75 years before the publication of KWQ. Why did Ricci say that the Europeans had "recently just got there", and hence, "the place has not been examined to see whether it is a big land or an island"? On the other hand, if the source maps of this region on KWQ are Chinese maps drawn between 1428 and the early 1430s (according to the annotation on KWQ, as explained in Section 3 of Chapter 1 in this book), this would imply that the Chinese were not sure then whether New Guinea was an island because they had not mapped the entire island yet. Is this correct?

The other possibility is the following: because the 1569 World Map and the 1570 World Map differ in their depictions of New Guinea from the 1594 World Map, did this difference make it difficult for Matteo Ricci in choosing how to draw this part of Mo Wa La Ni Jia (墨瓦蠟泥加)?

Apart from New Guinea, KWQ shows the outer contour of Mo Wa La Ni Jia (墨瓦蠟泥加) also in qualitatively closer match with that of *Terra Australis Incognita* in the 1594 World Map by Petrus Plancius, as can be seen from Figs. 2.1 and 2.4. Hence, my observation offers evidence that KWQ is *not,* as some scholars have suggested, based on the 1569 World Map by Gerardus Mercator or the 1570 World Map by Abraham Ortelius.

However, it remains a mystery at this point why these sixteenth century European cartographers drew *Terra Australis Incognita* with such a special geographical contour that Matteo Ricci and his Chinese collaborators later also followed suit, drawing Mo Wa La Ni Jia (墨瓦蠟泥加) like their maps. This is what I want to explore in what follows.

3. Extracting Detailed Geographical Information (Evidence) of Australia, New Zealand, Land of Fire and Antarctica from the Map of Mo Wa La Ni Jia (墨瓦蠟泥加)

When we compare the geographical information revealed in these four maps, the 1569 World Map by Gerardus Mercator shows many un-named rivers and a very simplified *Tierra del Fuego*; the 1570 World Map by Abraham Ortelius gives little geographical information and also shows a very simplified *Tierra del Fuego*; and the 1594 World Map by Petrus Plancius also gives little geographical information inside the map of the empty, vast and contiguous *Terra Australis Incognita* except in the *Tierra del Fuego* region. Hence, we can say that most of *Terra Australis Incognita/Magallanica/Magellanica* remained unknown to Europeans even in the late sixteenth century. However, we can see that KWQ contains much geographical information including bird names, mountains and rivers depicted in Mo Wa La Ni Jia (墨瓦蠟泥加), which should be closely examined. Is this information related to today's Australia, New Zealand, Land of Fire and Antarctica? If so, how? I shall attempt to analyse this in detail in three subsections: 3.1 for Australia, 3.2 for New Zealand, and 3.3 for Land of Fire and Antarctica.

3.1 *Australia*

I discovered that the section of KWQ shown in Fig. 2.5 reveals the mountain systems of the northern half of Australia. The mountains drawn in Mo Wa La Ni Jia (墨瓦蠟泥加) in this area remarkably follow the mountain systems of today's northwestern, northern and northeastern Australia, as shown by the one-to-one correspondence of the numbered mountain ranges in Fig. 2.5 with those in Fig. 2.6.

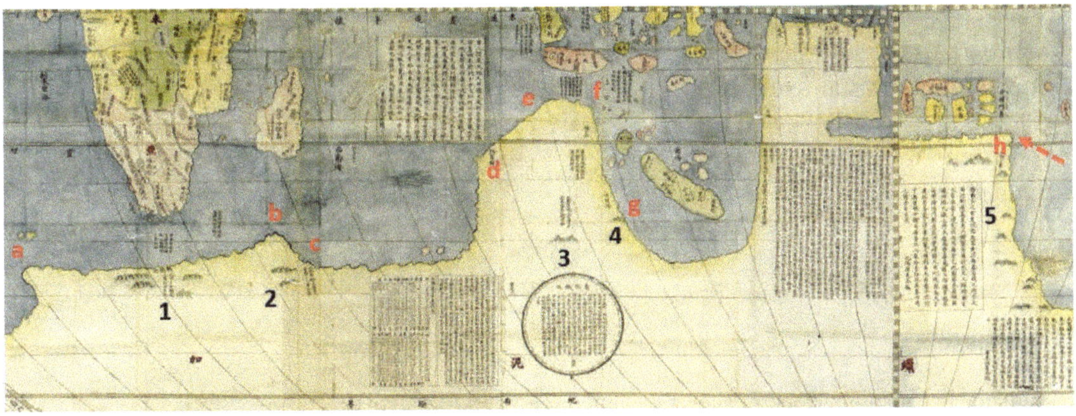

Fig. 2.5. The mountain systems labelled from 1 to 5 in this extract of Mo Wa La Ni Jia (墨瓦蠟泥加), extracted from Fig. 2.1, show a strong correlation with those in today's northwestern, northern, and northeastern Australia (see Fig. 2.6). The coast shape resembles its counterpart in today's northern half of Australia except in the region of Cape York Peninsula where the un-naturally shaped New Guinea (depicted as a square-shaped land area; it is the world's second-largest island) is mistakenly made to connect with the top portion of the Peninsula. There are many small islands to its east. Also indicated are Xian Ou Wu Si Ding He (仙歐吳私丁河) or St. Augustin River, Ping He (瓶河) or Ping River and two mountains to the east. To the southeast below these small islands is the northeastern region of Australia, characterised by many mountains. The over-simplified square corner pointed out here by a dashed red arrow near h, where the mountains are about to start, seems to correlate well with the protruding land area in the northeastern region of Australia pointed out by a dashed red arrow in Fig. 2.6.

Fig. 2.6. A display of mountain ranges (brown and yellow) in the northern half of Australia (drawn by Michel A. Van Hove). These mountains and coastal features labelled 1–5 and a–h, respectively, correlate well with those in Mo Wa La Ni Jia (墨瓦蠟泥加) on KWQ shown in Fig. 2.5. A dashed red arrow near h indicates where the mountains are about to start.

The shape of Mo Wa La Ni Jia (墨瓦蠟泥加) in this region seems to also match the shape of the northern half of Australia if the latter is stretched in the longitudinal direction, as we can see by comparing the coastal features labelled a–h in Figs. 2.5 and 2.6.

Near position 1

The annotation on KWQ is:

此地多有鸚哥之鳥故因名地

My translation is:

The place has many parrots; hence, it is named the Land of Parrots.

Ying Ge Di (鸚哥地) or Land of Parrots depicted in Mo Wa La Ni Jia (墨瓦蠟泥加) lies farther south and west of Kakadu National Park (the place proposed by S.L. Lee).[15] A *parrot* is named after its call. There are 56 parrot species in Australia. Parrots are found in many areas of Australia including the southwest corner of *Western Australia*,[16] which is relatively close to position 1 in Figs. 2.5 and 2.6.

From the above discussion, we know that it is erroneous to regard the longitude and latitude of Ying Ge Di (鸚哥地) or Land of Parrots as those of Australia (the smallest among the continents of the world, but it stretches in the south from latitude 10.60°S (Cape York, Queensland) to 43.64° South East Cape, Tasmania (South East Cape, Tasmania) and in terms of longitude from 113.15°E (Steep Point, Western Australia) to 153.64°E (Cape Byron, New South Wales). On KWQ, Australia is grossly expanded in longitude from its actual east–west extent.

Near position 2

The annotation on KWQ is:

此地香椒多端但人蠻猾不可同交易

My translation is:

This land produces spice and chili. But the residents here are cunning and like to fight for trivial matters, so trading with them should be avoided.

Due to Australia's huge size, the country has several different climate zones for spices. The tropical regions (see Fig. 2.7) lie in the north and across the east and west of Australia,[17] which is hot and humid in the summer, and quite warm and dry in the winter, suitable for people to sow cumin (a flowering plant; its seeds are used in the cuisines of many cultures)[18] during the dry season. Position 2 in Fig. 2.5 is in the tropical region of Australia; hence, the above passage correctly describes Australia's agricultural product.

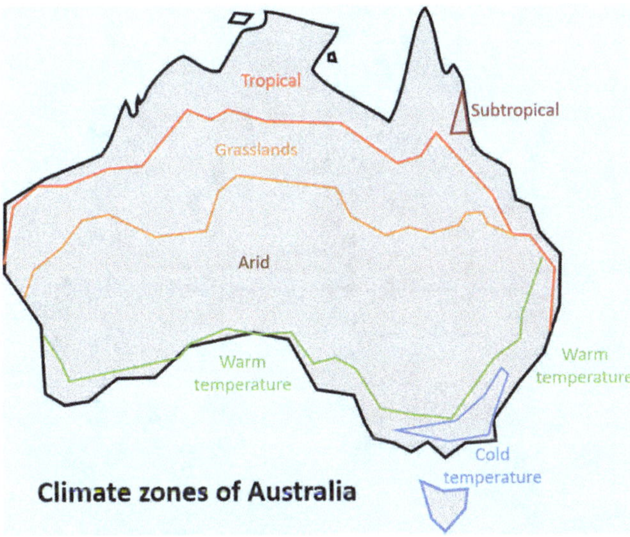

Fig. 2.7. A schematic drawing of climate zones of Australia (drawn by Michel A. Van Hove). The tropical regions of Australia lie in the north and across the east and west; position 2 in Fig. 2.5 lies inside the tropical climate zone.

"椒" can be interpreted as pepper or chili. However, mountain pepper only grows naturally in the cool temperate forests of Tasmania, Victoria and southern New South Wales in Australia. It is better not to interpret "椒" as pepper, since "椒" can also mean "chili". Chilies do grow well in most Australian climates, especially in their natural humid and hot habitat such as in tropical and subtropical regions of Australia near position 2.

Near position 3

The annotation is:

此南方地人至者少故未審其人物何如

My translation is:

This southern part of the land is rarely reached by humans. Hence, the place and people here have not been surveyed and studied.

Approximately 35 percent of the Australian continent receives very little rain and is effectively desert. The deserts in Australia are primarily distributed throughout the western plateau and interior lowlands of the country.[19]

The "southern part of the land" refers to the land in the regions labelled as 1, 3, 4, 7–10 in Fig. 2.8,[20] to the south-west of a place named "瑪力肚 (Maletur)" in Fig. 2.5. This region consists

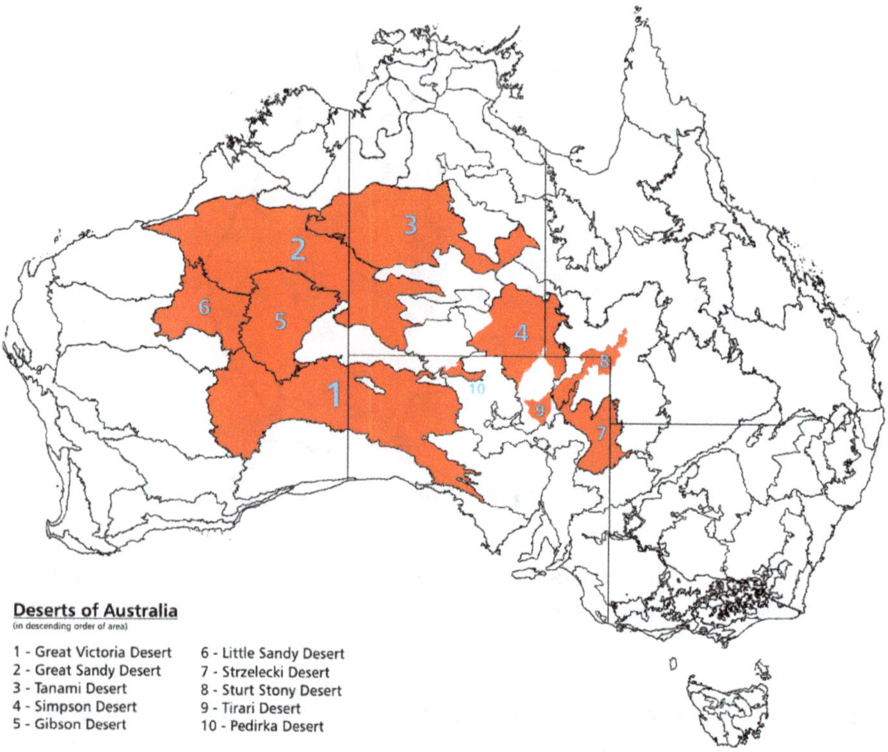

Fig. 2.8. Deserts of Australia (in red; see endnote 20; public domain).

of deserts and mountains and is hard to reach; hence, the place and people there had not been surveyed and studied.

Near position 4
The annotation is:

> 近年有被風浪加西良舶至此地惟言其廣闊無所產

My translation is:

> In recent years, there were Western ships coming by winds and currents to this area. They docked here and the sailors only said that the area was vast but had nothing to export.

This area is in the large and rectangular-shaped Gulf of Carpentaria between Australia, Indonesia and New Guinea. Across the Gulf is the northern tip of Australia — Cape York Peninsula — which is mistakenly shown in Fig. 2.5 as connected with New Guinea.

There are two places, "路客國 (Kingdom of Locach; the modern Thai city of Lopburi; but it is misplaced here, far to the south of Java and into the region of Mo Wa La Ni Jia (墨瓦蠟泥加))" and "伯亞祁 (Beach; an erroneous transcription of Locach)", on different sides of the mountainous area labelled as 3 in Fig. 2.5. As mentioned earlier, there is also a place named "瑪力肚 (Maletur; Kingdom of the Malays)", near the mountainous area 4. These three names are derived from Marco Polo's book which describes his journey by sea from China to India by way of Champa (Southern Vietnam), Java (which he called "Java Major"), Locach and Sumatra (which he called "Java Minor").[21] These names also appear on the 1569, 1570 and 1594 world maps inside the region of *Terra Australis Incognita*. Matteo Ricci simply followed the European cartographic practices at the time by *placing new lands to be part of hypothetical continents;* hence, he inevitably repeated the similar mistakes.

Near position 5

The eastern highlands of Australia are known as the Great Dividing Range.[22] The complex of mountain ranges starting from Xian Ou Wu Ding Feng (仙歐梧丁峰), labelled as "5" in Fig. 2.5, is called the East Australian Cordillera[23] which extends along the entire length of the country's eastern seaboard for about 3,500 km/2,175 mi from Cape York Peninsula in northern Queensland southward towards Tasmania. Only the northern part of the Great Dividing Range is shown in Fig. 2.5 of Mo Wa La Ni Jia (墨瓦蠟泥加), which matches closely the East Australian Cordillera labelled as "5" in Fig. 2.6.

Australia is the driest vegetated continent: 70 percent of its area is either arid or semi-arid land. It doesn't have many large rivers or large permanent lakes. But, since the Gulf of Carpentaria is tropical, it has a lot of rain and therefore many rivers flowing into it. But Mo Wa La Ni Jia (墨瓦蠟泥加) in Fig. 2.5 does not show any river.

3.2 *New Zealand*

We know that New Zealand consists of the largest islands to the southeast of Australia (see Fig. 2.9). The two places are separated by the Tasman Sea, which belongs to the western part of the South Pacific Ocean (Pacific Ocean, meaning Ning Hai (寧海) or a peaceful ocean on KWQ) and merges with the Coral Sea to the north. The Tasman Sea encloses a body of water about 2,250 km/1,400 mi wide and 2,300,000 km^2/900,000 mi^2 in area.[24]

We can also see the wide separation between the mountains in area 5 of Fig. 2.5 (and Fig. 2.10, extracted from KWQ) and the mountains which presumably are in North Island of New Zealand near the area named as Xiao Dao Bin (小島濱) or Near Island Coast (pointed out by the red arrow; could be the coast near today's Bay of Islands) in Fig. 2.10. But no geographical structure is entered in this huge and empty area, because the Tasman Sea is embedded entirely inside this expansive land space.

In addition, there is a group of small islands (pointed out by the red dashed arrow) in the north-eastern direction of Xiao Dao Bin (小島濱) or Near-Island Coast/coast near today's Bay of Islands, correctly depicted as on the present-day map of North Island, New Zealand. Mountains dominate

Fig. 2.9. Australia and New Zealand (drawn by Michel A. Van Hove).

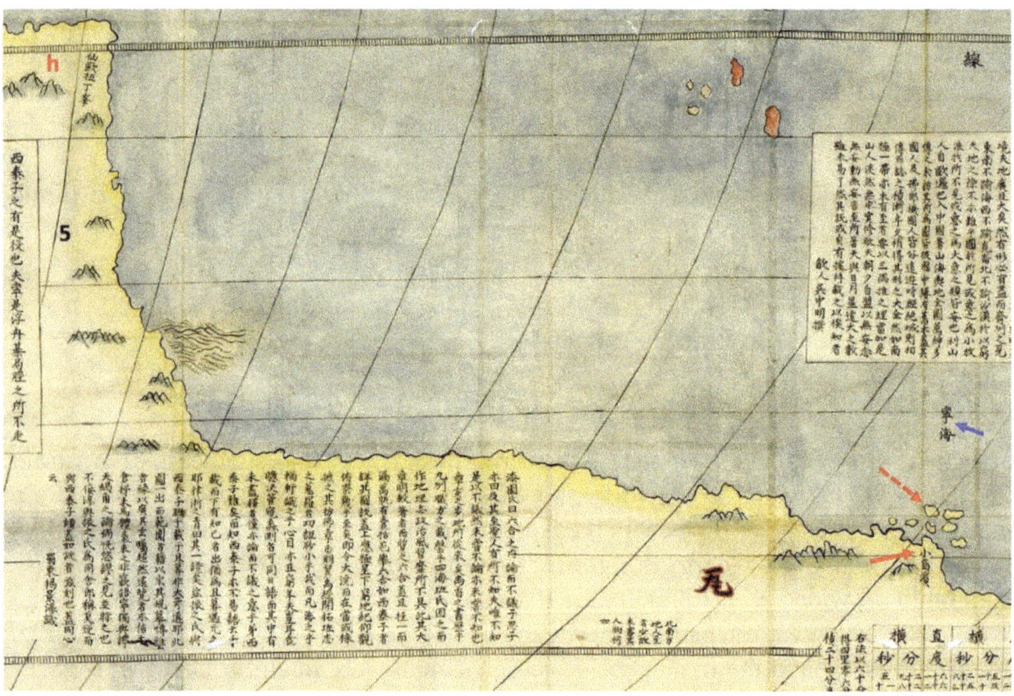

Fig. 2.10. The mountain system of Australia is on the left of this figure (near label 5); to the right, this part of Mo Wa La Ni Jia (墨瓦蠟泥加), extracted from Fig. 2.1 (public domain), can be correlated with today's mountainous New Zealand. The Tasman Sea is embedded between these two mountainous zones, however Ning Hai (寧海) or Pacific Ocean, pointed out by the blue arrow, is properly depicted at its correct geographical location. Xiao Dao Bin (小島濱) or Near-Island Coast could be the coast near today's Bay of Islands on the North Island of New Zealand and is pointed out by the red solid arrow; the islands nearby are pointed out by the red dashed arrow.

much of the New Zealand landscape,[25] because New Zealand straddles two tectonic plates. The plates push up the land and form the mountains. There are therefore many earthquakes in New Zealand.

Another interesting and very important feature is the dramatic difference in latitudes reflected between North Australia and Xiao Dao Bin (小島濱) or Near-Island Coast/coast near today's Bay of Islands. For example, Cape York Peninsula (the northernmost point in Australia) has a GPS-reported latitude of 10.68°S and Cape Reinga (the northwesternmost tip of the Aupouri Peninsula at the northern end of the North Island in New Zealand) has a GPS-reported latitude of 34.43°S. The difference is about 24°. From Fig. 2.1, we can estimate a much larger difference of over 35° between the two Capes. Similarly, their longitude difference is much larger on KWQ than on a modern map. Since New Guinea erroneously became part of the Cape York Peninsula and since the drawing of latitudes of geographical positions in Mo Wa La Ni Jia (墨瓦蠟泥加) is very inaccurate, all we can say is that KWQ only qualitatively depicts the relative latitude and longitude differences of Australia and New Zealand, but still allows us to differentiate Australia from New Zealand. Since there is no other major island between Australia and South America, the mountainous region near Xiao Dao Bin (小島濱) or Near-Island Coast/coast near today's Bay of Islands is most likely the northern part of New Zealand. The rest of New Zealand is embedded in Mo Wa La Ni Jia (墨瓦蠟泥加) on KWQ. However, none of the three European maps depicts this important geographical information, indicating that the *Europeans did not know the existence of New Zealand in the sixteenth century.*

In Section 4, I shall analyse the survey findings by T.C. Bell, a British surveyor specializing in Roman and Chinese engineering, to show Chinese presence in New Zealand since pre-B.C. days. Among the many surveys Bell conducted, I shall discuss an ancient Chinese walled city Waitangi (see Fig. 2.9), constructed in the pre-B.C. era in the coast near today's Bay of Islands (Xiao Dao Bin (小島濱) or Near-Island Coast), North Island, New Zealand, at the mouth of the Waitangi River (a river of the Northland region of New Zealand. It flows east from its origins, reaching the Bay of Islands close to the historic settlement of Waitangi). The site is where the Treaty was signed in 1840 between the European Settlers and the *Māori*, the indigenous Polynesian people of mainland New Zealand. The Chinese settlers knew that New Zealand is surrounded by water, that it consists of North Island and South Island, that its closest continent to the west across the Tasman Sea is Australia, and that its more distant continent to the east across the South Pacific Ocean is South America; Mo Wa La Ni Jia (墨瓦蠟泥加) should not be a contiguous and expansive land mass joining Australia, New Zealand, Land of Fire in South America and Antarctica together, as Matteo Ricci drew it on KWQ. Nonetheless, let us continue the map analysis of the last part: Land of Fire and Antarctica of Mo Wa La Ni Jia (墨瓦蠟泥加).

3.3 *Land of Fire and Antarctica*

As shown in Fig. 2.11, on the left side of Mo Wa La Ni Xia (墨瓦蠟泥峽) or Strait of Magellan, there is Huo Di (火地) or Land of Fire (*Tierra del Fuego*; actually, an archipelago at the southern extremity of South America), pointed out by the solid red arrow at the left.

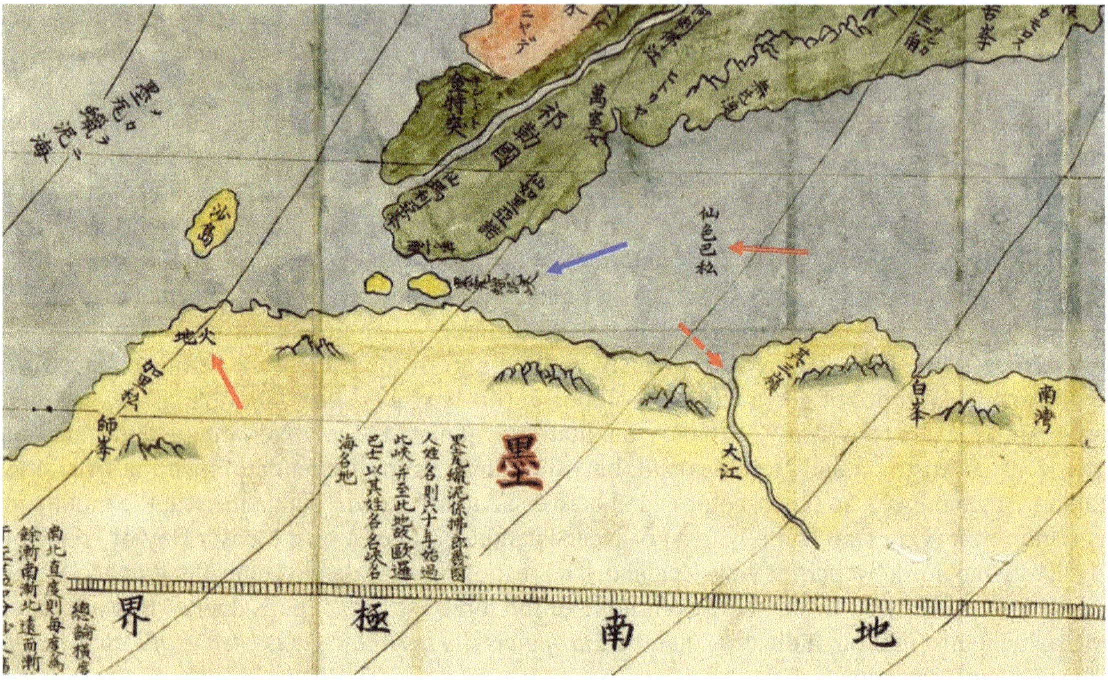

Fig. 2.11. This figure is extracted from KWQ in Fig. 2.1 (public domain); Huo Di (火地) or Land of Fire/*Tierra del Fuego*, pointed out by the solid red arrow, is an archipelago at the southern extremity of South America; the mountains nearby are a prolongation of the Andes. Xian Se Ba Si (仙色巴私) or *San Sebastián* Bay pointed out by the red double arrow, seems to mean *San Sebastián* Bay or Bahia San Sebastian ("bahia" is an ancient spelling of the Portuguese word "baia", meaning "bay") will be explained in the following text. To the right, Da Jiang (大江) (c. 62°S) or Big River, pointed out by the dashed red arrow, may be identified as today's Antarctic Sound (at a latitude of about 63.33°S on today's map; the value is very close to c. 62°S of Da Jiang (大江)) separating the northeastern end of the Antarctic Peninsula from the Joinville Island group. Mo Wa La Ni Xia (墨瓦蠟泥峡) or Strait of Magellan is pointed out by the blue solid arrow. 南灣 or Southern Bay at the far-right side of this figure should be deleted because this geographical item appearing on the Japanese copy of the KWQ is not depicted on the original KWQ.

The first European to reach this archipelago was Ferdinand Magellan (in 1520) when he sailed through the strait and called the region *Tierra del Fuego*.[26] The southern and western parts of the main island and the archipelago are a prolongation of the Andes, with peaks exceeding 2,234 m/7,000 feet. Most of the northern part of the main island, consisting of lakes and moraines, is under 180 m/600 ft in height. The eastern portion including the South Atlantic and Strait of Magellan coasts of the *Tierra del Fuego* archipelago is only a few hundred meters at most. This physiography of the Southern Andes is nicely depicted by the mountains drawn near Huo Di (火地) or Land of Fire in Fig. 2.11; the Strait of Magellan is pointed out by the blue solid arrow.

In Fig. 2.11, the ocean named Xian Se Ba Si (仙色巴私) or *San Sebastián* Bay, pointed out by the double red arrow, should be the South Atlantic Ocean, because *San Sebastián* Bay is part of Huo Di (火地) or Land of Fire depicted to the far left of Fig. 2.11 as pointed out by the solid red arrow. On today's map, *San Sebastián* Bay (it is semi-circular and about 30 km/19 mi across) is located south of Mo Wa La Ni Xia (墨瓦蠟泥峽) or Strait of Magellan in Argentina, and much smaller than it is depicted in Fig. 2.11. There are three major rivers in *Tierra del Fuego* in Argentina, and they all flow in the west–east direction. Hence, the large river named Da Jiang (大江) or Big River, pointed out by the dashed red arrow, depicted to the right of Fig. 2.11 in a mountainous region, must be flowing into the South Atlantic Ocean, not *San Sebastián* Bay. Instead, Da Jiang (大江) or Big River may be the 60 km/37 mi long and less than 15 km/9 mi wide at its narrowest stretch of water (flowing in a north–south direction) now called Antarctic Sound.[27] It was named by the Swedish Antarctic Expedition in 1902 under Otto Nordenskjöld[28] after the expedition ship *Antarctic* which was the first European vessel to navigate it. The first European to discover Antarctic Sound was British captain James Clark Ross during his expedition of 1839–1843.[29] The Antarctic Sound separates the northeastern end of the Antarctic Peninsula from the Joinville Island group and opens to the south into the wide bay now called Weddell Sea, see Fig. 2.12.[30] KWQ shows mountains all along these coasts: the Antarctic Peninsula is in fact quite mountainous.[31]

The appearance of Da Jiang (大江) or Big River/Antarctic Sound on KWQ is strong evidence showing that Zheng He's mariners explored a part of Antarctica most likely during their sixth voyage to the Western Ocean in the 1420s (according to the interpretation of the annotation on KWQ

Fig. 2.12. Map of Antarctica and the southern end of South America, with Strait of Magellan (pointed out by the blue dashed arrow), Land of Fire (blue solid arrow), Drake Passage (red dashed arrow), Antarctic Peninsula (labelled in black), Antarctic Sound (solid red arrow), the Joinville Island group (red double arrow), Weddell Sea (in blue lettering) and the mountains in Antarctica (in brown colour; see endnote 30; public domain).

as explained in Chapter 1 of this book), almost *five hundred years* before the European expedition ship *Antarctic* navigated through the Antarctic Sound. The Chinese mariners must have also discovered the open water south of South America — the Drake Passage — during this voyage.

We know that Drake Passage,[32] one of the most dangerous seas in the world, is located between Cape Horn in South America and the South Shetland Islands of Antarctica, which connects the South Atlantic Ocean with the South Pacific Ocean. It is supposed to lie to the south of Huo Di (火地) or Land of Fire. But it is totally missing or invisible on KWQ by being embedded inside the contiguous and expansive land space of Mo Wa La Ni Jia (墨瓦蠟泥加). The first European navigator to visit this passage was Francisco de Hoces[33] in 1526. Some Spanish, Argentinian and Chilean historians maintain that the "Drake Passage" should be named Mar de Hoces (*Hoces Sea*). But the Chinese got there about 100 years earlier, because, to explore the Antarctic Sound from the southern end of South America, sailors must first cross this open water.

In Fig. 2.11, the ocean north of Da Jiang (大江) or Big River/Antarctic Sound should be the South Atlantic Ocean, not Xian Se Ba Si (仙色巴私) or *San Sebastián* Bay. The Drake Passage which must be crossed from Huo Di (火地) or Land of Fire to Da Jiang (大江) or Big River/Antarctic Sound is embedded in the land space of Mo Wa La Ni Jia (墨瓦蠟泥加) on KWQ and is thus invisible.

As I have pointed out earlier, in the 1569 World Map by Gerardus Mercator (Fig. 2.2) there is an unnamed large river near position n, which may be the Antarctic Sound; and in the 1570 World Map by Abraham Ortelius (Fig. 2.3), there is also an unnamed river near position n, which may also be the Antarctic Sound. It seems that the cartographers of these two maps had seen some non-European maps which had Antarctic information on them. If these maps were not of Chinese origin, who else could have explored Antarctica before the Europeans?

3.4 *Uncovering the mystery of the land shape of Terra Australis Incognita and Mo Wa La Ni Jia* (墨瓦蠟泥加)

In Section 1, I mentioned that *when new lands were discovered, they were often considered to be part of hypothetical continents*. This practice seems to have dominated European cartography in drawing *Terra Australis Incognita* in the sixteenth century.

For example, on the 1569 World Map by Gerardus Mercator (Fig. 2.2), a portion of *Terra Australis Incognita* could correlate with Australia's northwest, north, and northeast; the Roper River can be identified; *Tierra del Fuego* is depicted at the south of South America (across the Strait of Magellan), and there are two rivers in that area. One might be the Antarctic Sound. But all the above are *part of Terra Australis Incognita.*

In the 1570 World Map by Abraham Ortelius (Fig. 2.3), very few geographical items are depicted in *Terra Australis Incognita*, but a portion of this region also correlates qualitatively with Australia's northwest, north, and northeast. In addition, *Tierra del Fuego* is depicted to the south of South America, across the Strait of Magellan in Fig. 2.3; and there is an un-named big river which might be the Antarctic Sound. Again, all the above are just *part of Terra Australis Incognita.*

In the 1594 World Map by Petrus Plancius (Fig. 2.4), the geographical information depicted on the map is similarly rare in comparison with the 1569 and 1570 maps.

Did these European cartographers simply draw the shape of their *Terra Australis Incognita* arbitrarily and accidentally happened to end up with similar land shapes? I doubt this very much, because the detailed shapes of their *Terra Australis Incognita* (especially near Australia) not only look mutually similar, but later even KWQ depicts Mo Wa La Ni Jia (墨瓦蠟泥加) qualitatively like these European maps.

Of course, by 1569, 1570 and 1594, European cartographers all knew that Ferdinand *Magellan had sailed through the Magellan Strait; hence, it is not surprising to see that* Huo Di (火地) or Land of Fire was added as part of *Terra Australis Incognita* at the end of the South America mainland. But what about the Australian part of *Terra Australis Incognita* in all these four maps? Why do they look so similar before the Europeans explored the region? There must be some reasons behind this similarity, and I wish to uncover them.

I decided to first focus my attention on ancient European maps to see since which year the European cartographers started to depict Australia as part of *Terra Australis Incognita.* Figure 2.13 is Ptolemy's world map,[34] reconstituted from Ptolemy's *Geography* (c. 150 A.D.) in the late fifteenth century.

On this map, the continents are given only as Europe, Asia and Libya (Africa). This ancient map does not closely resemble KWQ or any of the European maps researched in this chapter. Also, the bottom of the map looks uninterestingly structureless, which is distinctly different from its counterparts in the 1569, 1570 and 1594 European world maps and the 1602 KWQ.

Next, I direct my attention to maps which would show Australia for the first time. I found two maps shown in Fig. 2.14: (1) the left map is the De Virga World Map[35] made between 1411 and 1415 by Albertinus de Virga; he is also known for revising his 1409 map of Mediterranean by incorporating Arab and the 1418 Chinese World Map/Zheng He Map and published his map with amendments in 1419. Liu Gang (刘钢; Beijing lawyer, map collector and analyst) points out that the cartographer of the 1411–1415 De Virga World Map, shown in the left panel of Fig. 2.14, copied the Chinese map and left some remarks unchanged on his copied map[36]; and (2) the right map

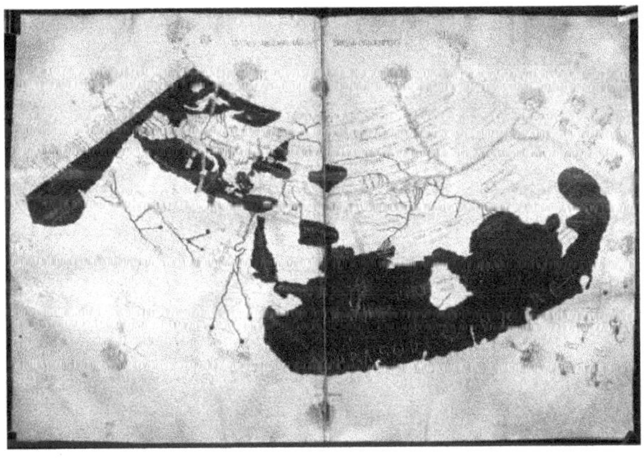

Fig. 2.13. Ptolemy's world map, reconstituted from Ptolemy's *Geography* (c. 150 A.D.) in the late fifteenth century (see note 34; public domain).

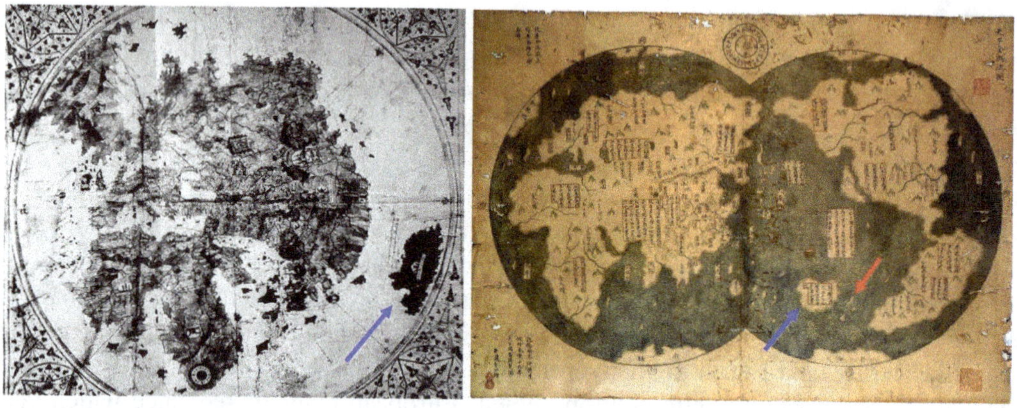

Fig. 2.14. The left map is the De Virga World Map (see endnote 35; public domain); Australia, pointed out by the solid blue arrow, is clearly shown at the far right of the map. The right map is a copy (see endnote 37; public domain) of the 1418 Chinese World Map/Zheng He Map made by Mo Yi-tong (莫易仝) in 1763; Australia, pointed out by the blue solid arrow, and New Zealand, consisting of two smaller islands, pointed out by the red solid arrow, are depicted. Here, the longitudinal positions of Australia and New Zealand are too far to the east.

is the 1763 copy of the 1418 Chinese World Map — Tian Xia Zhu Fan Shi Gong Tu 《天下诸番识贡图》 or Map of the Barbarians from All under Heaven Who Offer Tribute to the Court, or Zheng He Map[37] — made by Mo Yi-tong (莫易仝) and owned by Liu Gang.

At present, the 1763 copy is still controversial: there is a debate whether it is forged or authentic; I shall discuss both cases separately.

Case 1: Mo Yi-tong's 1763 map is a forgery. But this does not rule out the existence of the original 1418 Chinese World Map/Zheng He Map before its extinction. Otherwise, from whom did Albertinus de Virga receive information for him to draw Australia on his world map, if not from the Chinese? De Virga's world map was published nearly 200 years before the first European — Dutchman Willem Janszoon — landed on the western side of Cape York Peninsula and charted about 300 km/186 mi of coastline in 1606.[38] Hence, even if Mo Yi-tong's 1763 map is a fake, we may still suspect that Albertinus de Virga saw the original 1418 Chinese World Map and his map has influenced European drawing of the *Terra Australis Incognita*. However, since little geographical information is depicted inside the region of Australia on his map, this implies that other than the contour of the contiguous and extended Australia, the 1418 Chinese World Map/Zheng He Map (related to Zheng He's fifth voyage from 1417 to 1419), was not the source of the rich geographical information contained in Mo Wa La Ni Jia (墨瓦蠟泥加). That source may derive from Zheng He's sixth or seventh voyage. This will be investigated soon.

Case 2: Mo Yi-tong's 1763 copy is authentic. Both Liu Gang and Gunnar Thompson have strongly defended the authenticity of the 1418 Chinese World Map/Zheng He map.[39]

The 1418 Chinese World Map/Zheng He Map depicts both Australia and New Zealand, and their relative size and latitudinal differences seem reasonably represented. The date of the map drawing also overlaps with Zheng He's fifth voyage (1417–1419) to the Western Ocean.

Fig. 2.15. The left panel shows Australia (near bottom in this orientation) extracted from the De Virga World Map in Fig. 2.14. Its shape looks reasonably like today's Australia in the right panel.

In Fig. 2.15, a comparison of Australia extracted from the De Virga World Map is made with today's Australia: they look well correlated, and the former contains little geographical information inside the interior of the continent. This means that the European cartographers could have been influenced by De Virga's map to shape their *Terra Australis Incognita*, and only very limited geographical information will be depicted in the Australia portion of the 1569, 1570 and 1594 European world maps. This is what we see on these maps.

In the 1763 copy of the 1418 Chinese World Map/Zheng He Map, there is only one annotation inside the red frame pointed out by the solid blue arrow (Fig. 2.14, right panel):

土人膚亦黑色皆不著衣腰墜骨器皆有食人之習也

My translation is:

> The natives also have black skin, they wear no clothes and have bone artefacts hanging down from their waist; they all have the habit of cannibalism.

The annotation was not far off about the darker skin colour (on the Australian Aboriginal Flag, black represents the Aboriginal people of Australia) of the indigenous people of Australia in the fifteenth century.[40] This annotation does not appear on any of the European maps mentioned in this chapter (Figs. 2.2–2.4), nor on KWQ.

New Zealand is not depicted on the De Virga World Map, although it is depicted on the 1763 copy of the 1418 Chinese World Map/Zheng He Map. New Zealand is also completely missing from *Terra Australis Incognita* of the three European maps. A coincidence? Antarctica is embedded inside *Terra Australis Incognita* and Mo Wa La Ni Jia (墨瓦蠟泥加), instead of as a separate land in the 1418 Chinese World Map/Zheng He Map, which contains no other geographical information.

In summary, whether the 1763 copy of the 1418 Chinese World Map/Zheng He Map is authentic or not, the above discussion tells us that De Virga's world map apparently was influenced by the 1418 map, and Zheng He's mariners most likely had explored Australia, New Zealand and Antarctica no later than 1418, during their fifth voyage. However, it seems that they did not survey the interior of these lands as much as they did in their sixth voyage in the 1420s. Many new geographical data obtained from this voyage were depicted into the region of Mo Wa La Ni Jia (墨瓦蠟泥加) as can be seen on KWQ. But why the voyage in the 1420s, not the seventh voyage in the 1430s? The answer will be given in Section 4.

Including new southern lands as part of *Terra Australis Incognita* was the principle followed by the European cartographers even after Portuguese navigator Ferdinand Magellan became the first European to enter the Strait of Magellan in 1520.[41] The strait is located near the tip of South America, separating *Tierra del Fuego* from the continental mainland.[42] Since then, *Tierra del Fuego* was drawn into the sixteenth century *Terra Australis Incognita.* Magellan did not cross the water of Drake Passage to explore Antarctica. But why is the Antarctic Sound depicted on the European 1569 and 1570 world maps? The rich geographical information contained in Mo Wa La Ni Jia (墨瓦蠟泥加) are the results of Chinese explorations in the 1420s and the Antarctic Sound is clearly depicted on KWQ. This seems to indicate that after the Chinese discovered Antarctica in the 1420s during their sixth voyage to the Western Ocean, Mercator and Ortelius copied the Antarctic Sound from a Chinese map.

To explore the Antarctic Sound, the Chinese had to cross the body of water — Drake Passage — between the Land of Fire and the Antarctic Sound. This was about 100 years before the first European — *Spanish navigator Francisco de Hoces* — spotted that same body of water in 1525.

By the time Matteo Ricci was working on KWQ in China, he must have had the Chinese source maps of Australia, New Zealand, Land of Fire and Antarctica on hand. The Chinese mariners had explored all the southern continents no later than the 1420s, otherwise how could Matteo Ricci fill so much geographical information into the map of Mo Wa La Ni Jia (墨瓦蠟泥加)? However, he most likely was not willing to abandon the European concept of *Terra Australis Incognita* and their cartographic method of *including new lands as part of hypothetical continents.* He drew the shape of Mo Wa La Ni Jia (墨瓦蠟泥加) following the footsteps of the earlier European cartographers in the sixteenth century by inserting all the Chinese geographical information into the map of expanded Australia, New Zealand, Land of Fire and Antarctica to make it appear as a huge and contiguous Mo Wa La Ni Jia (墨瓦蠟泥加)/*Terra Australis Incognita.* Hence, KWQ looks very much like a European map at first instance. As a result, hundreds and thousands of years of Chinese marine history were lost into that void. And for more than 400 years, *no one else* seemed to have questioned and investigated the true meaning of that huge void on KWQ!

Finally, by the early seventeenth century, as soon as the first European reached Australia in 1606, this imaginary *Terra Australis Incognita* concept collapsed. Figure 2.16 left panel shows the 1658 Nicolaes Visscher World Map[43] in which a portion of Western Australia is depicted and there is no longer the huge, contiguous and empty *Terra Australis Incognita.* Figure 2.16 right panel

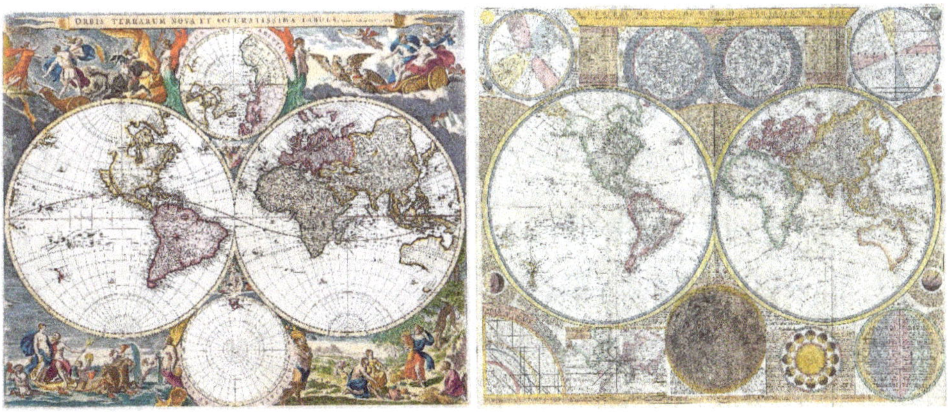

Fig. 2.16. Left: This double hemisphere map, *Orbis Terrarum Nova et Accuratissima Tabula*, showing no *Terra Australis Incognita*, was created by Nicolaes Visscher in 1658 in Amsterdam (see note 43; public domain). Right: The 1794 Samuel Dunn World Map shows all of Australia and New Zealand (see note 44; public domain).

shows that by the time of the late eighteenth century, the 1794 Samuel Dunn World Map[44] includes all of Australia and New Zealand, which are already very close to what we know today.

The first European to spot the northern end of the South Island in Golden Bay (the Island's West Coast) of New Zealand was Dutch navigator Abel Tasman on 13 December 1642. Tasman sketched sections of the two main islands' west coasts. The narrow islands of New Zealand did not appear on a world map until 1646. Nearly 130 years passed before British explorer James Cook circumnavigated and mapped New Zealand in October 1769.[45] These explorations finally put an end to the theory of a "great southern continent". But even when European settlers began arriving, most of the New Zealand interior remained uncharted.[46]

It was also known that the Māori, a group of Polynesian people, settled there sometime between 1200 and 1300.[47] By the time Abel Tasman reached New Zealand, the Māori had settled the land.

In the article titled "Who really discovered New Zealand?", author Alfie Shaw writes the following[48]:

> New Zealand lies at the bottom of Oceania, a region made up of thousands of islands scattered across the massive expanse of the Pacific Ocean, and these islands were explored and settled centuries before Europeans arrived by various Pacific peoples.

New Zealand can be identified on KWQ as part of Mo Wa La Ni Jia (墨瓦蠟泥加) in Figs. 2.1 and 2.10, but it is not depicted on any of the three major sixteenth century European maps discussed in this chapter. In Section 4, I shall show that the various "Pacific peoples" mentioned by Alfie Shaw *did* include the Chinese settlers, and their presence on New Zealand can be traced back to c. 1500 B.C. or earlier.

4. Archaeological Evidence of Chinese Settlement in New Zealand since pre-B.C. Era until Being Virtually Wiped Out in the Mid-1430s

My map analysis in Section 3 supports that the ancient Chinese explored New Zealand no later than in the 1420s. In fact, they also visited New Zealand in the 1430s, but something tragic happened, and the unfortunate event will be detailed in this section.

However, apart from the map evidence, do we have other evidence to support Chinese settlement in New Zealand long before the Europeans' arrival? According to T.C. Bell, his survey findings in New Zealand offer further support of Chinese presence in New Zealand since c. 1500 B.C. or earlier.

Mr. Bell[49] is a retired British marine engineer and production manager for Castrol. He has devoted his full time during the past 26 years surveying sites worldwide. Being an expert in Roman and Chinese engineering, he has sought evidence of ancient Chinese exploration of the world, especially the Roman remains in Britain, showing that Chinese and Romans collaborated in their engineering work in ancient days. Part of his effort also involves exhaustive surveys on Cape Breton Island in eastern Canada and on sites in New Zealand.[50] His survey findings have added valuable input to Chapter 1 of this book titled "Chinese Explored Cape Breton Island long before the Europeans",[51] as well as my 2019 book titled *The Last Journey of the San Bao Eunuch, Admiral Zheng He*,[52] and this chapter.

His multiple visits during 2003–2006 to New Zealand have convinced Bell that Chinese ships were there over 3,500 years ago. But due to space limitation, I was not able to present all his discoveries in New Zealand in my 2019 book. I am grateful to Bell again that he gives me the permission to analyse more of his valuable findings and publish the following results in this chapter.

(1) Bell located 50 Chinese sites including harbours, mining sites and cities/villages/towns in New Zealand. The three cities are Waitangi on North Island, Timaru and Christchurch on South Island (see these three cities in Fig. 2.9). In these cities, the foundations of the Chinese houses are still in place. Bell calls them cities due to their physical sizes. Carbon dates of samples collected by Bell from the Chinese sites have been determined (by Rafter Radiocarbon Laboratory and Waikato University in New Zealand)[53] to be around 190 B.C. and from the iron smelting site around 1066 A.D.; petroglyphs located on the east coast of North Island, inscribed on the massive gateway of a Chinese town, indicate Chinese presence no later than c. 1500 B.C.; it was examined by a Chinese expert in the National Library of Beijing, China, to belong to the Erlitou culture (二里头文化; c. 1750 to 1530 B.C.)[54] of the Shang Dynasty (the earliest ruling dynasty of China in recorded history; it ruled from 1600 to 1046 B.C.).

Weng Cheng (瓮城)[55] or Urn fort (see Fig. 2.17)[56] is a type of Chinese fort. New Zealand's *Weng Cheng* gateway site is the first *Weng Cheng* site Bell noticed in the world. The gateway in North Island formed the entrance to a massive Chinese city, now called Waitangi (a Māori-language name meaning "weeping waters"). This Chinese walled city was constructed in the Bay of Islands (may be identified as Xiao Dao Bin (小島濱) or Near-Island Coast/coast near today's Bay of Islands on KWQ), New Zealand, at the mouth of the Waitangi River. Interestingly, it was on the site revered by the Māori, where the Treaty of Waitangi was first signed on 6 February

Fig. 2.17. Left: A *Weng Cheng* (瓮城) or Urn fort (see endnote 56; public domain) in China. Right: Remains of *Weng Cheng* gateway to Chinese city of Waitangi; the Chinese walled city was constructed on North Island in the Bay of Islands, New Zealand, at the mouth of the Waitangi River. Photo by T.C. Bell 2003; courtesy of T.C. Bell.

1840 between the European Settlers and the Māori.[57] However, it is sad that this *obvious* Chinese *Weng Cheng* site has been overlooked.

A *Weng Cheng* (瓮城) or Urn fort is an additional semi-circular enclosure (half-round walls) built on the outside (or inside; or both; in New Zealand, the enclosure was built on the outside) of the city gate. The design and construction style are the hallmark of a Chinese development. External gateways of the *Weng Cheng* have one gate leading into the enclosure to watch and control arriving people, and one gate leading from there into the city/village/town. The design has been the same on the aforementioned two city sites located: Waitangi and Timaru. According to Bell, there are three more *Weng Cheng* sites in the world; all are in England,[58] as follows: one is in his local small town (Penrith, Cumbria) and controlled a small settlement; the second one is on Hadrian's Wall[59] at a site called Limestone Corner (in northern England); and the third one is on the Chinese fort guarding a barracks on the metallic ore quarrying site at Hardendale (in Cumbria), 32 km/20 mi south of Penrith.

The origin and development of *Weng Cheng* can be traced back to at least more than 2,200 years ago. The archaeological excavations of the ancient city of Zheng Han (郑韩故城) in Xinzheng City (新郑市), Henan Province (河南省), and of the ancient city of Erlong Lake (二龙湖古城) in Lishu County (梨树县), Jilin Province (吉林省), are good examples that *Weng Cheng* started to take its characteristic shape during the Spring and Autumn Period (770 B.C.–476 B.C. or according to some authorities until 403 B.C.)[60] and the Warring States period (481/403 B.C.–221 B.C.).[61] In the later dynasties, *Weng Cheng* matured from its initial function for basic combat to a comprehensive structure, integrating housing of soldiers and storage of grains, while serving as arrow towers and observation stations.[62] By the time of the Ming Dynasty, *Weng Cheng* was a necessary defensive structure for larger cities with complex internal structure and mature construction skills. The construction materials of *Weng Cheng* also developed from the original mud into externally coating with bricks, and then evolved into large stones, bricks bonded by lime and other building materials such as rice cohesive (to be discussed further in what follows) to form a strong structure.

(2) A rock which formed part of the outer gateway of the walled road to an "unknown" location[63] in the North Island of New Zealand was discovered. A photo of this rock was published in my 2019 book.[64] The rock has an "eye" carved into it, which kept watch on those entering the gate. Some writing has been carved into the rock face.[65]

(3) A total of about 78 Chinese shipwrecks on both North Island and South Island were discovered by Bell using magnetic anomaly and in a few corroborative cases by intrusive satellite searches carried out by Tim Akers, a UK-based marine archaeologist. In my 2019 book, I have that "Bell discovered in New Zealand that every Chinese vessel he has surveyed had a special kind of anchor in two sizes and they always identify a Chinese junk". They consist of two blocks approximately 3 m/9.8 ft long and 0.5 m/1.6 ft wide. The blocks are shaped to form a fluke at each end. A square hole was formed at the centre of each block and the blocks are connected by an eight-metre-long wooden stock. The anchor cable (rope) was attached to the centre of the stock. The design ensured that the two flukes always dug into the seabed. This design was adopted as the modern ships' anchors, and they are far superior to the Admiralty anchors.[66,67] With this important distinction, Bell was able to differentiate a Chinese shipwreck from a European one (Fig. 2.18).

Bell's surveys record that only two visible Chinese vessels, above shore level, have been located to date. The rest were sand-covered Chinese shipwrecks including 24 vessels of size of the order of c. 100 m × 50 m/328 ft × 164 ft (12 on North Island and 12 on South Island; see Table 2.1).

Table 2.1 lists all the Chinese junk wrecks at various locations, which Bell located in New Zealand and their approximate dimensions. His 2003 survey covered those on South Island,

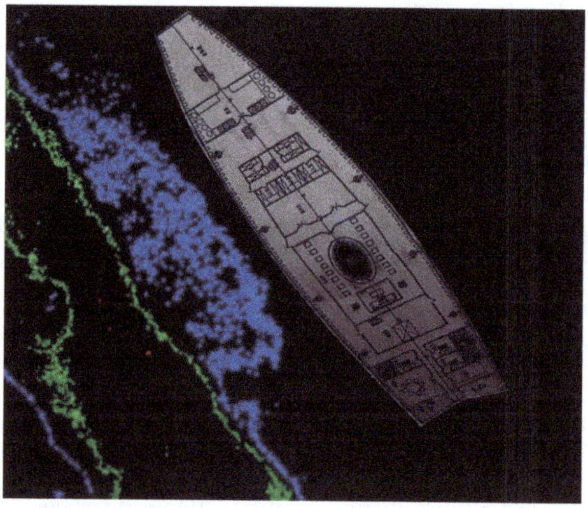

Fig. 2.18. Left (the blue-coloured shipwreck) is a wrecked Super Junk, c. 120 m × 48 m/394 ft × 157 ft, in New Zealand; sand covered and firstly located on shore at low tide by Magnetic Anomaly Survey (MAS) by T.C. Bell in 2006, then photographed by enhanced satellite by Tim Akers. The photo is owned by Bell. Note the *unusual extension* to the left of the bow. Similar bows were noticed on other wrecked junks, indicating possible anchors. It is a distinguishing feature of the Chinese Super Junks. Right (hand-drawn ship) is a drawing of a Super Junk from a Chinese shipyard (shipyard photo from Singapore 600-year Zheng He Exhibition taken by T.C. Bell, exhibitor). Courtesy of T.C. Bell.

Table 2.1. Summary of Total Junks located & their approximate dimensions.

2003 South Island February, June, and October/November Surveys				
			37 m × 15 m = 1	18 m × 7 m = 2
			34 m × 14 m = 1	16 m × 8 m = 2 Sampans = 4
100 m × 50 m	= 12		28 m × 11 m = 1	11 m × 8 m = 2
47 m × 11 m	= 36		28 m × 14 m = 1	20 m × 19 m =1 Raft
			22 m × 6.5 m =1	13 m × 10 m =1 Raft
			20 m × 7.0 m =1	
Over 40 m	= 48 (a)	20–40 m	= 6 (b)	Below 20 m = 8 (c)
TOTAL of		(a) = 48 Junks over 40 m	(c) 8 below 20 m, some possibly	
		and (b) 6 Junks 20 m–40 m	horse transports or mobile rafts	
TOTAL 2003		= 54 Junks over 20 m		
November 2006 Survey North & South Islands				
NORTH ISLAND				
Super Junks c. 100 m × 50 m (Treasure Ships) located				Harbour Bays
Waitangi	0			1
Dargaville	6	West Coast		6
Ahipara	1	West Coast (South end of 90 Mile Beach)		1
Makuta	1	East Coast (South of Tauranga)		
Coromandel	3	East Coast		3
Mangakuri Beach	1	East Coast (Dannevirke)		
Pourere Beach	0			2
North Island		12 Super junks		13
SOUTH ISLAND				
Moeraki		12		10
& Catlins			Further harbours anticipated at Catlins	
Total Super Junks all surveys North Island 12 + South Island 12 = 24				
Total Super Junk Harbour Bays North Island 13 + South Island 10 = 23				
Total wrecked junk sites located over 20 m in 2003/2006 surveys = 78				

Notes: These measurements are paces and to be taken as approximate.

Moeraki c. 120 m × 48 m and Wakanui (Maori for big ship!!) c. 47 m × 11 m visible.

and the 2006 survey included specifically Super Junks, of the order of c. 100 m × 50 m/328 ft × 164 ft, on both South Island and North Island.

There were 23 super-junk harbours (13 on North Island and 10 on South Island; also see Table 2.1), specifically designed for these large vessels with keel slots, flushed to remove debris and hull support pads at low tide.[68] The keels of these sailing ships are *not* one continual piece but lengths "scarfed" together (by angled jointing).

What has happened? Why were there so many Chinese shipwrecks including 24 Super Junks (Treasure Ships) in New Zealand when Bell surveyed the place during 2003–2006? What has drawn him there to carry out such an extensive archaeological survey? I wanted to find out more.

What surprised me was a report in 2004[69] in which it states that "More than five centuries ago, powerful tsunamis roared across parts of New Zealand, Australia and surrounding islands — waves that reached hundreds of feet up the mountainsides. Native peoples abandoned these regions, and legends of island-killing waves were born", and "The idea was born in 2003, when researchers discovered an impact crater on the sea floor near New Zealand. They named it Mahuika after a native god of fire. The crater is a dozen miles across and about 500 feet deep". I began to understand why Bell went to New Zealand in 2003.

Dr. Edward Bryant of the University of Wollongong carbon-dated the debris (some of it is 220 m high on the cliffs), washed onto the rocky shore by the tsunami, and estimated the impact to be around 1500 A.D.[70] A tsunami is a series of giant waves caused by earthquakes or volcanic eruptions under the sea, and as the waves travel inland, they build up to higher and higher heights. Other scholars suggested the impact around 1443 A.D.,[71] or 13 February 1491 A.D.[72] Dr. Bryant did further research and suggested that this disaster happened only shortly before European colonisation.[73] Despite the fact that there is debate about the origin of the Mahuika crater, scientists suggested that a meteorite (when a meteoroid — any fairly small natural object from interplanetary space — survives its passage through the atmosphere to reach the surface of our Earth, we call it a meteorite) striking out the sea floor near New Zealand, and the fire from the sky and the tsunami observed by the Aboriginals and in the Māori stories, could explain most plausibly the huge number of Chinese shipwrecks Bell has located.

One carbonised c. 47 m × 11 m/154 ft × 36 ft junk (see Fig. 2.19) was washed out of the ocean due to the same tsunami and entombed on a 23 m high cliff at Wakanui (see Fig. 2.9); hence the Māori name Wakanui, which translates as Big Ship.[74]

Despite some earlier concerns about the huge dimensions of the Chinese ships,[75] Bell thinks that the sizes of the Chinese ships are *not* exaggerated. He *did* discover a Chinese Super Junk, c. 120 m × 48 m/394 ft × 157 ft, arced over the tsunami-formed cliff at Moeraki (see Fig. 2.9) due to the same meteorite that crashed into the sea in the fifteenth century near New Zealand and wiped out all the ships in the harbour. Bell was able to see the concrete from the hull lining still appearing as the vessel is slowly breaking up (see Fig. 2.20), and the massive size is *fully* exposed (see Fig. 2.21).

Bell's discovery of the wreck of a Chinese Super Junk aroused my curiosity to ask another question: when did these Chinese Super Junks arrive at New Zealand? Which fleet did they belong to? I have mentioned earlier that the unfortunate event of a meteorite crashing into the Tasman Sea, south-west of New Zealand, was thought by scholars to occur in the fifteenth century — they suggested 1443 A.D. or 1491 A.D., a 48-year difference — but both dates are well before European colonisation.

Treasure Ships (a type of large wooden ships) were among Zheng He's fleet sent by the Ming Yongle Emperor (永乐皇帝; reigned from 1402 to 1424) and Xuande Emperor (宣德皇帝; reigned

Fig. 2.19. Wakanui (meaning Big Ship in Māori): Site of wrecked junks on the east coast of South Island, New Zealand. Here is the full view of a junk. This junk's hull was crushed to 50 mm/1.97 in and splayed out to 24 m/79 ft at its widest point. These are the remains of the overturned, splayed out and crushed hull of a c. 47 m × 11 m/154 ft × 36 ft Chinese junk, entombed in the 23 m/75 ft high tsunami-formed cliff. The tsunami was formed by a meteorite crashing into the sea near New Zealand. The entombed junk proves the presence of the Chinese and should also date the event (to be discussed in the following). This junk, and probably many others, were carbonised by heat from the low-flying meteorite crashing into the sea near New Zealand. The adjacent 12 m/39 ft wide canal linking Lakes Forsyth and Ellesmere to the Ashburton River at Hakatere suggests that a convoy of junks was in the canal when carbonised. Photo by T.C. Bell 2006; courtesy of T.C. Bell.

Fig. 2.20. Moeraki, New Zealand: Site of Chinese Treasure Ship, c. 120 m × 48 m/394 ft × 157 ft, blown out of the nearby harbour onto cliff by tsunami, following being carbonised by a low flying meteorite which caused the tsunami when it struck the Tasman Sea, south-west of New Zealand. Concrete from the hull lining is still appearing as the vessel is slowly breaking up and the cliff eroding. Bell's son is in the photo. Photo by T.C. Bell 2017; courtesy of T.C. Bell.

Fig. 2.21. Here is the full view of the same junk, c. 120 m × 48 m/394 ft × 157 ft in Fig. 2.20. Note the visible timbers of hull "T". From left (Paddy, Bell's wife in red), to right, there are three visible coated Moeraki balls breaking out of the hull of the junk. These irregular shaped concretion cores have man-made coatings applied to form a regular shape. The balls had a function within the junk, possibly as sail-balance weights. The concretions emanate from North Island, *not* South Island. Badger Bloomfield located both source and dumps of uncoated concretions. Photo by T.C. Bell 2006; courtesy of T.C. Bell.

from 1425 to 1435; grandson of the Yongle Emperor) to the Western Ocean for seven voyages, altogether. But soon after the seventh voyage, China decided in the mid-1430s to withdraw completely from the world stage. Long-distance sea voyages were prohibited, Treasure Ships were burned at anchor and the construction of new long-range ocean vessels was strictly forbidden, followed by imposing *Haijin* or Sea Ban. By 1525 all large types of ships were destroyed. Despite the fact that the Ming emperor partially lifted the *Haijin* starting from 1567, a series of related isolationist Chinese policies restricting private maritime trading and coastal settlement were imposed during most of the Ming Dynasty and part of the Qing Dynasty. For the next few hundred years, no Chinese Super Junks were sailing in the world oceans. *Hence, the Chinese Junks Bell located must belong to Zheng He's fleet in the early fifteenth century.* My task is to determine *which of Zheng He's fleets*.

In Section 3, I mentioned that all the geographical information depicted in the map of Mo Wa La Ni Jia (墨瓦蠟泥加) on KWQ were the work of mariners of Zheng He's Sixth Fleet, because ships from the previous five voyages all returned to their home port in China after the voyages ended; and ships reaching New Zealand in the 1430s from Zheng He's Seventh Fleet all perished in New Zealand in a tragic event as will be explained soon. It is a puzzle that Gavin Menezes previously suggested that of the Chinese ships in the southern ocean during their sixth voyage in the 1420s, all but one were destroyed by the above-mentioned unfortunate meteorite event.[76] In fact, such a catastrophic event could not spare any ship from misfortune and all the ships and survey data must have been carbonised, and lives lost, unless the survived ship can be *proven* to have been in an area very far away from the site. Hence, it is very unlikely that the Chinese Jumbo Junks located by Bell were from Zheng He's Sixth Fleet.

However, as I suggested in my 2019 book,[77] some members of Zheng He's Seventh Fleet, after exploring the inland of North America, could have taken a route to New Zealand on their way home and arrived there around mid-January to mid-May 1435 A.D.[78] There, they met with this unfortunate event and were virtually wiped out. This viewpoint can explain that the meteorite crashing into the Tasman Sea, south-west of New Zealand, took place around the mid-1430s, perhaps in 1435/1436 A.D., hence, the Chinese Super Junks Bell located in New Zealand were from Zheng He's Seventh Fleet.

(4) The carbonised junk (see Figs. 5.20 and 5.21) dumped at Moeraki by the tsunami was lined with concrete, bonded to the inside of the junk's hull with rice adhesive. The concrete lining the hull is not one continuous length but cast between the vessel's ribs. So, the hull is still flexible to some degree. The concrete took on the shape of the embossing of the Inspector's stamp, see Fig. 2.22.

Bell writes that a New Zealand geologist claimed that the hull's concrete was an igneous dyke.[79] There are no dykes recorded on the area's geology map and certainly never concrete ones. The analysis of the concrete by a New Zealand university showed it was identical to Roman waterproof concrete. It contained c. 28 percent volcanic rock and has very fine composition.[80] The concrete both stiffened the hull and provided the necessary ballast. Carbon dating of the hull timber was negated by the frequent exposure to salt water.

The Chinese lettering imprinted in the concrete was yet further confirmation of the origin of the vessel. When Bell first saw the arched Super Junk, there was a considerable amount of concrete from the hull lining lying on the sand. The concrete was cast in the shape of the spaces between the hull's timbers, plus some mortar or cannon balls, to be discussed soon in Item 5.

Fig. 2.22. Moeraki, New Zealand: Concrete lining from the hull of a Chinese Treasure Ship of size c. 120 m × 48 m/394 ft × 157 ft. The junk was carbonised by heat from a meteorite crashing into the Tasman Sea and blasted out of the nearby harbour by a tsunami. The solid red arrow points to the Chinese script; it is presumably the Inspector's approval stamp. Photo by T.C. Bell 2006; courtesy of T.C. Bell.

Rice adhesive or sticky rice mortar was invented by the ancient Chinese around the sixth century during the Southern and Northern Dynasties (南北朝; 420–589 A.D.). It was one of the greatest technological innovations of the time — using sticky rice soup mixed with slaked lime to make an inorganic–organic composite mortar that had more strength and water resistance than lime mortar.[81] During the Ming Dynasty (1368–1644 A.D.), techniques improved significantly, and the Great Wall sections used lime mortar and sticky rice to reinforce the bricks to resist earthquakes, see Fig. 2.23.[82]

The discovery of *rice adhesive used uniquely for binding concrete to the inside of the junk's hull is direct (hard) evidence of Chinese junks and Chinese presence in New Zealand* before a meteorite crashing into the sea near New Zealand in the mid-1430s.

(5) Bell discovered Chinese stone cannon and mortar balls (ammunition)[83] near the shipwrecks (see Fig. 2.24). Their sizes are 200 mm/8 in (weighing 10 kg/22 lb; near Shag Point and Moeraki, South Island, and Mangakuri, North Island), 300 mm/12 in (partially exposed on the sand below the Moeraki Chinese shipwreck; weighing c. 33 kg/73 lb) and 430 mm/17 in diameter (at North Island and near Shag Point in South Island shipwreck sites).

These cannon/mortar balls from Chinese vessels were carbonised by the same meteorite crashing into the sea near New Zealand in the mid-1430s, then either washed out of their harbour or washed ashore by the tsunami created by the meteorite. Later in 2017, an earthquake lifted the New Zealand seabed by around one metre and displayed several spherical stone balls and the base of the Chinese Harbour in Kaikoura.[84]

On his next visit, Bell discovered that the Government had swept the beach clear, cannon balls and all,[85] he could see nothing anymore.

Bell's record continues[86]:

> The Tsunami also covered most of the coastal settlements with debris from the seabed. Many of these vessels might have been part of the Pacific fleet gathering cargoes from the Pacific islands and might have involved some from the Chinese Seventh Fleet. Sadly, they chose the wrong time to meet in the fixed harbours, hence the demise of all the sailors. And we will never know how many residents and their Māori workers/slaves died.

Fig. 2.23. The bricks in the Great Wall of China are held together by sticky rice mortar (see endnote 82; public domain).

Fig. 2.24. Moeraki, New Zealand: Site of Chinese Treasure Ship, c. 120 m × 48 m/394 ft × 157 ft, blown out of nearby harbour onto a cliff by a tsunami, followed by being carbonised by a low-flying meteorite which caused the tsunami when it struck the Tasman Sea. Mortar balls from vessels are still being revealed as the cliff erodes. Photo by T.C. Bell 2017; courtesy of T.C. Bell.

The details of the sailing routes of some members of Zheng He's Seventh Fleet have been given in my 2019 book including those who sailed in 1434 from North America and reached New Zealand in 1435, but never made it home to China.[87] Hence, that tragic event vanished from history until Bell discovered the shipwrecks in New Zealand nearly 570 years later.

(6) Moeraki on South Island is the claimed home for the famous boulders — the "Moeraki balls"; so many were located because Moeraki is the site of the largest number of Chinese harbours in New Zealand, having 10 harbours designed for Treasure Ships (see Table 2.1). Bell explains the following (my comment is between square brackets)[88]:

> As the vessels in harbour disintegrated under the combined influence of the heat of the meteorite and the following tsunami, so the balls aboard, non-modified and the ships own infrastructure of modified balls spewed out into the bay, some can be seen today. Similar "modified balls" can be viewed in the North Island alongside Chinese wreck sites.
>
> Within the hull of the Moeraki shipwreck are two Moeraki boulders, both have a rough natural concretion core with a man-made outer, forming a perfect ball. Alongside the hull, is a similar one. Thus, proving the balls had a function within the vessel's hull. Geology maps of the area show no evidence of concretions. The presence of Moeraki balls is evidence of the wrecks of large Chinese vessels.
>
> In North Island I also located one large dump of uncoated concretions alongside a harbour ready for shipment (see Fig. 2.25); these concretions had modified ends to insert timber to roll them from the find source to the harbour ready for shipment. In North Island, massive concretions from

4 m/13 ft diameter were used by the Chinese as gateways to a large town. Chinese epigraphs on the concretions indicated a pre-B.C. origin.

(7) Bell located the barracks for the Chinese sailors, who slept ashore when in their home ports, indicating their crew size. This assumes that a nine-roomed barracks, with eight unpaved rooms, and one paved for the officer, each 4 m square, held eight sailors per room, a total of 64 men plus an officer. Roman barracks had 10 unpaved and one paved room, holding a total of 80 men plus an officer. The barracks alongside the harbour designed for a c. 120 m × 48 m/394 ft × 157 ft Treasure Ship held c. 768 men plus officers. The crew for a c. 47 m × 11 m/154 ft × 36 ft standard work Horse Junk,[89] by the same barrack block counting, numbered c. 360.

As stated in my 2019 book: "New Zealand would appear to have been the Chinese Southern Pacific Base during Chinese mariners' earlier voyages (we do not know when) as well as during their seventh voyage. Wear and tear of the Chinese ships required both repairs and replacement. Returning to China for repairs was out of the question due to the distance and with an unseaworthy vessel".[90] New Zealand's superb hardwoods and massive trees, including tall ones very suitable for masts, would have been a much-valued source of timber for China.

In summary, all the findings in this section reveal that from no later than c. 1500 B.C. Chinese knew that New Zealand consists of two main landmasses — the North Island and the South Island — that were surrounded by oceans from all sides.

New Zealand has been *deliberately* embedded into the contiguous and expansive imaginary land of Mo Wa La Ni Jia (墨瓦蠟泥加) on KWQ by Matteo Ricci. Did he try to make KWQ look

Fig. 2.25. A collection of natural cores in North Island, New Zealand, which the Chinese used as the cores (voidage) for the Moeraki balls. They were coated with concrete to form spheres for use on the Super Junks. Adjacent is a Chinese slot harbour. Presumably, these awaited shipment to Moeraki, prior to shipment to China. Bell's wife Paddy Bell is in the photo. Photo by T.C. Bell 2007; courtesy of T.C. Bell.

more like a European map because of his deeply held *Terra Australis Incognita* concept, or did he try to mislead the Chinese to believe that KWQ looked like a European map, hence, must be a European map? Then, who decided to insert all the Chinese geographical information of Australia, New Zealand, Land of Fire and Antarctica into the map of the contiguous and expansive imaginary land of Mo Wa La Ni Jia (墨瓦蠟泥加) to secretly retain the historical truth on the map? Were they his Chinese collaborators or Ricci himself? We may never know.

Once New Zealand is separated from the contiguous and expansive imaginary land to the west from Australia, about 2,000 km/1,200 mi across Tasman Sea, and to the east from South America, about 12,500 km/7,500 mi across the South Pacific Ocean, Australia will also separate from South America, because the De Virga World Map in Fig. 2.14 (left panel) shows Australia as a separate continent. That knowledge could only come from the Chinese.

Between about 1200 and 1300, Polynesians began to settle on the islands of New Zealand, and then developed a distinctive Māori culture. In 1642, Dutch explorer Abel Tasman became the first European to sight New Zealand, and in 1769 Captain James Cook was the first European to circumnavigate New Zealand. But Bell finds strong archaeological evidence that the Chinese were there no later than c. 1500 B.C., *more than 3,200 years before the Europeans*, and this chapter shows that the map of Mo Wa La Ni Jia (墨瓦蠟泥加) on KWQ uniquely depicts Xiao Dao Bin (小島濱) or Near-Island Coast/coast near today's Bay of Islands in New Zealand. New Zealand is not depicted on any of the three major sixteenth century European maps.

But before I close the report of this research, one big question remains: was Zheng He among the Chinese mariners and all the ships which were wiped out in the catastrophic tragedy which happened to New Zealand in the mid-1430s? Clearly, his fate remains a mystery, because he seems to have disappeared from history in the middle of his seventh voyage. There is *no* proof that he died at or near Calicut[91] on his journey returning to China, or that he returned to China, or that he died in North America (even though the medallion (宣德金牌) bestowed upon him by the Xuande Emperor was discovered in a remote town in North Carolina, USA, in the early 2000s).[92] In the latter case, Zheng He could have bestowed the medallion upon some of his mariners who decided to stay in North America, as a symbol of remaining with them, or to the aboriginals as a souvenir. This could have happened before he bade farewell to them and led the remaining fleet members back to China via New Zealand (then China's Pacific base) to complete his last world-wide voyage. This present chapter points to another possible fate of Zheng He, namely that, in New Zealand, he and his crew were the victims of an unexpected meteorite crashing into the Tasman Sea, which wiped them out in an instant, and buried them all together inside their beloved ocean waves, forever.

5. Conclusions

In this chapter, I have analysed in detail the geographical information contained in the map of the contiguous and expansive southern land of Mo Wa La Ni Jia (墨瓦蠟泥加) on Kunyu Wanguo Quantu《坤輿万国全图》(abbreviated as KWQ in this chapter), or Complete Geographical Map of

All the Kingdoms of the World. I was initially surprised to find that Mo Wa La Ni Jia (墨瓦蠟泥加) shares substantial commonality in its land shape with the three major world maps drawn by the European cartographers of the sixteenth century. Then, my in-depth analysis showed that Mo Wa La Ni Jia (墨瓦蠟泥加) contains much geographical information related to Australia, New Zealand, Land of Fire and Antarctica, which was mostly absent in the three major sixteenth century European world maps: (1) the 1569 World Map by Gerardus Mercator; (2) the 1570 World Map by Abraham Ortelius; and (3) the 1594 World Map by Petrus Plancius. My findings revealed that Chinese explored Australia, New Zealand, Land of Fire and Antarctica and obtained their geographical information long before the Europeans.

But I was still curious why the land shape of Mo Wa La Ni Jia (墨瓦蠟泥加) looks so much like *Terra Australis Incognita* on these European maps. Then I found out that the ancient Europeans once thought that there might be a huge continent in the southern hemisphere to balance the lands in the northern hemisphere, and *when new lands were discovered, they were often considered to be part of the hypothetical Terra Australis Incognita*. Matteo Ricci was a European. Hence, when he was making KWQ, he deliberately followed this European cartographic method by expanding and merging the actual Australia, New Zealand, Land of Fire and Antarctica on the Chinese source maps in his hands into the map of the huge and contiguous Mo Wa La Ni Jia (墨瓦蠟泥加), to make KWQ look almost like a European map. However, perhaps to show his map as more advanced than the other contemporary European maps or to show his indebtedness to the Chinese people, he and his collaborators embedded all the geographical information from the Chinese source maps — obtained by Zheng He's mariners during their sixth voyage in the 1420s — into the void of Mo Wa La Ni Jia (墨瓦蠟泥加) on KWQ. This void has misled the later generations for over 400 years until the truth was uncovered and presented in this chapter.

The Moeraki boulders, cannon/mortar balls, mining sites, *Weng Cheng* in cities/towns/villages, Chinese shipwrecks, harbours, sailor barracks and so on in New Zealand, discovered by Bell during his extensive survey work there from 2003 to 2006, reveal yet another aspect of the same story from Ricci's. It rekindles our memory of the epic voyages accomplished by Zheng He's mariners and the great Treasure Ships of the Ming Dynasty.

These mariners, during their expeditions, must have known well that not all lands were contiguous, and to them, *Terra Australis Incognita*, if it had ever existed in their mind, would have merely meant a possible continent, still to be explored in full. Led by that exploring spirit and courage, they continued their voyages and finally explored Australia, New Zealand, Land of Fire and Antarctica during their epic sixth voyage in the 1420s, long before the Europeans. But for some of them, I present strong evidence that their journey back home via New Zealand, after exploring North America during their seventh voyage, tragically ended in the mid-1430s, due to a meteorite crashing into the Tasman Sea, south-west of New Zealand, which wiped them out. Hence, the rich geographical information shown inside Mo Wa La Ni Jia (墨瓦蠟泥加) on KWQ was obtained by Zheng He's mariners during their sixth voyage, not their seventh voyage.

Today, when we begin to understand what really happened and what KWQ can truly reveal, we must be grateful for those Chinese mariners who made all the discoveries which paved the way for the "Age of Discovery".

Acknowledgements

The author thanks T.C. Bell for sending his precious survey photos and valuable survey findings for their publication in this chapter, for sharing his insights and advice and for approving the present chapter before its publication. The author also thanks Ian Hudson for reviewing this chapter prior to its publication. The author is grateful for Professor Emeritus Michel A. Van Hove's technical assistance in preparing all the figures in published form for this chapter. Special thanks go to M. Hayes-Shuptar (华铭叮) for the academic and editorial assistance with the endnotes.

Endnotes

[1]Library of Congress. "Kun yu wan guo quan tu". www.loc.gov, Library of Congress, 2022, loc.gov/item/2010585650.
[2]Siu-Leung Lee (李兆良). *Kun Yu Wan Guo Quan Tu Jie Mi: Ming Dai Ce Hui Shi Je*《坤輿萬國全圖解密: 明代測繪世界》or *Deciphering the Kunyu Wanguo Quantu, A Chinese World Map: Ming Chinese Mapped the World Before Columbus*. Taipei, Taiwan: Linking Publishing Company (聯經出版社), 2012; Sheng-Wei Wang. "Chapter nine: Secret revealed by ancient maps". *The Last Journey of the San Bao Eunuch, Admiral Zheng He*. Hong Kong, China: Proverse Hong Kong, 2019, pp. 290–301.
[3]See Chapter 1 of this book.
[4]*Ibid.*
[5]Stefan Zweig, *et al. Magellan*. London, UK: Pushkin Press, 2012.
[6]Aristotle and Jonathan Barnes. *The Complete Works of Aristotle: The Revised Oxford Translation, One-Volume Digital Edition (Bollingen Series Book 194)*. Princeton, NJ, USA: Princeton University Press, 2014.
[7]Claudius Ptolemy, *et al. Geography of Claudius Ptolemy.* New York, NY, USA: Cosimo Classics, 2011.
[8]Carlos Pedro Vairo. *Terra Australis — History of the Cartography of Tierra del Fuego, Patagonia & Antarctica*. Buenos Aires, Argentina: Zagier & Urruty Publications, 2011.
[9]"File:Mercator 1569 world map composite.jpg". *Wikimedia Commons: The Free Media Repository*, Wikimedia Foundation, 26 Nov. 2016, upload.wikimedia.org/wikipedia/commons/4/4b/Mercator_1569_world_map_composite.jpg.
[10]"File:OrteliusWorldMap1570.jpg". *Wikimedia Commons*, 17 May 2020, upload.wikimedia.org/wikipedia/commons/e/e2/OrteliusWorldMap1570.jpg.
[11]"File:1594 double hemisphere world map by Petrus Plancius.jpg". *Wikipedia Commons*, 25 Mar. 2023, upload.wikimedia.org/wikipedia/commons/0/0d/1594_double_hemisphere_world_map_by_Petrus_Plancius.jpg.
[12]"File:Kunyu Wanguo Quantu (坤輿萬國全圖).jpg". *Wikimedia Commons*, 10 Aug. 2021, upload.wikimedia.org/wikipedia/commons/7/71/Kunyu_Wanguo_Quantu_(坤輿萬國全圖).jpg.
[13]Sylvia Iparraguirre and Hardie St. Martin. *Tierra del Fuego*. Evanston, IL, USA: Curbstone Books, 2000.
[14]Paul H. Kratoska, ed. *South East Asia, Colonial History: Imperialism before 1800*. Oxfordshire, UK: Taylor & Francis, 2001, p. 56.
[15]Siu-Leung Lee (李兆良). *Kun Yu Wan Guo Quan Tu Jie Mi: Ming Dai Ce Hui Shi Je*《坤輿萬國全圖解密:明代測繪世界》or *Deciphering the Kunyu Wanguo Quantu, A Chinese World Map: Ming Chinese Mapped the World Before Columbus*, *op. cit.*, p. 192.
[16]Multiple photographers. "Australian Native Birds: Parrots, lorikeets and cockatoos, etc". www.waratah-software.com.au, Waratah Software, Alan & Dianne, 2022, waratahsoftware.com.au/wpr-birds-parrots.shtml.

[17]Ollie Jay and John R. Brotherhood. "Climate zones for climate and humidity". www.researchgate.net, ResearchGate, researchgate.net/figure/Climate-zones-of-Australia-Courtesy-of-Australian-Bureau-of-Meteorology_fig1_305696129.
[18]Dictionary. "Cumin". www.merriam-webster.com, Merriam-Webster, 2022, merriam-webster.com/dictionary/cumin.
[19]Australia Government. "Deserts of Australia". www.ga.gov.au, Australia Government, 2022, ga.gov.au/scientific-topics/national-location-information/landforms/deserts.
[20]"File:IBRA 6 Deserts legend.png". *Wikimedia Commons*, 16 Oct. 2020, upload.wikimedia.org/wikipedia/commons/a/aa/IBRA_6_Deserts_legend.png.
[21]Editor. "locach in a sentence". https://eng.ichacha.net, Native English, 2021, eng.ichacha.net/zaoju/locach.html.
[22]David Johnson. *The Geology of Australia*. Cambridge, UK: Cambridge University Press, 2009.
[23]*Ibid.*
[24]David Bateman Ltd. and New Zealand Department of Internal Affairs. *Bateman New Zealand Historical Atlas*. Auckland, New Zealand: David Bateman in association with Historical Branch, Dept. of Internal Affairs, 1997.
[25]*Ibid.*
[26]Francisco Coloane. Howard Curtis, trans. *Tierra del Fuego*. New York, NY, USA: Europa Editions, 2008.
[27]Staff writers. "Antarctic Sound dimensions and namesake". https://oceanwide-expeditions.com, Oceanwide Expeditions, 2022, oceanwide-expeditions.com/to-do/experiences/antarctic-sound.
[28]Otto Nordenskiold, *et al. The Swedish Antarctic Expedition*. Washington, DC, USA: Smithsonian Institution Annual Report, 1904.
[29]Ernest S. Dodge. *The Polar Rosses: John and James Clark Ross and Their Explorations*. London, UK: Faber, 1975.
[30]"File:Map of Transantarctic Mountains.jpg". *Wikipedia Commons*, 18 Mar. 2021, upload.wikimedia.org/wikipedia/commons/f/fc/Map_of_Transantarctic_Mountains.jpg.
[31]Gunter Faure and Teresa M. Mensing. *The Transantarctic Mountains: Rocks, Ice, Meteorites and Water.* Berlin, Germany: Springer, 2010.
[32]John Sugden. *Sir Francis Drake*. New York, NY, USA: Henry Holt and Company, 1991.
[33]Charles E. Nowell. "The Loaisa expedition and the ownership of the Moluccas". *Pacific Historical Review*, vol. 5, no. 4, Dec. 1936, pp. 325–336. https://www.jstor.org/stable/3632888.
[34]"File:PtolemyWorldMap.jpg". *Wikimedia Commons*, 29 Sept. 2022, commons.wikimedia.org/wiki/File:PtolemyWorldMap.jpg.
[35]"File:DeVirgaDetail.jpg". *Wikipedia Commons*, 12 Dec. 2014, upload.wikimedia.org/wikipedia/commons/7/7b/DeVirgaDetail.jpg.
[36]Liu Gang. "The Chinese inventor of bi-hemispherical world map". *e-Perimetron*, vol. 2, no. 3, summer 2007, pp. 185–193.
[37]"File:Zhenghemap.jpg". *Wikimedia Commons*, 11 Oct. 2022, upload.wikimedia.org/wikipedia/commons/b/bc/Zhenghemap.jpg.
[38]Mark Peel and Christina Twomey. *A History of Australia (Macmillan Essential Histories, 57)*. New York, NY, USA: Red Globe Press, 2018.
[39]Liu Gang (刘钢). *Gu Di Tu Mi Ma: Zhong Guo Fa Xian Shi Jie De Mi Tuan Xuan Ji* 《古地图密码: 中国发现世界的谜团玄机》 or *Secret Code of Ancient Maps: The Mystery of China's Discovery of the World.*

Guangxi, China: Guangxi Normal University Press Group (广西师范大学出版社集团), 2009; Gunnar Thompson. "Dr. Gunnar Thompson's opinion of the 1418 map". www.gavinmenzies.net, Gavin Menzies, 18 Aug. 2011, gavinmenzies.net/Evidence/12-dr-gunnar-thompsons-opinion-of-the-1418-map.
[40]Josephine Flood. *The Original Australians: Story of the Aboriginal People*. New South Wales, Australia: Allen & Unwin, 2019.
[41]Stefan Zweig, *et al. Magellan. Op. cit.*
[42]Rockwell Kent. *Voyaging: Southward from the Strait of Magellan*. Middletown, CN, USA: Wesleyan University Press, 2000.
[43]"File:Orbis Terrarum Nova et Accuratissima Tabula by Nicolaes Visscher, 1658.jpg". *Wikimedia Commons*, 23 July 2020, upload.wikimedia.org/wikipedia/commons/6/6c/Orbis_Terrarum_Nova_et_Accuratissima_Tabula_by_Nicolaes_Visscher%2C_1658.jpg.
[44]"File:1794 Samuel Dunn Wall Map of the World in Hemispheres — Geographicus — World2-dunn-1794.jpg". *Wikimedia Commons*, 6 Sept. 2020, upload.wikimedia.org/wikipedia/commons/f/ff/1794_Samuel_Dunn_Wall_Map_of_the_World_in_Hemispheres_-_Geographicus_-_World2-dunn-1794.jpg.
[45]William Frame and Laura Walker. *James Cook: The Voyages*. Quebec, Canada: McGill-Queen's University Press, 2018; Philippa Mein Smith. *A Concise History of New Zealand*. Cambridge, UK: Cambridge University Press, 2012.
[46]*Ibid*.
[47]New Zealand Government. "New Zealand Now". www.newzealandnow.govt.nz, Ministry of Business, Innovation & Employment, 2021, newzealandnow.govt.nz/live-in-new-zealand/history-government/a-brief-history.
[48]Alfie Shaw. "Who really discovered New Zealand?" www.bbcearth.com, BBC, 2022, bbcearth.com/news/who-really-discovered-new-zealand.
[49]T.C. Bell is a Chartered Engineer, with eight years at sea from Junior to Second Engineer, who has spent seven and a half years as a surveyor and around 18 years as a Works Engineer, two and a half years with Delta Metal, Birmingham, employing c. 1,000 employees, and 26 years with Castrol Oil, employing c. 450, the largest Lube Oil Factory in Europe, where he worked for around 15 years as Works Engineer and 11 years as Production Manager and Deputy Works Manager.
[50]From rammed earth dykes, to stone gateways and river navigation, etc., T.C. Bell spent c. 26 years working on Roman sites. He actually lives in an ex-Roman Town and is surrounded by still visible Roman sites. He has published a book on Roman Penrith and its region, and one on Roman Scotland; he has also spent six years training with the UK expert on Roman River navigation; he has given c. 150 site surveys to UK museums and libraries. In addition, he has published a paper on the comparison between Roman and Chinese engineering and another on Roman and Chinese river and canal engineering. He also carried out the original research for the Chinese canal which linked the Caribbean to the Pacific, surveyed on site by Col. John Blashford-Snell, Britain's best-known explorer. His surveying experience extends to one survey on Cape Breton Island in Canada, five surveys in New Zealand, one in Ireland, and hundreds in Scotland and the Isles as well as Northern England over a 30-year period. This includes the hiring of expert geophysics surveyors.
[51]See Chapter 1 of this book.
[52]Sheng-Wei Wang. "Chapter eight: Zheng He's generals force their way through the Fengdu gate, but later decide to return home". *The Last Journey of the San Bao Eunuch, Admiral Zheng He*, *op. cit.*, pp. 259–270.

[53]Private email communication with T.C. Bell; courtesy of T.C. Bell.
[54]Du Jinpeng (杜金鹏), ed. *Zhong Guo Zao Qi Qing Tong Wen Hua Er Li Tou Wen Hiua Zhuan Ti Yan Jiu* 《中国早期青铜文化 二里头文化专题研究》 or *A Special Study on Erlitou Culture in Early Chinese Bronze Culture*. Beijing, China: Science and Technology Press (北京科学出版社), 2008.
[55]*Weng Cheng* is an enclosure built on the outside (or inside; or both) of the city gate with defensive systems to protect towns and cities in China in pre-modern times; it is part of the walls of ancient Chinese cities; its shape is either semi-circular or square. The semi-circular one is called Weng Cheng (瓮城) or Weng City, or *Weng Cheng*; the square one is called Fang Cheng (方城) or Square City, Fang City or *Fang Cheng*; Sheng-Wei Wang. "Chapter eight: Zheng He's generals force their way through the Fengdu gate, but later decide to return home". *The Last Journey of the San Bao Eunuch, Admiral Zheng He*, *op. cit*., pp. 259–260.
[56]"File:XiAn CityWall SouthGate3.jpg". *Wikimedia Commons*, 3 Sept. 2020, commons.wikimedia.org/wiki/File:XiAn_CityWall_SouthGate3.jpg.
[57]Claudia Orange. *The Treaty of Waitangi*. New South Wales, Australia: Allen & Unwin, 1996.
[58]Private email communication with T.C. Bell; courtesy of T.C. Bell.
[59]David Breeze. "History of Hadrian's Wall". www.english-heritage.org.uk, English Heritage, 2022, www.english-heritage.org.uk/visit/places/hadrians-wall/hadrians-wall-history-and-stories/history.
[60]Chen Minzhen (陳民鎮) and Yuri Pines. "Where is King Ping? The history and historiography of the Zhou Dynasty's eastward relocation". *Asia Major (Academica Sinica)*, vol. 31, no. 1, 27 Jan. 2018. Retrieved 2022-06-15.
[61]Zhai Jiangyue. Yao Jiangyue, trans. *Records on the Warring States Period*. Guangxi, China: Guangxi Normal University Press (广西师范大学出版社), 2008. m
[62]Zeng Gongliang (曾公亮; 999–1078) and Ding Du (丁度; 990–1053). *Wu Jing Zhong Yao Qian Ji: Shou Cheng* 《武經總要前集·守城》 or *Collection of the Most Important Military Techniques, First Part: Defend the City*. Taipei, Taiwan: Waves Pacific Limited (河洛文化事業股份有限公司), 2009.
[63]T.C. Bell does not want to reveal the location of the evidence to protect it from being destroyed; courtesy of T.C. Bell.
[64]Sheng-Wei Wang. "Chapter eight: Zheng He's generals force their way through the Fengdu gate, but later decide to return home". *The Last Journey of the San Bao Eunuch, Admiral Zheng He*, *op. cit*., p. 260.
[65]*Ibid*.
[66]Betty Nelson Curryer. *Anchors: The Illustrated History*. Annapolis, MD, USA: Naval Institute Press, 1999.
[67]Sheng-Wei Wang. "Chapter eight: Zheng He's generals force their way through the Fengdu gate, but later decide to return home". *The Last Journey of the San Bao Eunuch, Admiral Zheng He*, *op. cit*., p. 264.
[68]T.C. Bell has published a booklet titled *A Chinese Boachuan (Treasure Ship): Shipyard to Wreck*; courtesy of T.C. Bell.
[69]Damond Benningfield. "Mahuika". www.scienceandthesea.org, The University of Texas Marine Science Institute, 2006–2020, scienceandthesea.org/program/201404/mahuika.
[70]Nick Sault. "The asteroid that hit New Zealand". *Investigate*, vol. 44, Mar. 2004, pp. 44–48.
[71]D. H. Abbott, *et al*. "Evidence from an Ice Core of a Large Impact Circa 1443 A.D". American Geophysical Union, Fall Meeting 2005, abstract #PP31C-05.
[72]James Goff, *et al*. "Analysis of the Mahuika comet impact tsunami hypothesis". *Marine Geology*, vol. 271, no. 3/4, 2010, pp. 292–296.
[73]Nick Sault. "The asteroid that hit New Zealand". *Op. cit*.
[74]Private email communication with T.C. Bell; courtesy of T.C. Bell.

[75]Sally K. Church. "Zheng He: An investigation into the plausibility of 450-ft treasure ships". *Monumenta Serica*, vol. 53, Dec. 2005, pp. 1–43. https://www.researchgate.net/publication/261905911_ZHENG_HE_AN_INVESTIGATION_INTO_THE_PLAUSIBILITY_OF_450-FT_TREASURE_SHIPS.
[76]Nick Sault. "The asteroid that hit New Zealand". *Op. cit.*
[77]Sheng-Wei Wang. "Chapter eight: Zheng He's generals force their way through the Fengdu gate, but later decide to return home". *The Last Journey of the San Bao Eunuch, Admiral Zheng He*, *op. cit.*, pp. 258–259.
[78]*Op. cit.*, p. 259.
[79]Private email communication with T.C. Bell; courtesy of T.C. Bell.
[80]*Ibid.*
[81]Yang Fuwei, *et al.* "Study of sticky rice–lime mortar technology for the restoration of historical masonry construction". *Accounts of Chemical Research*, vol. 43, no. 6, 2010, pp. 936–944. doi:10.1021/ar9001944.
[82]"File:2014.08.19.104149 Great Wall Badaling.jpg". *Wikimedia Commons*, 9 Sept. 2020, upload.wikimedia.org/wikipedia/commons/d/d0/2014.08.19.104149_Great_Wall_Badaling.jpg.
[83]Sheng-Wei Wang. "Chapter eight: Zheng He's generals force their way through the Fengdu gate, but later decide to return home". *The Last Journey of the San Bao Eunuch, Admiral Zheng He*, *op. cit.*, pp. 262–263.
[84]Private email communication with T.C. Bell; courtesy of T.C. Bell.
[85]*Ibid.*
[86]Sheng-Wei Wang. "Chapter eight: Zheng He's generals force their way through the Fengdu gate, but later decide to return home". *The Last Journey of the San Bao Eunuch, Admiral Zheng He*, *op. cit.*, p. 261.
[87]*Op. cit.*, p. 259.
[88]Private email communication with T.C. Bell; courtesy of T.C. Bell.
[89]Sheng-Wei Wang. "Chapter eight: Zheng He's generals force their way through the Fengdu gate, but later decide to return home". *The Last Journey of the San Bao Eunuch, Admiral Zheng He*, *op. cit.*, p. 252.
[90]*Op. cit.*, p. 261.
[91]Sheng-Wei Wang. "Chapter one: Mysteries surround the seventh voyage of Zheng He". *The Last Journey of the San Bao Eunuch, Admiral Zheng He*, *op. cit.*, pp. 39–40.
[92]Siu-Leung Lee (李兆良). *Xuan De Jin Pai Qi Shi Lu — Ming Dai Kai Tuo Mei Zhou*《宣德金牌啟示錄—明代開拓美洲》or *Revelation of the Xuande Medallion: Ming Chinese Explored the Americas*. Taipei, Taiwan: Linking Publishing Company (聯經出版社), 2013年。

Chapter 3

A Chinese-Based World Map Depicts Europe Between 1157 and 1166

Abstract

This chapter reports that a Chinese-based world map — the Kunyu Wanguo Quantu《坤舆万国全图》or Complete Geographical Map of All the Kingdoms of the World published by Matteo Ricci in 1602 in China — depicts Europe in the period between 1157 and 1166, during the Southern Song Dynasty (南宋; 1127–1279), and that a network of trade routes — the Maritime Silk Road routes connecting China and Europe — existed already before Christ. The findings are based on: (1) a comparison of key geographical features in the European portion of Kunyu Wanguo Quantu with major European and Arab maps from Antiquity to the sixteenth century; (2) a comprehensive examination of the geographical and historical information of each named European kingdom, principality, duchy, republic, state, confederation, province, county, region, autonomous or semi-autonomous region, city/town, peninsula, island, ocean, sea, lake and river depicted on the Kunyu Wanguo Quantu; (3) a historical record of China–Byzantine interactions during the rule of the Emperor Shenzong (神宗; 1048–1085) of the Northern Song Dynasty (北宋; 960–1127); (4) archaeological findings from the "Nanhai One (南海一号)" shipwreck dated around the 1160s of the Southern Song Dynasty and discovered in the South China Sea in 1987; and (5) the latest archaeological surveys made by T.C. Bell in Ireland and the United Kingdom, which reveal that the Chinese had actually operated in Western Europe as early as 2850 B.C., and suggesting the existence of the Sino-Europe maritime routes almost 5,000 years ago.

Keywords: China, Europe, Kunyu Wanguo Quantu, Nanhai One, Maritime Silk Road, Song Dynasty

1. Introduction

For over four hundred years, the world map — Kunyu Wanguo Quantu《坤舆万国全图》,[1] abbreviated here as KWQ — written with Chinese characters and having latitudinal and longitudinal lines, has been generally regarded as a map drawn in 1602 by Matteo Ricci (an Italian Jesuit priest and one of the founding figures of the Jesuit China missions) and his Chinese collaborators, based on the European maps which Ricci brought with him to China in 1582.[2] Recently, researchers have provided analysis and evidence[3] to show that geographical information on certain portions of KWQ is actually based on the knowledge obtained by the Ming (明代; 1388–1644) mariners in the 1420s during their sixth voyage to the Western Ocean led by Admiral Zheng He, implying that these portions of KWQ used Chinese maps as source maps (these no longer exist) and that they were not copied from other contemporary European maps. However, these analyses have not been thoroughly extended to the European portion of KWQ, abbreviated here as Europe-KWQ. Hence, it is important to examine the Europe-KWQ to determine the political era of Europe revealed by this map, to see whether the Europe-KWQ is also based on the knowledge of Zheng He's mariners.

2. The Europe-KWQ Is Not a Direct or Adapted Copy of the Two Major European Maps of the Sixteenth Century

Figure 3.1 shows the Europe-KWQ extracted from KWQ (a collection item at the United States Library of Congress's Geography & Map Division)[4]; Fig. 3.2 shows the map of Europe extracted from the 1569 World Map by Gerardus Mercator[5] abbreviated as Europe-Mercator (1569); and Fig. 3.3 shows the map of Europe extracted from the 1570 World Map by Abraham Ortelius[6]

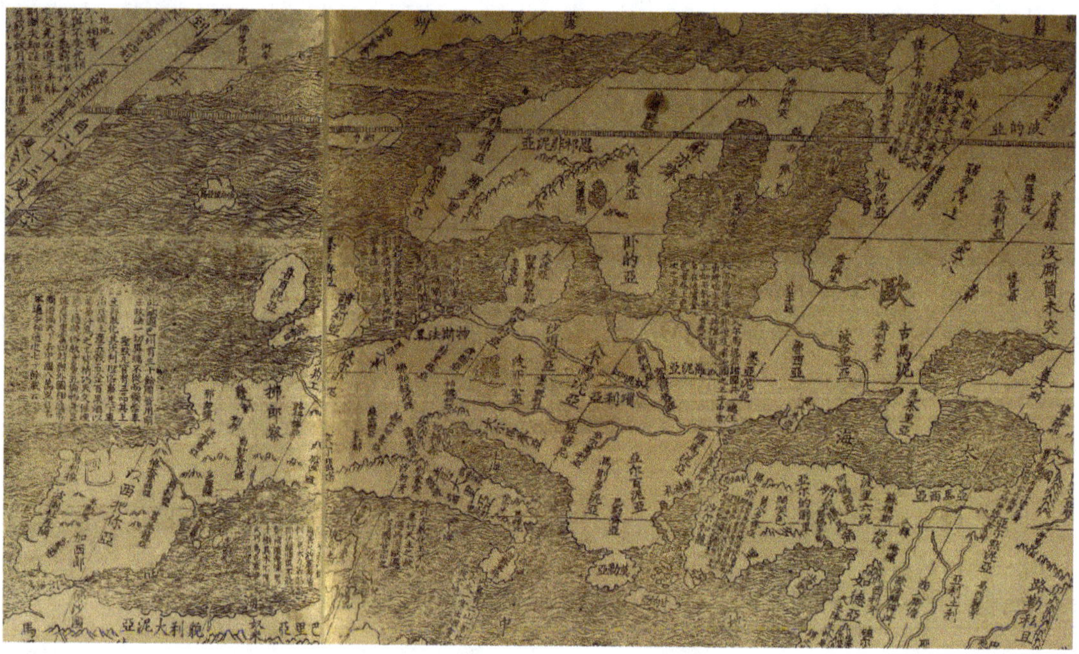

Fig. 3.1. The Europe-KWQ extracted from the 1602 KWQ (see endnote 7, public domain). The "toe" and "heel" of "Italy's boot" of the Italian Peninsula are poorly represented.

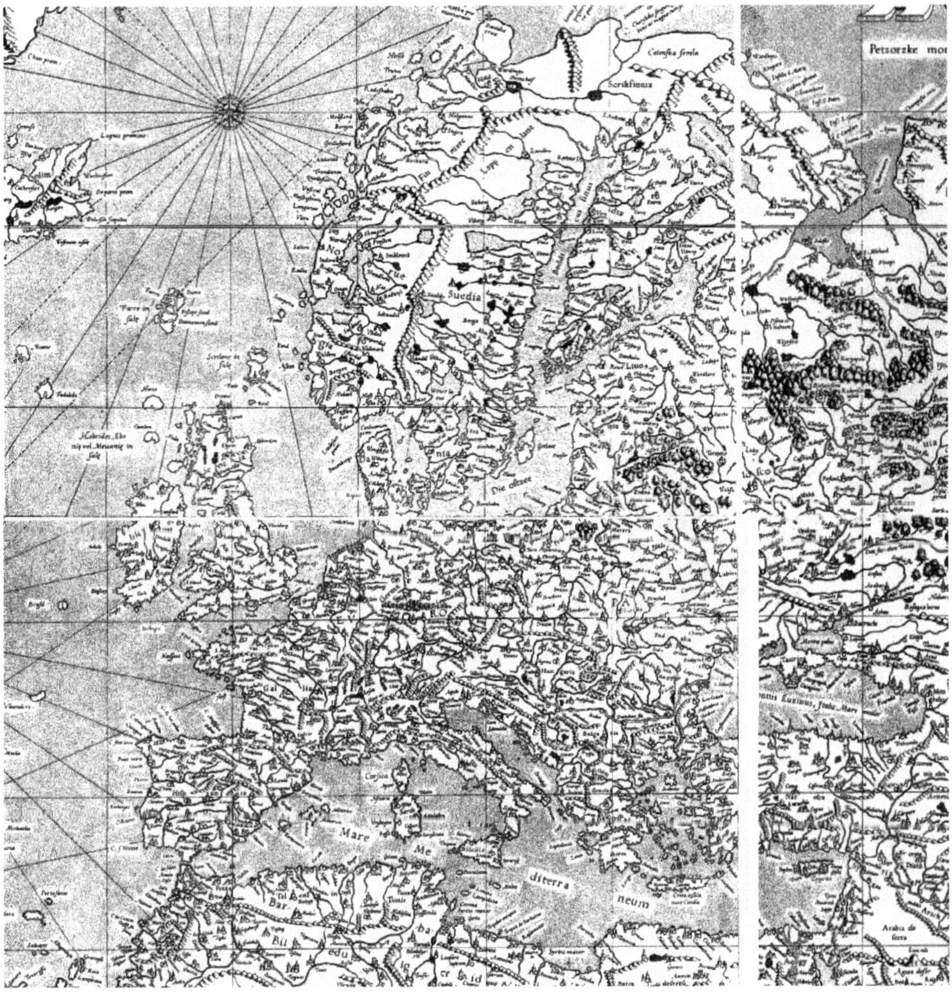

Fig. 3.2. The Europe-Gerardus (1569) extracted from the 1569 World Map by Gerardus Mercator (see endnote 8, public domain); the "toe" and "heel" of "Italy's boot" of the Italian Peninsula near the figure's bottom can be clearly identified.

abbreviated as Europe-Ortelius (1570). Both Europe Mercator (1569) and Europe-Ortelius (1570) were thought by some scholars to be the source maps of Europe-KWQ.

We can see very clearly that the "toe" and "heel" of "Italy's boot" of the Italian Peninsula are poorly depicted on the Europe-KWQ in Fig. 3.1, in comparison with the two European maps in Figs. 3.2 and 3.3. It seems that the Europe-KWQ is an older and less accurate map and not a direct or adapted copy of these two sixteenth century European maps.

In the Appendix of this chapter, I list all the geographical items shown on the Europe-KWQ in Fig. 3.1, and specifically denote those not appearing on the Europe-Gerardus (1569) or the Europe-Ortelius (1570). The denotation is mainly based on the Chinese book titled *Li Ma Dou Shi Jie Di*

Fig. 3.3. The Europe-Ortelius (1570) extracted from the 1570 World Map by Abraham Ortelius (see endnote 9, public domain); again, the "toe" and "heel" of "Italy's boot" can be clearly identified near the bottom of this figure.

Tu Yan Jiu《利玛窦世界地图研究》or Research on Matteo Ricci's World Map, written by Huang Shijian (黄时鉴) and Gong Yingyan (龚缨晏). My result shows that of the 117 geographical items analysed in this chapter, 30 (about 26 percent) do not appear on any of these two European maps.

Moreover, among the geographical items depicted on the Europe-KWQ, "黄魚島" (Yellow Fish Island) is of Chinese origin (Item 97 in the Appendix). The name is used to describe a kind of fish which lives around the island of Sardinia. This Chinese name has no connection with the pre-Latin root of the island.

"太海" (Tai Hai; Great Big Ocean; today's Black Sea) is also a Chinese-based name (Item 104 in the Appendix). The name "Black Sea" we know today was believed to be given by the Turkish during the Ottoman Empire period (1299–1922) when it was called Bahr-e Siyah or Karadeniz, which means the "Black Sea" in the Ottoman Turkish. The name "Black Sea" does not appear on the Europe-KWQ, implying that the era of the Europe-KWQ was earlier than 1299 A.D. and that Matteo Ricci has nothing to do with the name "太海" of this inland sea.

From the above analysis and simple comparisons, we can conclude that the Europe-KWQ cannot be a copy or adapted copy of the two prominent European maps of the sixteenth century shown in Figs. 3.2 and 3.3.

Since the Europe-KWQ seems to be an older and less accurate map in comparison with the sixteenth century European maps, I then decide to examine major European and Arab maps earlier than the sixteenth century to make further comparisons with the Europe-KWQ. The purpose is to see whether any different conclusion can be reached. This is done in Section 3.

3. The Europe-KWQ Is Very Different from the Major European and Arab Maps of the Twelfth to Fifteenth Centuries

I now compare the Europe-KWQ with major European maps drawn before the sixteenth century in reverse chronological order. The left panel in Fig. 3.4 shows the map of Europe extracted from the 1459 Fra Mauro map,[10] and the right panel is extracted from a modern reproduction of the 1375 Catalan Atlas.[11] Figure 3.5 shows the 1258–1291 map called Carta Pisane.[12] It appears that on all the maps of the thirteenth to fifteenth centuries, the European cartographers knew how to draw the "toe" and "heel" of "Italy's boot". Hence, there remains a puzzle: why did Matteo Ricci not know how to correctly draw the shape of the Italian Peninsula on the Europe-KWQ in 1602? It is also interesting to ask why and how the Europeans could already draw the Mediterranean region quite well in the thirteenth century as shown in Fig. 3.5.

History books record that Chinese developed navigational compasses as early as the eleventh or twelfth century (during the Song Dynasty; 960–1279), and that later, during the twelfth and thirteenth centuries, the compass arrived in Europe. After they learned how to use the compass, the

Fig. 3.4. At left is the map of Europe extracted from the 1459 Fra Mauro map (shown upside-down as the original had South at the top; see endnote 10; public domain). At right is the map of Europe extracted from the modern production of the 1375 Catalan Atlas (see endnote 11; public domain). Both show good depictions of the "toe" and "heel" of "Italy's boot".

Fig. 3.5. A picture of the Carta Pisane (see endnote 12; public domain), which was probably made around 1258–1291. We can recognise the "toe" and "heel" of "Italy's boot".

Europeans developed the Portolan charts which are nautical charts related to harbours. On these charts, "compass roses", also called "wind roses" or "roses of the wind" are drawn at various significant geographical locations with a network of rhumb lines that focused on the coasts and islands.[13] Sailing along any of these lines can lead ships to a harbour; hence, the maps were important guides for the sailors. A drawback of the Portolan charts is that they did not consider the spherical shape of the Earth; instead, they regarded the sea as a flat surface over which sailors could sail in a straight-line direction to reach their destination. As a result, the Portolan charts cannot provide effective sailing directions on a curved surface such as a vast ocean. But in the small and nearly flat Mediterranean Sea, the charts were very useful. The Carta Pisana (a Portolan chart) shows high accuracy in presenting the Mediterranean and is regarded as the cumulative work of the local fishermen and coastal merchants over the years.

However, when stepping back to the twelfth century, we notice that the Arab cartographers were doing much better in their overall depiction of the world than the European cartographers. The Moroccan geographer, Muhammad al-Idrisi, created in 1154 the *Tabula Rogeriana*, which was the most accurate map of the world at the time. He incorporated into his map the knowledge of Africa, the Indian Ocean and the Far East gathered by Arab merchants and explorers with the information inherited from the classical geographers.[14] The right panel of Fig. 3.6 shows the Italian Peninsula extracted from his world map.[15] Although the *Tabula Rogeriana* was regarded as the most accurate world map for the next three centuries, the Italian Peninsula is so distorted that it looks more like an elephant.

From the above comparisons, we know that the Europe-KWQ is also not a direct or adapted copy of the major European and Arab maps during the twelfth to the fifteenth centuries.

Since the Europe-KWQ contains many country names, I sensed that a comprehensive understanding of the history and geography of Europe would most likely lead me to figure out the

Fig. 3.6. At left is the Italian Peninsula extracted from the Europe-KWQ (see endnote 4; public domain). At right is the Italian Peninsula extracted from the *Tabula Rogeriana*, drawn by al-Idrisi in 1154, shown upside-down as the original had South at the top (see endnote 15; public domain).

political landscape embedded in the Europe-KWQ, and from there I should be able to determine the era of the map's content. With this goal in mind, I started to make a full analysis of the history and geography of each named geographical item depicted on the Europe-KWQ, and from there to extract the era of the map. This era corresponds to the date when the region was explored. This is an arduous process requiring thorough examination of the complicated European history and geography. But the effort paid off. My results are presented in the following sections and in the Appendix.

4. Medieval European History and Geography Are Keys in Determining the Era Represented by the Europe-KWQ

The Middle Ages or medieval period of Europe lasted from around the third to the late fifteenth century. After the fall of the Western Roman Empire[16] (286–476/480) came the tumultuous Dark Ages of the tenth and eleventh centuries, which later transitioned into the Renaissance and the Age of Discovery in the fifteenth and sixteenth centuries (or even the early seventeenth century). Here, I give some details:

After the western part of the Roman Empire fell, it was replaced by a series of kingdoms ruled by the Germans, but Roman identity and culture had spread to the Germanic tribes. The eastern part continued to exist as the Byzantine Empire[17] (c. 500–1492; Eastern Roman Empire) for about 1,000 years.

The above brief review is evidenced by Europe in Fig. 3.7[18] which shows that by 814, a unified Frankish Empire (or the Kingdom of Franks, Francia or Frankland; 481–843) already existed in Western and Central Europe. However, the Europe-KWQ in Fig. 3.1 does not show the same political landscape as in Fig. 3.7. The two maps clearly represent Europe in different eras. Also notice that the Europe-KWQ does not draw the territorial boundaries of a kingdom (the only exception is Sweden, which will be discussed in detail later). Figures 3.7, 3.8[19] and 3.9[20] show historical maps formed by drawing countries that existed in 814, 1142 and 1190 into modern European maps, respectively.

The Kingdom of Franks remained united until 843 when the Treaty of Verdun partitioned it. Later in 987, West Francia became the Kingdom of France; however, the territory remained known as Francia and its ruler, as "King of the Franks" well into the High Middle Ages (from around 1000 to 1250). The first king to call himself "King of France" was Philip II, in 1190.[21] The Kingdom of the Franks was the largest post-Roman kingdom in Western Europe, and the Carolingian Empire (800–888; the Carolingian Dynasty had ruled as kings of the Franks since 751) is considered the first phase in the history of the Holy Roman Empire[22] (800/962[23]–1806), leading later to the modern states of France and Germany.[24] We should also be aware that the largest territory of the Holy Roman Empire after 962 was the Kingdom of Germany, while the Holy Roman Empire also included the Kingdoms of Bohemia, of Italy, of Burgundy and numerous other territories.

Fig. 3.7. Europe in 814 (see endnote 18; public domain).

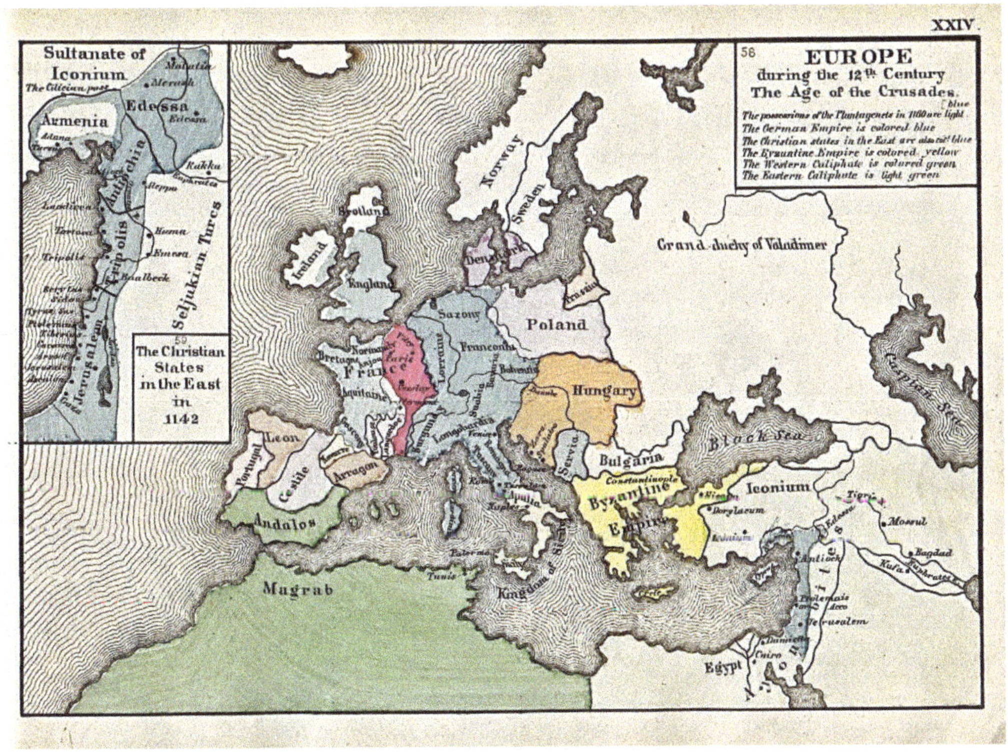

Fig. 3.8. Europe in 1142 (see endnote 19; public domain).

From this history, we conclude that the political landscape of Europe in 814 must belong to an older era than that of the Europe-KWQ, because the Europe-KWQ depicts Francia (拂郎察; the future France) and Germania (入尔馬泥亞; the future Germany) as separate kingdoms, which was the situation after the unified Frankish Empire ended.

After the fall of the Frankish Empire in the ninth century, Europe entered the Central and High Middle Ages[25] and experienced the political era of the Crusades and the Papal monarchy. By the mid-to-late twelfth century, a new political landscape of Europe gradually stabilised as shown in Fig. 3.8 for Europe in 1142 and Fig. 3.9 for Europe in 1190.

Generally speaking, on these two maps there are several kingdoms on the Iberian Peninsula, the Kingdom of France (or France) in Western Europe, the Holy Roman Empire ruling over much of Western and Central Europe (and extended to the northern Italian Peninsula) characterised by strong Papal authority, several independent city-states on the Northern and Southern Italian Peninsula, many countries in Eastern and Northern Europe, and the Byzantine Empire taking over the Balkan Peninsula and areas beyond. This political landscape seems to correlate better with the Europe-KWQ in Fig. 3.1, yet there are still distinct differences which after investigation have enabled me to determine the political era revealed by the Europe-KWQ. This investigation is detailed in Section 5 and in the Appendix.

Fig. 3.9. Europe in 1190 (see endnote 20; public domain).

5. The Europe-KWQ Depicts Europe Between 1157 and 1166

In the following, I shall list my key analyses which led me to determine the political era of Europe as depicted on the Europe-KWQ:

I. On the Europe-KWQ, Sweden's territory (蘇亦齊界; Item 77 in the Appendix) extends to the entire coastal region of southern and central Finland (沸你刪突; Item 76). The Swedish colonisation of Finland started in c. 1150,[26] and the year 1362, just six years before the Yuan Dynasty (1271–1368) of China ended, is generally considered the year of incorporation of Finland into the Kingdom of Sweden. Such an extended Swedish territorial extent depicted in Fig. 3.1 is not seen in Fig. 3.8 of Europe in 1142 but is seen in Fig. 3.9 of Europe in 1190. This difference confirms that the Europe-KWQ depicts a Europe after c. 1150.

II. At the western end of the Iberian Peninsula in Figs. 3.8 and 3.9, the name "Portugal" (葡萄牙) is denoted, not the old name 波爾杜瓦爾 (Portucale, Portugale or Portugália; Item 1 in the Appendix) shown on the Europe-KWQ in Fig. 3.1. The Kingdom of Portugal[27] originated from the County of Portugal (1096–1139) which was a semi-autonomous county of the Kingdom of León (910–1230). Independence from León (利昂; Item 11 in the Appendix) took place in three stages: (1) in 1139, Portugal's independence was claimed internally; (2) in 1143, Portugal's independence was recognised by the Kingdom of León and the Kingdom of Castile (1065–1230; 加西郎; Item 5 in the Appendix); and (3) in 1179, Portugal's independence was recognised by Pope Alexander III. In Figs. 3.8 and 3.9, the name Portugal is used. However, on the Europe-KWQ, the use of the ancient name 波爾杜瓦爾 also makes sense, because the independent status of 波爾杜瓦爾 had not yet received the Pope's confirmation which came only in 1179.[28] Combining this date with the date of Sweden's territorial expansion into Finland in c. 1150, we can infer that the Europe-KWQ depicts the political era of Europe between c. 1150 and 1179, which overlaps with the early Southern Song Dynasty of China (1127–1279).

III. Castile (加西郎; Item 5), which originally was an eastern county of the Kingdom of León (利昂; Item 11; however, its location is misplaced to the far northeast of Castile on the Europe-KWQ), became an independent realm from 1065. However, between 1072 and 1157, it was again united with the Kingdom of León. Confusingly, in Fig. 3.8, the map of Europe in 1142 seems to still draw the two kingdoms in two different colours, as if they were not united. Later, from 1230 this union became permanent, and the Kingdom of Castile and León came to be known collectively as the Crown of Castile.[29] In Fig. 3.9, we do see that there is separately the Kingdom of Castile and the Kingdom of León, because the year 1190 falls between 1157 and 1230, when the two kingdoms were not united. Since 加西郎 (the Kingdom of Castile; Item 5) and 利昂 (the Kingdom of León; Item 11) also appear as two separate kingdoms on the Europe-KWQ, this would support an era of Europe between 1157 and 1230. Combining this finding with the previous inference in II that the Europe-KWQ depicts an era between c. 1150 and 1179, leads to a new era of Europe between 1157 and 1179, which still overlaps with the era of the Southern Song Dynasty of China.

IV. Granada (厄辣捺達; Item 3) is in southern Spain and in the eastern and middle part of Andalusia, but on the Europe-KWQ, 厄辣捺達 is depicted too westward on the Iberian Peninsula and too close to the Gibraltar Strait. In Fig. 3.9, it is better located. Granada was under the rule of the Almohad Empire (1121–1269; a North African Berber Muslim empire)[30] from 1166[31] (the Almoravids ruled Granada from 1090, but their last king was killed in April 1147 by the Almohad Caliphate). Since the Almohad Empire is not depicted in Fig. 3.1 (as opposed to Fig. 3.9), this is the main reason that the Europe-KWQ depicts Europe in an era before 1166. Combining this finding with the previously inferred era between 1157 and 1179 (discussed in III) gives a final era of Europe revealed by the Europe-KWQ between 1157 and 1166.

V. The Europe-KWQ gives the names of the kingdoms, but does not draw their territorial borders, except for Sweden. Although the names of the Holy Roman Empire and the Byzantine Empire are not explicitly written, you can still notice several kingdoms on the Iberian Peninsula, Francia (拂郎察; Item 18; future France) in Western Europe, and the territories over much of Western and Central Europe (and extended to the northern Italian Peninsula) were ruled by the Holy Roman Empire and characterised by strong Papal authority; there were several independent city-states on the Northern and Southern Italian Peninsula; there were also many countries in Eastern and

Northern Europe, and the Balkan Peninsula and areas beyond were taken over by the Byzantine Empire (Eastern Roman Empire; the terms “Byzantine Empire” and “Eastern Roman Empire” were coined after the end of the realm). Such a political landscape represented by the Europe-KWQ is consistent with the Europe between 1157 and 1166.

VI. However, it is odd that 苏亦微亚 (Switzer; Item 23) and 十三郡 (the thirteen cantons; also included in Item 23) show up on the Europe-KWQ.

As explained in the Appendix, the Old Swiss Confederacy was a loose confederation of independent small states within the Holy Roman Empire. In 1351, a pact between the cantons of Uri, Schwyz and Unterwalden became the precursor of the modern state of Switzerland. Only by 1513 did the federation expand to thirteen members (the thirteen cantons). The name Schwyz over time became dominant and spawned the modern German Schweiz, French Suisse and English Swiss, while the derivative Switzer (苏亦微亚) was mainly used during the sixteenth to nineteenth centuries to denote a Swiss person, and especially a Swiss soldier or mercenary; it is still used in the modern English name “Switzerland”.

On the Europe-KWQ, “Switzerland” is denoted by three names: 赫尔勿妻亞 (Helvetia), 苏亦微亚 (Switzer) and 十三郡 (the thirteen cantons). 赫尔勿妻亞 is the transliteration of the Latin term *Helvetia* which was used to name the ancient region of central Europe between the Alps and the Jura Mountains before the Roman conquest, encompassed by modern Switzerland; the name is derived from the Helvetii tribe. It was the only name used before the sixteenth century to describe that region.

Since the Europe-KWQ is already shown to depict the political era of Europe between 1157 and 1166, *Helvetia*, transliterated as 赫尔勿妻亞, should be the only name used to depict that area. It appears that it was Matteo Ricci who added 苏亦微亚 (mainly used during the sixteenth to nineteenth centuries) and 十三郡 (not appeared before the sixteenth century) next to 赫尔勿妻亞 on the Europe-KWQ, to describe the same region, *Helvetia*, to make the Europe-KWQ look more updated. The presence of these two geographical items are incompatible with the rest of the political landscape revealed by the Europe-KWQ. The Latin term *Confoederatio Helvetica*, still in use today, means Swiss Confederation and is the nation’s full Latin name; here *Helvetica* is the adjective of the noun *Helvetia*.

In summary, from the analysis presented in Sections 2–5 and in the Appendix, I can now conclude that the Europe-KWQ is not a direct or adapted copy of major European and Arab maps from the twelfth to the sixteenth centuries. Instead, it depicts the political landscape of Europe in the era between 1157 and 1166, which overlaps with the early Southern Song Dynasty (1127–1279) of China. Europe-KWQ provides many additional geographical items (about 26 percent) not present in the two major sixteenth century European maps. These additional geographical items are highlighted in the Appendix.

Given that Song China was the world’s strongest nation (its GDP accounted for 80 percent of the world’s GDP), far surpassing Europe at that time[32]; and given its much advanced maritime capability (compasses were first used by Chinese for navigation during the Song Dynasty in the eleventh century[33]), it is not surprising that the Chinese mariners in the Southern Song Dynasty had explored Europe. It seems very logical for the Chinese cartographers of the Song era to depict Europe on map(s) with information they knew of. However, here is the puzzle; why did Matteo Ricci publish the Europe-KWQ in 1602 using an old source map of the Southern Song Dynasty?

Was he not able to locate a Ming map of Europe? Were he really the cartographer of the Europe-KWQ, why did he not instead simply use the Europe-Mercator (1569) or the Europe-Ortelius (1570) of his own era and drawn by European cartographers?

In Section 6, I shall list other evidence which gives further support from different angles, showing that the content of the Europe-KWQ is based on maritime information in the early Southern Song Dynasty as well as eras long before that.

6. China–Europe Interactions Can Be Traced to the Millennia Before Christ

In this section, I shall list a few examples of the China–Europe interactions in diplomacy and trade in the Song Dynasty, and Chinese exploration of Europe in the millennia before Christ:

I. There exist historical records (*Wenxian Tongkao* 《文献通考》[34] or *Comprehensive Examination of Literature* and *Song Shi* 《宋史》[35] or *History of the Song Dynasty*) of China–Byzantine interactions during the era of Emperor Shenzong (神宗; 1048–1085) of the Northern Song Dynasty (960–1127)[36]: in October 1078, the Byzantine Emperor Nikephoros III Botaneiates sent his envoys to seek an alliance with Taugus (唐家子); Chinese people in the Tang Dynasty were known to the Europeans as Taugus, and Christianity in China appeared in the seventh century during the Tang Dynasty. He was hoping that China in the east would be willing to form an alliance with him to fight against the Turkic people. The envoys took the Silk Road routes by land along the Aegean Sea and the ancient Anatolian Plateau to finally arrive at the Northern Song capital Kaifeng (开封) in late 1081. However, at that time Emperor Shenzong had just been defeated by the neighbouring Western Xia (西夏; 1038–1227); hence, he was not in the mood to heed the request of these ambassadors. When the envoys returned home, they found that Nikephoros III Botaneiates had already been overthrown by Alexios I Komnenos.[37]

II. Chinese wooden ships had sailed the Maritime Silk Road from Southeast China to the Indian Ocean, the Arab countries, the east coast of Africa and most likely Europe in the Southern Song Dynasty (1127–1279): a wooden shipwreck — Nanhai One (南海一号)[38] — was discovered in 1987 in the South China Sea. Some archaeologists date it to c. 1160s. This type of ship, 41.8 m × 11 m × c. 4 m/137 ft × 36 ft × c. 13 ft,[39] is a sharp-bottomed sea-going ship, which is famous for sailing in the South China Sea and the open sea. The wreck contains about 180,000 artefacts; most of the porcelains were from the famous Jingdezhen (景德镇) and other Chinese kiln families; many were made in Western and Arab cultural styles to satisfy foreign customers. Moreover, through the copper coins found on the Nanhai One, it can be roughly inferred that China's copper alloy coins were the main currency for regional trade at that time.

A photo in endnote 40 shows copper alloy coins[40] collected from Nanhai One. It is also worth noting that Chinese coins have a smooth round shape because they were manufactured by being cast in molds; European coins were typically cut and hammered manually or in later times milled, hence they were not easy to form with a smooth round shape.

In short, the discovery of Nanhai One is testimony of the development of maritime trade in the Song Dynasty, especially in the Southern Song Dynasty. After China's northwestern transportation link with Central Asia was blocked by the Western Xia, Uyghur and Liao regimes, the Song imperial court had to turn to sea to develop trade relations with many countries and regions in South Asia, Central Asia, Africa, the Middle East and Europe. The widespread use of compasses in the Song Dynasty and the new breakthroughs in the technology of manufacturing sea-going ships made all these maritime trades possible and successful, hence, China became a great maritime power in the Song Dynasty. The Belt and Road Initiative currently implemented by China to connect the economies of various countries and regions is a modern form of the Maritime Silk Road routes in the Southern Song Dynasty.

III. British surveyor T.C. Bell has discovered that ancient Chinese had operated on the British Isles since Antiquity, based on the following evidence he collected[41]:

(1) In Ireland, the Chinese presence on Innis Mor, an island in Galway Bay, was confirmed using magnetic anomaly (a magnetic anomaly is the change in magnitude of the Earth's magnetic field with respect to the expected value for that location) which located sand-covered wrecks of two Chinese Junks of size c. 47 m × 11 m/154 ft × 36 ft as well as their unique anchors (explained in great detail in Section 4 of Chapter 2 of this book) and two matching adjacent Chinese standard slot harbours.
(2) In Scotland, T.C. Bell discovered in Rhins of Galloway a c. 130 m × 38 m/427 ft × 125 ft Chinese catamaran and, flanking it, a c. 70 m × 27 m/230 ft × 89 ft Chinese Junk. Both are sand-covered and tidal, and both were located by magnetic anomaly at low tide. Adjacent was a large sand-covered Roman harbour and several Chinese slot harbours, capable of taking c. 70 m × 28 m/230 ft × 92 ft vessels. There was a massive 240 m/787 ft harbour wall behind which the massive Chinese catamaran and c. 70 m/230 ft junk were anchored. Moreover, nearly every bay down the Rhins of Galloway held Chinese slot harbours, some with shipwrecks. There was even one with stone cannon balls on the beach. All had barracks for the Chinese crews, some had Chinese towns, and several had small cliff-top forts overlooking the bays.
(3) In Scotland, one of the three bays Bell surveyed in Wester Ross had a large double-walled field with visible wall foundations and Chinese barrack foundations, overlooking a six-bay Chinese harbour (with stonework in place). There was also a farmer's excavation, c. 1 m/c. 3 ft deep, where drainage revealed burnt brush wood; this was carbon dated by a major Scottish laboratory to 2850 B.C.
(4) Again, in Scotland, a fleet of four shipwrecks located in Gruinard Bay on the west coast, and one wreck confirmed by satellite, were apparently part of the first Chinese fleet sent out to survey the world c. 5,000 years ago.
(5) In Cumbria (the most northwesterly county of England), T.C. Bell located a large metallic ore exploitation site, and in southern England, what he considers to be a Chinese shipyard.

In summary, T.C. Bell found that the Chinese appear to have entered the British Isles in two waves: the first occurred in the millennia before Christ; the second took place during the Roman period c. 44–410[42] (he found Roman forts alongside Chinese barracks and harbour sites to defend them). He has located sand-covered Chinese shipwrecks from both periods. Bell is currently writing a book titled *The Chinese Discovery of Britain* to give more details of his investigations.

IV. Ancient Sino-Europe maritime routes must have long existed. This is not only evidenced by T.C. Bell's surveys in Ireland and the United Kingdom, but also by some of the sea names depicted on the Europe-KWQ. These names are in ancient Latin (from 75 B.C. to 300 A.D.) such as 上海 (*Mare Superum*; Item 105) and 下海 (*Mare Inferum*; Item 106), also known in more recent medieval Latin (from late third to late fifteenth century) as *Mare Adriaticum* (today's Adriatic Sea) and *Mare Tyrrhenum* (today's Tyrrhenian Sea), respectively. Also, the Chinese name 太海 (great big sea or very big ocean; Item 104) is used in place of the much later name "Black Sea" during the Ottoman Empire period (1299–1922), and 入尔馬尼海 (*Mare Germanicum*; Item 112; before 75 B.C.) is used to depict the southern arm of the Baltic Sea (this name was first used in the eleventh century). All these observations indicate that the Chinese had sailed along the waters surrounding the European continent before and after Christ and used the ancient Latin names or the Chinese-invented names to denote the waters they had sailed through.

8. Conclusions

In Sections 2 and 3, I have shown that an ancient world map depicting Europe. This map is Kunyu Wanguo Quantu《坤舆万国全图》or Complete Geographical Map of All the Kingdoms of the World, published by Matteo Ricci in 1602 in China, and it is of Chinese origin; it is not a direct or adapted copy of the prominent European and Arab maps from the twelfth to the sixteenth century. Then in Sections 4 and 5, based on the knowledge of the Medieval-European history and geography, I was able to analyse each named geographical item depicted on that map — abbreviated as the Europe-KWQ in this chapter — to extract the political era of Europe being presented. My conclusion is that the Europe-KWQ depicts Europe between 1157 and 1166, a period overlapping with the early Southern Song Dynasty (1127–1279). This conclusion is consistent with the known historical records in Section 6 of the Northern Song Dynasty and the latest archaeological findings of the Nanhai One shipwreck of the Southern Song Dynasty. Furthermore, the survey findings made by T.C. Bell in Ireland and the United Kingdom reveal that the Chinese had operated on the western coast of Europe as early as 2850 B.C., which suggests the existence of Sino-Europe maritime routes almost 5,000 years ago.

Acknowledgements

The author thanks T.C. Bell for communicating his valuable survey results in Ireland and the United Kingdom and allowing their publication in this chapter, and for approving the present chapter before publication. The author also thanks Ian Hudson and Professor Klaus Hermann of the Max Planck Society in Berlin, Germany, for reviewing this chapter prior to publication. Special thanks go to Professor Jimin Sun (孙继敏), Institute of Geology and Geophysics, Chinese Academy of Sciences, for reviewing my chapter despite his busy schedule. Finally, I am grateful to Professor Emeritus Michel A. Van Hove for his technical assistance in preparing all the figures in this chapter in published form, and M. Hayes-Shuptar (华铭叮) for the academic and editorial assistance with the endnotes.

Appendix

Items preceded by "•" are geographical items which do not appear on the Europe-Mercator (1569) or the Europe-Ortelius (1570). All the geographical items are categorised into 13 different zones. The Europe-KWQ uses almost entirely Chinese traditional characters; the author of this chapter uses modern Chinese simplified characters except in cases when the geographical items used by the Europe-KWQ are mentioned. Chinese pinyin for pronunciation is listed for each item in this Appendix.

Europe-KWQ: Europe depicted on Kunyu Wanguo Quantu				
Item #	**Name**	**Pinyin**	**Etymology**	**History and geography and my comments**
Zone I: Southwest Europe (Iberian Peninsula)				
1	波爾杜瓦爾	Bo Er Du Wa Er	The old name of Portugal; Latin: *Portucale*; it evolved into *Portugale* (波尔杜瓦尔) during the seventh and eighth centuries. During the ninth century, *Portucale*, *Portugale* and *Portugália* were used to refer to the region between the rivers Douro and Minho (parts of today's border of Spain and Portugal).[43]	The Kingdom of Portugal (1139–1910) originated from the County of Portugal (868–1139) which was a semi-autonomous county of the Kingdom of León (Item 11).[44] Independence from León took place in three stages: (1) in 1139, Portugale's independence was claimed internally; (2) in 1143, Portugale's independence was recognised by León and Castile (Item 5) through the Treaty of Zamora; and (3) in 1179, Portugale's independence was recognised by Pope Alexander III.[45] The geographical location of 波爾杜瓦爾 (*Portucale*, *Portugale* or *Portugália*) is correctly depicted on the Europe-KWQ. The use of the old name on the Europe-KWQ reveals a political era before the independence of Portugale (波爾杜瓦爾) into Portugal (葡萄牙); the earliest date could be regarded as 1139 and the latest date could be regarded as 1179, depending on how the cartographer viewed the situation.

Item #	Name	Pinyin	Etymology	History and geography and my comments
2	曷利擦	He Li Ca	Spanish: *Galicia*[46]; Portuguese: *Galiza*; the term was derived from the Latin toponym *Callaecia*.[47]	During the Middle Ages, the Kingdom of Galicia (409–1833)[48] was occasionally ruled by its own kings, but most of the time it was leagued to the Kingdom of León (910–1230), and later to the Kingdom of Castile (1065–1230),[49] while maintaining its own legal and customary practices and culture. From 1230 on, Galicia was under the control of the Crown of Castile (1230–1715),[50] because that year León united with Castile and formed the Crown of Castile. The Europe-KWQ correctly depicts the geographical position of 曷利擦 (Galicia) and the co-existence of the kingdoms of Galicia, León and Castile between 1157 and 1166, when Galicia was leagued to León or Castile. However, the Europe-KWQ misplaced León to the far northeast of Castile.
3	厄辣捺達	E La Na Da	Granada; Latin: *Granātum*	This was the region surrounding today's city of Granada in the community of Andalucía. The Almoravid Dynasty (1040–1147) ruled Granada from 1090 to 1147, and the Almohad Dynasty (1121–1269; a North African Berber Muslim empire) ruled from 1166.[51] By 1171, all of the Muslim Iberian Peninsula was under the Almohad.[52] The Emirate of Granada (1230–1492) became a tributary state to the Kingdom of Castile from 1238.[53] On the Europe-KWQ, Granada is depicted too westward on the Iberian Peninsula and too close to the Gibraltar Strait. In Fig. 3.9 (Europe in 1190), it is correctly located. Since the Almohad Caliphate is not depicted on the Europe-KWQ (as opposed to Fig. 3.9), this is the main reason that the Europe-KWQ depicts Europe in an era before 1166.
• 4	多勒篤	Duo Le Du	Toledo; Latin: *Toletum*	In the early eleventh century, Toledo became an independent Taifa kingdom. It was conquered by Castile in 1085, and the Kingdom of Toledo (1085–1833) was created in the central Iberian Peninsula. The Muslim-led Kingdom of Toledo became a subordinate to the Christian-led southern realm of the Crown of Castile after 1230 but had its own court and rulers.[54]

(*Continued*)

(*Continued*)

● 5	加西郎	Jia Xi Lang	Castile; Spanish: *Castilla*	Castile, originally an eastern county of the Kingdom of León in the eleventh century, became an independent realm (1065–1230). Between 1072 and 1157, it was again united with León, and after 1230, this union became permanent. It added Toledo in 1085 and integrated the Kingdom of León in the Crown of Castile in 1230.[55] Both 利昂 (Item 11) and 加西郎 (Castile; Item 5) appear on the Europe-KWQ, indicating an era after 1157, but before 1230. Hence, it is consistent with the Europe-KWQ revealing an era between 1157 and 1166. However, the location of 加西郎 (Castile) is too far south, and it also occupies an area that should belong to 俺大魯西亞 (Item 6; Andalusia), such that the latter's geographic location is grossly misplaced (see discussion in Item 6).
● 6	俺大魯西亞	An Da Lu Xi Ya	Andalusia; Spanish: *Andalucía.* The name of Andalusia is derived from the Arabic name "*Andalus*", Arabic: الأندلس, transliteration: *al-Andalus*, during Muslim rule.[56]	Andalusia is the southernmost region of the Iberian Peninsula. After the abrupt end of the rule of the Kingdom of the Visigoths (an early Germanic people) in 711, the region was under Muslim rule.[57] The name "Al-Andalus" was originally applied to the Iberian Peninsula, and later it referred to the parts not controlled by the Gothic states in the North of the Peninsula until January 1492.[58]  Fig. A3.1. Map of Andalusia (in red).[59] The geographical location of Andalusia on the Europe-KWQ differs substantially from historically known Andalusia.[60] If Matteo Ricci drew it in 1602, it would be quite inconceivable that he made such an error.

Item #	Name	Pinyin	Etymology	History and geography and my comments
7	曷剌甕	He La Weng	The Kingdom of Aragon; Latin: *Regnum Aragoniae*; Spanish: *Reino de Aragón*	The Kingdom of Aragon existed from 1035 to 1516. The Kingdom gave the name to the Crown of Aragon (1162–1716)[61] and became a member of the Holy Roman Empire. In 1479, the Crowns of Aragon and Castile were united to form modern Spain.[62]  Fig. A3.2. Map of Aragon (in red).[63] On the Europe-KWQ, the Kingdom of Aragon (曷剌甕) is depicted at its correct geographical location in the era between 1157 and 1166.
8	葛荅籠亞	Ge Da Long Ya	Catalonia; Spanish: *Cataluña.* Catalan: *Catalunya*; Medieval Latin: *Cathalonia* or *Cathalauni*	The earliest known use of "Catalonia" for the counties in the northeast of Iberia was in 1117. The Count of Barcelona ruled Catalonia from the end of the ninth century until 1410.[64] During most of the Principality of Catalonia's history, it was in dynastic union with the Kingdom of Aragon, constituting together the Crown of Aragon.[65] 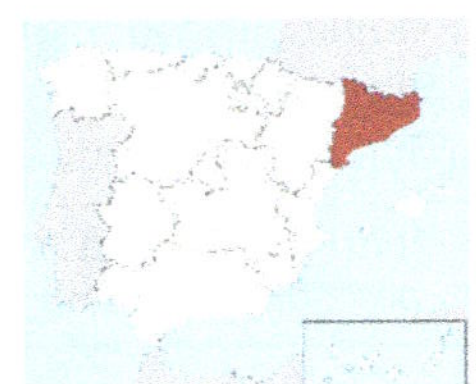 Fig. A3.3. Map of Catalonia (in red).[66]

(*Continued*)

(*Continued*)

				葛沓籠亞 (Catalonia) is incorrectly placed to the south of 曷剌甕 (Aragon) on the Europe-KWQ (it should be to the east).
9	迷施葛亞	Mi Shi Ge Ya	Biscay; Basque: *Bizkaia*. Spanish: *Vizcaya*	Biscay is a historical region in Spain, lying on the south shore of the eponymous bay. It was a dependency of the Kingdom of Pamplona in the eleventh century, then became autonomous[67] and finally a part of the Crown of Castile (1230–1715).[68] 迷施葛亞 (Biscay) showing up as an autonomous region on the Europe-KWQ is consistent with the era of Europe between 1157 and 1166.
10	那勿蠟	Na Wu La	Navarre; Spanish: *Navarra*	The Kingdom of Navarre (824–1941) was a Basque kingdom.[69] In 1200, Navarre lost the key western Basque districts to Castile, leaving the kingdom landlocked.[70] 那勿蠟 (Navarre) is not land-locked on the Europe-KWQ. This is consistent with the era between 1157 and 1166 (before 1200) of Europe, revealed by this region of the map.
11	利昂	Li Ang	León; Latin: *Leō*	The Kingdom of León (910–1230) was an independent kingdom situated in the northwest region of the Iberian Peninsula.[71] It was the most important kingdom of the Iberian Peninsula until the two kingdoms of León and Castile were split in 1157; already in 1139, the County of Portugal had won independence to become the Kingdom of Portugal. In 1230, the kingdoms of León and Castile had the first joint sovereign.[72] On the Europe-KWQ, 利昂 (Kingdom of León) is incorrectly placed in the northeast region of 以西把你亞 (the Iberian Peninsula; Item 12), being too far away from 加西郎 (Kingdom of Castile; Item 5). However, the fact that they are two separated kingdoms on the Europe-KWQ is consistent with the history of Europe between 1157 and 1166 revealed by the Europe-KWQ.

Item #	Name	Pinyin	Etymology	History and geography and my comments
12	以西把你亞	Yi Xi Ba Ni Ya	Spain; Latin: *Hispania*. Hispania was the Roman name for the Iberian Peninsula and its provinces, which was often used during Antiquity and the Low Middle Ages (476 A.D.–1000 A.D.) as a geographical name,[73] and the collective names of the Iberian Peninsula kingdoms of the Middle Ages. By the late twelfth century, the whole Iberian Peninsula became known as "Spain".[74]	In the eighth century, nearly all of the Iberian Peninsula was conquered by largely Moorish Muslim armies from North Africa. Later, the Reconquista (Reconquest) was the centuries-long period in which Christian rule was re-established over the Iberian Peninsula. The Kingdom of León was the strongest Christian kingdom for centuries. In 1492, the combined forces of Castile and Aragon ended the last remnant of a 781-year presence of Islamic rule in Iberia. The term 以西把你亞 (Hispania; Item 12) shown on the Europe-KWQ is used to denote the collective names of the Iberian Peninsula kingdoms. This implies that the era of the Iberian Peninsula on this map cannot be later than the last years of the twelfth century when the term "Spain" began to replace "Hispania". The era between 1157 and 1166 is not later than the last years of the twelfth century.
Summary A1: Items 1–12 present the history and geography of 以西把你亞 (Iberian Peninsula; Item 12) on the Europe-KWQ, which are consistent with what we know of Europe between 1157 and 1166. During this short period, there existed on the peninsula six kingdoms: 喝利擦 (Galicia; Item 2), 多勒篤 (Toledo; ruled by Muslims; Item 4), 加西郎 (Castile; Item 5), 曷剌甕 (Aragon; Item 7), 那勿蠟 (Navarre; Item 10), 利昂 (León; Item 11); a region: 俺大魯西亞 (Andalusía; Item 6); an autonomous region: 迷施葛亞 (Biscay; Item 9); a county 波爾杜瓦爾 (Portugale; Item 1) whose independence was not yet recognised by Pope Alexander III; a principality: 葛苔籠亞 (Catalonia; Item 8) and a city: 厄辣捺達 (Granada; Item 3). The Kingdom of Aragon was the only member of the Holy Roman Empire on the Iberian Peninsula. On the Europe-KWQ, territorial borders are not drawn.				

(*Continued*)

(*Continued*)

Zone II: Western Europe				
13	多羅薩	Duo Luo Sa	Toulouse; Latin: *Tolosa*	The city Toulouse was the capital of the County of Toulouse during the Carolingian era (800–888).[75] In the twelfth century, the city obtained considerable autonomy.[76]
● 14	曷計荅尼亞	He Ji Da Ni Ya	Aquitaine; Latin: *Aquitania*	It is a historical region in southwestern France. In the Middle Ages (from around the third to the fifteenth century), Aquitaine was a duchy (602–1453), whose boundaries fluctuated considerably. In 1058, the Duchy of Vasconia (Gascony; 瓦斯工; Item 17) and Aquitaine merged under the rule of the Duke of Aquitaine. Aquitaine passed to Francia in 1137,[77] hence Fig. 3.8 shows both Aquitaine and Gascony as part of France in 1142. But from 1154, the area became an English possession, a part of the Angevin Empire,[78] until the end of the Hundred Years' War in 1453, when it was annexed by France.[79] During the twelfth and thirteenth centuries, the Angevin Empire was the most powerful vassal in France. Figure 3.9 no longer shows Aquitaine and Gascony since they were possessed by the Angevin Empire from 1154. But the 1157–1166 Europe-KWQ still shows 曷計荅尼亞 (Aquitaine; Item 14) and 瓦斯工 (Gascony; Item 17), because both were duchies possessed by the Angevin Empire and lasted until 1453 before they were annexed by France.

Item #	Name	Pinyin	Etymology	History and geography and my comments
• 15	羅尺刺 and 路得棲亞	Luo Chi La and Lu De Qi Ya	Lutetia[80]; Latin: *Lutetia Parisiorum*. Both 罗尺刺 and 路得棲亞 are the transliterations of the Latin *Lutetia*, which means "marsh" or "swamp". The term was used to indicate the small island in the middle of the Seine River where modern Paris originated.[81]	Around 1 and 2 A.D., the Gallo-Roman city became Lutetia. In 360 A.D., Julian the Apostate was proclaimed Emperor of the Gauls and Lutetia was rebaptised Paris.[82] By the end of the twelfth century, Paris had become the political, economic, religious and cultural capital of France. On the Europe-KWQ, both 罗尺刺 (at left) and 路得棲亞 (at right) denote Lutetia, today's Paris, because the Seine River cuts the city into two parts: the Rive Gauche (the Left Bank, south of the Seine) and the Rive Droite (the Right Bank, north of the Seine). The Europe-KWQ correctly reflects the geographical condition around Lutetia/Paris between 1157 and 1166.
16	波尔卧尼	Bo Er Wo Ni	Burgundy; French: *Bourgogne*; Latin: *Burgundia*. It takes its name from the Burgundians, an East Germanic people who moved westwards beyond the Rhine during the late Roman period.[83]	During the fourth century, the Burgundians founded the Kingdom of the Burgundians (411–534),[84] which was conquered in the sixth century by the Franks. The Duchy of Burgundy (918–1482) emerged in the tenth century and lasted until 1428.[85] The original territory of the Duchy of Burgundy was in what is now eastern France (see Fig. 3.8). But on the Europe-KWQ, 波尔卧尼 (the Duchy of Burgundy) in the era between 1157 and 1166 is incorrectly located near the North Sea. Only in the fourteenth century did the territory of the Duchy of Burgundy expand further north in the Low Countries; by then the entire duchy was called Burgundian Netherlands (1384–1482).[86]

(Continued)

(*Continued*)

17	瓦斯工	Wa Si Gong	Gascony; French: *Gascoigne*	Gascony is a historical region in southwestern France. The history of Duchy of Vasconia/Gascony (602–1453) from 1058 is already detailed in Item 14 (曷計荅尼亞; Aquitaine). However, on the Europe-KWQ, 瓦斯工 (Gascony; Item 17) is depicted erroneously in northeast Francia, next to 波尔卧尼 (Burgundy), although 曷計荅尼亞 (Aquitaine) is correctly depicted in southwest Francia.
18	拂郎察	Fu Lang Cha	France; Latin: *Francia*; it means "land of the Franks". The Franks were a group of Germanic peoples. Originally *Francia* was applied to the whole Empire of the Franks, from southern France to eastern Germany.	The Kingdom of West Francia (843–987) is also known as The Kingdom of the West Franks.[87] France originated from the Kingdom of West Francia in 987, which was the western half of the Carolingian Empire (the empire of the Franks). However, the territory remained known as *Francia* and its ruler as "King of the Franks" well into the High Middle Ages. The first king calling himself "King of France" was Philip II, in 1190.[88] In the early-mid twelfth century, the French kings controlled little more than Paris and the surrounding region.[89] Indeed, on the Europe-KWQ, 拂郎察 seems to cover only Paris and the surrounding region; it correctly reflects the political landscape of the Kingdom of Francia in the era between 1157 and 1166.
Summary A2: Items 13–18 present the history and geography of Western Europe including a kingdom: 拂郎察 (Francia); a city: 多羅薩 (Toulouse; Item 13); three duchies: 曷計荅尼亞 (Aquitaine; Item 14), 波尔卧尼 (Burgundy; Item 16) and 瓦斯工 (Gascony; Item 17); as well as the geographical information surrounding 羅尺剌 (Lutetia; Item 15) and 路得棲亞 (Lutetia; Item 15). The Europe-KWQ correctly reflects the political landscape of Europe in this region between 1157 and 1166. But 波尔卧尼 (Burgundy; Item 16) is depicted erroneously near the North Sea in the era between 1157 and 1166 A.D. On the Europe-KWQ, territorial borders are not drawn.				

Item #	Name	Pinyin	Etymology	History and geography and my comments
Zone III: Central and Southern Europe (including Northern and Central Italian Peninsula)				
• 19	麻尔西里亞	Ma Er Xi Li Ya	Marseille; Latin: *Massilia*, *Marsilia*	Marseille, located on the Mediterranean coast, has been a trading port since antiquity.[90] It became part of the County of Provence[91] (part of the Holy Roman Empire[92]) during the tenth century.
20	突尔蜚諾	Tu Er Fei Nuo	Dauphiny; French: *Dauphiné*; Latin: *Delphinus*	Dauphiny is a historical territory in southeastern France.[93] The Dauphiné was originally the Dauphiné of Viennois. Later, it took the name of Dauphiné. It became a state of the Holy Roman Empire in the eleventh century.[94] After 1461, Dauphiné united with France.[95]
21	百尔入革	Bai Er Ru Ge	Belgium; Latin: *Belgium*. "Belgium" comes from Gallia Belgica (a Roman province in Gaul, where the Belgae, a mix of Celtic and Germanic peoples, lived).[96]	After the fall of the Roman Empire and the beginning of the Middle Ages, the territories which presently form Belgium became part of the Holy Roman Empire and would remain as such until the eleventh and twelfth centuries.[97]
• 22	赫尔勿妻亞	He Er Wu Qi Ya	赫尔勿妻亞 is the transliteration of *Helvetia*. The Latin term *Helvetia* was an ancient region in central Europe between the Alps and the Jura Mountains.	*Helvetia* is derived from the ethnonym Helvetii, the name of the Gaulish tribe inhabiting the Swiss Plateau before the Roman conquest.[98] This Latin name is still used in today's Switzerland.

(*Continued*)

(*Continued*)

23	蘇亦微亚 and 十三郡	Su Yi Wei Ya and Shi San Jun	The English name "Switzerland" is a compound containing *Switzer*, an obsolete term for the Swiss, which was in use during the sixteenth to nineteenth centuries. 苏亦微亚 can be considered as the transliteration of *Switzer*. The Latin term *Confoederatio Helvetica* (Swiss Confederation) is used officially.[99]	In 800, Switzerland was part of Charlemagne's empire. After his death, the eastern part of Switzerland became part of the Holy Roman Empire together with Germany, Austria and Italy, while the western part became part of Burgundy.[100] In 1291, three provinces (called cantons) joined together to form the nucleus of the Old Swiss Confederacy (OSC). In the sixteenth century, the OSC had 13 members (the thirteen Cantons),[101] and between 1353 and 1481, it had only eight members.[102] Switzer (苏亦微亚) was used during the sixteenth to nineteenth centuries, and the use of thirteen Cantons (十三郡) started from 1513 when the OSC had thirteen members. But the Europe-KWQ reveals the era of this region between 1157 and 1166, which is much earlier than the sixteenth century. It appears that 苏亦微亚 and 十三郡 were added onto the Europe-KWQ to denote the country named 赫尔勿妻亞 (Item 22) (Helvetia) by Matteo Ricci in 1602. See detailed discussion in subsection VI of Section 5 in the main text.
• 24	沙勿牙	Sha Wu Ya	Savoy; Italian: *Savoia*. The name *Savoy* stems from the Late Latin *Sapaudia*, referring to a fir forest in the Western Alps.	The County of Savoy (1003–1416) was a State of the Holy Roman Empire. It emerged in the eleventh century, along with the free communes of Switzerland, from the collapse of the Kingdom of the Burgundians.[103]

Item #	Name	Pinyin	Etymology	History and geography and my comments
• 25	別蒙突	Bie Meng Tu	Piedmont; Italian: *Piemonte*. The name comes from medieval Latin *Pedemontium* or *Pedemontis*, i.e. *ad pedem montium*, meaning "at the foot of the mountains" (referring to the Alps).	It is a historical region in northwest Italy. After the fall of the Western Roman Empire, the region was successively invaded by the Burgundians, Ostrogoths, East Romans, Lombards and Franks. In the ninth–tenth centuries, Piedmont was part of the Kingdom of Italy within the Holy Roman Empire, subdivided into several marches and counties. In the eleventh century, Piedmont was added to the County of Savoy (1013–1416), which was a State of the Holy Roman Empire.[104]
26	惹怒襪	Re Nu Wa	Genoa; Italian: *Genova*; Latin: *Genua*. The place is on the Gulf of Genoa in the Ligurian Sea.	Before 1100, Genoa was an independent city-state. Later, the Holy Roman Emperor was the overlord; however, the actual power was wielded by several "consuls" annually elected by popular assembly.[105]
27	隆拔勒地亞	Long Ba Le Di Ya	Lombardy; Italian: *Lombardia*	Lombardy is a historical region in northern Italy. During the early Middle Ages, "Lombardy" referred to the Kingdom of the Lombards (568–774).[106] In the tenth century, Lombardy, under the rule of the Holy Roman Empire, was divided into small, autonomous city-states.[107] Hence, the name "Lombardy" was used to denote the whole of northern Italy until the fifteenth century and sometimes later. By the end of the fourteenth century, Milan and Mantua emerged as rival hegemons in Lombardy. The duchies of Milan[108] and Mantua[109] were two centres of the European Renaissance culture, trade and banking activities in northern Europe. However, neither is shown on the Europe-KWQ; this is consistent with the era between 1157 and 1166, revealed by the map, which was long before the European Renaissance.

(*Continued*)

(*Continued*)

28	勿耨茶	Wu Ru Cha	Venice; Italian: *Venezia*; Latin: *Venetia*	The city is in northeast Italy. It is composed of many small islands that are separated by canals and linked by hundreds of bridges. The city was the capital of the Republic of Venice (697–1797).[110] Of all the northern Italian city-states, Venice is the only one that was not a part of the Holy Roman Empire.
29	羅馬	Luo Ma	Rome; Latin: *Roma*	From the fourth century, the Lateran Palace in southeast Rome was the principal residence of the popes and continued so for about a thousand years. However, from 1309 to 1377, the popes lived at Avignon in France,[111] and on their return to Rome, they chose to live at the Vatican. Later, they moved back to the Quirinal Palace in Rome in 1583. But on the Capture of Rome (the final event of the long process of Italian unification) in 1870,[112] they retired to the Vatican. Note that the Treaty of Venice made official the independence of Papal States from the Holy Roman Empire in 1177.[113] While the Pope is elected by the College of Cardinals, the Holy Roman Emperor was first elected by a small body of the greatest princes of the Empire, then followed by his coronation as Emperor by the Pope. The original annotation on the Europe-KWQ has this text: 此方教化王不娶專行天主之教在邏馬國歐邏巴諸國皆宗之 My translation and comments (between square brackets) are as follows: The enlightened king — the Pope — at this place [Rome], does not get married, but only practices the religion of God. In the Papal States [邏馬國; the State of the Church],[114] all the territories were under the direct sovereign rule of the Pope. Although the Europe-KWQ does not show the borders of the Papal States, the map does depict 羅馬 (Rome; the residence of the Pope) and 邏馬國 (the Papal States) to show their difference.

Item #	Name	Pinyin	Etymology	History and geography and my comments
• 30	麻勒葛	Ma Le Ge	Marche[115]	Marche is a historical region in central Italy. Marche[116] was part of the Papal States, but most of the territory was under local lords, while the major cities ruled themselves as free communes.
31	意大里亞	Yi Da Li Ya	*Italia* was the Latin and Italian name for the Italian Peninsula. A nineteenth century map of Ancient Sicily and Southern *Italia*, illustrating the homeland of the Romans and metropole of Rome's empire in classical Antiquity.[117]	*Italia* was the ancient name of the Italian Peninsula, originally applied only to today's Southern Italy. In the eighth century, the Franks helped the formation of the Papal States in Central Italy. Until the thirteenth century, Italian politics was dominated by the relations between the Holy Roman Emperors[118] and the Papacy.[119] Major renaissance cities in Italia such as Florence and the region Tuscany are not depicted on the Europe-KWQ, indicating that the map depicts Europe before its renaissance, consistent with the era between 1157 and 1166 deduced in the present chapter.

Summary A3: Items 19–31 present the history and geography of Central and Southern Europe (including the northern and central Italian Peninsula) under mostly the rule of the Holy Roman Empire in the Middle Ages. The geographical items on the Europe-KWQ include trading port: 麻尔西里亞 (Marseille; Item 19); a county: 沙勿牙 (Savoy; Item 24); five regions: 突尔蜚諾 (Dauphiny; Item 20), 百尔入革 (Belgium; Item 21), 赫尔勿妻亞 (Helvetia; Item 22), 別蒙突 (Piedmont; Item 25) and 隆拔勒地亞 (Lombardy; Item 27); a confederation: 蘇亦微亚 and 十三郡 (Switzerland; Item 23); a semi-independent city-state: 惹怒襪 (Genoa; Item 26); an independent city-state: 勿耨茶 (Venice; Item 28); the Pope's residence: 羅馬 (Rome; Item 29); 邏馬國 (the Papal States; mentioned in the annotation), 麻勒葛 (Marche; Item 30) which was part of the Papal States, and 意大里亞 (Italia; Item 31). Among all the city-states in northern Italy, only 勿耨茶 (Venice; Item 28) was not part of the Holy Roman Empire. The Europe-KWQ correctly reflects the political landscape of this part of Europe between 1157 and 1166, although the name of the Holy Roman Empire is not depicted (the exact term "Holy Roman Empire" was not used until the thirteenth century). The only exception is the addition of the sixteenth century names 蘇亦微亚 and 十三郡 (Switzerland; Item 23) to the Europe-KWQ by Matteo Ricci in 1602. On the Europe-KWQ, no country border is drawn.

(*Continued*)

(*Continued*)

Zone IV: Southern Europe (Southern Italian Peninsula)				
32	那波里	Na Bo Li	Naples; Italian: *Napoli*	The city was the capital of the Duchy of Naples (661–1139). In 1137, the Duchy of Naples was under Norman control and thus joined the Kingdom of Sicily (1130–1816).[120] The Kingdoms of Naples and Sicily were ruled by the Staufer Dynasty between 1194 and 1254.[121] The Europe-KWQ correctly depicts 那波里 (Naples) during the era between 1157 and 1166.
33	步尔牙	Bu Er Ya	Apulia; Italian: *Puglia*	Apulia is a historical region located in the Southern Italian Peninsula. In the north, the Gargano promontory extends out into the Adriatic like a "spur", while in the south, the Salento peninsula forms the "heel" of "Italy's boot". Apulia became an autonomous duchy in 1059.[122] But Apulia was disestablished in 1130 to become part of the Kingdom of Sicily (1130–1816).[123] After 1282, when the Kingdom lost the island of Sicily itself, Apulia remained part of the remnant Kingdom of Naples, and remained so until the unification of Italy in 1861.[124]  Fig. A3.4. Apulia on the Italian Peninsula (in red).[125] The nickname given to Italy is the "boot" and Apulia is the region that stretches as the "heel" of "Italy's boot". 步尔牙 (Apulia) is poorly depicted on the Europe-KWQ, so it is hard to imagine that the map was drawn by Matteo Ricci in 1602.

Item #	Name	Pinyin	Etymology	History and geography and my comments
34	葛辣比	Ge La Bi	Calabria; Calabrian: *Calàbbria*	Calabria is a historical region in the Southern Italian Peninsula. In the seventh century, the Byzantine Empire created the Duchy of Calabria[126] and continued to use the name *Calabria* for its territory in ancient Bruttium.[127] Calabria is the "toe" of "Italy's boot". On the Europe-KWQ, 葛辣比 (Calabria) does not seem to have toes (see left figure in Fig. 3.6).  Fig. A3.5. Calabria on the Italian Peninsula (in red).[128]
Summary A4: Items 32–34 present the history and geography of Southern Europe (Southern Italian Peninsula) including the city 那波里 (Naples, Item 32), as well as the "heel" 步尔牙 (Apulia; Item 33) and "toe" 葛辣比 (Calabria; Item 34) of "Italy's boot". All are consistent with the era of Europe between 1157 and 1166, but the Europe-KWQ does not correctly depict the shape of "Italy's boot".				
Zone V: Western and Central Europe (including regions near the Baltic Sea and the North Sea)				
35	拂朗製泥亞	Fu Lang Ke i Ya	Franconia; German: *Franken*, the term comes from the plural form of *Franke*, a member of the Germanic tribe known as the Franks.[129]	At the beginning of the tenth century, a Duchy of Franconia was established, covering what is now northern parts of the state of Bavaria, a small part of southern Thuringia and a region in north-eastern Baden-Württemberg.[130] Franconia remained a royal domain nurturing no strong ducal dynasty and supported German kings and Holy Roman emperors. By the twelfth century, the name had come to refer only to East Franconia.[131]

(*Continued*)

(*Continued*)

36	帕襪利亞	Pa Wa Li Ya	Bavaria; New Latin: *Bavaria*	Bavaria (a landlocked region) became a stem duchy (c. 555–1623) in the sixth century and a part of the Holy Roman Empire from the tenth century.[132] It became an independent kingdom after 1805.[133]
37	物斯法畧	Wu Si Fa Lue	Westphalia; German: *Westfalen*	It is a historical region of northwestern Germany. Westphalia first appears as the name of the western third of the duchy of Saxony (Item 45) in the tenth century and survived the breakup (1180) of the Saxon duchy which was part of the Holy Roman Empire.[134] The creation of the Duchy of Westphalia expanded its territories.[135] On the Europe-KWQ, 物斯法畧 (Westphalia) is depicted along the North Sea; this is incorrect; it is landlocked.
38	非里西亞	Fei Li Xi Ya	Frisia[136]	Frisia is a historical coastal region in Germany and the Netherlands, along the southeastern corner of the North Sea. Frisia was inhabited by the Frisians (a West Germanic ethnic group). The Frankish Empire subjugated Frisia in the eighth century.[137] After the fall of the Frankish Empire, the Frisia lands owed their allegiance to the Holy Roman emperor.[138] On the Europe-KWQ, 物斯法畧 (Westphalia; Item 37) is depicted along the North Sea, whereas 非里西亞 (Frisia; Item 38) becomes a landlocked region. The opposite is true.
39	虎西亞	Hu Xi Ya	Prussia[139]; Latin: *Borussia*	Prussia is a historical region on the southeastern shore of the Baltic Sea. From 1147 to 1166, the Polish tried to subdue Prussia to no avail.[140] The Old Prussia was under the rule of the Holy Roman Empire.[141] The Europe-KWQ incorrectly depicts 虎西亞 (Prussia) as a land-locked region. However, in Figs. 3.8 and 3.9 of the main text, Prussia is depicted correctly near the Baltic Sea and bordering 波羅尼亞 (Poland; Item 51).

Item #	Name	Pinyin	Etymology	History and geography and my comments
● 40	肥良的亞	Fei Liang Di Ya	Flanders; Latin: *Flandria*	Flanders was a county and an ancient Dutch-speaking northern territory in Belgium[142] that existed from 862 until its absorption by the French First Republic in 1795.[143] Like Belgium (百尔入革; Item 21), Flanders (肥良的亞; Item 40) was under the rule of the Holy Roman Empire during the eleventh and twelfth centuries. On the Europe-KWQ, 百尔入革 (Belgium; Item 21) and 肥良的亞 (Flanders; Item 40) are depicted at widely separated geographical locations. This kind of error shows that the cartographer who drew the Europe-KWQ was not very familiar with Europe's internal geographical information, hence, was not a European.
41	波亦米亞	Bo Yi Mi Ya	Bohemia[144]; Latin: *Bohemia*	The Duchy of Bohemia was established in c. 870 and promoted to the Kingdom of Bohemia (the predecessor of the modern Czech Republic) in 1198. From 1004 to 1806, Bohemia was part of the Holy Roman Empire.[145]
42	喎蘭地	Wai Lan Di	Holland; Latin: *Hollandia*. It is a region and former province on the western coast of the Netherlands.	From the tenth to the sixteenth century, Holland proper was a unified political region within the Holy Roman Empire as a county ruled by the counts of Holland. The name "Holland" was first used for the region around Haarlem, and by 1064, it was used as the name of the entire county.[146]  Fig. A3.6. North and South Holland (in orange) shown together within the Netherlands.[147]

(*Continued*)

(Continued)

43	則闌地	Ze Lan Di	Zeeland	The County of Zeeland (1012–1796) was a county of the Holy Roman Empire in the Low Countries. It roughly corresponded to the modern Dutch province of Zeeland.[148] The original annotation next to 喎闌地 (Holland) and 則闌地 (Zeeland) on the Europe-KWQ has this: 西洋布此二島最妙 My translation is: It is most interesting that these two islands are placed in the Western Ocean. However, Holland (Item 42) is not an island, and Zeeland included many islands in the past (they are now connected by dikes). The annotation does not describe the geographical conditions of these two places correctly. 西 means "west". 西洋 means the "Western Ocean", because Europe is to the west of China and the ocean there was named 西洋 or Western Ocean. 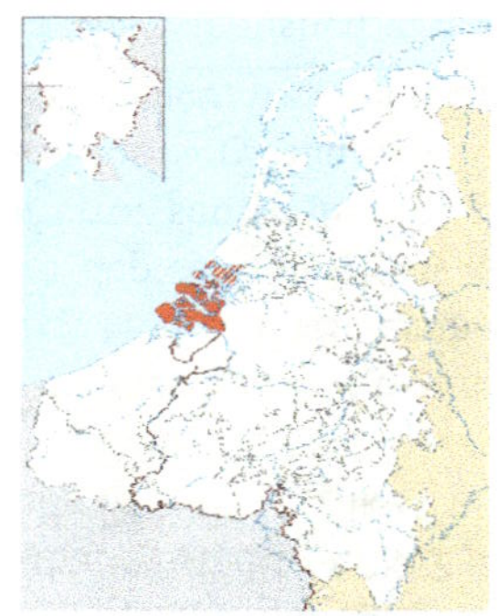 Fig. A3.7. County of Zeeland around 1350 (in red).[149]

Item #	Name	Pinyin	Etymology	History and geography and my comments
44	噢失突利亚	O Shi Tu Li Ya	Austria; German: *Österreich*. The word "Austria" is a Latinisation of the German name and was first recorded in the twelfth century.	The name "Ostarrîchi" first appeared in 996.[150] Austria initially emerged as a margraviate around 976 and the Duchy of Austria (1156–1453) was a medieval principality of the Holy Roman Empire.[151]  Fig. A3.8. The first appearance of the word "Ostarrîchi", circled in red and magnified. Modern Austria honours this document, dated 996, as the founding of the nation (see endnote 150; public domain).
45	沙瑣泥亞	Sha Suo Ni Ya	Saxony; Latin: *Saxonia*	The Saxons were originally a small tribe living near the North Sea between the Elbe and Eider Rivers. Saxony has a long history as a duchy (804–1296), an electorate of the Holy Roman Empire (the Electorate of Saxony). In 1180, the Duchy of Saxony was divided, and its territory sharply reduced.[152]
46	入尔馬泥亞	Ru Er Ma Ni Ya	Germania; Latin: *Germania*. *Germani* (for the people) and *Germania* (for the area where they lived) became the common Latin words for Germans and Germany.	Germania was a large historical region east of the Rhine in Central Europe, which was associated with the Germanic peoples.[153] The Kingdom of Germany (843–911) denotes the mostly Germanic-speaking Eastern Frankish Kingdom. After 962, East Francia became part of the Holy Roman Empire, which also included the Kingdom of Italy and, after 1032, the Kingdom of Burgundy as well as the various states located in Western Europe during the Middle Ages.[154] By the twelfth century, East Francia was called the Kingdom of the Germans by the historian Otto of Freising.[155] The Europe-KWQ depicts 入尔馬泥亞 (Germania) in Central Europe and 拂郎察 (Francia; Item 18) in Western Europe, which clearly shows the east–west division, enforced by the German and

(*Continued*)

(Continued)

				Latin language split since the ninth century, in consistence with the Europe of 1157–1166 depicted on the Europe-KWQ. The original annotation at the far right of 入尔馬泥亞 on the Europe-KWQ is: 入尔馬泥亞諸國共一總王也世繼者七國之王於中常共推一賢者為之 My translation is: The kingdoms of Germania are ruled by a single chief king. In the hereditary monarchy, the chief king is often the one considered to have attained wisdom among the kings of the seven kingdoms.
Summary A5: Items 35–46 give the history and geography of Western and Central Europe (including regions near the Baltic Sea and the North Sea). The geographical items under analysis include a huge area consisting of many small kingdoms collectively being called 入尔馬泥亞 (Germania; Item 46); four duchies: 拂朗殼泥亞 (Franconia; Item 35), 波亦米亞 (Bohemia; Item 41), 噢失突利亚 (Austria; Item 44) and 沙瑣泥亞 (Saxony; Item 45); a stem duchy: 帕襪利亞 (Bavaria; Item 36); three counties: 肥良的亞 (Flanders; Item 40), 喎蘭地 (Holland; Item 42) and 則蘭地 (Zeeland; Item 43); and three regions: 物斯法畧 (Westphalia; Item 37), 非里西亞 (Frisia; Item 38) and 虎西亞 (Prussia; Item 39). All were under the rule of the Holy Roman Empire, consistent with the depiction of the Europe-KWQ in the era between 1157 and 1166. On the Europe-KWQ, no border is shown, nor does it indicate the name of the Holy Roman Empire, because the exact term "Holy Roman Empire" was not used until the thirteenth century.				
Zone VI: Central and Southeast Europe				
47	翁阿利亞	Weng A Li Ya	Hungary; Latin: *Hungaria*	The Kingdom of Hungary existed in Central Europe from 1000 A.D. into the 20th century. By the twelfth century, it became a European middle power within the Western world.[156] On the Europe-KWQ, the geographical location of the Kingdom of Hungary is incorrectly depicted to the northwest, instead of directly south, of 波羅尼亞 (Kingdom of Poland; Item 51).
• 48	瑣利亞	Suo Li Ya	Croatia; Medieval Latin: *Croātia*	Croatia is in the Balkan Peninsula and at the crossroads of Central and Southeast Europe. The Croats arrived in the sixth century and Croatia became a kingdom in 925. The Kingdom of Croatia (and Dalmatia) was a province within the Kingdom of Hungary (1102–1527).[157]

Item #	Name	Pinyin	Etymology	History and geography and my comments
				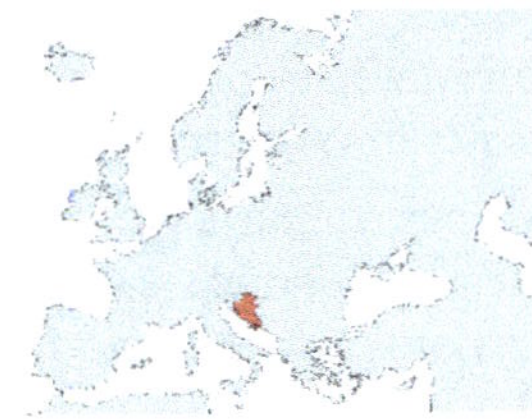 Fig. A3.9. Kingdom of Croatia and Dalmatia (in red).[158] The Europe-KWQ erroneously depicts Croatia as a landlocked kingdom.
• 49	大尔馬齊亞	Da Er Ma Qi Ya	Dalmatia; Latin: *Dalmatia*	Dalmatia is a narrow belt along the east shore of the Adriatic Sea, and one of the four historical regions of the Kingdom of Croatia (925–1102)[159] from the eighth century. Dalmatia was located at the very south of Croatia in the Balkan Peninsula. The Kingdom of Croatia (and Dalmatia) was a province within the Kingdom of Hungary from 1102 to 1526.[160]
50	突浪西尔襪尼亞	Tu Lang Xi Er Wa Ni Ya	Transylvania; Medieval Latin: *Ultrasylvania*, later *Transylvania*	Transylvania is a historical region in today's central Romania.[161] Between 1003 and 1526, Transylvania was a voivodeship in the Kingdom of Hungary, led by a voivode appointed by the King of Hungary.[162]
Summary A6: Items 47–50 give the history and geography of Central and Southeast Europe (including a portion of the Balkan Peninsula). This contains a kingdom: 翁阿利亞 (Hungary; Item 47) and three provinces: 瑣利亞 (Croatia; Item 48), 大尔馬齊亞 (Dalmatia; Item 49) and 突浪西尔襪尼亞 (Transylvania; Item 50). The Europe-KWQ correctly reflects the political landscape of this part of Europe between 1157 and 1166. But the geographical locations of 翁阿利亞 (Hungary; Item 47) and 瑣利亞 (Croatia; Item 48) are erroneous. The Europe-KWQ shows no borders.				
Zone VII: Eastern Europe				
51	波羅尼亞	Bo Luo Ni Ya	Poland; Latin: *Polonia*	The Kingdom of Poland started from 1025 and formed a union with Lithuania in 1385. The union bound the two countries together for the next four centuries.[163]

(*Continued*)

(Continued)

				Although the Europe-KWQ correctly depicts 波羅尼亞 (the Kingdom of Poland; Item 51) to the east of 波亦米亞 (the Kingdom of Bohemia became the predecessor of the modern Czech Republic; Item 41), the two realms are drawn too far apart.
• 52	墨亞尼亞	Mo Ya Ni Ya	Moravia; Czech: *Morava*	Moravia was a historical region in the east of the Czech Republic and one of three historical Czech lands, with Bohemia and Czech Silesia. Moravia had varying degrees of autonomy from the ruler of Bohemia and reached its height of autonomy in 1182.[164] 墨亚尼亚 (Moravia; Item 52) should be located between 波亦米亞 (Duchy of Bohemia; Item 41) and 波羅尼亞 (Kingdom of Poland; Item 51). But on Europe-KWQ, it was wrongly depicted in the eastern part of 波羅尼亞 (Kingdom of Poland; Item 51). As a result, 墨亚尼亚 (Moravia; Item 52) and 波亦米亞 (Duchy of Bohemia; Item 41) are far apart.
Summary A7: Items 51 and 52 give the history and geography of 波羅尼亞 (Kingdom of Poland; Item 51) and one of the three historical Czech lands — 墨亞尼亞 (Moravia; a semi-autonomous region; Item 52) — in Eastern Europe. The Europe-KWQ correctly reflects the political landscape of this part of Europe between 1157 and 1166, but mistakenly depicts the geographical location of 墨亞尼亞 (Moravia) to the east of 波羅尼亞 (Kingdom of Poland). The Europe-KWQ shows no borders.				
Zone VIII: Southeast Europe (mainly the Balkan Peninsula)				
53	羅馬泥亞	Luo Ma Ni Ya	Romania; Latin: *romanus*, meaning "Roman" or "of Rome".	Romanians (or Vlachs of Germanic origin) established their presence on the Balkan Peninsula, in the lands to the north of the Lower Danube, from the eleventh century.[165]
54	比產齊何	Bi Chan Qi He	Byzantium; Greek: *Byzantion*	Byzantium was an ancient Greek city. The Byzantine Empire (395–1453), also referred to as the Eastern Roman Empire or Kingdom of the Romans or simply Byzantium, in its own time was the continuation of the Roman Empire in its eastern provinces.[166] The capital city was Constantinople founded in 324 on the site of an already-existing city, Byzantium, and now called Istanbul.
55	蘭被礼	Lan Bei Li	Yambol; old name: *Hiambouli*	Yambol was a town in southeastern Bulgaria under Byzantine rule (1018–1185).[167] Its name derived from a nearby ancient city Kabile or Kabyle, whose name evolved from Diospolis, Hiambouli, etc., to become Yambol.

Item #	Name	Pinyin	Etymology	History and geography and my comments
56	亞尔百泥亞	Ya Er Bai Ni Ya	Latin: *Albania*: the region where the Albanians lived	The first undisputed mention of Albanians who lived in the territory of Albania on the Balkan Peninsula dates to historical records from 1079 or 1080.[168] The Albanians called their country *Arbëri or Arbëni*. However, the Kingdom of Albania (medieval) was created in 1272 and lasted until 1368.[169] The Europe-KWQ depicts 亞尔百泥亞 (Albania; Item 56) as facing the Black Sea, and 步尔葛利亞 (Item 60; Bulgaria) facing the Adriatic Sea. The reality is just the opposite.
• 57	厄勒齊亞	E Le Qi Ya	Greece; Latin: *Graecia* and *Graecus*	厄勒齊亞 (Graecia; Greece) was under the rule of the Byzantine Empire from the fourth century until 1453. Greece suffered from the dislocation of barbarian invasions and retained formal control of only the islands and coastal areas. During the eleventh and twelfth centuries, Greece was able to retain stability and strong economic growth.[170]
58	馬則多泥亞	Ma Ze Duo Ni Ya	Macedonia; Greek: *Makedonía*	Macedonia is a historic region that spans parts of northern Greece and the Balkan Peninsula. From the beginning of the sixth century, Macedonia — then part of the Byzantine Empire — became a subject of frequent raids by Slavic tribes. Little is known of the provincial organisation in the twelfth century.[171]
• 59	班諾尼	Ban Nuo Ni	Pannonia[172]	Pannonia was a province of the Roman Empire bounded on the north and east by the Danube.[173] The Kingdom of Hungary by the eleventh century comprised the entire Pannonian basin.[174] Hungary was not under the rule of the Holy Roman Empire, and peace was concluded with the Byzantine war of 1127–1129 A.D.[175] 班諾尼 (Pannonia) is a landlocked region, but the Europe-KWQ mistakenly depicts it as facing the Adriatic Sea.
60	步尔葛利亞	Bu Er Ge Li Ya	Bulgaria: the name is derived from Bulgars, a Turkic tribe which founded the country.[176]	The First Bulgarian Empire (681–1018) was a medieval Bulgar-Slavic state in Southeast Europe.[177] From 1018, the local Slavic inhabitants were called Bulgarians by the Byzantines. But the Bulgarians kept their nationality until the Second Bulgarian Empire was formed in 1185.[178] The Europe-KWQ incorrectly depicts the geographical location of Bulgaria as facing the Adriatic Sea.

(*Continued*)

(*Continued*)

• 61	莫勒亞	Mo Le Ya	Morea: the name is first recorded in the tenth century in the Byzantine chronicles.[179]	Morea was the name of the Peloponnese Peninsula in southern Greece during the Middle Ages and the early modern period.[180]
Summary A8: Items 53–61 give the history and geography of Southeast Europe (mainly the Balkan Peninsula). The geographical items include a kingdom: 步尔葛利亞 (Bulgaria; Item 60); a province: 馬則多泥亞 (Macedonia; Item 58); three regions: 亞尔百泥亞 (Albania; Item 56), 厄勒齊亞 (Greece; Item 57) and 班諾尼 (Pannonia; Item 59); a peninsula: 莫勒亞 (Morea; Item 61); a city: 比產齊何 (Byzantium; Item 54); a town: 蘭祓礼 (Yambol; Item 55); and a semi-autonomous region: 羅馬泥亞 (Romania; Item 53). The Europe-KWQ correctly reflects the political landscape of this part of Europe under the rule of the Byzantine Empire in the era between 1157 and 1166. However, the geographical locations of 亞尔百泥亞 (Albania), 步尔葛利亞 (Bulgaria) and 和班諾尼 (Pannonia) are incorrect. The Europe-KWQ does not show borders. The name of Byzantine Empire is also not indicated, perhaps because the terms "Byzantine Empire" and "Eastern Roman Empire" were coined after the end of the realm in 1453; its citizens continued to refer to their empire simply as the Roman Empire, and to themselves as Romans. Hence, when this region of Europe was explored between 1157 and 1166, the explorers would not know the names "Byzantine Empire" and "Eastern Roman Empire".				
Zone IX: British Isles				
62	諳厄利亞	An E Li Ya	England; Latin: *Anglia*; it was the medieval Latin name for England that is part of today's United Kingdom.	England became a unified state in the tenth century.[181] Later, the Kingdom of Great Britain (1707–1801), officially called Great Britain, was a sovereign state formed by uniting the kingdoms of England (which includes Wales) and Scotland (思可齊亞; Item 64) to form a single kingdom. The original annotation next to 諳厄利亞 (Anglia) on the Europe-KWQ has this: 諳厄利亞無毒蛇等蟲雖別處攜去者到其地即無毒性 My translation is: Anglia has no venomous snakes or other poisonous insects; even those which were brought in from elsewhere become non-toxic once they arrive there. Among the three types of snakes native to Britain, only one native reptile, the adder, is venomous. However, the adder's bite generally does not cause death.[182]

Item #	Name	Pinyin	Etymology	History and geography and my comments
63	婆林日	Po Lin Ri	Belerion; Latin: *Belerion* or *Bolerium*	Bolerium or Balerion (婆林日) is the ancient Celtic name for the westernmost tip of Cornwall (the westernmost part of England). After a period of Roman rule, by the middle of the ninth century, Cornwall had fallen under the control of Wessex.[183] Later in 1337, the title Duke of Cornwall was created by the English monarchy.[184]
64	思可齊亞	Si Ke Qi Ya	Scotland; Late Latin: *Scotia*	The name Scotia was initially used to refer to Ireland. From the ninth century, Scotia referred to only the part of Britain — the Kingdom of Scotland (843–1707)[185] — lying north of the River Forth.
65	喜百泥亞	Xi Bai Ni Ya	Ireland; Classical Latin: *Hibernia*, which was taken from Greek geographical accounts.	Beginning from the seventh century, all the Irish kingdoms had their own kings, but were nominally subject to the high king. Later the Kingdom of Ireland was created in 1542 and lasted until 1800.[186]
Summary A9: Items 62–65 give the history and geography of the British Isles in the North Atlantic. The geographical items include a state: 諳厄利亞 (England; Item 62); a kingdom: 思可齊亞 (Scotland; Item 64); many small kingdoms collectively being called 喜百泥亞 (Ireland; Item 65); and a region: 婆林日 (Belerion; Item 63). The Europe-KWQ correctly depicts their political landscape and geographical locations between 1157 and 1166. However, ancient Latin names or Greek names were used for these places, indicating that this part of the world may have been known to the Chinese in the millennia before Christ, consistent with the findings of T.C. Bell's surveys. The Europe-KWQ does not show borders.				
Zone X: Northern Europe				
• 66	玉良氐	Yu Liang Di	Jutland; Latin: *Lutlandia*; Danish: *Jylland*; German: *Jütland*	Jutland is a peninsula in Northern Europe which forms the continental portion of Denmark and a part of Northern Germany.

(*Continued*)

(*Continued*)

				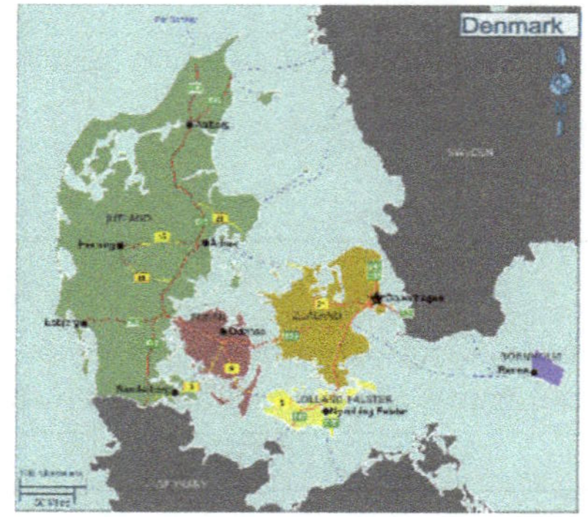 Fig. A3.10. Regions of Denmark; Jutland is highlighted in green.[187]
67	大泥亞 or 第那瑪尔加	Da Ni Ya or Di Na Ma Er Jia	Denmark; Latin: *Dania* (named for the Dani tribe) Denmark in Spanish is *Dinamarca* (transliterated as 第那瑪尔加), the name shown in an annotation on the Europe-KWQ.	Throughout the High and Late Middle Ages (1000–1250 and 1250–1500), the Kingdom of Denmark (eighth century?–1397) consisted of a peninsula, Jutland, and an archipelago of 443 named islands, and included present-day south Sweden.[188] The original annotation to the right of 玉良氏 (Jutland; Item 66) on the Europe-KWQ is: 大泥亞即第那瑪尔加 My translation is: Dania is also known as Dinamarca. The Europe-KWQ depicts Denmark as only a peninsula, without islands.
68	诺尔物入亚	Nuo Er Wu Ru Ya	Norway; Latin: *Northuagia*	The Kingdom of Norway was established in 872 as a merger of many petty kingdoms and has existed continuously for 1149 years.[189]
• 69	羅多里	Luo Duo Li	Lutuoli	This place cannot be identified on a modern map.
70	非馬祁亞	Fei Ma Qi Ya	Finnmark; Latin: *Finmarchia*	Finnmark was a county in the northern part of Norway. Coastal areas of Finnmark were colonised by Norwegians beginning in the tenth century.[190]

Item #	Name	Pinyin	Etymology	History and geography and my comments
71	思祁非尼亞	Si Qi Fei ni Ya	Scricfinnia[191]	Scricfinnia lies between Lapland (the largest and northernmost region of Finland) and the Arctic Ocean. It was a region in the northern part of Norway.
72	臥的亞	Wo Di Ya	Götaland, also *Geatland*, *Gothia*, *Gothland*, *Gothenland* or *Gautland*	Gothia lies in today's southern Sweden. Parts of the area were at that time, between 1157 and 1166, either Danish or Norwegian.[192]
73	蠟皮亞	La Pi Ya	Two possibilities: (1) Lapland; Latin: *Lapponia*; or (2) Lapia in Southern Finland.	Case 1: Lapland includes the northern parts of Scandinavia and Finland with the Kola Peninsula in Russia, where Sami people, also called Laplanders, live.[193] Case 2: Lapia is in Southern Finland[194] with a small population. But the Europe-KWQ wrongfully depicts 蠟皮亞 in Southern Sweden (蘇亦齊; Item 75). This has made the determination of the precise location of 蠟皮亞 uncertain.
74	比葛謎亞	Bi Ge Mi Ya	Bjarmaland or Bjarmland; Latin: *Biarmia*	Until the sixteenth century, Bjarmaland usually referred to the southern shores of the White Sea in northwestern Russia and the basin of the Northern Dvina River in Finland as well as some of the surrounding areas.[195]
75	蘇亦齊	Su Yi Qi	Kingdom of Sweden; Swedish: *Sverige*	The Swedish Viking Age lasted roughly from the eighth century to the eleventh century.[196] An independent Swedish state emerged during the early twelfth century. However, it is not known when and how the Kingdom of Sweden was born. The Swedish colonisation of the coastal areas of Finland started during the twelfth century. During the seventeenth century, Sweden emerged as a European great power.[197]
76	沸你刪突	Fei Ni Shan Tu	Finland; old Latin name: *Finnia*	The earliest known written mention of Finland is around the eleventh century.

(*Continued*)

(*Continued*)

				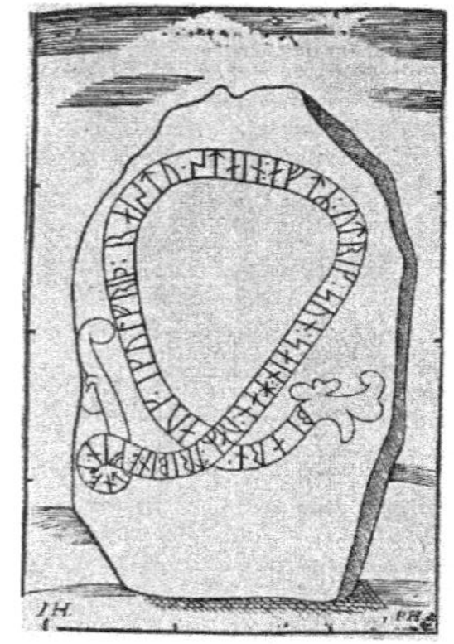 Fig. A3.11. Drawing of Runestone U 582 (see endnote 197; public domain). Finland became a common name for the whole country around the twelfth century. Swedish colonisation of Finland took place during the Northern Crusades from the twelfth century. Colonisation focused on the Finnish archipelago and some of its coastal regions.[199] Finland was part of Sweden for almost 700 years from around 1150 until the Finnish War of 1809 that saw Finland becoming an autonomous part of the Russian Empire.[200] The year 1362 is often considered to be the time when Finland was incorporated into the Kingdom of Sweden.[201]
• 77	蘇亦齊界	Su Yi Qi Jie	Jie (界) means "boundary". 蘇亦齊 is Sweden (see Item 75).	On the Europe-KWQ, Finland is poorly drawn. The Europe-KWQ generally does not contain territorial boundaries, but Sweden is an exception: the Europe-KWQ clearly depicts a region across the Baltic Sea as Sweden's territorial boundary which extends to the entire coastal region in the south and middle portion of Finland. This fact by itself reveals the political landscape of this part of Europe between c. 1150 and 1362 (when Finland was completely incorporated into Sweden). However, after analysing other regions on the map, the date of the map is narrowed to between 1157 and 1166.

Item #	Name	Pinyin	Etymology	History and geography and my comments
Summary A10: Items 66–77 give the history and geography of Northern Europe including four kingdoms: 大泥亞/第那瑪尔加 (Denmark; Item 67), 诺尔物入亚 (Norway; Item 68), 蘇亦齊 (Sweden; Item 75) and 沸你刪突 (Finland; Item 76); and five regions: 非馬祁亞 (Finnmark; Item 70), 思祁非尼亞 (Scricfinnia; Item 71), 臥的亞 (Gothia; Item 72), 蠟皮亞 (Lapland or Lapia; Item 73) and 比葛謎亞 (Bjarmaland; Item 74); and a peninsula: 玉良氏 (Jutland; Item 66). The Europe-KWQ correctly reflects the political landscape of this part of Europe between 1157 and 1166. However, the map erroneously depicts 大泥亞/第那瑪尔加 (Denmark; Item 67) as only a peninsula without islands; it also places 蠟皮亞 (Lapland or Lapia; Item 73) in Southern Sweden, instead of Lapland in the northern parts of Scandinavia and Finland or Lapia in Southern Finland, while 羅多里 (Lutuoli; Item 69) cannot be identified. The Europe-KWQ shows no borders except for 蘇亦齊 (Sweden; Item 75) with its territorial border extending to the coastal area in the southern and central parts of 沸你刪突 (Finland; Item 76).				
Zone XI: Eastern Europe				
• 78	葛勒施葛	Ge Le Shi Ge	Kvenland; Latin: *Caienska Semla*, which is a synonym of the non-Finnic name Kvenland.	Caienska (Kvenland) is an ancient name for an area in Fennoscandia and Scandinavia.[202]
79	諾尔京	Nuo Er Jing	Nordenborg[203]	There is a Nordenborg in today's Norway; the Europe-KWQ depicts 諾尔京 (Nordenborg) to the north of 葛勒施葛 (Kvenland/Caienska *Semla*; Item 78).
• 80	矮人國	Ai Ren Guo	Kingdom of Dwarfs	In the Norse mythology, homeland of the dwarfs could mean Svartalfaheim[204] (also known to its natives as Niðavellir). But the Europe-KWQ has a realistic description of a 矮人國 (the Kingdom of Dwarfs). The original annotation to the left of 矮人國 (Kingdom of Dwarfs; Item 80) on the Europe-KWQ is: 國人男女長止尺餘五歲生子八歲而老常為鸛鷂所食其人穴居以避每候夏三月出壞其卵雲以羊為騎 My translation and comment (between square brackets) are as follows:

(*Continued*)

(*Continued*)

				The men and women of the Kingdom of Dwarfs are only a little more than a *Chi* [about 1/3 m] tall. At age five, they give birth to children and at age eight, they are old. They often become prey of stork harriers and must live in burrows to avoid them. Every year they wait for the three summer months to come out to destroy the birds' egg nests. They ride on sheep. The dwarfs are not as powerful as the storks, but fortunately, they are smarter, knowing to dig out the birds' eggs to maintain ecological balance.
81	諾勿瓦的亞	Nuo Wu Wa Di Ya	Novgorod; Latin: *Novogardia*	The Novgorod Republic was a medieval East Slavic state (1136–1478)[205] stretching from the Gulf of Finland in the west to the northern Ural Mountains in the east, including the city of Veliky Novgorod (usually shortened to Novgorod) in modern Russia. However, on the Europe-KWQ, Novgorod is mistakenly drawn as a landlocked state.
82	葛尔曷波利	Ge Er Ge Bo Li	Cargapowl[206] or Kargopol	Kargopol, located north of Moscow, was first chronicled in the twelfth century or even earlier.[207] It was the most significant trade station of the Novgorod Republic and one of the most northerly permanent Slavic settlements.[208]
83	札勿泥亞	Zha Wu Ni Ya	Zawonia	Zawonia is a village in the administrative district of Gmina Chlewiska, within Szydłowiec County, Masovian Voivodeship, in east-central Poland.[209] On the Europe-KWQ, 札勿泥亞 (Zawonia) is erroneously placed near the coast, instead of at its correct geographical location in landlocked southwestern Poland.

Item #	Name	Pinyin	Etymology	History and geography and my comments
• 84	臥尔丁然	Wo Er Ding Ran	Region in today's Lithuania: The Chinese name is not a transliteration of Lithuania, but the geographical location of 臥尔丁然 on the Europe-KWQ matches Lithuania.	For centuries, the southeastern shores of the Baltic Sea were inhabited by various Baltic tribes. The region of today's Lithuania is first mentioned in 1009. Later, the Grand Duchy of Lithuania was a European state that lasted from the thirteenth century to 1795.[210]
• 85	孛漏生	Bei Lou Sheng	Belarus: The name is closely related with the term *Belaya Rus'*, i.e. White Rus'.[211]	The region that is now Belarus was first settled by Baltic tribes in the third century. Around the fifth century, the area was taken over by Slavic tribes. In the ninth century, the territory of modern Belarus became part of Kievan Rus', a vast East Slavic state ruled by the Rurikid Dynasty. The lands of modern Belarus avoided the brunt of the Mongol invasion in the thirteenth century invasion and eventually joined the Grand Duchy of Lithuania (1236–1795) in 1242.[212]
86	魯西亞	Lu xi Ya	Rus'; Latin: *Ruthenia*; it is derived from Rus', a medieval state populated primarily by the East Slavs.	The establishment of the first East Slavic states was in 862. In the tenth to eleventh centuries, Kievan Rus' (879–1240) became one of the largest and most prosperous states in Europe.[213] But Kievan Rus' disintegrated in the Mongol invasion of 1237–1240. The invaders, later known as Tatars, formed the state of the Golden Horde, which ruled Southern and Central Russia for over two centuries.[214]
87	波多里亞	Bo Duo Li Ya	Podolia	Podolia is a vast historic region in East Europe, located in the west-central and southwestern parts of modern Ukraine and in northeastern Moldova (Moldova is the Romanian name; Moldavia is the correct English name; Item 88). During the Great Migration Period (a period during and after the decline of the Western Roman Empire), many nationalities passed through this territory or settled within it for some time,[215] until the Polish colonisation beginning in the fourteenth century.[215]

(Continued)

(Continued)

88	莫大未亞	Mo Da Wei Ya	Moldavia[217]	Moldavia lies in the northeastern corner of the Balkan region of Europe. The inhabitants of the Moldavian region were Christians. The Principality of Moldavia existed from 1359 to 1859.[218] The Europe-KWQ erroneously depicts 莫大未亞 (Moldavia) as a peninsula.
89	契利未牙	Qi Li Wei Ya	Crimea[219]	Crimea is a peninsula located on the northern coast of the Black Sea in Eastern Europe. The Byzantine emperors controlled the southern shores of the Crimea peninsula (the theme of Cherson) until the thirteenth century.[220] The Europe-KWQ depicts 契利未牙 (Crimea; Item 89) in the inland region and 瓦荼里亞 (Gazaria; Item 90) as a peninsula. This has caused great confusion. As explained in Item 90, 瓦荼里亞 (Gazaria) refers to an area on 契利未牙 (Crimea; a peninsula lying between the Black Sea and the Sea of Azov), which was initially occupied by the Khazars.
90	瓦荼里亞	Wa Cha Li Ya	Gazaria[221]	Gazaria is a word derived from Khazaria, which refers to an area on Crimea initially occupied by the Khazars[222] in the eighth century.[223] Later, different parts of Crimea were colonised or occupied by various empires or invaded by steppe nomads. In 1783, the Russian Empire annexed all of Crimea.[224]
91	古馬泥	Gu Ma Ni	The name Cumania originated as the Latin exonym for the Cuman–Kipchak confederation (tenth century–1241).[225]	Cumania was a Turkic confederation from the tenth century in the western part of the Eurasian Steppe. It ended its existence in the middle of the thirteenth century, with the Great Mongol Invasion of Europe.[226]
• 92	兀失丁入	Wu Shi Ding Ru	Usting[227]	The place cannot be properly identified. There is a "Usting" in today's Austria.[228] But the location is too far away from 兀失丁入 (Usting; Item 92) depicted on the Europe-KWQ.

Item #	Name	Pinyin	Etymology	History and geography and my comments
• 93	勒贊	Le Zan	Ryazan	There is the so-called Old Ryazan which was the capital of the Ryazan principality. It was burned in the thirteenth century during the Mongol invasion of the Batu Khan.[229] There is a museum on that site now. Modern Ryazan before the decree of Empress Catherine II of 1778 was called Pereyaslavl-Ryazansky.[230] Ryazan became an independent Russian princedom early in the twelfth century to the early sixteenth century.
Summary A11: Items 78–93 give the history and geography of Eastern Europe including the Balkan Peninsula and the Black Sea coastal area. The geographical items include a kingdom: 矮人国 (the Kingdom of Dwarfs; Item 80); a republic: 諾勿瓦的亞 (Novgorod; Item 81); six regions: 葛勒施葛 (Caienska/Kvenland; Item 78), 臥尔丁然 (Lithuania; Item 84), 孛漏生 (Belarus; Item 85), 波多里亞 (Podolia; Item 87), 莫大未亞 (Moldavia; Item 88) and 瓦茶里亞 (Gazaria; Item 90); a trade station: 葛尔曷波利 (Cargapowl; Item 82); a village: 札勿泥亞 (Zawonia; Item 83); a city: 勒贊 (Ryazan; Item 93); a state: 魯西亞 (Rus'; Item 86); a peninsula: 契利未牙 (Crimea; Item 89); and a confederation: 古馬泥 (Cumania; Item 91). The Europe-KWQ correctly reflects the political landscape of this part of Europe in the era between 1157 and 1166. However, the geographical locations of 諾勿瓦的亞 (Novgorod; Item 81), 札勿泥亞 (Zawonia; Item 83) and 莫大未亞 (Moldavia; Item 88) are erroneously depicted. Also, 瓦茶里亞 契利未牙 (Crimea; Item 89) is not correctly depicted as a peninsula, and 諾尔京 (Nordenborg; Item 79) and 兀失丁入 (Usting; Item 92) cannot be properly identified. The Europe-KWQ shows no borders.				
Zone XII: Islands (other than the British Isles)				
94	干的亞	Gan Di Ya	Candia	Crete is the largest among the Greek islands. After 900 years as a Roman and then Eastern Roman (Byzantine) island, Andalusian Muwallads established the Emirate of Crete (820s–961),[231] and made the city Crete its new capital. Both the city and the island became known as *Chandax* or *Chandakas*, which gave Latin, Italian and Venetian *Candia*, from which were derived French *Candie* and English *Candy* or *Candia*. In 961, the island returned to Byzantine rule after expelling the Arabs until 1205, when the Duchy of Candia became an overseas colony of the Republic of Venice.[232]
• 95	漚白亞	Ou Bai Ya	Euboea is the ancient and current name of the second-largest Greek island.	As the second largest Greek island, this ancient name Euboea (漚白亞) remained in use by classic authors until the sixteenth century. After Constantinople was founded in 330, Greece became part of the Eastern Roman Empire.[233] The Byzantine Empire was in existence from 330 to around 1453.[234]

(*Continued*)

(Continued)

96	西齊里亞	Xi Qi Li Ya	Sicily; Italian: *Sicilia*	Sicily is the largest island in the Mediterranean Sea. The Kingdom of Sicily (1130–1816)[235] was a successor state of the County of Sicily which had been founded in 1071. The original annotation next to 西齊里亞 (Sicily) on the Europe-KWQ is: 此岛有二山一常出大火一常出烟昼夜不绝 My translation and comments (between square brackets) are as follows: There are two [volcanic] mountains on this island: one [Mount Etna which dominates the eastern third of Sicily] often emits big fires and the other often emits smoke day and night without stopping. In fact, aside from Etna, other Sicilian volcanoes that are always active are Stromboli and Vulcano.[236]
97	黃魚島	Huang Yu Dao	黃魚島 (Yellow Fish Island) is the Island of Sardinia. Sardinia has pre-Latin roots coming from the pre-Roman ethnonym *s(a)rd-*, later Romanised as *sardus*.	The sea that surrounds the island produces small, oily forage fish which have a light yellow shiny longitudinal band on their upper body, hence, the Chinese called them "yellow fish", and the island 黃魚島 (Yellow Fish Island). The term "sardine" (fish name) may come from the Island of Sardinia during the early fifteenth century, around which sardines were once abundant; this name was first used in English during the early fifteenth century, not between 1157 and 1166. The Chinese name 黃魚島 (Yellow Fish Island) is not connected with this European pre-Latin roots. The original annotation next to the island is: 此島生珊瑚樹長 My translation is: This island produces corals, and trees are tall. The Riviera del Corallo is a coastal stretch of northwest Sardinia, which includes the town of Alghero where the Macor Coral Museum is located,[237] which is famous for producing the red coral used to make jewelry. Sardinia also has vast quercus suber (commonly called the cork oak) forests (Sardinia supplies approximately 85 percent of corks made in Italy). The cork oak usually grows to 20 m/66 ft in height but can reach 25 m/82 ft.[238]

Item #	Name	Pinyin	Etymology	History and geography and my comments
98	瑪兒大	Ma Er Da	Malta	Malta is the largest of the three major islands that constitute the Maltese archipelago. Malta was ruled by the Phoenicians, Carthaginians, Romans, Byzantines and Arabs before it was occupied by the County of Sicily in 1091.[239] The island then became part of the Kingdom of Sicily until it was given to the Order of Saint John (also known as Knights Hospitaller) along with Gozo and Tripoli (North Africa) in 1530.[240] The Malta Island should be some 80 km/50 mi south of the island of Sicily across the Malta Channel. However, on the Europe-KWQ, 瑪兒大 (Malta) is displaced to the far west of Sicily and to the south-east of the island of Sardinia.
99	哥尔西克	Ge Er Xi Ke	Corsica	Corsica is an island in the Mediterranean Sea. Corsica was ruled by the Popes from 1077. The Pope yielded civic administration to Pisa in 1090, but contention between the Pisans and Genoese soon engulfed Corsica.[241] Corsica was finally removed from the fighting by annexation to the Papal States in 1217.[242]
100	米諾里革	Mi Nou Li Ge	Menorca	Menorca is one of the Balearic Islands located in the Mediterranean Sea. Menorca was annexed to the Caliphate of Córdoba in 903, until the rule of the last Muslim ra'îs in 1287.[243]
101	馬岳里革	Ma Yue Li Ge	Mallorca	Mallorca is the largest island in the Balearic Islands located in the Mediterranean Sea. The island came under rule as an independent Taifa from 1087 to 1114 and was later ruled from North Africa until 1176. The ruling was replaced by the Almohad Dynasty until 1229, and finally annexed by the Crown of Aragon in 1230 until the eighteenth century.[244]

(Continued)

(*Continued*)

102	伊加里亞島	Yi Jia Li Ya Dao	Iceland[245]; Latin: *Islandia* (directly from Icelandic language *Ísland*). The oldest known record of the name "Iceland" is an eleventh-century rune carving from Gotland, which can be traced back to the ninth century.[246]	Iceland was founded more than 1,000 years ago during the Viking age of exploration beginning in the second half of the ninth century. The place was settled by a mixed Norse and Celtic population.[247] The Kingdom of Norway (872–1397) owned Iceland until 1415, when the kingdoms of Norway, Denmark and Sweden were united.
103	都力	Du Li	Thule	Thule is the northernmost location mentioned in ancient Greek and Roman literature and cartography.[248] It has been confusing where Thule is located. On the Europe-KWQ, 都力 (Thule) is depicted as an island off the coast of Norway. Today's Thule is a military base in northwest Greenland.
Summary A12: Items 94–103 give the history and geography of the major islands in Europe (other than near Britain). These islands are 甘的亞 (Candia: Item 94), 滬白亞 (Euboea; Item 95), 西齊里亞 (Sicily; Item 96), 黃魚島 (Yellow Fish Island; Item 97), 瑪兒大 (Malta; Item 98), 哥尔西克 (Corsica; Item 99), 米諾里革 (Menorca; Item 100), 馬岳里革 (Mallorca; Item 101), 伊加里亞島 (Iceland; Item 102) and 和都力 (Thule; Item 103). The Europe-KWQ correctly depicts their names, geographical locations, and characteristics in the era between 1157 and 1166, except the geographical position of 瑪兒大 (Malta; Item 98) is far away from its actual location.				
Zone XIII: Seas, oceans, lake, river and the name of Europe				
104	太海	Tai Hai	"太" meaning "great" or "big"; 太海 (today's Black Sea) is the world's largest inland body of water.[249] In Europe, until	The Black Sea was known around 400 B.C., based on the dating of a Greek merchant vessel discovered on the floor of the sea, which is the oldest known intact shipwreck ever located.[250] The sea was first named by the ancient Greeks who called it "Inhospitable Sea", because it was difficult to navigate, and hostile tribes inhabited its shores. Later, the name "Black Sea" was believed to be given by the Turkish during the Ottoman Empire period (1299–1922) when it was called Bahr-e Siyah or Karadeniz,

Item #	Name	Pinyin	Etymology	History and geography and my comments
			1570, the map *Asiae Nova Descriptio* from Abraham Ortelius's *Theatrum Orbis Terrarum* labels this sea *Mar Maggior* (Great Sea).	which means "the Black Sea" in the Ottoman Turkish.[251] The name also referred to a system of colour symbolism representing the cardinal directions, with black or dark for north. The Black Sea is to the north of the Ottoman Empire, hence, "Black Sea" meant "Northern Sea".[252] For sailors, the sea was black due to severe storms in the winter. For Chinese, the Black Sea is named as 太海 (Great Big Sea) on the Europe-KWQ, implying that the Chinese knew that this sea was a very big inland body of water, most likely because they had explored it to realise its size, hence used a suitable Chinese name 太海 (Great Big Sea) to describe it.
105	上海	Shang Hai	In the Latin of the Romans (Ancient or Classical Latin used from 75 B.C. to 300 A.D.) the sea was *Mare Superum* (the "upper sea", meaning 上海; here 上 means "upper" and 海 means "sea")[253]; in the medieval Latin (from around late third to late fifteenth century), the sea was *Mcre Hadriaticum* or *Mare Adriaticum*.	Both *Mare Superum* and *Mare Adriaticum* (Adriatic Sea) are names for designating the body of water separating the Italian Peninsula from the Balkans.[254] The name *Mare Superum* existed in Ancient Latin, and Mare Adriaticum, in the later medieval Latin. On the Europe-KWQ, 上海 (*Mare Superum*) is used, implying that the knowledge of this sea came from the period of around 75 B.C. to 300 A.D. (during the era of the Western Roman Empire). Today's large city Shanghai, located on the Chinese coast, bears the same Chinese name with a little different meaning: 上 (Shang; "upon") and 海 (Hai; "sea") together mean "Upon the Sea"[255]: the Chinese historians have suggested that during the Tang or the Song Dynasty, the area of modern-day Shanghai was under the sea level so the land appeared to be literally "on the sea".

(*Continued*)

(Continued)

• 106	下海		The Romans often designated the sea to the West of Italy as "the lower sea", or *Mare Inferum* (下海; 下 meaning "lower").[256]	On the Europe-KWQ, 下海 (Mare Inferum; the "lower sea") is used to depict the Tyrrhenian Sea (which name was originally used by the Greeks),[257] indicating that this geographical information came from the period of around 75 B.C. to 300 A.D.
107	地中海	Di Zhong Hai	Mediterranean Sea: The earliest proven use of the term *Mare Mediterrāneum* (Late Latin; from the third to the sixth centuries and continuing into the seventh century in the Iberian Peninsula), meaning "in the middle of land", is in the sixth century.	The original annotation for Mediterranean Sea on the Europe-KWQ is this: 此海生有一種咽機那魚長尺許周身皆刺 而有大力若貼船後雖順風不能動海濱產蠟里千樹其木不畏火可為屯寨 My translation and comment (between square brackets) are as follows: This sea has a kind of pharyngeal fish. The fish is about one *chi* [about 1/3 of a meter] in length with thorns all over the body and is very powerful. If it attaches its body to the back of the ship, even sailing downwind cannot make the ship move. The seashores grow Laliqian trees which produce wax and can burn easily. The wood can be stored for use in the fort. The Europe-KWQ shows what 地中海 (the Mediterranean Sea) could produce, indicating that the Chinese mariners might have explored this sea.
108	巴尔德峡	Ba Er De Xia	Strait of Gibraltar; Spanish: *Estrecho de Gibraltar*	The Strait of Gibraltar comes from the Rock of Gibraltar, which name is derived from the Arabic words "Jabal Ṭāriq (Mountain of Ṭāriq)", honouring Ṭāriq ibn Ziyād, who captured the peninsula during the early eighth century.[258]

Item #	Name	Pinyin	Etymology	History and geography and my comments
109	大西洋	Da Xi Yang	大西洋 literally means Great Western Ocean; it is today's Atlantic Ocean. The oldest known mention by the Greeks of an "Atlantic" sea was around mid-sixth century B.C.,[259] which refers to "the sea beyond the pillars of Heracles"; these are said to be part of the sea that surrounds all land.[260] However, to early Greek sailors and in Ancient Greek mythological literature, this all-encompassing ocean was instead known as Oceanus, the gigantic river that encircled the world, instead of the enclosed seas.	The term "Atlantic" originally referred specifically to the Atlas Mountains in Morocco and the sea off the Strait of Gibraltar and the North African coast. It then expanded to the ocean beyond Gibraltar. During the Age of Discovery (early fifteenth century to early seventeenth century), the Atlantic was also known to English cartographers as the Great Western Ocean (大西洋).[261] But in China, the geographical item "西洋", literally "Western Ocean", originally has different meanings during different periods of Chinese history. From the Five Dynasties (五代; 907–960),[262] through the Song[263] and Yuan[264] dynasties, and until the Ming Hongwu (明洪武) period,[265] it always meant a specific region or country to the west of China, not an ocean. By the time of the Yongle (永乐) era, the term 西洋 — literally the Western Ocean — had developed to include the meaning of many overseas countries.[266] In the early Ming Dynasty, the term also had a new meaning: the middle of Borneo (婆罗洲) was the boundary — the east was called 东洋 — the Eastern Ocean, and the west was called 西洋 — the Western Ocean. In the Ming Dynasty, the place called the South China Sea in the past contained both 东洋 (the Eastern Ocean) and 西洋 (the Western Ocean). Hence, from the Western Region of the Han Dynasty to the Western Ocean of the Ming Dynasty, the change reflects the expansion of travel between China and foreign countries from land to the ocean.[267] Matteo Ricci introduced the term 大西洋 (Great Western Ocean on the Europe-KWQ, meaning the ocean was the ocean of the Great Westerners) on the Europe-KWQ; this term had not existed on Chinese maps before 1602.

(*Continued*)

(*Continued*)

			During the Age of Discovery, the Atlantic was also known to English cartographers as the Great Western Ocean, literally translated as 大西洋, because in Latin, *Oceanus Occidentalis* means "Western Ocean", and the ocean was Great Westerners' Ocean.	
• 110	河摺亞諸滄	He Zhe Ya Nuo Cang	河摺亞諸 is the transliteration of the first Latin word of *Oceanus Atlanticus* or *Oceano Occidentale*; Latin: *Oceanus* (a water deity) is known as the gigantic river that encircled the world; 滄 means "ocean" or "wave".	Matteo Ricci placed 大西洋 (Great Western Ocean) off the Strait of Gibraltar and placed 河摺亞諸滄 (Ocean) off the west coast of Africa, seemingly to indicate that 大西洋 (Great Western Ocean) was the smaller water area off the Strait of Gibraltar, whereas 河摺亞諸滄 was the gigantic river that encircled the world in the Greek mythological literature. The Europe-KWQ in Fig. 3.1 does not show 河摺亞諸滄, because it extends beyond the coastal areas of Europe. Readers can find 河摺亞諸滄 near the western coast of Africa by viewing the original KWQ listed in endnote 1. It will be discussed in Chapter 4 of this book when Africa on KWQ is analysed.

Item #	Name	Pinyin	Etymology	History and geography and my comments
111	肥良的海	Fei Liang Di Hai	肥良的海 refers to the sea north of *Flandria*; it is today's North Sea.	The North Sea has had various names throughout history. One of the earliest recorded names was "Northern Ocean".[268] The name "North Sea" probably came into English, via the Dutch "Noordzee", because the sea is generally to the north of the Netherlands. Before the adoption of "North Sea", the Latin words *Mare Frisicum* or the English name "Frisian Sea" were commonly used. The Latin name of Flanders, a Dutch-speaking county from 862 to 1795 in northern Belgium, is *Flandria*; it is transliterated as 肥良的亞 in Item 40. Hence, 肥良的海 (the Chinese pronunciation of 肥良的 is close to Flandria; 海 is "ocean") refers to the sea north of Flandria, which is today's North Sea. The cartographer of the Europe-KWQ used a Chinese expression 肥良的海, which literally means "the sea off northern Flandria".
112	入尔馬尼海	Ru Er Ma Ni Hai	On the Europe-KWQ, it is today's southern arm of the Baltic Sea. The sea was known in ancient Latin language in the period before 75 B.C. as *Mare Germanicum* (入尔馬尼海).	On the Europe-KWQ, 入尔馬尼海 (Latin: *Mare Germanicum*, before 75 B.C.) becomes the southern arm of the Baltic Sea; 窩窩所德海 (Item 113), 波的海 (name used from the eleventh century; Item 114) are derived from different languages, and on the Europe-KWQ they demote the northernmost and northeasternmost arm, respectively, of today's Baltic Sea (in Modern English). I shall give further explanations in Items 113 and 114. The original annotation to the left of 入尔馬尼海 (Item 112) states the following: 入尔馬尼海出琥珀生石上如石乳然多在海濱金色者為上藍次之赤最下 My translation is: In the southern arm of the Baltic Sea, amber often grows like stalactite on raw stones near the seashore. The golden-coloured ones are the best, the blue-coloured ones are the next, and the red-coloured ones have lower quality.

(*Continued*)

(*Continued*)

• 113	窩窩所德海	Wo Wo Suo De Hai	On the Europe-KWQ, it is today's Gulf of Bothnia.	The name "Gulf of Bothnia" may have originated in the ninth century. It is the northernmost arm of the Baltic Sea and is situated between Finland's west coast and Sweden's east coast. 窩窩所德海 is literally a transplantation of Oostzee in Dutch or Ostsee in German. The meaning is "East Sea", which corresponds to the Gulf of Bothnia. On the Europe-KWQ, 窩窩所德海 represents the Gulf of Bothnia.
114	波的海	Bo Di Hai	Gulf of Finland.	On the Europe-KWQ, 波的海 is the Gulf of Finland, the eastern arm of the Baltic Sea (波的海; Item 114). The first to name the Baltic Sea (*Mare Balticum*) was the eleventh century German chronicler Adam of Bremen.[269] Hence, the Europe-KWQ cannot be a map drawn before the eleventh century. This is consistent with the Europe-KWQ revealing an era of Europe between 1157 and 1166.
115	大努毘河江	Da Nu Pi He Jiang	Danube River; Latin: *Danubius*, *Danuvius*; German: *Donau.* 河江 means "river".	Before the Romans had reached the river, it was little known. The Greeks called the western part Danube and the eastern part *Ister* or *Hister*.[270] Later, the Romans called it *Danubius*, or *Danuvius* (in Latin), the name of the Roman god of rivers. The Danube is Europe's second-longest river. It is in Central and Eastern Europe. It is also the only named river on the Europe-KWQ.
116	勿匿尔湖	Wu Ni Er Hu	On the Europe-KWQ, it is today's Lake Vanern.	Lake Vanern (Vänern or Wener) is in the southwest of Sweden. It is the largest lake in Sweden and the third-largest lake of all Europe.
117	歐邏巴	Ou Luo Ba	Europe; Latin: *Europa*	The first recorded usage of *Eurṓpē* (Greek *Ευρώπη*) as a geographic term refers to the western shore of the Aegean Sea. It is first used in the sixth century B.C.[271]
Summary A13: Items 104–117 give the history and geography of the oceans, seas, a lake and a major river in Europe. They are: 太海 (Great Big Sea or today's Black Sea; Item 104), 上海 (*Mare Superum*; Item 105), 下海 (*Mare Inferum*; Item 106), 地中海 (Mediterranean Sea; Item 107), 大西洋 (part of Atlantic Ocean; Item 109), 河摺亞諾滄 (part of Atlantic Ocean; Item 110), 肥良的海 (North Sea; Item 111), 入尔馬尼海 (southern arm of the Baltic Sea; Item 112), 窩窩所德海 (northern arm of the Baltic Sea; Item 113), 波的海 (on the Europe-KWQ, it is the eastern arm of the Baltic Sea; Item 114), 巴尔德峡 (Strait of Gibraltar; Item 108), 大努毘河江 (Danube River; Item 115), and 勿匿尔湖 (Lake Vanern; Item 116). The Europe-KWQ correctly shows their locations, names and characteristics. These indicate that the ancient names and water transportations, which already existed before and after Christ, were still in use between 1157 and 1166 on the Europe-KWQ. The last item gives the origin of the name of the European continent 歐邏巴 (Europe; Item 117) as the sixth century B.C.; the name was still in use between 1157 and 1166.				

Endnotes

[1]Library of Congress. "Kun yu wan guo quan tu." www.loc.gov, Library of Congress, 2022, loc.gov/item/2010585650.

[2]Ronnie Po-Chia Hsia. *Matteo Ricci and the Catholic Mission to China, 1583–1610: A Short History with Documents (Passages: Key Moments in History)*. Indianapolis, IN, USA: Hackett Publishing Company, Inc., 2016.

[3]Siu-Leung Lee (李兆良). *Kun Yu Wan Guo Quan Tu Jie Mi: Ming Dai Ce Hui Shi Je*《坤輿萬國全圖解密:明代測繪世界》or *Deciphering the Kunyu Wanguo Quantu, A Chinese World Map: Ming Chinese Mapped the World Before Columbus*. Taipei, Taiwan: Linking Publishing Company (聯經出版社), 2012; Sheng-Wei Wang. "Chapter nine: Secret revealed by ancient maps." *The Last Journey of the San Bao Eunuch, Admiral Zheng He*. Hong Kong, China: Proverse Hong Kong, 2019, pp. 290–301; Chapters 1 and 2 of this book.

[4]"File:Kunyu Wanguo Quantu by Matteo Ricci Plate 1-3.jpg." *Wikipedia Commons: The Free Media Repository*, Wikimedia Foundation, 23 Oct. 2020, upload.wikimedia.org/wikipedia/commons/b/b8/Kunyu_Wanguo_Quantu_by_Matteo_Ricci_Plate_1-3.jpg; "File:Kunyu Wanguo Quantu by Matteo Ricci Plate 4-6.jpg." *Wikipedia Commons*, 23 Oct. 2020, commons.wikimedia.org/wiki/File:Kunyu_Wanguo_Quantu_by_Matteo_Ricci_Plate_4-6.jpg.

[5]"File:Mercator 1569 world map composite.jpg." *Wikimedia Commons*, 26 Nov. 2016, upload.wikimedia.org/wikipedia/commons/4/4b/Mercator_1569_world_map_composite.jpg.

[6]"File:OrteliusWorldMap1570.jpg." *Wikimedia Commons*, 19 Nov. 2020, upload.wikimedia.org/wikipedia/commons/e/e2/OrteliusWorldMap1570.jpg.

[7]"File:Kunyu Wanguo Quantu by Matteo Ricci Plate 1-3.jpg." *Op. cit.*

[8]"File:Mercator 1569 world map composite.jpg." *Op. cit.*

[9]"File:OrteliusWorldMap1570.jpg." *Op. cit.*

[10]"File:FraMauroDetailedMap.jpg." *Wikipedia Commons*, 10 Sept. 2022, upload.wikimedia.org/wikipedia/commons/1/1b/FraMauroDetailedMap.jpg.

[11]"File:Atles català (full 4, ca000004).jpg." *Wikipedia Commons*, 10 Feb. 2022, upload.wikimedia.org/wikipedia/commons/c/c3/Atles_català_%28full_4%2C_ca000004%29.jpg.

[12]"File:Carte Pisane Portolan.jpg." *Wikipedia Commons*, 29 Oct. 2020, upload.wikimedia.org/wikipedia/commons/b/bb/Carte_Pisane_Portolan.jpg.

[13]Roel Nicolai. *The enigma of the origin of Portolan charts: a geodetic analysis of the hypothesis of a medieval origin*. Leiden, Netherlands: Brill, 2016.

[14]S. Maqbul Ahmad. "Cartography of al-Sharif al-Idrisi." *The History of Cartography Vol. 2, Book 1*, Chicago, IL, USA: University of Chicago Press, 1992, pp. 156–174.

[15]"File:Tabula Rogeriana 1929 copy by Konrad Miller.jpg." *Wikipedia Commons*, 28 Sept. 2022, upload.wikimedia.org/wikipedia/commons/2/21/Tabula_Rogeriana_1929_copy_by_Konrad_Miller.jpg.

[16]Peter Heather. *The Fall of the Roman Empire: A New History of Rome and the Barbarians*. Oxford, UK: Oxford University Press, 2005.

[17]Jonathan Shepard. *The Cambridge History of the Byzantine Empire c. 500–1492*. Cambridge, UK: Cambridge University Press, 2019.

[18]"File:Europe 814.svg." *Wikipedia Commons*, 3 Aug. 2022, upload.wikimedia.org/wikipedia/commons/thumb/9/90/Europe_814.svg/2560px-Europe_814.svg.png.

[19]"File:Europe 1142.jpg." *Wikipedia Commons*, 1 July 2021, upload.wikimedia.org/wikipedia/commons/8/83/Europe_1142.jpg.

[20]"File:Europe mediterranean 1190.jpg." *Wikipedia Commons*, 1 May 2022, upload.wikimedia.org/wikipedia/commons/7/75/Europe_mediterranean_1190.jpg.
[21]Jim Bradbury. *Philip Augustus: King of France 1180–1223 (The Medieval World)*. Oxfordshire, UK: Routledge, 2015.
[22]Peter H. Wilson. *Heart of Europe: A History of the Holy Roman Empire.* Cambridge, MA, USA: Belknap Press, 2016.
[23]Some historians consider the crowning of Frankish king Charlemagne in 800 as the first Holy Roman Emperor as the beginning of the Holy Roman Empire (Terry Breverton. *Everything You Ever Wanted to Know About the Tudors but Were Afraid to Ask*. Gloucestershire, UK: Amberley Publishing, 2014, p. 104); others consider the coronation of Otto I in 962 as the beginning of the Holy Roman Empire (Edward Peters. *Europe: The World of the Middle Ages*. London, UK: Pearson, 1977).
[24]Rosamond McKitterick. *The Frankish Kingdoms under the Carolingians, 751–987*. Oxfordshire, UK: Routledge, 1983.
[25]The Early Middle Ages are also sometimes referred to as Late Antiquity. This time period is usually viewed as beginning in the third century and stretching to the seventh century or as late as the eighth; the High Middle Ages is the time period from around 1000 to 1250 A.D.; the Central Middle Ages are between the Early Middle Ages and the High Middle Ages; the Late Middle Ages begin in 1300, though some scholars look at the mid- to late-fifteenth century as the beginning of the end. Once again, the end of the end is debatable, ranging from 1500 to 1650; Chris Wickham. *Medieval Europe*. New Haven, CN, USA: Yale University Press, 2017.
[26]T. K. Derry. *A History of Scandinavia: Norway, Sweden, Denmark, Finland, and Iceland.* Minneapolis, MN, USA: University of Minnesota Press, 2000.
[27]A. R. Disney. *A History of Portugal and the Portuguese Empire: From Beginnings to 1807, Vol. 1 and 2: Portugal*. Cambridge, UK: Cambridge University Press, 2009.
[28]Ministry of Foreign Affairs (Portugal). "Permanent Mission of Portugal to the United Nations." https://onu.missaoportugal.mne.gov.pt, Ministry of Foreign Affairs (Portugal), onu.missaoportugal.mne.gov.pt/en/about-portugal/history.
[29]John Stevens. *The General History of Spain from the First Peopling of It by Tubal, till the Death of King Ferdinand, Who United the Crowns of Castile and Aragon: ... to the Death of King Philip III (1699).* Ann Arbor, MI, USA: ProQuest, 2011.
[30]Amira K. Bennison. *The Almoravid and Almohad Empires*. Edinburgh, UK: Edinburgh University Press, 2016.
[31]Colum Hourihane, ed. "Granada." *The Grove Encyclopedia of Medieval Art and Architecture*. Oxford, UK: Oxford University Press, 2012.
[32]William Guangling Liu. *The Chinese Market Economy 1000–1500*. New York, NY, USA: State University of New York Press, 2016.
[33]William Lowrie. *Fundamentals of Geophysics*. London, UK: Cambridge University Press, 2007, p. 281.
[34]As found in the electronic version on the Internet at https://www.burnook.com/book/item/c6c5d64bbc1f47ed9793183636d53682.html.
[35]As found in the electronic version on the Internet at https://www.burnook.com/book/item/52e1ef44a6224ca0b63cb828c7d5fa8d.html.
[36]Romain Thurin. "China and the Two Romes. The 1081 and 1091 'Fulin' Embassies to the Song Empire." *Journal of the Economic and Social History of the Orient*, vol. 64, no. 1–2, 16 Mar. 2021.
[37]Margaret Mullett and Dion Smyth, eds. *Alexios I Komnenos*. Belfast, UK: Belfast Byzantine Enterprises, 1996.

[38]Li Yan (李岩) and Chen Yiqin (陳以琴). *Nan Hai I Hao Fu Chen Ji Ji Wang Kai Lai De Hang Cheng* 《南海I號浮沉記—繼往開來的航程》 or *The Story of the Nanhai One: The Voyage of the Past and the Future*. Beijing, China: Cultural Relics Publishing House (文物出版社), 2009; LI Qing Xin. *Nanhai 1 and The Maritime Silk Road*. Beijing, China: China Intercontinental Press, 2010.
[39]National Conservation Center for Underwater Cultural Heritage (国家文物局水下文化遺産保護中心). *Nan Hai I Hao Chen Chuan Kao Gu Bao Gao Zhi Yi 1989–2004 Nian Diao Cha* 《南海I号沉船考古报告之一 1989–2004年调查》 or *One of the Archaeological Reports on the Shipwreck of Nanhai One: 1989–2004 Investigation*. Beijing, China: Cultural Relics Publishing House (文物出版社), 2017.
[40]Lin Junying (林君穎). "Shi Da Kao Gu Xin Fa Xian Zhi Nan Hai Yi Hao Nan Song Gu Chuan Zhui Yi Hai Shang Si Lu Hui Huang 十大考古新發現之「南海一號」南宋古船追憶海上絲路輝煌 or 'Nanhai One', one of the top ten new archaeological discoveries: The ancient ship of the Southern Song dynasty rekindles the glory of the Maritime Silk Road." www.hk01.com, Hong Kong 01 (藝文中國), 7 May 2020, hk01.com/藝文中國/470171/十大考古新發現之-南海一號-南宋古船追憶海上絲路輝煌.
[41]T.C. Bell, private email communication.
[42]T.C. Bell, private email communication; Raoul McLaughlin. *Rome and the Distant East: Trade Routes to the Ancient Lands of Arabia, India and China*. London, UK: Continuum, 2010.
[43]A. R. Disney. *A History of Portugal and the Portuguese Empire: From Beginnings to 1807, Vol. 1 and 2: Portugal*. Cambridge, UK: Cambridge University Press, 2009.
[44]Ângelo Ribeiro and José Hermano. *História de Portugal I — A Formação do Território [History of Portugal: The Formation of the Territory] (in Portuguese)*. Lisbon, Portugal: QuidNovi, 2004.
[45]A. R. Disney. *A History of Portugal and the Portuguese Empire: From Beginnings to 1807, Vol. 1 and 2: Portugal*. *Op. cit*.
[46]William D. Phillips Jr. and Carla Rahn Phillips. *A Concise History of Spain*. Cambridge, UK: Cambridge University Press, 2016.
[47]Juan J. Moralejo. "Callaica nomina: estudios de onomástica gallega (PDF)." *A Coruña: Fundación Pedro Barrié de la Maza*, 2008, pp. 113–148. http://ilg.usc.es/agon/wp-content/uploads/2010/09/Callaica_Nomina.pdf.
[48]Marc Lodewijckx. *Archaeological and Historical Aspects of West-European Societies: Album Amicorum André Van Doorselaer*. Leuven, Belgium: Leuven University Press, 1996.
[49]Bernard F. Reilly. *The Contest of Christian and Muslim Spain*. Hoboken, NJ, USA: Blackwell Publishers, 1995.
[50]John Stevens. *The General History of Spain from the First Peopling of It by Tubal, till the Death of King Ferdinand, Who United the Crowns of Castile and Aragon: ... to the Death of King Philip III (1699)*. *Op. cit*.
[51]Amira K. Bennison. *The Almoravid and Almohad Empires*. *Op. cit*.
[52]*Ibid*.
[53]Ambroxio Huici Miranda. "The Iberian Peninsula and Sicily." In: P. M. Holt, *et al*., eds. *The Cambridge History of Islam Vol. 2A*. Cambridge, UK: Cambridge University Press, 1970, p. 429.
[54]Bernard F. Reilly. *The Medieval Spains*. Cambridge, UK: Cambridge University Press, 1993.
[55]James J. Todesca, ed. *The Emergence of León-Castile c. 1065–1500. Essays Presented to J.F. O'Callaghan*. Oxfordshire, UK: Routledge, 2015.
[56]Reinhart Anne Pieter (1820–1883). Dozy, creater. *Recherches Sur L'Histoire Et la Littérature de L'Espagne Pendant Le Moyen Age*. Charleston, SC, USA: BiblioBazaar, 2009, p. 303.
[57]Dario Fernandez-Morera. *The Myth of the Andalusian Paradise: Muslims, Christians, and Jews under Islamic Rule in Medieval Spain*. Wilmington, DE, USA: Intercollegiate Studies Institute, 2016.

[58]The Arabs and Muslims from the Middle Ages used the name of Al-Andalus for all those lands that were formerly part of the Visigothic kingdom: the Iberian Peninsula and Septimania.
[59]"File:Andalucia in Spain (including Canarias) (special marker).svg." *Wikipedia Commons*, 23 Feb. 2022, upload.wikimedia.org/wikipedia/commons/thumb/d/d0/Andalucia_in_Spain_(including_Canarias)_(special_marker).svg/2387px-Andalucia_in_Spain_(including_Canarias)_(special_marker).svg.png.
[60]*Ibid.*
[61]Thomas N. Bisson. *The Medieval Crown of Aragon: A Short History*. Oxford, UK: Clarendon Press, 1991.
[62]William D. Phillips Jr. and Carla Rahn Phillips. *A Concise History of Spain. Op. cit.*
[63]"File:Localización de Aragón.svg." *Wikipedia Commons*, 6 Sept. 2020, upload.wikimedia.org/wikipedia/commons/thumb/0/08/Localización_de_Aragón.svg/2560px-Localización_de_Aragón.svg.png.
[64]Charles River Editors. Catalonia*: The History and Legacy of Spain's Most Famous Autonomous Community.* Ann Arbor, MI, USA: Charles River Editors, 2019.
[65]*Ibid.*
[66]"File:Cataluna in Spain (including Canarias).svg." *Wikipedia Commons*, 26 Mar. 2022, upload.wikimedia.org/wikipedia/commons/thumb/a/a0/Cataluna_in_Spain_%28including_Canarias%29.svg/2387px-Cataluna_in_Spain_%28including_Canarias%29.svg.png; author TUBS, date 7 March 2011, titled "Location of Catalonia in Spain."
[67]Mark Kurlansky. *The Basque History of the World: The Story of a Nation*. London, UK: Penguin Books, 2001.
[68]John Stevens. *The General History of Spain from the First Peopling of It by Tubal, till the Death of King Ferdinand, Who United the Crowns of Castile and Aragon: ... to the Death of King Philip III (1699). Op. cit.*
[69]Roger Collins. *The Basques*. Oxford, UK: Basil Blackwell, 1990.
[70]*Ibid.*
[71]R. A. Fletcher. *The Episcopate in the Kingdom of León in the Twelfth Century.* Oxford, UK: Oxford University Press, 1978; as found in the electronic version on the Internet at https://libro.uca.edu/ekl/ekl.htm.
[72]*Ibid.*
[73]John Stevens. *The General History of Spain from the First Peopling of It by Tubal, till the Death of King Ferdinand, Who United the Crowns of Castile and Aragon: ... to the Death of King Philip III (1699). Op. cit.*
[74]*Ibid.*
[75]Heinrich Fichtenau. Peter Munz, trans. *The Carolingian Empire*. Toronto, Canada: University of Toronto Press, 1978.
[76]H. Spencer Lewis. *Toulouse The Mystic City of France*. Whitefish, MT, USA: Literary Licensing, LLC, 2014.
[77]Marion Meade. *Eleanor of Aquitaine: A Biography.* London, UK: Penguin Books, 1991.
[78]John Gillingham. *The Angevin Empire*. London, UK: Bloomsbury Academic, 2000.
[79]Desmond Seward. *The Hundred Years War: The English in France, 1337–1453*. New York, NY, USA: Atheneum, 1978.
[80]Philippe Velay. *From Lutetia to Paris: The Island and the Two Banks*. Paris, France: Presses du CNRS, 1992.
[81]Elaine Sciolino. *The Seine: The River That Made Paris*. New York, NY, USA: W. W. Norton & Company, 2020.

[82]Colin Jones. *Paris: The Biography of a City*. London, UK: Penguin Books, 2006.
[83]Herwig Wolfram. *The Roman Empire and Its Germanic Peoples*. Berkeley, CA, USA: University of California Press, 1997.
[84]Charles River Editors. *The Burgundians: The History of the Early Kingdoms of Burgundy in the Middle Ages Kindle Edition*. Ann Arbor, MI, USA: Charles River Editors, 2021.
[85]Percy Allen. *Burgundy: The Splendid Duchy Volume 1*. Norderstedt, Germany: Book on Demand Ltd., 2013.
[86]Walter Prevenier and Wim Blockmans. *The Burgundian Netherlands*. Cambridge, UK: Cambridge University Press, 1986.
[87]Joshua J. Mark. "Kingdom of West Francia." www.ancient.eu, Ancient History Foundation, 16 Nov. 2018, ancient.eu/Kingdom_of_West_Francia.
[88]Jim Bradbury. *Philip Augustus: King of France 1180–1223 (The Medieval World)*. Oxfordshire, UK: Routledge, 2015.
[89]Colin Jones. *Paris: The Biography of a City. Op. cit.*
[90]William Kornblum. *Marseille, Port to Port*. New York, NY, USA: Columbia University Press, 2022.
[91]Lawrence Durrell. *Provence*. New York, NY, USA: Arcade, 2014.
[92]Peter H. Wilson. *Heart of Europe: A History of the Holy Roman Empire. Op. cit.*
[93]Felix Vernay. *Petite Histoire Du Dauphine*. Cresse, France: Regionalismes, 2014.
[94]Peter H. Wilson. *Heart of Europe: A History of the Holy Roman Empire. Op. cit.*
[95]Roger Price. *A Concise History of France*. Cambridge, UK: Cambridge University Press, 2014.
[96]Samuel Humes. *Belgium: Long United, Long Divided*. London, UK: Hurst, 2014.
[97]Emile Cammaerts. *Belgium from the Roman Invasion to the Present Day*. Open Library: Loki's Publishing, 2019.
[98]Clive H. Church and Randolph C. Head. *A Concise History of Switzerland*. Cambridge, UK: Cambridge University Press, 2013.
[99]*Ibid.*
[100]*Ibid.*
[101]*Ibid.*
[102]*Ibid.*
[103]John Dormandy. *A History of Savoy: Gatekeeper of the Alps*. Oxford, UK: Fonthill Media, 2018.
[104]Giovanni Tabacco. *The Struggle for Power in Medieval Italy: Structures of Political Rule*. Cambridge, U. K.: Cambridge University Press, 1990; Edoardo Manarini. *Struggles for Power in the Kingdom of Italy: The Hucpoldings, c. 850-c. 1100*. Amsterdam, Netherlands: Amsterdam University Press, 2022.
[105]Steven A. Epstein. *Genoa and the Genoese, 958–1528*. Chapel Hill, NC, USA: University of North Carolina Press, 2022.
[106]Icon Group International. *Lombardy: Webster's Timeline History, 730 BC-2007*. Las Vegas, NV, USA: ICON Group International, Inc., 2009.
[107]*Ibid.*
[108]"File:Location of the Duchy of Milan-it.svg." *Wikipedia Commons*, 5 Aug. 2021, commons.wikimedia.org/wiki/File:Location_of_the_Duchy_of_Milan-it.svg.
[109]The Duchy of Mantua was a duchy in Lombardy, Northern Italy. Its first duke ruled Mantua since 1328; UNESCO World Heritage Center. "Mantua and Sabbioneta." https://whc.unesco.org/, UNESCO World Heritage Center, 2022, whc.unesco.org/en/list/1287/.
[110]John Julius Norwich. *A History of Venice*. New York, NY, USA: Vintage, 1989.

[111]Edwin Mullins. *The Popes of Avignon: A Century in Exile*. New York, NY, USA: BlueBridge, 2011; Joëlle Rollo-Koster. *Avignon and Its Papacy, 1309–1417: Popes, Institutions, and Society*. Lanham, MD, USA: Rowman & Littlefield Publishers, 2015.
[112]R. De Cesare. Helen Zimmern, trans. *The Last Days of Papal Rome: 1850–1870 (1909)*. Whitefish, MT, USA: Kessinger Publishing, LLC, 2010.
[113]Ernest F. Henderson. "The Avalon Project: The Peace of Venice; 1177." https://web.archive.org/, The Avalon Project at Yale Law School, 1996–2005, web.archive.org/web/20060825040541/; http://www.yale.edu/lawweb/avalon/medieval/venice.htm.
[114]John Miley. *The History of the Papal States: From Their Origin to the Present Day (in Three Volumes)*. London, UK: Forgotten Books, 2018.
[115]Huang Shijian (黄时鉴) and Gong Yingyan (龚缨晏). *Li Ma Dou Shi Jie Di Tu Yan Jiu*《利玛窦世界地图研究》or *Research on Matteo Ricci's World Map*. Shanghai, China: Shanghai Chinese Classics Publishing House (上海古籍出版社), 2004, p. 202.
[116]Dana Facaros and Michael Pauls. *Italy: Umbria and the Marche*. Buckinghamshire, UK: Bradt Travel Guides, 2021.
[117]"File:1814 Thomson Map of Southern Italy (Naples ^ Sicily) - Geographicus - NaplesSicily-t-1814.jpg." *Wikipedia Commons*, 6 Sept. 2020, commons.wikimedia.org/wiki/File:1814_Thomson_Map_of_Southern_Italy_(_Naples_^_Sicily_)_-_Geographicus_-_NaplesSicily-t-1814.jpg.
[118]Peter H. Wilson. *Heart of Europe: A History of the Holy Roman Empire. Op. cit.*
[119]John Miley. *The History of the Papal States: From Their Origin to the Present Day. Op. cit.*
[120]"Kingdom of Naples." www.zum.de, World History at KMLA, 2022, zum.de/whkmla/region/italy/xnaples.html.
[121]*Ibid.*
[122]Graham Loud. *The Age of Robert Guiscard: Southern Italy and the Northern Conquest*. Oxfordshire, UK: Routledge, 2000.
[123]Hubert Houben. Graham A. Loud and Diane Milburn, trans. *Roger II of Sicily: A Ruler between East and West*. Cambridge, UK: Cambridge University Press, 2002.
[124]Charles River Editors. *The Unification of Italy: The History of the Risorgimento and the Conflicts That Unified the Italian Nation*. Ann Arbor, MI, USA: Charles River Editors, 2019; Desmond Seward. *An Armchair Traveller's History of Apulia*. London, UK: Haus Publishing, 2013.
[125]"File:Apulia in Italy.svg." *Wikipedia Commons*, 14 Feb. 2021, upload.wikimedia.org/wikipedia/commons/thumb/1/1a/Apulia_in_Italy.svg/1630px-Apulia_in_Italy.svg.png; author TUBS, date 7 March 2011, titled "Location of Apulia in Italy."
[126]George Ostrogorsky. "The Byzantine Empire in the world of the seventh century." *Dumbarton Oaks Papers*, vol. 13, 1959, pp. 1–21. https://doi.org/10.2307/1291126 or https://www.jstor.org/stable/1291126.
[127]Icon Group International. *Calabria: Webster's Timeline History, 730 BC-2007*. Las Vegas, NV, USA: ICON Group International, Inc., 2009.
[128]"File:Calabria in Italy.svg." *Wikipedia Commons*, 25 Oct. 2021, upload.wikimedia.org/wikipedia/commons/thumb/9/91/Calabria_in_Italy.svg/1630px-Calabria_in_Italy.svg.png; author TUBS, date 7 March 2011, titled "Location of Calabria in Italy."
[129]Staff writer. "The Frankish Empire." https://resources.saylor.org/, The Salor Foundation, 2022, resources.saylor.org/wwwresources/archived/site/wp-content/uploads/2012/10/HIST201-1.1.1-FrankishEmpire-FINAL1.pdf.
[130]Staff writer. "Duchy of Franconia." www.britishmuseum.org, The British Museum, 2022, britishmuseum.org/collection/term/x105678.

[131]Örjan Martinsson. "Dukes of Franconia." www.tacitus.nu, Tacitus Historical Atlas, 2022, tacitus.nu/historical-atlas/regents/germany/franconia.htm.
[132]Staff writer. "The History of Bavaria." https://germanamericansociety-spokane.org/, German American Society-Spokane, 2022, germanamericansociety-spokane.org/hx_of_bav.htm.
[133]J. Augusteijn and H. Storm, eds. *Region and State in Nineteenth-Century Europe: Nation-Building, Regional Identities and Separatism*. London, UK: Palgrave Macmillan, 2012.
[134]Robert Zagolla. *Saxony: A Short History*. Open Library: Bebra Verlag, 2008.
[135]*Ibid.*
[136]Huang Shijian (黄时鉴) and Gong Yingyan (龚缨晏). *Op. cit.*, p. 196.
[137]James Minahan. *Encyclopedia of the Stateless Nations: Ethnic and National Groups Around the World Volume II D-K*. Santa Barbara, CA, USA: Greenwood, 2002, p. 614.
[138]Peter H. Wilson. *Heart of Europe: A History of the Holy Roman Empire. Op. cit.*
[139]Huang Shijian (黄时鉴) and Gong Yingyan (龚缨晏). *Op. cit.*, p. 195.
[140]Hans W. Koch. *A History of Prussia.* Oxfordshire, UK: Routledge, 1978; Sebastian Haffner. Ewald Osers, trans. *The Rise and Fall of Prussia*. Lexington, MA, USA: Plunkett Lake Press, 2012.
[141]*Ibid.*
[142]Samuel Humes. *Belgium: Long United, Long Divided. Op. cit.*
[143]John Julius Norwich. *A History of France*. New York, NY, USA: Grove Press, 2019; William Doyle. *The Oxford History of the French Revolution*. Oxford, UK: Oxford University Press, 2018.
[144]Huang Shijian (黄时鉴) and Gong Yingyan (龚缨晏). *Op. cit.*, p. 196.
[145]Nora Berend, *et al. Central Europe in the High Middle Ages: Bohemia, Hungary and Poland, c. 900-c.1300.* Cambridge, UK: Cambridge University Press, 2014.
[146]James C. Kennedy. *A Concise History of the Netherlands*. Cambridge, UK: Cambridge University Press, 2017.
[147]"File:Holland position.svg." 24 Sept. 2020, upload.wikimedia.org/wikipedia/commons/thumb/6/69/Holland_position.svg/1736px-Holland_position.svg.png.
[148]James C. Kennedy. *A Concise History of the Netherlands. Op. cit.*
[149]"File:Locator County of Zeeland (1350).svg." *Wikipedia Commons*, 8 Jan. 2021, upload.wikimedia.org/wikipedia/commons/thumb/4/45/Locator_County_of_Zeeland_%281350%29.svg/2818px-Locator_County_of_Zeeland_%281350%29.svg.png.
[150]"File:Ostarrîchi Document (zoomed).jpg." *Wikimedia Commons*, 25 May 2022, commons.wikimedia.org/wiki/File:Ostarrîchi_Document_(zoomed).jpg.
[151]Gunter Bischof. *Austrian Historical Memory and National Identity*. Oxfordshire, UK: Routledge, 1996; U.S. Library of Congress. "The Holy Roman Empire and the Duchy of Austria." http://countrystudies.us/, U.S. Library of Congress, 2022, countrystudies.us/austria/6.htm.
[152]Robert Zagolla. *Saxony: A Short History*. *Op. cit.*; James Westfall Thompson. *Feudal Germany*. Chicago, IL, USA: University of Chicago Press, 1928.
[153]Simon Winder. *Germania: In Wayward Pursuit of the Germans and Their History*. London, UK: Picador, 2011.
[154]Charles River Editors. *The Burgundians: The History of the Early Kingdoms of Burgundy in the Middle Ages Kindle Edition. Op. cit.*
[155]Milwaukee Public Museum. "Medieval Germany (c. 481–1350 CE)." www.mpm.edu, Milwaukee Public Museum, 2022, mpm.edu/index.php/[menu-link-parent-path-raw]/title-raw-120; Admin. "European Royal History: Exploring the History of European Royalty." https://europeanroyalhistory.wordpress.com/, Blog at WordPress.com, 2022, europeanroyalhistory.wordpress.com/2019/09/10/history-of-germany-part-vi-kingdom-of-east-francia-and-the-kingdom-of-germany/.

[156]Pal Engel. *Realm of St. Stephen: A History of Medieval Hungary, 895–1526*. London, UK: I. B. Tauris, 2001.
[157]The Miroslav Krleža Institute of Lexicography. "Croatia in union with Hungary (1102–1527)." https://croatia.eu/, The Miroslav Krleža Institute of Lexicography, 2022, croatia.eu/index.php?view=article&lang=2&id=21.
[158]"File:Kingdom of Croatia and Dalmatia.png." *Wikipedia Commons*, 19 May 2021, upload.wikimedia.org/wikipedia/commons/6/63/Kingdom_of_Croatia_and_Dalmatia.png.
[159]Harold B. Lee Library. "Croatia: Kingdom (925–1102)." https://eudocs.lib.byu.edu/, Brigham Young University, 2022, eudocs.lib.byu.edu/index.php/Croatia:_Kingdom_(925–1102).
[160]The Miroslav Krleža Institute of Lexicography. "Croatia in union with Hungary (1102–1527)." *Op. cit.*
[161]Ion Grumeza. *Dacia: Land of Transylvania, Cornerstone of Ancient Eastern Europe*. Falls Village, CN, USA: Hamilton Books, 2009.
[162]Radu Marza. "Transylvania and Hungary or Transylvania in Hungary as Viewed by the Romanian Historiography." www.academia.edu, 2022, academia.edu/5915626/Transylvania_and_Hungary_or_Transylvania_in_Hungary_as_Viewed_by_the_Romanian_Historiography.
[163]de Robert I. Frost. *The Oxford History of Poland-Lithuania: Volume I: The Making of the Polish-Lithuanian Union, 1385–1569*. Oxford, UK: Oxford University Press, 2018.
[164]Will S. Monroe. *Bohemia and the Cechs the History, People, Institutions, and the Geography of the Kingdom, Together with Accounts of Moravia and Silesia*. London, UK: Forgotten Books, 2012.
[165]Keith Hitchins. *A Concise History of Romania*. Cambridge, UK: Cambridge University Press, 2014.
[166]Jonathan Shepard, ed. *The Cambridge History of the Byzantine Empire c. 500–1492*. *Op. cit.*
[167]Department of information and analytics of the Izhevsk City Administration. "Yambol (the Republic of Bulgaria)." www.izh.ru, Department of information and analytics of the Izhevsk City Administration, 1998–2022, www.izh.ru/en/info/14207.html; European National Networks for the Enhancement of VET. "Yambol, a city in the south-eastern region of Bulgaria." www.enneproject.eu, European National Networks for the Enhancement of VET, 2022, enneproject.eu/yambol-a-city-in-the-south-eastern-region-of-bulgaria; R. J. Crampton. *A Concise History of Bulgaria*. Cambridge, UK: Cambridge University Press, 2006.
[168]Alexandru Madgearu. *The Wars of the Balkan Peninsula: Their Medieval Origins*. Lanham, MD, USA: Scarecrow Press, 2007.
[169]Bernd J. Fischer and Oliver Jens Schmitt. *A Concise History of Albania, New Edition*. Cambridge, UK: Cambridge University Press, 2022.
[170]Richard Clogg. *A Concise History of Greece*. Cambridge, UK: Cambridge University Press, 2021.
[171]Andrew Rossos. *Macedonia and the Macedonians: A History*. Stanford, CA, USA: Hoover Institution Press, 2008.
[172]Huang Shijian (黄时鉴) and Gong Yingyan (龚缨晏). *Op. cit.*, p. 200.
[173]András Mócsy. *Pannonia and Upper Moesia: A History of the Middle Danube Provinces of the Roman Empire*. Oxfordshire, UK: Routledge, 2015.
[174]GeoERA researchers. "Pannonian Basin." https://geoera.eu/, Geological Service for Europe (GeoERA), 2022, geoera.eu/projects/geoconnect3d6/pannonian-basin/.
[175]János B. Szabó. "Forgotten Wars: Hungarian-Byzantine wars in the 10th–13th centuries." www.academia.edu, Academia, 2022, academia.edu/4141996/FORGOTTEN_WARS._HUNGARIAN_BYZANTINE_WARS_IN_10th_13th_CENTURY.
[176]R. J. Crampton. *A Concise History of Bulgaria*. *Op. cit.*
[177]Dennis P. Hupchick. "The First Bulgarian Empire." In: *The Balkans*. London, UK: Palgrave Macmillan, 2002. https://doi.org/10.1057/9780312299132_3; Steven Runciman. *A History of the First Bulgarian Empire*. Morrisville, NC, USA: lulu.com, 2018.

[178]Florin Curta. "Chapter 31: Second Bulgarian Empire." *Eastern Europe in the Middle Ages (500–1300)*. Leiden, Netherlands: Brill, 2019, pp. 671–672.
[179]Teresa Shawcross. *The Chronicle of Morea: Historiography in Crusader Greece*. Oxford, UK: Oxford University Press, 2009.
[180]*Ibid.*
[181]Sarah Foot. *Aethelstan: The First King of England*. New Haven, CN, USA: Yale University Press, 2012.
[182]Emily Osterloff. "Should we be scared of British snakes?" www.nhm.ac.uk, The Natural History Museum, 2021, nhm.ac.uk/discover/should-we-be-scared-of-british-snakes.html.
[183]F. E. Halliday. *A History of Cornwall*. Cornwall, UK: House of Stratus, 2008.
[184]The National Archives. "A Charter of 1337." www.legislation.gov.uk, The National Archives on behalf of HM Government, 2022, legislation.gov.uk/aboutus.
[185]Alistair Moffat. *Scotland: A History from Earliest Times*. Edinburgh, UK: Birlinn, 2017.
[186]Thomas Bartlett. *Ireland: A History*. Cambridge, UK: Cambridge University Press, 2011.
[187]"File:Denmark regions map1.png." *Wikipedia Commons*, 22 Oct. 2020, upload.wikimedia.org/wikipedia/commons/9/9e/Denmark_regions_map1.png.
[188]Paulle Lauring. David Hohnen, trans. *A History of Denmark*. Copenhagen, Denmark: Host & Son, 2004.
[189]Stale Dyrvik, *et al*. Rolf Danielsen, ed. *Norway: A History from the Vikings to Our Own Times*. Oslo, Norway: Scandinavian University Press, 1995.
[190]*Ibid.*
[191]Huang Shijian (黄时鉴) and Gong Yingyan (龚缨晏). *Op. cit*., p. 199.
[192]T. K. Derry. *A History of Scandinavia: Norway, Sweden, Denmark, Finland, and Iceland. Op. cit.*
[193]Giuseppe Acerbi. *Travels Through Sweden, Finland, and Lapland, to the North Cape, in the Years 1798 and 1799; Volume 1*. Open Library: Andesite Press, 2017.
[194]A Google map of Lapia can be found on the Internet at https://www.google.com.hk/maps/place/18600+L%C3%A4pl%C3%A4,+Finland/@61.1438143,26.0808003,1306m/data=!3m2!1e3!4b1!4m5!3m4!1s0x46902be37838569f:0x279259c3f168ef25!8m2!3d61.1438047!4d26.0895551?hl=en.
[195]C. S. Christensen. "Bjarmaland — The forgotten history of the vikings in Northern Russia: A description of the Vikings, the Bjarmians, the area around the delta of Northern Dvina and the wealth of nature in the Russian forests in the years 890–1250." https://st-hum.ru/, Studia Humanitatis, 2022, st-hum.ru/content/christensen-cs-bjarmaland-forgotten-history-vikings-northern-russia-description-vikings.
[196]T. K. Derry. *A History of Scandinavia: Norway, Sweden, Denmark, Finland, and Iceland. Op. cit.*
[197]Neil Kent. *A Concise History of Sweden*. Cambridge, UK: Cambridge University Press, 2008.
[198]"File:U 582, Söderby-Karl.jpg." *Wikipedia Commons*, 24 Feb. 2021, upload.wikimedia.org/wikipedia/commons/e/e7/U_582%2C_Söderby-Karl.jpg.
[199]Raita Merivirta, *et al*. "Finns in the Colonial World." In: Raita Merivirta, *et al*., eds. *Finnish Colonial Encounters: From Anti-imperialism to Cultural Colonialism and Complicity*. London, UK: Palgrave Macmillan, 2021. https://doi.org/10.1007/978-3-030-80610-1_1.
[200]David Kirby. *A Concise History of Finland*. Cambridge, UK: Cambridge University Press, 2006.
[201]*Ibid.*
[202]Lars Elenius. "The dissolution of ancient Kvenland and the transformation of the Kvens as an ethnic group of people. On changing ethnic categorizations in communicative and collective memories." www.tandfonline.com, Informa UK limited, 2022, tandfonline.com/doi/full/10.1080/08003831.2019.1681225.
[203]Huang Shijian (黄时鉴) and Gong Yingyan (龚缨晏). *Op. cit*., p. 207.
[204]Charles Russell Coulter and Patricia Turner. *Ancient Deities: An Encyclopedia*. Jefferson, NC, USA: McFarland & Company, 2000, p. 447.

[205]Joel Raba. "The fate of the Novgorodian Republic." *The Slavonic and East European Review*, vol. 45, no. 105, 1967, pp. 307–323. JSTOR, http://www.jstor.org/stable/4205876; Charles J. Halperin. "Novgorod and the 'Novgorodian Land'." *Cahiers du Monde russe*, vol. 40, no. 3, Jul.–Sept. 1999, pp. 345–363. JSTOR, https://www.jstor.org/stable/20171136.
[206]Huang Shijian (黄时鉴) and Gong Yingyan (龚缨晏). *Op. cit.*, p. 203.
[207]Uilyam Brumfild. *Kargopol. Arhitekturnoe nasledie v fotografiyah (Russian edition)*. Moscow, Russia: Tri Kvadrata, 2007.
[208]Charles Henry Coote. Edward Delmar Morgan, ed. *Early Voyages and Travels to Russia and Persia by Anthony Jenkinson and Other Englishmen: With Some Account of the First Intercourse of the English ... Caspian Sea (Hakluyt Society, First Series)*. Oxfordshire, UK: Routledge, 2017.
[209]Damian Dabrowski. *Palace Dolnego Slaska Powiat trzebnicki Gmina Trzebnicca i Zawonia.* Warszawa, Poland: CM Jakub Jagiello, 2020.
[210]Saulius Sužiedėlis. *Historical Dictionary of Lithuania (2nd ed.)*. Lanham, MD, USA: Scarecrow Press, 2011, p.119; Office of the Seimas of the Republic of Lithuania. "Coat of arms of the Republic of Lithuania." www.lrs.lt, Office of the Seimas of the Republic of Lithuania, 2022, lrs.lt/sip/portal.show?p_r=38110&p_k=2.
[211]Serhii Plokhy. *The Origins of the Slavic Nations: Premodern Identities in Russia, Ukraine, and Belarus*. Cambridge, UK: Cambridge University Press, 2010.
[212]*Ibid.*
[213]George Vernadsky. *Kievan Russia*. New Haven, CN, USA: Yale University Press, 1973.
[214]Charles Halperin. *Russia and the Golden Horde: The Mongol Impact on Medieval Russian History*. Bloomington, IN, USA: Indiana University Press, 1987.
[215]Columbia University Press. "Podolia." www.infoplease.com, Columbia University Press, 2022, infoplease.com/encyclopedia/places/baltic-cis/countries/Podolia.
[216]*Ibid.*
[217]Huang Shijian (黄时鉴) and Gong Yingyan (龚缨晏). *Op. cit.*, p. 200.
[218]Europe Direct Information Center. "Moldavia." www.strasbourg-europe.eu, Information Center on the European Institutions (CIIE), 2022, strasbourg-europe.eu/moldavia.
[219]Huang Shijian (黄时鉴) and Gong Yingyan (龚缨晏). *Op. cit.*, p. 197.
[220]Neil Kent. *Crimea: A History*. London, UK: Hurst, 2016.
[221]Huang Shijian (黄时鉴) and Gong Yingyan (龚缨晏). *Op. cit.*, p. 186.
[222]Bernard D. Weinryb. "The Khazars: An annotated bibliography." *Studies in Bibliography and Booklore*, vol. 6, no. 3, 1963, pp. 111–129. Hebrew Union College — Jewish Institute of Religion. https://www.jstor.org/stable/27943361.
[223]Neil Kent. *Crimea: A History. Op. cit.*
[224]M. S. Anderson. "The Great Powers and the Russian annexation of the Crimea, 1783–4." *The Slavonic and East European Review*, vol. 37, no. 88, 1958, pp. 17–41. JSTOR, 4205010.
[225]Bratislava Jaroslav Drobný. "Cumans and Kipchaks: Between Ethnonym and Toponym." Zborník Filozofickej Fakulty Univerzity Komenského Ročník XXXIII — XXXIV Graecolatina et Orientalia Bratislava, 2012, stránku 205–217. https://fphil.uniba.sk/fileadmin/fif/katedry_pracoviska/kksf/kf/GLO-2012_Drobny.pdf.
[226]*Ibid.*
[227]Huang Shijian (黄时鉴) and Gong Yingyan (龚缨晏). *Op. cit.*, p. 185.
[228]Accu weather reporter. "Usting." www.accuweather.com, AccuWeather, 2022, accuweather.com/zh/at/usting/946271/satellite-enhanced/946271.

[229]The Presidential Library. "Ryazan Region. History of 'isolated lands' in the collection of the Presidential Library." www.prlib.ru, Boris Yeltsin Presidential Library, 2022, prlib.ru/en/news/1165339.
[230]*Ibid.*
[231]Christos G. Makrypoulias. "Byzantine Expeditions against the Emirate of Crete c. 825–949." *Graeco-Arabica*, vol. 7–8, 2000, pp. 347–362. https://www.scribd.com/document/16196011/Ch-G-Makrypoulias-Byzantine-Expeditions-against-the-Emirate-of-Crete-c-825–949.
[232]Chris Moorey. *A History of Crete*. London, UK: Haus Publishing, 2020.
[233]Richard Clogg. *A Concise History of Greece. Op. cit.*
[234]Timothy E. Gregory. *A History of Byzantium*. Hoboken, NJ, USA: Wiley-Blackwell, 2009.
[235]Louis Mendola. *The Kingdom of Sicily 1130–1860*. Chicago, IL, USA: Trinacria Editions LLC, 2017.
[236]The Earth Watching project. "Stromboli Volcano." https://earth.esa.int/, ESA, 2022, earth.esa.int/web/earth-watching/historical-views/content/-/article/stromboli-volcano/; Smithsonian Institution. "Vulcano." https://volcano.si.edu/, Global Volcanism Program of the Department of Mineral Sciences at National Museum of Natural History of Smithsonian Institution, 2022, volcano.si.edu/volcano.cfm?vn=211050.
[237]MACOR Coral Museum. "MACOR Coral Museum." www.algheroexperience.it, MACOR Coral Museum, 2022, algheroexperience.it/en/macor-coral-museum.html.
[238]T. Houston Durrant, *et al*. "Quercus suber in Europe: Distribution, habitat, usage and threats." https://forest.jrc.ec.europa.eu/, Tree species | European Atlas of Forest Tree Species, 2022, forest.jrc.ec.europa.eu/media/atlas/Quercus_suber.pdf.
[239]Charles Dalli. "The Siculo-African Peace and Roger I's Annexation of Malta in 1091." In Toni Cortis and Timothy Gambin, eds. *De Triremibus: Festschrift in honour of Joseph Muscat (PDF)*. Publishers Enterprises Group (PEG) Ltd. p. 273. ISBN 9789990904093. Archived from the original (PDF) on 2 January 2018. https://web.archive.org/web/20180102191402/http://melitensiawth.com/incoming/Index/The%20Arabs%20in%20Malta/2005%20Dalli.pdf.
[240]John Car. *The Knights Hospitaller: A Military History of the Knights of St John*. South Yorkshire, UK: Pen and Sword Military, 2017.
[241]L. H. Caird. *The History of Corsica*. Marrickville, Australia: Wentworth Press, 2019; Editor. "Corsica." www.historyfiles.co.uk, History Files, 2022, historyfiles.co.uk/KingListsEurope/ItalyCorsica.htm.
[242]*Ibid.*
[243]Icon Group International. *Menorca: Webster's Timeline History, 1235–2007*. Las Vegas, NV, USA: ICON Group International, Inc, 2010.
[244]Icon Group International. *Mallorca: Webster's Timeline History, 902–2007*. Las Vegas, NV, USA: ICON Group International, Inc, 2010
[245]Andrew Evans. "Is Iceland Really Green and Greenland Really Icy?" www.nationalgeographic.com, National Geographic Society, 2022, nationalgeographic.com/science/article/iceland-greenland-name-swap.
[246]Axel Kristinsson. "Is there any tangible proof that there were Irish monks in Iceland before the time of the Viking settlements?" https://www.why.is/, The Icelandic Web of Science, 2022, why.is/svar.php?id=4802.
[247]Gunnar Karlsson. *The History of Iceland*. Minneapolis, MN, USA: University of Minnesota Press, 2000.
[248]"File:Thule carta marina Olaus Magnus.jpg." *Wikimedia Commons*, 13 June 2019, commons.wikimedia.org/wiki/File:Thule_carta_marina_Olaus_Magnus.jpg.
[249]Jason Daley. "Black Sea." www.Smithsonian.com, Smithsonian Magazine, 14 Oct. 2018, smith sonianmag.com/smart-news/oldest-intact-shipwreck-discovered-black-sea-180970622/.

[250]Tasos Kokkinidis. "The Oldest Intact Shipwreck 'Odysseus' was an Ancient Greek Vessel." https://greekreporter.com/, Greek Reporter, 18 May 2022, greekreporter.com/2022/05/18/2400-year-old-odysseus-greek-vessel-discovered-intact-at-the-bottom-of-the-black-sea.
[251]Shamseer Mambra. "8 Amazing Facts about the Black Sea." https://www.marineinsight.com, marine Insight, 24 Nov. 2020, marineinsight.com/know-more/8-amazing-facts-about-the-black-sea/.
[252]*Ibid.*
[253]Gregory R. Crane (Editor-in-Chief). "Dictionary of Greek and Roman Geography (1854)." http://www.perseus.tufts.edu/, Perseus Digital Library of Tufts University, 2022, www.perseus.tufts.edu/hopper/text?doc=Perseus:text:1999.04.0064:id=tyrrhenum-mare-geo.
[254]Igor S. Zonn, *et al. The Adriatic Sea Encyclopedia.* Cham, Switzerland: Springer, 2021.
[255]Eric N. Danielson. *Shanghai and the Yangzi Delta: The New Yangzi River Trilogy*, Volume 1. Independently published, 2023.
[256]Gregory R. Crane (Editor-in-Chief). "Dictionary of Greek and Roman Geography (1854)." *Op. cit.*
[257]H. M. Denham. *The Tyrrhenian Sea.* London, UK: John Murray, 1969.
[258]Ajay Menon. "10 Interesting Facts about the Straits of Gibraltar." https://www.marineinsight.com, Marine Insight, 2022, marineinsight.com/know-more/10-interesting-facts-about-the-straits-of-gibraltar/.
[259]Domingo Plácido Suárez. Julio Mangas Manjarrés, *et al.*, eds. *La Península Ibérica en los autores griegos: de Homero a Platón.* Ann Arbor, MI, USA: Fundación de Estudios Romanos (Roman Studies Foundation), 1998.
[260]S. Thomas. "How the oceans got their names." Oxford Dictionaries by Oxford University Press, 2015. Retrieved 1 October 2016.
[261]*Ibid.*
[262]Rong Liang (荣良). *Zheng He Xia Xi Yang Yan Jiu Wen Xuan* 《郑和下西洋研究文选》 or *Selected Works of Zheng He's Voyages to the Western Ocean.* Beijing, China: China Ocean Press (海洋出版社), 2005, p. 73.
[263]*Op. cit.*, p. 73.
[264]*Op. cit.*, p. 74.
[265]*Op. cit.*, p. 106.
[266]*Op. cit.*, p. 107; Xiang Da (向达), ed. *Liang Zhong Hai Dao Zhen Jing* 《两种海道针经》 or *Two Types of Nautical Reference Books Based on Compass Reading and Sailing Mealiage.* Beijing, China: Zhonghua Book Company (中华书局), 2000, p. 7.
[267]Rong Liang (荣良). *Zheng He Xia Xi Yang Yan Jiu Wen Xuan* 《郑和下西洋研究文选》 or *Selected Works of Zheng He's Voyages to the Western Ocean. Op. cit.*, p. 104.
[268]Duane W. Roller. "Roman exploration." *Through the Pillars of Herakles: Greco-Roman Exploration of the Atlantic.* Oxfordshire, UK: Taylor and Francis, 2006, p. 119. Retrieved 8 December 2008. "Footnote 28. Strabo 7.1.3. The name North Sea — More properly 'Northern Ocean'. Septentrionalis Oceanus – probably came into use at this time; the earliest extant citation is Pliny, Natural History 2.167, 4.109."
[269]Adam is most famous for his chronicle *Gesta Hammaburgensis ecclesiae pontificum (Deeds of Bishops of the Hamburg Church)*; Rasma Lazda. "Adam of Bremen." In: Graeme Dunphy and Cristian Bratu, eds. *Encyclopedia of the Medieval Chronicle.* Consulted online on 27 May 2022. http://dx.doi.org/10.1163/2213-2139_emc_SIM_00031. First published online: 2016.
[270]Radoslav Katicic. *Ancient Languages of the Balkans, Part One.* Paris, France: Mouton, 1976, p. 144.
[271]As a name for a part of the known world, it is first used in the sixth century B.C. by a pre-Socratic Greek philosopher Anaximander as shown in his map: "File:Anaximander world map-en.svg." *Wikipedia Commons*, 26 Apr. 2022, commons.wikimedia.org/wiki/File:Anaximander_world_map-en.svg.

Chapter 4

A Chinese-Based World Map Depicts Africa in 1433

Abstract

This chapter shows that the African portion of a Chinese-based world map Kunyu Wanguo Quantu《坤輿萬國全圖》, or the Complete Geographical Map of All the Kingdoms of the World published by Matteo Ricci in 1602 in China, is not a direct or adapted copy of the three major sixteenth century European maps. This chapter also presents strong evidence which challenges the viewpoint that the Treasure Fleets led by Admiral Zheng He (郑和) to the Western Ocean during their seven epic voyages only reached as far as the east coast of Africa. A thorough analysis of the eleven original annotations and the 185 geographical items — kingdoms/empires, city-states, regions, cities/towns, rivers, lakes, mountains, islands, oceans and seas — depicted on the map of Africa in Kunyu Wanguo Quantu, and their comparison with the corresponding geographical items on the three major sixteenth century European maps, lead me to draw the following conclusions: (1) the Ming mariners drew the map of Africa in 1433 — the year when they explored Africa during their seventh (and last) voyage to the Western Ocean — decades before the Europeans were able to make a similar level of exploration in Africa; this Chinese map later became the source map for Matteo Ricci to draw the African portion of Kunyu Wanguo Quantu; and (2) Kunyu Wanguo Quantu shows correctly the relative positions of the southernmost tip of Africa and the Cape of Good Hope, and strongly supports the claim that Zheng He's mariners most likely rounded that tip into the Atlantic Ocean on the western side of the continent, enabling them to eventually reach the New World beyond.

Keywords: Africa, Kunyu Wanguo Quantu, Matteo Ricci, Treasure Fleets, Zheng He

1. Introduction

The world map — Kunyu Wanguo Quantu《坤輿萬國全圖》, abbreviated here as KWQ[1] — written with Chinese characters and having latitudinal and longitudinal lines, has been generally regarded as drawn in 1602 by Matteo Ricci — an Italian Jesuit priest and one of the founding figures of the Jesuit China missions — and his Chinese collaborators, based on the European maps which Ricci brought with him to China in 1582.[2]

In Chapters 1–3 of this book, I have presented with analysis and evidence to show that (1) on KWQ, the map of Cape Breton Island (in today's southeastern Canada), and the map of Australia, New Zealand, Land of Fire and Antarctica embedded inside Mo Wa La Ni Jia (墨瓦蠟泥加) on KWQ, are actually based on geographical information obtained in the 1420s by the Ming (明代; 1388–1644) mariners no later than their sixth voyage to the Western Ocean; and (2) surprisingly, the map of Europe on KWQ actually reflects the political landscape of Europe between 1157 and 1166 in the era of the Southern Song Dynasty (南宋; 1127–1279). In all these cases, I have shown that KWQ is of Chinese origin and the map is not a direct or adapted copy of other contemporary European maps for the above-mentioned map sections.

The question now is whether a similar analysis can be extended to the African part of KWQ, abbreviated here as Africa-KWQ, to also find its true origin. To answer this question requires a good knowledge of Africa's history, political landscape and geography from both the Chinese and the European perspectives; this will be the focus of Sections 2–7 of this chapter.

Africa is the second largest continent in the world. At its peak, prior to European colonialism, it is estimated that Africa had up to 10,000 different states and autonomous groups. In the absence of written sources about many of them, my analysis basically relies on the writing of early geographers and travellers, or limited data derived from oral tradition. Due to this limitation, it is inevitable that some small African kingdoms/city-states presented in the Appendix of this chapter do not have more detailed information other than their names transliterated into Chinese. But it will become obvious at the end of the present analysis that this limitation does not seem to have any obvious impact on my conclusions.

2. Chinese Drew Maps of Africa since Antiquity

Chinese knowledge about Africa started early. The earliest surviving Chinese map relating China with foreign countries — the Huayi Tu《華夷圖》or Map of China and Barbaric Countries — was engraved in the year 1136 as a stone stele of the Southern Song Dynasty (1127–1279), see Fig. 4.1. It is important to know that Chinese scientists already developed navigational compasses as early as the eleventh or twelfth century during the Song Dynasty (960–1279). With the use of compasses, China's marine navigation expanded considerably. Western Europeans soon followed at the end of the twelfth century.

Africa is shown at the lower left corner of Fig. 4.1. The shape of the continent, with its triangular southernmost region, correlates closely with today's geography of Africa except with an eastward-rotated orientation. The oceans on the east and west sides of the African continent are connected.

The next prominent Chinese map which contains Africa is shown in Fig. 4.2, at the left panel. It is the Da Ming Hunyi Tu《大明混一圖》[5] or Composite Map of the Ming Empire, painted in colour on 17 m^2 (183 sq ft) of stiff silk in 1389 of the Ming Dynasty. The year is determined by the political landscape revealed in this map. The original text on the map was written in Classical Chinese, but on the surviving copy Manchu labels were later superimposed. The coast of Africa is surrounded by the Indian Ocean, showing the area of the Cape of Good Hope and the southern tip

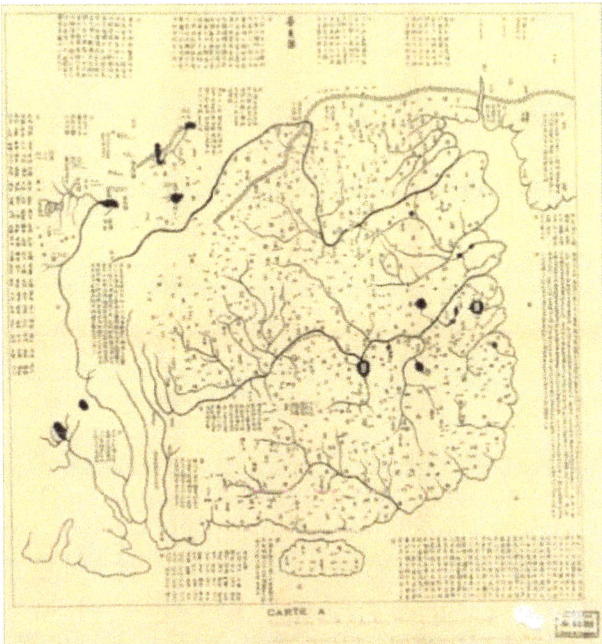

Fig. 4.1. This map shows the 1136 Huayi Tu《華夷圖》or Map of China and Barbaric Countries; see endnote 3, public domain). It is the earliest surviving map of China that relates China with other foreign states. It has very accurate latitudes and longitudes for the interior part of China, which may indicate the use of spherical trigonometry projection.[4] Africa is at the lower left corner of the figure. Its shape resembles today's Africa, but it has an eastward-rotated orientation.

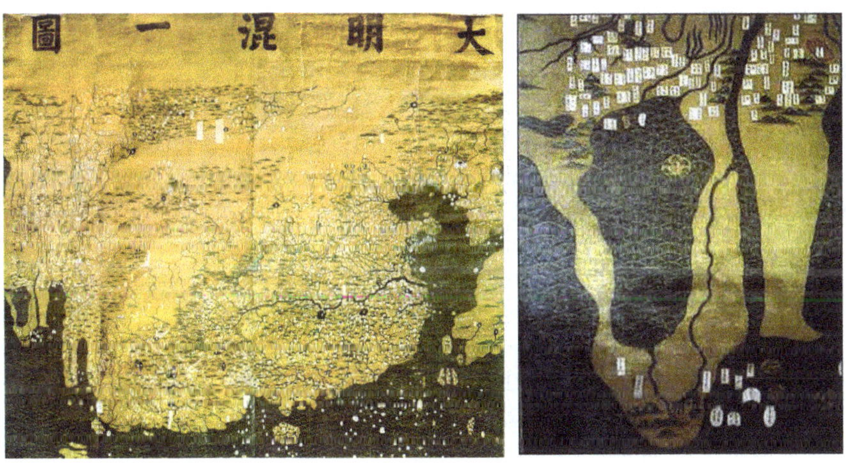

Fig. 4.2. At left is the map of the Da Ming Hunyi Tu《大明混一圖》or Composite Map of the Ming Empire (endnote 5; public domain) created around 1389 in the early Ming Dynasty; at right is a blow-up of this map's Africa: the large dark area may be Lake Victoria, because its wavy shading is like other water surfaces, and it connects to apparent rivers. The surviving copy of the map shows later revisions, and it is uncertain how closely it matches the original.

of Africa. It appears that sailors could round the southernmost tip of Africa to get to the western side of the continent. Maps of this type were made since about the 1320s, but they were all lost; Da Ming Hunyi Tu is the earliest survivor.

On this map, Sub-Saharan Africa (geographically, the area of the continent of *Africa* that lies south of the *Sahara Desert*) is depicted in a reasonable approximation of the correct shape, complete with mountains near the southern tip (see Fig. 4.2, right panel), which is close to reality. We can see that the interior of the continent is extraordinary, as well as confusing and inaccurate (also in Fig. 4.2, right panel): (1) the large dark area may be Lake Victoria (because its wavy shading is similar to other water surfaces and it connects to apparent rivers); (2) presumably the long and wide waterway to the right of the huge lake is the Red Sea and the long and narrower waterway, presumably the Nile River, flows into it. (The Mediterranean Sea is far to the north beyond a huge land area, see Fig. 4.2, left panel.) This Nile stops somewhere in the far south, does not flow into any ocean, and is not connected with the huge lake. Such a river system is the common depiction in Classical and Islamic maps of the Nile; and (3) many place names in the north and the south are marked. Again, the oceans on the eastern and the western sides of Africa are connected.

From these two maps, we know that by the end of the fourteenth century, Chinese knew the general shape of Africa and had some knowledge about its interior, albeit not all correct. The knowledge was the fruit of obtaining frequent reports of the topography — coastal and internal — of Africa by Chinese and non-Chinese merchants and mariners who for centuries had frequent and extensive contacts with cities along the southern coast of China.[6] After the Mongols' rule over China (1271–1368) and much of Eurasia,[7] the Islamic maps of Africa could also have found their way to China for Chinese cartographers to incorporate new information into their maps.

3. Zheng He's Seven Voyages (1405 to mid-1430s) Unmatched in World History

As stated in Section 2, the Ming Chinese knew the general shape of Africa and the many places depicted on the 1389 Da Ming Hunyi Tu《大明混一圖》or Composite Map of the Ming Empire even before Admiral Zheng He sailed to the Western Ocean in the early fifteenth century. Most importantly, they knew that in the Western Ocean — defined in the early Ming Dynasty as the ocean to the west of the middle of Borneo (婆罗洲) — there were many countries worthy of their diplomatic and trade contacts. Hence, soon after Zhu Di (朱棣; 1360–1424)[8] became the third Emperor — Emperor Yongle (who reigned from 1402 to 1424) of the Ming Dynasty — he decided to expand China's tributary system to these countries and to project China's strength to the rest of the world.

He ordered his favourite Admiral Zheng He (郑和; 1371–1435)[9] to lead Treasure Fleets (a Chinese *Treasure Ship* is a type of large wooden ship) to sail to the Western Ocean six times. Later, after Zhu Di's death, his grandson Zhu Zhanji (朱瞻基; 1399–1435)[10] — the fifth Emperor (Emperor Xuande reigned from 1425 to 1435) — resumed this activity and ordered Zheng He's seventh (and last) voyage. These voyages were carried out with hundreds of huge ships and tens of thousands of sailors and other passengers.[11] They are unmatched in world history.

With Zheng He's frequent trips to the Western Ocean, a steady stream of foreign envoys came from far away, offering all kinds of rare and precious gifts. The Ming emperors also treated these envoys well and generously gave gifts to them in return. According to the historical records, the Ming mariners reached Africa in their fifth through seventh voyages.[12]

But soon after the seventh voyage around the mid-1430s, the Ming rulers halted these activities. Due to the enormous cost of these voyages (since they were not really trade excursions, the government recovered little of the cost), and due to a growing threat to China's land borders in the north and west, the Ming rulers wanted to focus on domestic needs and secure the country's inland borders. They either burned the great fleet in the harbours or allowed it to rot. Most of the valuable navigational records and maps were destroyed, vanished in fires, or hidden, hence were lost to history. It became illegal to talk or write about Zheng He's voyages.

Due to these difficulties, modern historians have been struggling to pick up the pieces from the residual historical records to reconstruct that missing chapter of China's remarkable marine navigation in the Zheng He era. Among the historians, the extent of Zheng He's sailing is still hotly debated. Some hold the viewpoint that Zheng He's fleets did not travel beyond the east coast of Africa. But this viewpoint has faced challenges including my own work presented in my 2019 book titled *The Last Journey of the San Bao Eunuch, Admiral Zheng He*. The book suggests that during the seventh voyage, Zheng He's mariners made a trans-Atlantic crossing via southern Africa to arrive at the North American coast in the autumn of 1433 and later plausibly explored the North American heartland in the Central Mississippi Valley.[13]

This chapter is my fourth in a series of investigations, aiming at shedding light on the amazing story of Zheng He's seven epic voyages. It analyses and checks the extent and era of Chinese mariners' exploration of Africa accomplished by their seventh (and last) voyage.

Is it true that Zheng He's fleets only visited the east coast or Swahili coast[14] of Africa? This will be investigated in the following sections and in the Appendix. Before that, I present the European knowledge of Africa in Section 4, for comparison with the Chinese knowledge.

4. Major European Maps of Africa since Antiquity

The African continent is separated from the European continent by the Mediterranean Sea. Their narrowest separation is the Gibraltar Strait which has a minimum width of only 13 km/8.1 mi. Due to the proximity, the geography of northern Africa was reasonably well known among Europeans during Classical Antiquity (from circa eighth century B.C. to fifth century A.D.) in Greco-Roman geography: northwestern Africa (the Maghreb) was known as either *Libya* or *Africa*, while Egypt was considered part of Asia. The activities of the Romans and later the Muslims (beginning in the seventh century) took place mainly along the northern and eastern coasts of Africa. They did not explore the western, southern and internal lands of Africa.

(1) Ptolemy's map of Africa

Ptolemy's world map, reconstituted from Ptolemy's *Geography*[15] (c. 150 A.D.) in the late fifteenth century, does not show the correct shape of Africa, as evidenced in Fig. 4.3.

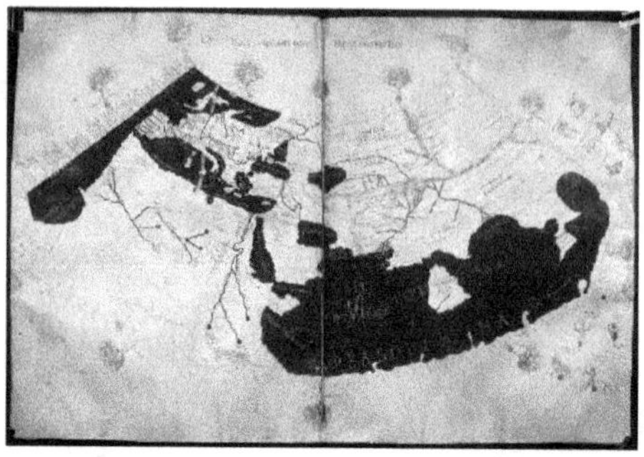

Fig. 4.3. Ptolemy's world map (see endnote 16, public domain), reconstituted from Ptolemy's *Geography* (c. 150 A.D.) in the late fifteenth century.

Africa is depicted as an extended continent in the south and there is no sea passage from the Mediterranean Sea to the Indian Ocean via the southern end of the African continent.

The European exploration of Sub-Saharan Africa begins from the fifteenth century, pioneered by the Kingdom of Portugal under Henry the Navigator (1394–1460).[17] Maps of Africa became more precise. But the interior of Africa was not mapped in any detail before the second half of the nineteenth century.

In Section 2, I have shown that by the twelfth century Chinese cartographers already drew the entire outline of Africa in the 1136 Huayi Tu《華夷圖》or Map of China and Barbaric Countries. And by the late fourteenth century, they also drew Lake Victoria in the African interior and wrote many place names in northern and southern Africa on the 1389 Da Ming Hunyi Tu《大明混一圖》or Composite Map of the Ming Empire.

Then in Section 3, I have shown that Zheng He's seven voyages (1405 to mid-1430s) are unmatched in world history. Although the extent of his expeditions is still under debate, most scholars agree that his mariners reached Africa during their fifth to seventh voyages.

Europeans reached the south of Africa in the late fifteenth century: Portuguese navigator and explorer Bartolomeu Dias (c. 1450–1500)[18] was the first European to round the Cape of Good Hope in 1488; another Portuguese explorer, Vasco da Gama (c. 1460s–1524)[19] reached India from the Atlantic Ocean by that route in 1497–1499.

(2) Major European maps of Africa in the fifteenth century

But surprisingly, in Fig. 4.4, the De Virga World Map (1411–1415) published by Albertinus de Virga, a fifteenth century Venetian cartographer, depicts the entire continent of Africa (and even Australia) in its modern shape long before the Europeans explored Sub-Saharan Africa.

Yet, there exist no credible academic sources to answer the following questions: (1) how did De Virga know the modern shape of Africa decades before the European explorers rounded the Cape of Good Hope in 1488 and reached India from the Atlantic Ocean in 1497 to 1499? (2) Could De Virga have heard about the shape of the African continent from Africans or Arabs? In Section 3 of Chapter 1 and Sections 3.4 of Chapter 2, I have mentioned that De Virga is known for

Fig. 4.4. De Virga World Map (1411–1415; see endnote 20, public domain) shows entire Africa.

revising his 1409 map of the Mediterranean by incorporating information from the Arab map and the 1418 Chinese World Map/Zheng He Map[21] (it is no longer in existence). Did he copy the Chinese map?

Liu Gang (刘钢) has provided some evidence which may help to clarify the above matters: He is a Beijing lawyer, map collector and analyst, and author of the Chinese book titled *Gu Di Tu Mi Ma: Zhong Guo Fa Xian Shi Jie De Mi Tuan Xuan Ji* 《古地图密码:中国发现世界的谜团玄机》 or Secret Code of Ancient Maps: The Mystery of China's Discovery of the World.[22] He points out that a bi-hemispherical world map — Yu Di Tu 《舆地图》 or Terrestrial Map — was created in 1320 by the Chinese cartographer Zhu Siben (朱思本; 1273–14th Century) in the Yuan Dynasty (1279–1368); *the cartographer of the De Virga world map (1411–1415) copied the Chinese map and revised his own map, but left some remarks unchanged on his revised map published (as amendment) in 1419.*[23]

What was equally if not more inspiring to the Europeans is the Fra Mauro map of the world (see Fig. 4.5, left panel), made around 1450s, after the Zheng He era. It shows a more detailed picture of Africa as a continent and it annotates (summarised and translated into English) that around 1420 a ship, or junk, from India crossed the Sea of India towards the Island of Men and the Island of Women, off Cape Diab; it sailed for 40 days in a south-westerly direction without ever finding anything other than wind and water (see Fig. 4.5, right panel). Today's Cape of Good Hope was named by Fra Mauro as *Cap de Diab*.

On the map, Fra Mauro also annotates (summarised and translated into English) the construction of the junks (a type of Chinese sailing ship with fully battened sails) that navigated in the Indian Ocean as carrying four masts or more, some of which can be raised or lowered, and have 40–60 cabins for the merchants and only one tiller. They can navigate without a compass, because they have an astronomer, who stands on the side and, with an astrolabe in hand, gives orders to

Fig. 4.5. At left is the Fra Mauro map of the world (see endnote 24, public domain), which is inverted to match the modern North-South orientation; at right is a blow-up of a portion of the Fra Mauro map which describes the travels around 1420 of a junk beyond the southernmost tip of Africa (he named it *Cap de Diab*; it is today's Cape of Good Hope); the ship is also illustrated above the text (see endnote 25, public domain).

the navigator. Fra Mauro was then convinced that the Indian Ocean was not a closed sea, and that Africa could be circumnavigated by her southern tip.

This knowledge together with the De Virga World Map (1411–1415) probably encouraged the Portuguese to intensify their effort to round the tip of Africa in the late fifteenth century.

(3) Major European maps of Africa in the sixteenth century

In the sixteenth century, Europeans made more explorations: the Cantino Planisphere (1502; see Fig. 4.6)[26] is the earliest surviving map showing Portuguese discoveries in eastern and western Africa, including capes, mountains, islands, rivers, etc. But much of Africa's interior was yet to be explored.

The Caverio Map (c. 1506)[27] is like the Cantino Planisphere (1502). It was probably either made in Lisbon by the Genoese Canveri or copied by him in Genoa from the very similar Cantino map.

Other prominent European maps in the early sixteenth century are: (1) the Ruysch world map (1507),[28] which contains the discoveries the Portuguese had made along the African coast and the Nile (Fig. 4.7, left panel); (2) the Waldseemüller map (1507),[29] which includes more African rivers and lakes, and information on the most recent discoveries (Fig. 4.7, right panel); (3) the Map of Pietro Coppo (1520),[30] which includes many African mountains in addition (Fig. 4.8, left panel); (4) the Diego Ribero world map (1529),[31] which depicts Africa's animals, kingdoms, mountains, rivers and forests, and is considered the first scientific world map based on empirical latitude observations (Fig. 4.8, right panel); (5) the Mercator world map (1569),[32] which uses a cylindrical map projection invented by Gerardus Mercator and was for the use of marine navigation (Fig. 4.9, left panel); (6) the *Africae Tabula Nova* (New Map of Africa) by Abraham Ortelius (1570),[33] which is considered to be the first true modern atlas; many of his maps were based upon sources that no longer exist or are extremely rare (Fig. 4.9, right panel); (7) the *Orbis Terrarum* (Map of the World)

Fig. 4.6. The Cantino Planisphere (1502; endnote 26; public domain) is a manuscript Portuguese world map preserved at the Biblioteca Estense in Modena, Italy.

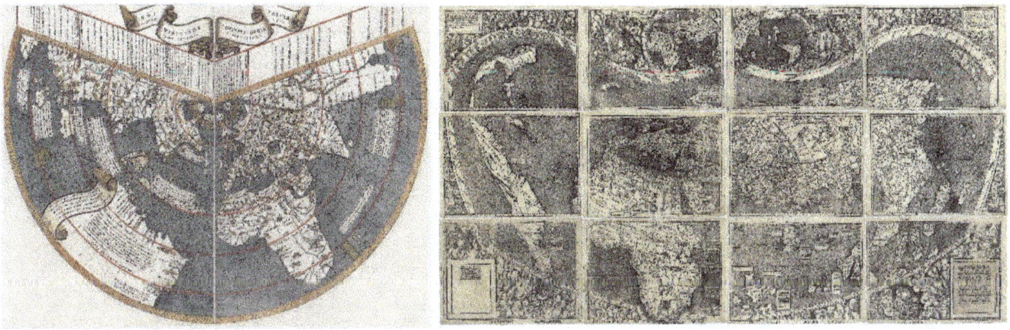

Fig. 4.7. At left is the Ruysch world map (1507; endnote 28; public domain), which shows the Nile; at right is Waldseemüller's world map (1507; endnote 29; public domain), which includes more African rivers and lakes.

Fig. 4.8. At left is the Map of Pietro Coppo (1520; endnote 30; public domain) which includes many African mountains; at right is the Diego Ribero world map (1529; endnote 31; public domain), which contains Africa's animals, kingdoms, mountains, rivers and forests, and is considered the first scientific world map.

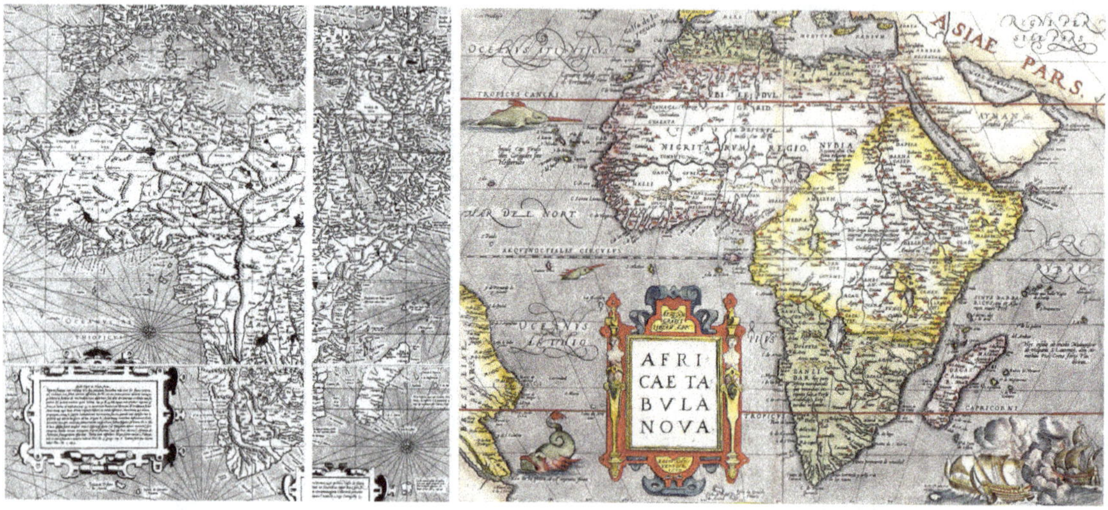

Fig. 4.9. At left is Africa extracted from the Mercator world map (1569; endnote 32; public domain); at right is the *Africae Tabula Nova* by Abraham Ortelius (1570; endnote 33; public domain).

Fig. 4.10. At left is Africa extracted from the *Orbis Terrarum* (Map of the World) by Petrus Plancius (1594; endnote 34; public domain); at right is Africa extracted from Kunyu Wanguo Quantu《坤輿萬國全圖》or the Complete Geographical Map of All the Kingdoms of the World published by Matteo Ricci (1602; see endnote 36, public domain).

by Petrus Plancius (1594)[34] (Fig. 4.10, left panel); Plancius is considered to be one of the notable figures in the Golden Age of Netherlandish cartography (c. 1570s–1670s).

However, despite their separate merit, the eight European maps still show that their geographical knowledge of the interior of Sub-Saharan Africa was very limited during the sixteenth century.[35] The situation lasted until the beginning of the nineteenth century.

5. Africa Depicted on Kunyu Wanguo Quantu to be Compared with Three European Maps

After reviewing the Chinese and European knowledge about Africa, in the following sections and in the Appendix, I shall investigate the Africa-KWQ published by Matteo Ricci in China in 1602 (Fig. 4.10, right panel). The map surprisingly contains many geographical items of Chinese origin, which should not appear on a "European" map; it also contains many more geographical items and much information absent from the above-mentioned eight European maps. All these have given me a strong motivation to establish that the Africa-KWQ is not a direct or adapted copy of the following three major European maps: (1) the map of Africa extracted from the Mercator world map (1569), abbreviated as Africa-Mercator (1569); (2) the *Africae Tabula Nova* by Abraham Ortelius (1570), abbreviated as Africa-Ortelius (1570); and (3) the map of Africa extracted from *Orbis Terrarum* (Map of the World) by Petrus Plancius (1594), abbreviated as Africa-Plancius (1594). These three maps are referred to as "the three major sixteenth-century European maps" in the remaining part of this chapter, and they have been considered the most likely maps referenced by Matteo Ricci in making the Africa-KWQ.

My goals also include exploring the following: how different is the Africa-KWQ from the three major sixteenth century European maps? What geographical items on the Africa-KWQ are of Chinese origin? What is the most important geographical item missing on the three major sixteenth century European maps when comparing them with the Africa-KWQ? Was Matteo Ricci really the cartographer of this map or were the Chinese? If he was not, did he change the Chinese map and, if so, in what way or to what extent? And what is the true era of Africa reflected by the Africa-KWQ? That era would correspond to the date when Africa was explored.

These topics will be thoroughly investigated and answered in Sections 6 and 7 as well as in the Appendix of this chapter.

6. The Africa-KWQ Is of Chinese Origin, Not a Direct or Adapted Copy of European Maps

I present in the Appendix the results of my analysis of all the geographical items depicted on the Africa-KWQ after investigating their histories and geographical locations as best as I can. Also included in the main text and in the Appendix are comparisons of the Africa-KWQ with the three major sixteenth century European maps whenever possible to show their differences. As I have written in Section 1 of this chapter, in the absence of written sources about some of the small kingdoms, my analysis basically relies on the writing of early geographers and travellers, or limited data derived from oral tradition. Due to this limitation, it is inevitable that some small African kingdoms/city-states presented in the Appendix do not have more detailed information other than their names transliterated into Chinese. Nevertheless, it will become obvious at the end of my analysis that this limitation does not seem to have obvious impact on my conclusions.

My analysis has enabled me to conclude that the Africa-KWQ is a Chinese-based map which includes Chinese geographical and political knowledge of Africa obtained by Ming mariners in 1433 during their seventh (and last) voyage.

Here are the main bases for me to draw the above conclusions:

(1) Eleven Chinese-based annotations on the Africa-KWQ are absent on the three major sixteenth century European maps.

These 11 annotations are realistic descriptions of the African geography, natural environment, human races and habits, in addition to animals, plants and produce on land and seas. The descriptions most likely were witnessed by Chinese explorers who had visited these places. Some of these annotations reflect specifically and uniquely their connections with the Chinese culture or philosophy: for example, annotation @1 next to 馬拿莫大巴 (Item 17; Monomotapa/Kingdom of Mutapa) about qilin — a gentle, kind and benevolent creature which carries a mystical good omen in Chinese mythology — signifies the unique Chinese passion about this creature. Such a passion was absent from Europeans' minds; and @8 uses the Chinese "Yin and Yang" concept (representing two opposite forces in the universe) to describe an African habit, which uniquely reflects the long- and deeply-held Chinese philosophy: every matter, tangible or non-tangible, in the universe must maintain a good balance. All the annotations are listed in the Appendix with my English translations and comments.

(2) Many geographical items on the Africa-KWQ are of Chinese origin.

On the Africa-KWQ, there are geographical items which have no connection with names given by the Europeans after their arrival in Africa from the late fifteenth century. This is strong evidence showing the older age of the information source of the Africa-KWQ. These Chinese-originated geographical items are specifically listed in Table 4.1; each item is followed by a second number (such as "Item 4") showing its listing position in the Appendix under Item #. Items preceded by "•" are geographical items which do not appear on any of the Africa-Mercator (1569), Africa-Ortelius (1570) or Africa-Plancius (1594) maps.

(3) Among the 185 geographical items on the Africa-KWQ, 70 (about 38 percent) are not depicted in any of the three major sixteenth century European maps.

They are listed in Table 4.2 separately in nine zones (Item numbers refer to the Appendix).

(4) The Africa-KWQ depicts the correct southernmost tip of Africa.

In Fig. 4.11, the three major sixteenth century European maps all mistakenly show the Cape of Good Hope as the southernmost tip of Africa, whereas the Africa-KWQ correctly depicts 羅經正峰 (Luo Geng Zheng Feng; Item 4) — today's Cape Agulhas — as the geographic southernmost tip of the African continent, see Fig. 4.12 for a comparison.

The only other European map which correctly depicts Cape Agulhas as the southernmost tip of Africa is the later drawing of the whole of Africa by Italian cartographer, engineer and astronomer Giacomo Gastaldi (c. 1500–1566), engraved by Fabricius Licinus (c. 1521–1565).[49] This rare map from 1564 printed on eight copperplates is a large-scale map (108 cm × 142 cm/3.54 ft ×

Table 4.1. Geographical items of Chinese origin on the Africa-KWQ.

#	**Geographical items of Chinese origin**
1 羅經正峰 Item 4	The Africa-KWQ correctly depicts 羅經正峰 (Luo Geng Zheng Feng; Item 4) — today's Cape Agulhas — as the geographic southernmost tip of the African continent. In the Chinese nomenclature, "羅經" means "compass", "正" here means that the direction of magnetic north (pointed at by the compass needle) coincides with true north, and "峰" means "mountain or mountain peak". There are many mountains near the southern end of Africa. This Chinese nomenclature reveals not only precisely the meaning of the direction of magnetic north coinciding with true north, but also the cape's surrounding environment. The cape was visited by Portuguese Bartolomeu Dias in 1488, who called it *Cabo das Agulhas* (after St. Agulhas) or "Cape of Needles", after noticing that in the region the direction of magnetic north pointed by the compass needle coincided with true north. But his nomenclature lacks an indication of the mountainous environment near the cape.
2, 3 大浪山 and 大浪山角 Items 5, 6	The Africa-KWQ depicts 大浪山 (Da Lang Shan; Item 5) and 大浪山角 (Da Lang Shan Jiao; Item 6), which can be identified as today's Cape of Good Hope and the nearby coastal region. In that region the currents of the Indian Ocean and the Atlantic Ocean meet at the point where the warm-water Agulhas current meets the cold-water Benguela current. That oceanic meeting point fluctuates between Cape Agulhas and Cape Point.[37] Hence, huge waves are generated in that neighbourhood. 大 means "huge", 浪, "wave", 山, "mountain" and 角, "cape". Portuguese explorer Bartolomeu Dias in 1488 was the first European to round the Cape of Good Hope and reach the East African coast. On the journey back, he observed the southernmost point of Africa, later called *Cabo das Agulhas*, or Cape of Needles. Bartolomeu Dias named the rocky second cape *Cabo das Tormentas* (Cape of Storms) to signify the tempestuous storms and strong Atlantic–Antarctic currents that made ship travel so dangerous. This "Cape of Storms" was later renamed as "Cape of Good Hope" (*Cabo da Boa Esperança*) because of the great optimism engendered by the opening of a sea route to India and the East. Later, Vasco da Gama also recorded a sighting of the Cape of Good Hope in 1497. While Cape of Storms reflects the tempestuous storms, the Chinese names 大浪山 (Da Lang Shan; Item 5) and 大浪山角 (Da Lang Shan Jiao; Item 6) signify, instead, the strong Atlantic–Antarctic currents, the big waves.[38] These prove that Chinese explorers had experienced the high waves at 大浪山 and 大浪山角 and gave suitable names to the region to warn future navigators.
• 4 泥羅河泉 Item 35	On the Africa-KWQ, the source or spring of the White Nile is depicted specifically as 泥羅河泉 (Ni Luo He Quan; Item 35). 泥羅河 means Nile River and 泉, "source or spring". This nomenclature is Chinese based, although 泥羅河 is a direct transliteration of Nile River known at the time. None of the three major sixteenth century European maps uses this name to denote the source of the Nile.

(*Continued*)

Table 4.1. (*Continued*)

#	**Geographical items of Chinese origin**
● 5 齊歷湖 Item 36	On the Africa-KWQ, the source lake of the White Nile is depicted as 齊歷湖 (Qi Li Hu; Item 36; Lake of Cichlids; today's name is Lake Victoria). Cichlids (齊歷) are fish from the family *Cichlidae* (a large family of chiefly tropical freshwater percoid fishes).[39] Nearly all African cichlids live in one of the great lakes in East Africa: Lake Malawi, Lake Victoria and Lake Tanganyika. The first known European use of *cichlid* was in 1884[40]; and the word's origin may be from the base of New Latin *Cichlidae*. However, the Chinese name 齊歷 (Qi Li), which may be regarded as the transliteration of the local name of the fish, is already depicted on the Africa-KWQ centuries before the first use of *cichlid* by the Europeans. None of the three major European maps uses this fish name to name the source lake of White Nile.
6 黑湖 Item 54	On the Africa-KWQ, 黑湖 (Hei Hu; Item 54) is depicted as the source lake — Lake Tana — of today's Blue Nile. The water's transparency being quite low may explain the Chinese name of the lake: 黑 means "black" and 湖 "lake", whereas none of the three major sixteenth century European maps depicts this lake as Black Lake.
7 入匿 Item 102	On the Africa-KWQ, 入匿 (Ru Ni; Item 102) depicts the region of today's Guinea. The English term Guinea comes from the Portuguese word *Guiné*, which emerged in the mid-fifteenth century to refer to the lands inhabited by the *Guineus*, a generic term for the black African peoples south of Senegal.[41] The name "Guine" is written on the Africa-Plancius (1594). But the Chinese name 入匿 (Ru Ni) is not a transliteration of Guinea, rather a geographical term of Chinese origin.
8 入蠟河 Item 158	On the Africa-KWQ, 入蠟河 (Ru La He; Item 158) is depicted in northwestern Africa, which corresponds to River Ghir. But 入蠟 (Ru La) is not a transliteration of "Ghir"; 河 means "river". On the three major sixteenth century European maps, River Ghir is depicted. The name 入蠟河 is of Chinese origin.
9 得莫甑 Item 159	得莫甑 (De Mo Zheng; Item 159) on the Africa-KWQ, is depicted as Tebesse[42] on the Africa-Ortelius (1570). 得莫甑 is not a transliteration of Tebesse. The Africa-Mercator (1569) and the Africa-Plancius (1594) do not depict Tebesse, but the Africa-Ortelius (1570) does.

10 福島 Item 168	On the Africa-KWQ, 福島 (Fu Dao; Item 168) is today's Fuerteventura,[43] the second largest of the Canary Islands[44] in the Atlantic Ocean. The island's name may have been derived from *Fortunatae Insulae* (The Isles of the Blessed), the name by which the Romans knew the Canary Islands. In the fourteenth century, European explorers and traders reached this island. In Spanish, "Fuerteventura" can refer to the strong winds around the island and the resulting danger to nautical adventurers or, instead, wealth, luck, or destiny, because "fuerte" means "strong" and "ventura" "fortune". The three major sixteenth century European maps use this island name; hence, its meaning is ambivalent. The Chinese nomenclature of 福 means "blessed" and 島, "island". The meaning unambiguously matches the Roman meaning of the island name, implying that Chinese and Romans might have had contacts as early as during the Roman era (27 B.C.–476 A.D.). In Chapter 3 of this book, I have mentioned T.C. Bell's survey findings which show that Chinese appear to have entered the British Isles in two waves: the first wave occurred in the millennia before Christ; and the second wave took place during the Roman period c. 44–410[45] (he found Roman forts alongside a Chinese barracks and harbour sites to defend them).
11 綠峰島 Item 170	On the Africa-KWQ, 綠峰島 (Lu Feng Dao; Item 170) is today's Cape Verde archipelago. *Cabo Verde* means "Green Cape", *Verde* means "green"; it is the name that Portuguese navigators originally gave to a cape on mainland Africa in c. 1444, a few years before they discovered the islands — the Cape Verde archipelago — to its west; the name "Cape Verde archipelago" was not originally meant to describe the green mountains on these islands. The Chinese name is not an exact transliteration of "Cape Verde archipelago" which does not signify the "green mountain" on the islands. Because of their proximity to the Sahara, most of the Cape Verde islands are dry. But on islands with high mountains and farther away from the African coast, the much higher humidity provides a rainforest habitat on the mountains. This has precisely given the source of the Chinese name 綠峰島; here 綠 means "green", 峰, "mountain or mountain peak", 島, "islands or archipelago".
12 利未亞 Item 179	The name "Africa" is used to denote this continent on the European maps, but on the Africa-KWQ, present-day Africa is named as 利未亞 (Li Wei Ya; Item 179). This Chinese name is not connected with "Africa Proconsularis", given to the Roman province established in 146 B.C. following the defeat of Carthage,[46] nor "Alkebulan" which was the original ancient name of Africa.[47]
13 利未亞海 Item 182	On the Africa-KWQ, 利未亞海 (Li Wei Ya Hai; Item 182) is depicted as the ocean to the west of the African continent; it is today's South Atlantic Ocean. On all the three major sixteenth century European maps, the Latin expression *Oceanus Æthiopicus* is the name given to the southern half of the Atlantic Ocean. This name appeared on European maps from ancient times up to the turn of the nineteenth century. The Chinese nomenclature has no connection with this Latin expression.

(*Continued*)

Table 4.1. (*Continued*)

#	**Geographical items of Chinese origin**
14 西南海 Item 183	On the Africa-KWQ, 西南海 (Xi Nan Hai; Item 183) is today's Indian Ocean. Since Antiquity, Chinese already grasped the technique of measuring latitude on the Earth by observing the height of the Polaris star in the Northern Hemisphere; hence they knew well the direction of north. North, south, east and west are the four cardinal directions. On the compass, which was widely used in the Song Dynasty (960–1279), each of the four cardinal directions are separated by 90 degrees, and secondarily divided by four ordinal (intercardinal) directions — northeast, southeast, southwest and northwest — each located halfway between two cardinal directions. The Chinese also used these directions to name the oceans surrounding their territory and elsewhere in the world. 西 means "west", 南, "south" and 海, "sea or ocean". 西南海 is the ocean to the southwest of China, which is the Indian Ocean. On the 1570 *Theatrum Orbis Terrarum* by Ortelius,[48] the Latin expression *Mar Di India* is used to denote the Indian Ocean; but the Africa-Mercator (1569) and the Africa-Ortelius (1570) do not depict the Indian Ocean. On the Africa-Plancius (1594), *Oceanus Indicus* is displaced at where today's Arabian Sea is.
15 西紅海 Item 184	On the Africa-KWQ, 西紅海 (Xi Hong Hai; Item 184) is today's Red Sea. 西 means "west", 紅, "red", and 西紅海, "Western Red Sea". "West" is used, because on KWQ, there is 东紅海 (Eastern Red Sea) in North America. This will be explained when the map of North America is analysed in Chapter 5 of this book. On the Africa-Mercator (1569), the Red Sea is named *Sinus Arabicus* (literally "Arabian Gulf"); on the Africa-Ortelius (1570), it is depicted as *Mare Rosso, ol Sinus Arabicus*; and on the Africa-Plancius (1594), as *Mare Rubrum* (in Latin; *Rubrum* means "Red"). The name of the sea may be due to a seasonal change of the red-coloured water surface or refer to the direction south, just as the Black Sea's name may refer to north. The Chinese nomenclatures of 西紅海 (Western Red Sea) and 东紅海 (Eastern Red Sea) refer specifically to the red-coloured water surfaces and their east–west cardinal directional difference. The Chinese nomenclatures have different meanings from the European names.
Summary: The above are the 15 Chinese-based geographical items depicted on the Africa-KWQ. Matteo Ricci has kept them "as is" when he edited the Chinese source map to make the final Africa-KWQ. This list provides strong evidence showing that the Africa-KWQ is not a copy of the three major sixteenth century European maps discussed in this chapter.	

Table 4.2. Thirty-eight percent of geographical items not depicted in the three European maps.

Zone I Southern Africa There are 7 items.	1. 未惹荅 (Ueigeta; Item 10); 2. 步都牙 (Putuia; Item 11); 3. 只伐 (Scioa; Item 15); 4. 馬拿莫大巴 (Monomotapa/Kingdom of Mutapa; Item 17); 5. 伯那人 (Péna man; Item 18); 6. 月山 (Moon Mountain; Item 19); 7. 長山 (Long Mountain; Item 24)
Zone II Area between the Equator and southern Africa There are 10 items.	1. 黑人國 (Kingdom of People with Black-coloured Skin; Item 27); 2. 布都曷 (Putuho; Item 30); 3. 古西媽 (Cusima; Item 32); 4. 尼羅河泉 (the source of the Nile; Item 35); 5. 齊歷湖 (Lake of Cichlids; Item 36); 6. 假佛尔 (Cafri; Item 37); 7. 仙勞冷祖濱 (San Lorenzo Beach; Item 39); 8. 兀兒曷 (Urho; Item 41); 9. 岸辨叔旺 (Ambiancativa; Item 45); 10. 干只入閦 (Canzegemen Item 49)
Zone III Eastern and northeastern Africa (north of the Equator) There are 10 items.	1. 黑湖 (Lake Tana; Item 54); 2. 入曷謎的里 (Bagamidri; Item 56); 3. 吾入勒卯翁 (Tigremahon; Item 57); 4. 下黑地阿皮亞 (Ethiopia Inferiore; Item 60); 5. 多亞剌 (Doara; Item 62); 6. 党各哩 (Dangali; Item 65); 7. 多拔西 (Dobas; Item 66); 8. 黑地兀皮亞 (Ethiopia; Item 68;); 9. 巴皮羅泥亞 (Babylon in Egypt; Item 71); 10 薩係得 (Sahid; Item 72)
Zone IV Central Africa (north of the Equator; Region I) There are 8 items.	1. 革剌漫的亞 (Calamat; Item 78); 2. 帝曷辣耶 (Tigrai, Item 79); 3. 訝德 (Iaté; Item 80); 4. 雷地 (*Terra dei lampi*, Item 83); 5. 默大入剌 (Media Guilan; Item 85); 6. 伯恁 (Benin City/*Bini*; Item 89); 7. 這阿安 (Cheogane; Item 90); 8. 巴瓦米 (Bagamidu; Item 92)
Zone V Western Africa (Region I) There are 4 items.	1. 七力日突 (Meleghete; Item 96); 2. 色匿客 (Scenico; Item 99); 3. 匆突伊 (Vodei; Item 106); 4. 曷噩特 (Hongoté; Item 108)
Zone VI Western Africa (Region II) There are 7 items.	1. 訝喜黄野 (Iasci, deserto; Item 113); 2. 熱土利亞 (Getulia; Item 114); 3. 葛那葛 (Conaco; Item 115); 4. 巴瓦諾 (Pauano; Item 116); 5. 息歪瓦河 (Senegal River; Item 120); 6. 刻剌可 (Colaco; Item 121); 7. 黑牙數 (Heasus; Item 130)
Zone VII Central Africa (north of the Equator; Region II) There are 8 items.	1. 辣葛 (Lacca; Item 135); 2. 摩色嶺 (Muslin; Item 136); 3. 小利未亞 (Libia minore; Item 137); 4. 呉沙兒瓦蠟上 (Usciaru-alasciam; Item 138); 5. 色則尔没 (Secelme; Item 139); 6. 荒野 (Huan Ye; Item 140); 7. 葛那 (Ge Na; Item 141); 8. 亞入新巴 (Agisymba; Item 143)
Zone VIII North Africa There are 10 items.	1. 小亞非利加 (Africa Minor; Item 149); 2. 訝沙登 (*Iasciatem*; Item 150); 3. 馬尔馬利加 (Marmarica; Item 151); 4. 沙尔 (Sar; Item 152); 5. 曷热尔 (Hoger; Item 153); 6. 奴米德 (Numidia; Item 154); 7. 大兒瓦國 (Tarua; Item 157); 8. 貌利大泥亞 (Mauretania; Item 161); 9. 亞無音 (Abuinam; Item 162); 10. 德利非 (Trisidis; Item 164)
Zone IX Islands, oceans and seas of Africa There are 6 items.	1. 匆自島 (*Vuiza, isola di*; Item 173); 2. 未尔曷六剌 (Ueirholieula; Item 176); 3. 檀香樹嶺 (literally means "Sandato odoro-so, picco de"; Item 177); 4. 曷叭布剌 (Hopapulal; Item 178); 5. 利未亞 (today's Africa; Item 179); 6. 利未亞海 (today's South Atlantic Ocean; Item 182).
Summary: The Africa-KWQ shows that 70 of its 185 depicted geographical items (about 38 percent) do not appear on any of the three major sixteenth century European maps. We may also include the Iron Island (鐵島, mentioned in @10 on the Africa-KWQ, but the island itself is not depicted explicitly). The places represented by these geographical items spread over all parts of the African continent, implying that *the Chinese explorers had explored or had gained knowledge (or a combination of both) about much of the African continent, not only the east coast of Africa.*	

Fig. 4.11. The incorrect southernmost tip of Africa — Cape of Good Hope — shown in the Africa-Mercator (1569) at top left, in the Africa-Ortelius (1570) at top right, in the Africa-Plancius (1594) at lower left, and the correct southernmost tip of Africa — 羅經正峰 (today's Cape Agulhas; Item 4) — in the Africa-KWQ at lower right. All figures are extracted from the maps shown in Figs. 4.9 and 4.10 (public domain).

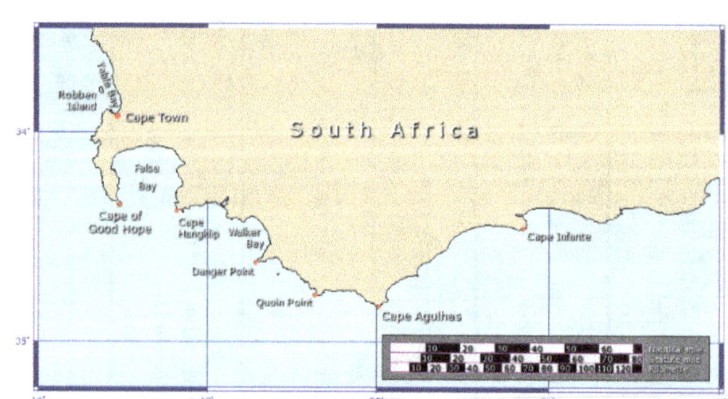

Fig. 4.12. A modern map (see endnote 53, public domain) shows the southernmost tip — Cape Agulhas — of Africa and the Cape of Good Hope to its northwest.

4.66 ft); placenames are abundant; cape and bay names are from Portuguese and Arab sources; hence, no Chinese-based geographical item appears on this map. Gastaldi, who was cosmographer to the Republic of Venice, is regarded by many scholars as the greatest cartographer of the Italian school.[50] This map was used as an important source by later sixteenth and seventeenth century cartographers, including Mercator, Ortelius, etc.[51] But surprisingly, none of them, Mercator, Ortelius or Plancius, followed Gastaldi to draw Cape Agulhas as the correct southernmost tip of Africa. Did Matteo Ricci draw the Africa-KWQ based on the 1564 Gastaldi map? This does not seem likely, because at that time in China Matteo Ricci only had a small Italian wall map (made in 1570) in his possession,[52] not the large-scale Gastaldi map. Besides, the rest of the Gastaldi map differs substantially from the Africa-KWQ. (Due to the scope limitation of this chapter, I cannot make detailed comparisons here. It will be the content of another book). Since Matteo Ricci was in China and published KWQ in Beijing, this leaves only the possibility that he was using a Chinese map as his source map to correctly draw the southernmost tip of Africa on the Africa-KWQ.

The latest date for the Chinese mariners to round this southernmost tip, before Matteo Ricci published the KWQ in 1602, would be during the Ming Treasure Fleets' seventh (and last) voyage in the early 1430s, before the Hăijìn (海禁) or sea ban was imposed. By then, the Chinese mariners already mastered the astronomical navigation described in the *Zheng He Hang Hai Tu*《郑和航海图》[54] or *The Charts of Zheng He's Voyages* for determining the latitudes with high accuracy. Since they knew how to measure time, they could also obtain the longitude of their ships' location.[55] If the Chinese mariners had not mapped the two capes accurately, and if Matteo Ricci had not used their map as source map, then how could he correctly show the geographical locations of these two capes on the Africa-KWQ?

(5) The river and lake systems depicted on the Africa-KWQ show complicated differences from the three major sixteenth century European maps.

The river and lake systems are analysed in two categories: to the south of the Equator and to the north of the Equator.

Subsection (1): To the south of the Equator, all four maps show a river (today's White Nile) flowing northward from a big lake located south of the Equator. More details are given as follows:

Lake Victoria and rivers flowing into or out of it

On the Africa-Mercator (1569), the lake is made into a black-coloured area in the middle of the figure and the river flowing northward is unnamed (Fig. 4.13, top left panel); there are four named rivers flowing into the southern or southeastern side of the lake from mountains further south. On this map, the Zaire River (today's Congo River) is depicted correctly as not flowing into the abovesaid lake, but into a smaller unnamed lake (near the left side of the figure).

On the Africa-Ortelius (1570), there is also an unnamed lake (Fig. 4.13, top right panel) and it is made into a beige-coloured dotted area; the river flowing northward is named as *Nilus fl.* (Nile River); there is a river named *Zuama fl.*, which seems to wrongfully flow out from the southern side of the lake and into today's Indian Ocean as two rivers — *Zuama fl.* and *R. di Spirito Sant.*

Fig. 4.13. At top left is the lake area near the source of the White Nile extracted from the Africa-Mercator (1569; public domain); at top right is the similar area extracted from the Africa-Ortelius (1570; public domain); at bottom left is the lake area near the source of the White Nile extracted from the Africa-Plancius (1594; public domain; note, there is some duplication because of overlapping map segments); and at bottom right is the similar area extracted from the Africa-KWQ (public domain).

This map also wrongfully depicts the Congo River/Zaire River as flowing out of the above-said lake.

On the Africa-Plancius (1594), the lake is made into a blue-coloured area with a few islands in it (Fig. 13, bottom left panel). The river flowing northward is named *Nilus fl.* (it is not visible on this map but can be seen on the larger-scale map in Fig. 4.10), and there is an unnamed river flowing from the southern side of the lake into the ocean (today's Indian Ocean). The Congo River/Zaire River is depicted correctly as not being connected to the above-said lake, but it does not correctly show a loop-like river path.

On the Africa-KWQ, the lake is a shady area named 齊歷湖 (Qi Li Hu; Item 36; meaning Lake of Cichlids; today's Lake Victoria) in Fig. 4.13 at the bottom right panel. There is a river named 尼羅河泉 (Ni Luo He Quan; Item 35; the source of the Nile) flowing out of the southern side of the lake and flows into 西南海 (Xi Nan Hai; Item 183; today's Indian Ocean; but this is incorrect, since Nile does not flow into the Indian Ocean) as two branches; one branch is not named, and the other is named 金河 (River of Gold; Item 13). The Congo River/Zaire River is depicted incorrectly as connected with the Lake of Cichlids; a loop-like river path is drawn. (Congo River/Zaire River is the world's deepest recorded river and the only major river to cross the Equator twice. Lake Tanganyika near Bujumbura in Burundi is thought to be the source of the Congo River/Zaire River; the Lake was first visited by Europeans in 1858, in their quest for the source of the Nile.[56]) The map shows that the Congo River/Zaire River has two branches leaving the lake in the western and southwestern side of the lake: one direct branch and one looping to the southwest. The southwestern branch is interrupted by the mountain range seemingly blocking its course, and then this branch flows in a northwesterly direction to join the direct branch. This is incorrect in comparison with today's Congo River/Zaire River shown in Fig. 4.14.

In summary, it seems that only the Africa-Mercator (1569) and the Africa-Plancius (1594) are correct in disconnecting the Congo River/Zaire River from Lake Victoria, while the Africa-Ortelius (1570) and the Africa-KWQ incorrectly depict the White Nile and the Congo River/Zaire River as sharing the same source lake. But all these four maps depict the Congo River/Zaire River completely south of the Equator. Hence, the similar mistakes and the many differences in river details among the four maps and their comparison with today's courses of the White Nile and the Congo River/Zaire River send a clear message: the mapping of the African river and lake systems was a

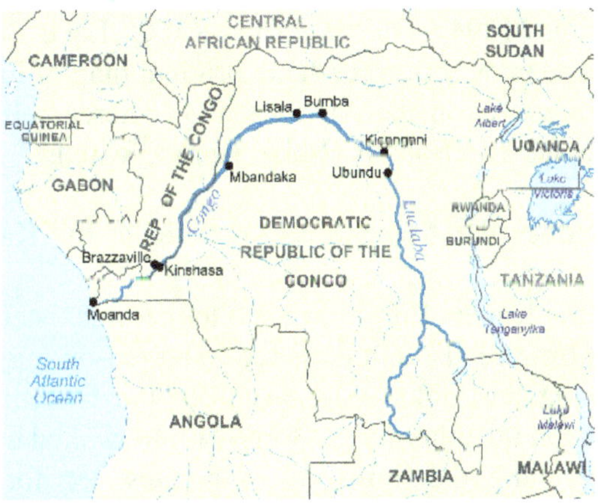

Fig. 4.14. Map of the Congo River/Zaire River in West-Central Tropical Africa, showing national boundaries (see endnote 57, public domain).

huge challenge for the earlier Chinese as well as the later Europeans. The interior of Africa was not mapped correctly in any detail before the second half of the nineteenth century.

Subsection (2): To the north of the Equator, the four maps also show different river and lake systems to be discussed as follows:

Lake Borno

Today, the White Nile meets the Blue Nile near Khartoum in Sudan, which is a rain-poor city. But on the three major sixteenth century European maps, the White Nile meets the Blue Nile, then splits, then meets again, before joining to become the Nile which flows northward into the Mediterranean Sea. The ring-shaped land area formed between their first and second meeting places has cities/towns drawn into it, as can be seen on the Africa-Mercator (1569) in Fig. 4.9, left panel and the Africa-Ortelius map (1570) in Fig. 4.9, right panel on the larger scale of the original maps. A similar depiction is shown on the Africa-Plancius (1594) in Fig. 4.10, left panel, but no city/town is drawn into that ring-shaped land area.

However, on the Africa-KWQ, the ring-shaped land area (Fig. 4.10, right panel) is replaced by a lake named 波尔諾湖 (Bo Er Nuo Hu; Item 67; Lake Borno). This is incorrect.

Lake Borno is related to the Kanem-Bornu Empire (Kanem Empire: c. 700–1380; Bornu Empire: 1380s–1893).[58] Kanem-Bornu was an African trading empire that controlled the area around its territory at various times, including what is now southern Chad, northern Cameroon, north-eastern Nigeria, eastern Niger and southern Libya.[59] The rulers of the old Bornu Empire had the title Shehu of Borno/Shehu of Bornu/Sultan of Borno. The Kanem-Bornu Empire was centred first in Kanem in western Chad, and after 1380, in Borno (in today's north-eastern Nigeria), which is very close to today's Lake Chad.[60]

Both the Africa-Mercator (1569) and the Africa-Plancius (1594) depict the Bornu Empire as *Borno*. *Borno* is absent from the Africa-Ortelius (1570), but *Borno Lacus* (Lake Borno; presumably Lake Chad) is depicted on the same map, and as an unnamed lake on both the Africa-Mercator (1569) and the Africa-Plancius (1594). On the Africa-KWQ, Lake Borno is clearly placed at a wrong location which can be seen on the larger-scale original map of Fig. 4.10, right panel. Nither Kanem nor Borno/Bornu shows up on the Africa-KWQ, whereas all the three major sixteenth century European maps depict most likely Lake Borno, whether named (*Borno Lacus*) or unnamed.

You may wonder: why does the Africa-KWQ not depict the Kanem, or the Borno/Bonru Empires?

It is understandable that the Kanem Empire was no longer in existence in the fifteenth century; but why is the Borno/Bornu Empire also not depicted? The reason may be that Borno/Bornu was in turmoil and unstable during the first three-quarters of the fifteenth century and, hence, not included in the Africa-KWQ. On the other hand, Borno/Bornu peaked during the reign of Mai Idris Alooma (c. 1564–1596), reaching the limits of its greatest territorial expansion.[61] Hence, it deserves to be depicted as *Borno* on the Africa-Mercator (1569) and the Africa-Plancius (1594). But the Africa-Ortelius (1570) fails to depict it.

Did the Nile have branches flowing east–west across Africa in ancient days? (1) River Ghir

Today's Nile does not have any branch oriented in the east–west direction across northern Africa through the Sahara Desert. The climate of the Sahara Desert has been essentially unchanged for the past 13,000 years.[62] So there was no significant climate change and no forest as it is shown on Sebastian Munster's 1554 map of Africa![63]

Before analysing the rivers and lakes on ancient maps, we must be aware of the complication that their flow directions are not depicted. Nevertheless, we can assume that rivers with one end at an ocean coast must flow towards that ocean, while rivers with one end in mountains must flow away from those mountains; lakes do not indicate flow directions, and in deserts, a river can end in a lake that evaporates.

On the Africa-KWQ, the Nile has a branch (Fig. 4.15, top left panel) which points in the south-westerly direction before it splits into two further branches: one branch continues mostly westward, then north-westward, and stops near a location where an annotation appears (close to the central vertical seam in the map in Fig. 4.15, top left panel). The annotation (see @7 in the Appendix) says that this river appears then disappears three times at about two hundred *li* (one *li* is about ½ km) distance apart. Indeed, another river, labelled 入蠟河 (Ru La He; Item 158: River Ghir), appears to the southwest of that disappearance, and its course continues, through several other branch points, until reaching mountains in the territory of 馬邏可國 (Ma Luo Ke Guo; Item 166; Kingdom of Morocco), which are probably the Atlas Mountains in northwest Africa. In short, on the Africa-KWQ, the River Ghir does not seem to relate to the Nile, nor does it flow into the North Atlantic Ocean.

The Africa-Mercator (1569) depicts a similar Nile branch (with the name *Gtras* or *Giras flu*) as that on the Africa-KWQ, but this branch is connected with a few lakes and splits into one or more branches before it is at a location where the river disappears and another river starts on the other end of the mountain (later it is named *Ghir flu*); this river flows through several complicated branch points until reaching mountains in 馬邏可國 (Ma Luo Ke Guo; Item 166; Kingdom of Morocco); one branch flows into a lake, one branch flows into the North Atlantic Ocean, and the rest disappear in the desert or mountains. On the Africa-Mercator (1569), River Ghir does not seem to relate to the Nile but does flow into the North Atlantic Ocean.

The Africa-Ortelius (1570; Fig. 4.15, lower left panel) shows no branch of the Nile flowing across northern Africa, and River Ghir is a much shorter southward flowing river (in what is now Algeria) and stops in a lake, together with many other rivers in north and northwestern Africa flowing into the desert or the Mediterranean Sea or the North Atlantic Ocean.

The Africa-Plancius (1594; Fig. 4.15, lower right panel) shows a branch of the Nile in the southwesterly direction originating from some mountains. There is an unnamed river (it has a few branching points) which originates from other mountains and flows into an unnamed lake (presumably Lake Borno) near *Borno*. This river has two branches (or perhaps three) and passes through the Tropic of Cancer. River Ghir lies to the west of this unnamed river and flows into an unnamed lake, not the North Atlantic Ocean. In short, River Ghir is not connected with the Nile and does not flow into the Atlantic Ocean.

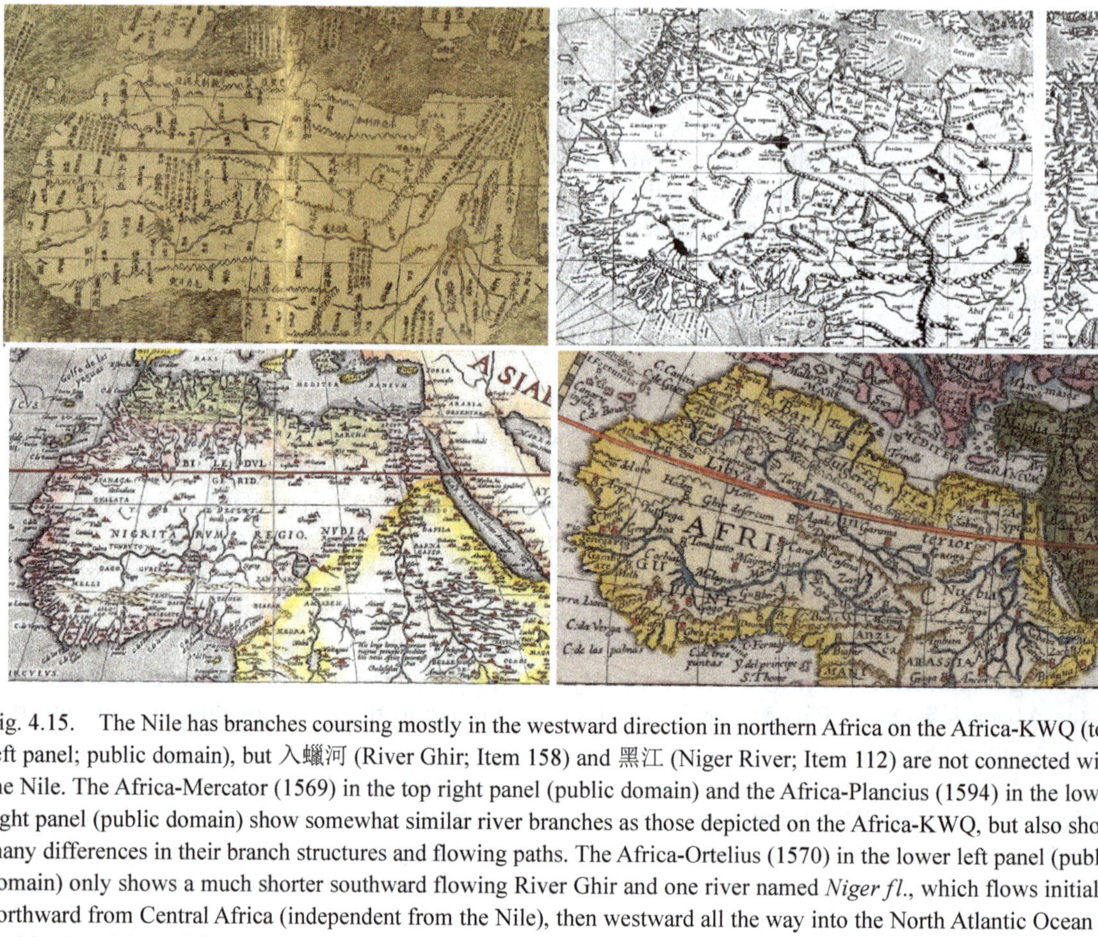

Fig. 4.15. The Nile has branches coursing mostly in the westward direction in northern Africa on the Africa-KWQ (top left panel; public domain), but 入蠟河 (River Ghir; Item 158) and 黑江 (Niger River; Item 112) are not connected with the Nile. The Africa-Mercator (1569) in the top right panel (public domain) and the Africa-Plancius (1594) in the lower right panel (public domain) show somewhat similar river branches as those depicted on the Africa-KWQ, but also show many differences in their branch structures and flowing paths. The Africa-Ortelius (1570) in the lower left panel (public domain) only shows a much shorter southward flowing River Ghir and one river named *Niger fl.*, which flows initially northward from Central Africa (independent from the Nile), then westward all the way into the North Atlantic Ocean in a wide river delta. All figures are extracted from the maps shown in Figs. 4.9 and 4.10 (public domain).

In summary, the river and lake systems of these four maps share some similarities, but also show many differences. All these cartographers may imagine rivers that may or may not be connected and are generally wrong in comparison with today's map of Africa. We will see more of their mistakes in the process of making progress in enhancing our world knowledge about the river and lake systems in Africa, starting from the earlier Chinese to the later European explorers.

(2) Niger River

On the Africa-KWQ, the other branch of the Nile appears to be oriented first south-westerly, and then mostly westward, but it seems to terminate, without any annotation, close to the central vertical seam in the map (Fig. 4.15, top left panel). Then a long river named 黑江 (Hei Jiang; Item 112; the Niger River,[64] today's third-longest river in Africa, exceeded only by the Nile and the Congo River/Zaïre River) appears near that central vertical seam and flows westward and ends in the

Fig. 4.16. This is a modern-day map of the Niger River, with national boundaries included (see endnote 67, public domain).

North Atlantic Ocean as three branches including the Senegal River. This course is very different from today's Niger River shown in Fig. 4.16: it takes its source near the Atlantic in the west, flows eastward into Central Africa and ends up there in the Gulf of Guinea of the Atlantic Ocean; it never connects with the Nile.

On the Africa-Mercator (1569; Fig. 4.15, top right panel), there is also a second Nile branch flowing in the southwest direction, then disconnected from the Nile by a mountain range, but an unnamed river (presumably the Niger River) appears to flow in a different direction from today, ending up (as three unnamed branches) in the Atlantic in western Africa, rather than in the Gulf of Guinea.

But the Africa-Ortelius (1570) does show a river named *Niger fl.*, which flows initially westward from Borno lacus (presumably Lake Chad) in Central Africa, and it is independent from the Nile. The lake is the remnant of a former inland sea, paleolake Mega-Chad.[65] The Africa-Ortelius (1570) then shows the Niger continuing to flow westward into Lake Guber, and on to western Africa to discharge in a wide river delta into the North Atlantic Ocean at four estuaries. Lake Guber is also known as the Great Lake or Lake Débo.[66] Indeed, today's Niger receives water from Central Africa through tributaries, but then veers southward into the Gulf of Guinea.

On the Africa-Plancius (1594; Fig. 4.15, lower right panel), there is an unnamed long river, disconnected from the Nile, flowing across northern Africa in an east–west direction. It starts from the mountains in central Africa and flows like the 黑江/Niger River depicted on the Africa-KWQ; finally, it discharges into the North Atlantic Ocean at three estuaries in western Africa. But unlike the Africa-Ortelius (1570), on this map, the Niger River is not connected with Lake Chad/Lake Borno.

The river and lake systems of the Africa-Ortelius (1570) are least correlated with those on the Africa-KWQ, and it is the only map that depicts the name of *Lacus Guber* (Lake Guber; also known as the Great Lake or Lake Débo) among the three major sixteenth century European maps. This lake is transliterated as 呉巴湖 (Wu Ba Hu; Item 142; Lake Guber) on the Africa-KWQ. The Africa-KWQ, the Africa-Mercator (1569) and the Africa-Plancius (1594) seem to have been influenced by the early Arab geographers who believed the upper Senegal River and the upper Niger

River were connected and formed a single river flowing from east to west, which they called the "Western Nile". This is incorrect, but some of the headwaters of the Senegal River are in fact near the Niger River in Mali and Guinea.

In short, all these four maps have serious errors and great uncertainties in their depiction of the interior rivers and lakes of Africa. This indicates the difficulty of obtaining accurate and complete information about Africa's interiors during that time, and that the Europeans, despite being closer to Africa and coming a hundred years later, still made mistakes.

A summary of Item 5

In summary, the analysis in Items (1) to (5) clearly provides strong evidence to support the important conclusion that the Africa-KWQ is of Chinese origin, and not a direct or adapted copy of the three major sixteenth century European maps.

My next goal is to find the date of the Chinese source map used by Matteo Ricci to make the Africa-KWQ. This can be done by analysing the political landscape revealed by this map as I did for the European portion of the KWQ — the Europe-KWQ.[68] Once this is done, we will know the role played by the Ming mariners during their last voyage in the 1430s and their contribution to the making of the Africa-KWQ. My analysis is presented in Section 7.

7. The Political Era Revealed by the Africa-KWQ

There are several key kingdoms or empires on the Africa-KWQ, which can reveal the political landscape of Africa, hence help to determine the date when Africa on the Africa-KWQ was explored. Here are the key facts:

(1) 馬拿莫大巴 (Item 17; Monomotapa or Kingdom of Mutapa) is on the Africa-KWQ.

This kingdom was established in c. 1430.[69] Although the Portuguese entered direct relations with the kingdom in the 1560s, Monomotapa/Kingdom of Mutapa is not depicted on any of the three major sixteenth century European maps. There was only a sixteenth century Portuguese map of Monomotapa lying in the interior of southern Africa[70]; the map was from the late 1500s/early 1600s. However, this map does not correctly depict the southernmost tip of Africa and contains only the southern portion of the African continent with limited geographical information. Moreover the date of the map is too close to 1602 to be in the hands of Matteo Ricci, making it less likely to be the source map of the Africa-KWQ. Hence, this geographical item and its location on the Africa-KWQ can only come from the Chinese source. *This is the most important geographical item missing on the three major sixteenth century European maps when comparing them with the Africa-KWQ.*

Zheng He's seventh voyage to the Western Ocean was ordered by the Xuande Emperor in 1430. The complete navigational routes and timelines have been worked out in Chapter 3 of my book titled *The Last Journey of the San Bao Eunuch, Admiral Zheng He* by extracting navigational information from Luo Maodeng's 1597 epic work titled *San Bao Tai Jian Xi Yang Gi* 《三宝太监西洋记》 or *An Account of the Western World Voyage of the San Bao Eunuch*. The earliest date for the Ming mariners to arrive at Africa has been extracted to be in 1433[71] and will be further explained here.

We may be curious about who sent these mariners to Africa. There are two possibilities: (1) Assistant Envoy Hong Bao (洪保) ordered a squadron to sail to Africa to explore the African continent soon after the fleet he led arrived at Calicut (in today's India) on 10 December 1432[72]; or (2) the fleet led by Principal Envoy Zheng He (郑和) received information about the Kingdom of Mutapa after his arrival at Mogadishu (in today's Somalia in eastern Africa) in January or February 1433.[73] I shall soon discuss which event is more likely the case.

Both fleets must have a map of Africa on hand based on their previous voyages (according to the historical records, the Ming Treasure Fleets reached Africa in their fifth through seventh voyages),[74] and now they could update their map with the latest information. This means not only that the geographical location of the Kingdom of Mutapa can only come from the Chinese source, but more precisely, from the Treasure Fleet's seventh (and last) voyage, due to the date of this newly established kingdom in c. 1430.

After this last voyage, the Ming rulers no longer sent sea-faring ships to the rest of the world; hence the date of the Ming mariners' last visit of Africa determines the date of the political landscape reflected in the Chinese source map of the Africa-KWQ and vice versa.

Moreover, the annotation (@1 in the Appendix) next to 馬拿莫大巴 (Item 17; Monomotapa/ Kingdom of Mutapa) about qilin — a gentle, kind and benevolent creature which carries a mystical good omen in Chinese mythology — reveals and signifies the unique Chinese passion about this creature. This kind of passion was absent from Europeans' minds.

The political era of the source map of the Africa-KWQ is determined by the political landscape of Africa as reflected by that map, and it was a period between the moment the Ming mariners arrived in Africa to explore the continent in early 1433 and the moment they left that continent. Since they had only a few months to stay in Africa before returning to China in the same year, the political era revealed by the Africa-KWQ should be 1433. This date is also consistent with the interpretation of an annotation on the Europe-KWQ, which I have discussed near the end of Section 3 in Chapter 1, and which will be further explained in this chapter.

(2) The year 1433 overlaps with the eras of the political entities depicted on the Africa-KWQ.

In the year of 1433, the following 17 political entities existed in Africa and they are properly depicted on the Africa-KWQ: 馬泥工哥 (Item 42; Kingdom of Kongo; 1390–1940), 亞毘心域 (Item 55; Abyssinia; 1270–1974), 下黑地阿皮亞 (Item 60; Ethiopia Inferiore; 1270–1974), 訝德兒 (Item 63; Kingdom of Adal; 1415–1577), 黑地兀皮亞 (Item 68; Ethiopia; 1270–1974), 巴皮羅泥亞 (Item 71; Babylon in Egypt; c. fourth century–fifteenth century), 皮亞法尒 (Item 88; Biafar; from the fifteenth to the nineteenth centuries), 必寧 (Item 94; Kingdom of Benin, also known as the Kingdom of Edo; 1180–1897), 道馬 (Item 95; Dauma; a dependency of the Kingdom of Allada; c. 1100–1724; a coastal western African kingdom in what is now southern Benin), 墨利國 (Item 100; Mali Empire, also known as the Kingdom of Melli; c. 1230–1670), 鑾定曷 (Item 109; 1235–1901), 側側 (Item 111; Zegzeg; the kingdom existed since the eleventh century), 息匿瓦國 (Item

118; Jolof Empire of Senegal; 1350–1549), 怒皮亞 (Item 133; the last Nubia kingdom collapsed in 1504), 亞入新巴 (Item 143; Agisymba; a kingdom already mentioned by Ptolemy in second A.D., which was near today's Central African Republic), 堵泥素 (Item 156; Tunis; in the early fifteenth century, Tunis was ruled by Arabic-speaking Muslims) and 馬邏可國 (Item 166; Morocco; 788–present).

(3) The Songhai Empire (c. 1464–1591) does not appear on the Africa-KWQ.
The Songhai Empire (c. 1464–1591) does not appear on the Africa-KWQ, because it was not yet established. This points to the fact that the date of the Africa-KWQ must be earlier than c. 1464. The year 1433 is before the year 1464 and is consistent with the above fact. Bornu/Borno is also not depicted on the Africa-KWQ, but for a different reason: Bornu/Borno was in turmoil and unstable during the first three-quarters of the fifteenth century.

(4) There is a long annotation near the location of Spain on KWQ.
There is a long annotation near the location of Spain on KWQ, which may be related to the date when KWQ was drawn:

> 此歐邏巴州有三十 餘國皆用前王政法一切異端不從而獨崇奉天主上帝聖教凡官有 三品其上主興教化其次判理俗事其下專治兵戎土產五穀五金百果酒以葡萄汁為之工皆精巧天文性理無不通曉俗敦實重五倫物匯甚盛君臣康富四時與外國相通客商遊遍天下去中國八萬里自古不通今相通近七十餘載雲
>
> My translation of the above passage and my comments (between square brackets) are as follows:
>
> The European continent has more than 30 countries. They all implement political and legal systems of their previous kings. They do not follow any heretics but revere only the Catholic God. All their government officials are categorised into three ranks: the top rank must address education and enlightenment; the next rank must deal with justice and mundane affairs; and the lowest rank must specialise in military affairs. The local products include five kinds of grain, five kinds of metal, 100 kinds of fruit, and wines made of grapes. Their craftsmanship is very elaborate, and they thoroughly master astronomy and all theories. Their cultural customs are honest, and they care about human relationships. Their mercantile exchanges are prosperous, and the countries are affluent. They communicate with other countries all the time and their people and merchants travel all over the world. The continent is about 80,000 *li* [1 *li* is about ½ km] away from China. Since ancient times, the place had no interaction with China. Now they have had diplomatic relations [with China] for more than 70 years and less than 80 years.

Siu-Leung Lee (李兆良) suggests[75] that the last sentence of this annotation determines the date of the Chinese source map of KWQ to be after Zheng He's sixth voyage but before his seventh voyage (during the Ming era) started in 1430. There are two problematic issues in his arguments: (1) in Section 3 of Chapter 1, I have shown that the correct interpretation of the above annotation in fact gives the date when KWQ was drawn between 1428 and 1433, not between 1428 and 1430; there is no reason to narrow the date range by three years to exclude 1433, especially given that the Treasure Fleets did visit Africa during their seventh (and last) voyage in the 1430s; and (2) we should not confuse the date when the map was drawn with the date when the places were explored or the date when the geographical data were collected. For example, the map of Europe on KWQ reveals a political era between 1157 and 1166, implying a much earlier date than the Zheng He era[76] when the region was explored by the Chinese, although the map of the European region on KWQ was drawn between 1428 and 1433 according to the above annotation. Hence, this annotation gives the date when the map was drawn, rather than the dates when the different regions were explored.

After I fully analyse the Africa-KWQ in this chapter, it becomes clear that there is strong evidence to show that the Ming mariners must have had the time and opportunity in 1433 to remap Africa and to depict the Kingdom of Mutapa on their African map to reflect the political landscape of Africa of that year.

I mentioned earlier in this section that there were at least two possibilities for this remapping of Africa to take place: (1) Assistant Envoy Hong Bao ordered a squadron to sail to Africa to explore the continent soon after the fleet he led arrived in Calicut on 10 December 1432[77]; and (2) the fleet led by Principal Envoy Zheng He received information about the Kingdom of Mutapa after their arrival at Mogadishu in January or February 1433[78] and updated their map of Africa; however, his fleet may have only briefly visited Africa with no time to remap the rest of the continent, because they later sailed non-stop from Mecca to North America by rounding the southernmost tip of Africa, according to my previous analysis.[79] Hence, this second possibility was less likely to happen. Besides, there is still the major mystery surrounding whether Zheng He himself returned to China, or stayed in North America, or his fleet was tragically destroyed in New Zealand by a meteorite hitting the Tasman Sea.[80]

This leaves the more plausible case (1) for the Ming mariners to reexplore Africa in 1433. It would only take a little over 20 days and nights with tailwind,[81] based on similar sailing distances to sail from Calicut to Mogadishu (2,134 nautical miles/nm; 1 nm = 1.15 mile) vs. from Kollam (old name Quilon) to Mogadishu (2,073 nm). The squadron sent out by Hong Bao could reach the east coast of Africa in early 1433 to remap Africa and record the location and the founding of the Kingdom of Mutapa, etc., while they conducted their diplomacy in Africa. Afterwards, these mariners would rejoin Hong Bao and Wang Jinghong (王景弘; the other Principal Envoy) at Calicut on 9 April 1433, and all the squadrons would be led by Wang Jinghong back to China (they arrived at Nanjing in July 1433) with their updated map of Africa.

8. Conclusions

In Section 2 of this chapter, I have shown that Chinese mastered marine navigation and drew maps of Africa since Antiquity. In Section 3, I have shown that, by the time Admiral Zheng He set sail to the Western Ocean in the early fifteenth century, his seven voyages (1405 to mid-1430s) were unmatched in world history. But the Ming rulers halted these epic expeditions soon after the seventh voyage ended to focus on domestic needs and to secure the country's inland borders. They either burned the great fleet in the harbours or allowed it to rot. Most of the valuable navigational records and maps vanished in fire or were lost in history.

However, evidence given in Section 4 shows that some of the Chinese maps and records must have gone to the Europeans for them to draw maps of the entire continent of Africa and ignited their zeal to find a new direct route from the Atlantic Ocean to the Indian Ocean and China for trade and exploration.

Hence, for nearly 600 years after the abrupt termination of maritime exploration by the Ming rulers, modern historians have been struggling to pick up the pieces from the residual historical records to reconstruct that missing chapter of China's epic maritime navigation. Especially, the extent of Zheng He's sailing is still hotly debated. Even though some historians hold the viewpoint that Zheng He's fleets did not travel beyond the east coast of Africa, others have doubts and are actively seeking evidence or have obtained some evidence to topple that conventional thinking.

Among the various approaches is the reinvestigation, from a fresh angle, of Kunyu Wanguo Quantu《坤輿萬國全圖》or the Complete Geographical Map of All the Kingdoms of the World — published by Matteo Ricci in 1602 in China, abbreviated as KWQ in this book. The KWQ, with Chinese characters, may be considered by some to be crucial in expanding Chinese knowledge of the world, but by others to be a map of Chinese origin, which must have enhanced the European knowledge of the world and laid the foundation of the Age of Discovery. The latter also begin to think that Matteo Ricci's contribution to KWQ may be *mainly editorial work involving mostly changing the Chinese-based names into Chinese transliterated European names to make the resulting map of Africa — abbreviated as the Africa-KWQ — looks more European.* Despite his claim to be the cartographer of the map, there are still 15 Chinese-based geographical items Matteo Ricci could not replace; the many other important facts shown on the map also seem to be telling a different story.

In Section 5, I set up my goal of investigating all the geographical items on the Africa-KWQ and comparing them with those on the three major sixteenth century European maps to resolve the issues. These European maps are abbreviated as: (1) the Africa-Mercator (1569); (2) the Africa-Ortelius (1570) and (3) the Africa-Plancius (1594).

In Section 6, my analyses have shown that the Africa-KWQ is indeed a Chinese-based map, which is based on a thorough examination of the pre-colonial African history and geography presented in the Appendix of this chapter. The key points are: (1) there are 11 detailed annotations of Chinese origin on the Africa-KWQ, which are absent on the three major sixteenth century Europeans maps; they are realistic descriptions of the African geography, natural environment, human races and their habits, animals, plants and natural products on land and seas, and they can only have been witnessed by Chinese explorers who had actually visited these places; (2) among the 185 geographical items depicted on the Africa-KWQ, 15 are of Chinese origin and they are listed in Table 4.1; (3) among the 185 geographical items, 70 (about 38 percent; 70/185) are not depicted on any of the three major sixteenth century European maps; these places are spread all over the African continent as listed in Table 4.2; this implies that the Chinese had basically explored or known about (or both) the whole of Africa when they made the final update of their map (the source map of the Africa-KWQ); (4) the Chinese had correctly located the Cape of Good

Hope and the southernmost tip of Africa and depicted the latter's geographical position with a Chinese-based name 羅經正峰 (today's Cape Agulhas) on the Africa-KWQ; this is a strong indication that they had rounded the southernmost tip of Africa to sail into the Atlantic Ocean on the western side of Africa on their way to the Americas (to be further discussed in Chapter 5); (5) the Africa-KWQ depicts its river and lake systems not only differently from the three major sixteenth century European maps in gross features and in detail, but also differently from those on today's map of Africa. The latter finding indicates that the source map of the Africa-KWQ may imagine rivers that may or may not be connected and are generally wrong.

Then in Section 7, I present my analysis and evidence of the political landscape of the major kingdoms and empires reflected on the Africa-KWQ to conclude that the Chinese source map of the Africa-KWQ reveals the geographical and political condition of Africa in 1433, after the Kingdom of Mutapa was founded in c. 1430. This year overlaps with Zheng He's seventh (and last) voyage to the Western Ocean. The Ming mariners sent by Hong Bao to Africa remapped Africa during their seventh (and last) expedition and updated their map of Africa on hand in the hope of preparing for their next voyage, albeit it was never launched again.

Acknowledgements

The author thanks Ian Hudson and T.C. Bell for reviewing this chapter prior to publication. The author is also grateful to Professor Emeritus Michel A. Van Hove for his critical reading and valuable comments, as well as technical assistance in preparing most of the figures in this chapter in published form. Finally, special thanks go to M. Hayes-Shuptar（华铭叮）for the academic and editorial assistance with the endnotes.

Appendix

This appendix analyses all the geographical items on the Africa-KWQ in nine categories (zones). The Africa-KWQ uses almost entirely Chinese traditional characters; this chapter uses modern Chinese simplified characters in the text except in quoting some ancient Chinese maps and the geographical items depicted on the Africa-KWQ. The non-Chinese-based geographical items are European-based names or African names transliterated by Matteo Ricci into Chinese names. It is my viewpoint that Matteo Ricci most likely has replaced the original Chinese names of the geographical items by the names given by the European explorers to make the Africa-KWQ look more European. But there are still many Chinese-based geographical items and annotations remaining on the map to reveal that the true cartographer of the map must be Chinese. Items preceded by "•" are geographical items which do not appear on the Africa-Mercator (1569), or the Africa-Ortelius (1570), or the Africa-Plancius (1594). The sign "@" means "annotation" and it is numbered in sequence in this Appendix. Chinese pinyin is listed for the pronunciation of each geographical item.

Africa-KWQ				
Item #	**Name**	**Pinyin**	**Etymology**	**Analysis of history, politics, geography; comparison or comment**
Zone I: Southern Africa				

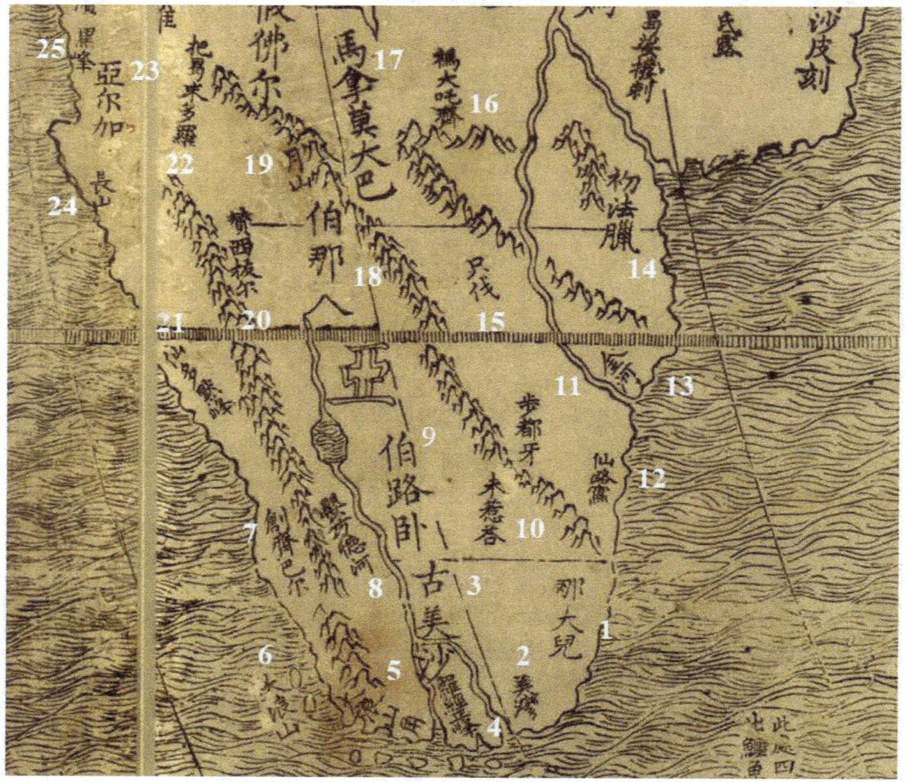

Fig. A4.1. Extracted from Fig. 4.10, right panel, public domain.

1	那大兒	Na Da Er	那大兒 (Na Da Er) is the transliteration of Nadal. On the Africa-Ortelius (1570), it is depicted as *Terra, de nadal*, meaning “the land of Nadal”.	

2	美灣	Mei Wan	In the Chinese name, 美 means "beautiful" and 灣 "bay"; 美灣 means the Beautiful Bay. On the Africa-Ortelius (1570), there is *Baia hermosa y pesqueria*, which means the Beautiful Bay for Fishing.
3	古美沙	Gu Mei Sha	The Chinese name 古美沙 (Gu Mei Sha) is the transliteration of "Cumissa". This place is also depicted on the Africa-Ortelius (1570).
4	羅經正峰	Luo Geng Zheng Feng (here 經 pinyin is the same as 耕 Geng; the normal pinyin is Jing)	The Africa-KWQ correctly depicts 羅經正峰 as the geographic southernmost tip of the African continent, which is today's Cape Agulhas. In the Chinese nomenclature, "羅經" means "compass", "正" means the direction of magnetic north (pointed at by the compass needle) coinciding with true north, and "峰" means "mountain or mountain peak". Cape Agulhas is a rocky headland, like the Cape of Good Hope; it is near where the Atlantic Ocean meets the Indian Ocean and forms the beginning of the dividing line between the Atlantic and Indian Oceans. The sea off Cape Agulhas (and the Cape of Good Hope) is notorious for gigantic waves and winter storms, which can sink even large ships. These geographical conditions are caused by the west-to-east strong winds of the Roaring Forties (areas of latitudes 40°–50° south in the Southern Hemisphere, where the prevailing winds blow persistently throughout the year), and the cold Antarctic Circumpolar Current flowing in the same direction, meeting the warmer Agulhas Current in the region of the Cape.[82] However, the cartographers of the three major sixteenth century European maps did not seem to know where to place Cape Agulhas; they incorrectly placed the southernmost tip of Africa at the Cape of Good Hope. For example, on the Africa-Ortelius (1570), it is depicted as *C. de Buona Speransa*, here *Buona* means "good" and *Speransa* "hope". In 1487, the Portuguese Bartolomeu Dias led an expedition in search of a trade route around the southern tip of Africa. In 1488, on the journey back, he encountered two capes: the Cape of Good Hope, originally named by him as the Cape of Storms (*Cabo das Tormentas*); it was renamed later to the Cape of Good Hope (*Cabo da Boa Esperança*) to celebrate the opening of a route to the east[83]; and the second cape he named *Cabo das Agulhas* (after St. Agulhas), or Cape of Needles (the direction of magnetic north, and therefore the compass needle, coincided with true north in the region). The Africa-Mercator (1569) does not depict Cape Agulhas.

(*Continued*)

(*Continued*)

Item #	Name	Pinyin	Etymology	Analysis of history, politics, geography; comparison or comment
				On the Africa-Ortelius (1570), there is *Costa Dritta*, literally "straight coast", meaning the coast where the direction of magnetic north, pointed at by the compass needle, coincided with true north. But this map fails to show *Costa Dritta* as the geographic southernmost tip of Africa. On the Africa-Plancius (1594), Cape Agulhas is depicted as *C. das Aguilhas*, but it still fails to show it as the geographic southernmost tip of Africa. The only other sixteenth century map to correctly depict Cape Agulhas as the geographic southernmost tip of Africa is the modern geography of the whole of Africa by Italian cartographer, engineer and astronomer Giacomo Gastaldi. But this map was too big for Matteo Ricci to bring with him to China, and the content is also different from the Africa-KWQ.
5	大浪山	Da Lang Shan	Cape of Good Hope; Portuguese: *Cabo da Boa Esperança*; Spanish: *Cabo de Buena Esperanza*	The Cape of Good Hope is a rocky headland on the Atlantic coast of the Cape Peninsula in South Africa. The currents of the Indian Ocean and the Atlantic Ocean meet at the point where the warm-water Agulhas current meets the cold-water Benguela current and turns back on itself. That oceanic meeting point fluctuates between Cape Agulhas and Cape Point.[84] The Cape of Good Hope is not the southernmost tip of Africa, but it is at the southern tip of the Cape Peninsula. The city centre of Cape Town is about 40 km/25 mi to the north of the cape. Cape Agulhas is approximately 150 km/93 mi to the east-southeast of Cape Point and the Cape of Good Hope. See more of the discussion in Item 4.
6	大浪山角	Da Lang Shan Jiao	On the Africa-KWQ, both 大浪山 and 大浪山角 denote the neighbourhood of the Cape of Good Hope. More detailed information can be read from Items 4 and 5.	
7	創其巴尔	Chuang Qi Ba Er	Zanzibar; Arabic: *zanjibār*, which is in turn from Persian *zangbâr*;	創其巴尔 (Chuang Qi Ba Er) is the transliteration of Zanzibar which is an Archipelago in the Indian Ocean, off today's Tanzania's coast. Its major city is also named Zanzibar. But on the Africa-KWQ, it is depicted as a region in southern Africa. A similar

			Zhang means "black" and *bâr* "coast"; it is the "land of the blacks", in reference to the dark skin of the inhabitants.	mistake also appears on the Africa-Ortelius (1570). However, t*he island is correctly depicted off the east coast of Africa on the Africa-Mercator (1569) and the Africa-Plancius (1594). Such a mistake on the Africa-KWQ simply implies the older age of this map compared to these two European maps.* In 1503 or 1504, Zanzibar became part of the Portuguese Empire and remained so for almost two centuries.[85] It was one of the many autonomous city-states along the East African coast for long-distance traders from both the hinterland and the Indian Ocean.[86] Hence, it is more surprising to see the Africa-Ortelius (1570) making a mistake like that of the Africa-KWQ, after the Europeans had known the island for almost 70 years.  Fig. A4.2. Zanzibar's location (see endnote 87, public domain).
8	婴方德河	Ying Fang De He	*Infante, fl.*	The Chinese character 婴 means "infant". 婴方德 is a transliteration of *infant* (infant). 河 means "river"; in Spanish, it is abbreviated as "fl."
9	伯路卧	Bo Lu Wo	Belugaras	Kingdom of Belugaras
• 10	未惹荅	Wei Re Da	Ueigeta[88]	

(*Continued*)

(*Continued*)

Item #	Name	Pinyin	Etymology	Analysis of history, politics, geography; comparison or comment
● 11	步都牙	Bu Du Ya	Putuia[89]	
12	仙路寫	Xian Lu Xie	*P. de S. Lucia Praia*	仙路寫 (Xian Lu Xie) is a transliteration of "S. Lucia". The place is close to the Indian Ocean. "Praia" means "beach".
13	金河	Jin He	*Fl: del oro*	"Oro" means "gold". 金 also means "gold". *Fl: del oro* means "River of Gold." On the Africa-KWQ, 金河 flows into the Indian Ocean and is one of the branches of 泥羅河泉 (Item 35). On the Africa-Ortelius (1570), *Fl: del oro* is an independent river flowing into the Indian Ocean. However, neither the Africa-Mercator (1569) nor the Africa-Plancius (1594) depicts a river named *Fl: del oro.*
14	初法蠟	Chu Fa La	Cefala	The place was a town on the eastern coast of southern Africa facing the Indian Ocean.[90]
● 15	只伐	Zhi Fa	Scioa[91]	
16	禡大吒齊	Ma Da Zha Qi	Maitachasi[92]	
● 17	馬拿莫大巴	Ma Na Mo Da Ba	Portuguese: *Monomotapa*; it is a transliteration of the African royal title *Mwenemutap.* Later, this monarch's royal title was applied by the Portuguese to the kingdom.[93]	馬拿莫大巴 (Ma Na Mo Da Ba) is the transliteration of Monomotapa (the Kingdom of Mutapa; c. 1430–1760)[94] or its ruler's title *Mwenemutapa*. The kingdom was established around 1430 by the Prince from the Great Zimbabwe[95] (an ancient city in the southeastern hills of Zimbabwe). Portuguese explorer and spy Pêro da Covilhã (c. 1460–after 1526) was the first European known to have visited Sofala (the chief seaport of the Kingdom of Mutapa) in 1489 and to know the Kingdom of Mutapa. The Portuguese entered direct relations with the kingdom in the 1560s.[96] The Africa-KWQ uniquely depicts the existence of the Kingdom of Mutapa to reflect the political landscape of Africa in 1433, whereas surprisingly none of the three major sixteenth century European maps depicts this kingdom. This finding further supports that the Africa-KWQ was drawn by the Ming mariners in early 1433 when they explored Africa during their seventh voyage.

@1	馬拿莫有獸首似馬額上有角皮極厚遍身皆鱗其足尾如牛疑麟雲。(This annotation is shown in Fig. A4.3.)	Ma Na Mo You Shou Shou Si Ma E Shang You Jiao Pi Ji Hou Bian Shen Jie Lin Qi Zu Wei Ru Niu Yi Lin Yun	My translation and comments (between square brackets) are as follows: In Monomo [tapa] there is a kind of beast whose head is like that of a horse but with horns on the forehead. They have thick skins which are covered with a sheet of scales [or scale-like patterns]; their feet and tails are like oxen. There is a suspicion that they may be qilins. The qilin's modern name is giraffe. Giraffes occur across east and southern Africa. The Kingdom of Mutapa is in southern Africa. Today's Namibia and South Africa have some of the largest populations of giraffes. On the fourth voyage (1413–1415), Zheng He met in Bengal (in today's India) with envoys from Melindi (now part of Kenya). These men had brought with them giraffes as a tribute, and they gave one of those giraffes to the Chinese who took it home.[97] In Chinese mythology, qilin is a gentle, kind and benevolent creature which carries a mystical good omen. This annotation is absent on any of the three major sixteenth century European maps. It signifies the unique connection of the Africa-KWQ with the Chinese mythology, hence, the Chinese origin of the Africa-KWQ.	
• 18	伯那人	Bo Na Ren	*Péna, uomini*[98]	The meaning of "uomini" is "man". 伯那人 (Bo Na Ren) is the transliteration of "Péna man".
• 19	月山	Yue Shan	*Luna, monte della*[99]	"*Luna*" means "moon", and "monte" means "mountain". 月 means "moon" and 山, "mountain"; 月山 (Yue Shan) is also Moon Mountain.
20	贊西拔爾	Zan Xi Ba Er	This is a repetition of 創其巴尔 (Chung Qi Ba Er; Item 7, Zanzibar). On the Africa-KWQ, 贊西板爾 should be corrected as 贊西拔爾.[100] None of the three major sixteenth century European maps shows Zanzibar twice. The location of the second Zanzibar (贊西拔爾; Item 20) on the Africa-KWQ is still incorrect.	
21	仙多默峰	Xian Duo Mo Feng	*G di S Tomo*	仙多默 (Xian Duo Mo) is the transliteration of *Saint Tomo*; 峰 means "mountain peak". "Di" means "of". 仙多默峰 (Xian Duo Mo Feng) means the Saint Tomo Mountain Peak.

(*Continued*)

(Continued)

Item #	Name	Pinyin	Etymology	Analysis of history, politics, geography; comparison or comment
22	把曷末多羅	Ba He Mo Duo Luo	Bagamidri	把曷末多羅 (Ba He Mo Duo Luo) is the transliteration of Bagamidri, which was said to be a large and fertile kingdom, watered by many rivers; silver was found in its mountains.[101] On the Africa-KWQ, there are mountains near 把曷末多羅 (Bagamidri; Item 22).
23	亞尔加	Ya Er Jia	*C. de Arcas*	Arcas is facing the Atlantic Ocean.
• 24	長山	Zhang Shan	*Montagna, grande*[102]	"Montagna, grande" means "great mountain". 長山 means Long Mountain, implying a great mountain.
25	黑峰	Hei Feng	*C. Negro*	"*C. Negro*" means "Black Mountain Peak". 黑 means "black" and 峰 "mountain or mountain peak". 黑峰 means Black Mountain or Black Mountain Peak.

Zone II: Area between the Equator and southern Africa

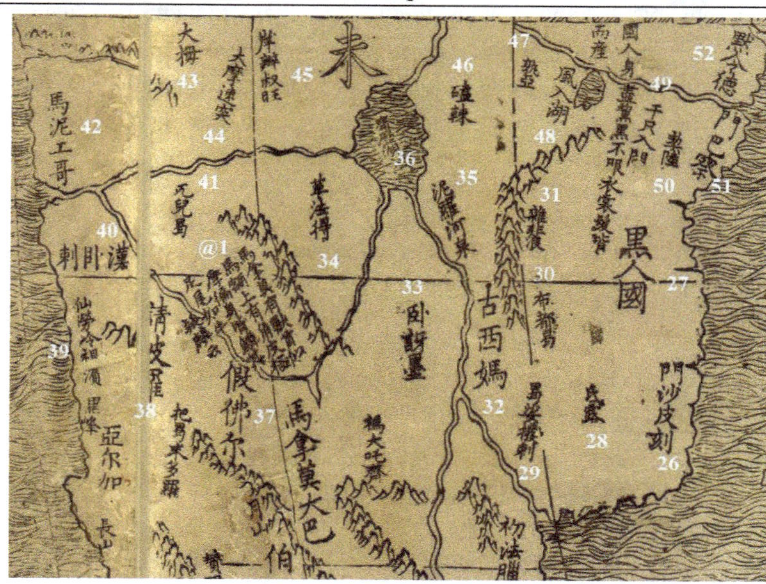

Fig. A4.3. Extracted from Fig. 4.10, right panel, public domain.

26	門沙皮刻	Men Sha Pi Ke	Mozambique; Portugese: *Moçambique*	門沙皮刻 (Men Sha Pi Ke) is the transliteration of Mozambique. The country was named Moçambique by the Portuguese after the small Island of Mozambique, which is located on the southeast coast of Africa[103]; a town was founded there no later than the fourteenth century.[104] This name was subsequently applied to the mainland country, and the island was renamed *Ilha de Moçambique* (Island of Mozambique). Portuguese explorer Vasco da Gama (c. 1460–1526) reached the coast of Mozambique in 1498,[105] and Portuguese Mozambique lasted from 1498 to 1975. Many coastal towns became regular ports of call on the new European sea route to the east.[106]
• 27	黑人国	Hei Ren Guo	*Negrita Rum Regio*[107]: literally means "Kingdom of People with Black-coloured Skin".	黑人國 (Hei Ren Guo) was the Kingdom of People with Black-coloured Skin. It was in the region corresponding to today's Tanzania. 黑 means "black", 人, "people" and 國, "kingdom". "Tanzania" was created as a clipped compound of the names of the two states that unified to create the country: Tanganyika and Zanzibar (combining "Tan" and "Zan" and the suffix, "ia" to form Tanzania). "Zanzibar" comes from "Zengi", the name for a local people (said to mean having "black-coloured skin").
@2	此地倶近日故國人身盡黧黑不服衣裳髮皆捲短土不產鐵而產金銀象牙犀角寶貝之類		Ci Di Ju Jin Ri Gu Guo Ren Shen Jin Li Hei Bu Fu Yi Chang Fa Jie Juan Duan Tu Bu Chan Tie Er Chan Jin Yin Xiang Ya Xi Jiao Bao Bei Zhi Lei	My translation and comment (between square brackets) are as follows: This region is close to the Sun [near the Equator], hence, the country's inhabitants have dark-black skins. They do not wear clothes and their hair is short and curly. Their lands do not produce iron, but produce gold, silver, ivory, rhino horns and other valuable objects which can be regarded as treasures.
28	氐露	Di Lu	Tirut	Kingdom/City-state of Tirut
29	曷娑撥剌	He Suo Bo La	*Casabele*	Kingdom/City-state of Casabele
• 30	布都曷	Bu Dou He	*Putuho*[108]	

(*Continued*)

(*Continued*)

Item #	Name	Pinyin	Etymology	Analysis of history, politics, geography; comparison or comment
31	雜非浪	Za Fei Long	*Gefala*	
• 32	古西媽	Gu Xi Ma	*Cusima*[109]	
33	卧訝墨	Wo Ya Mo	*Goyame*	Kingdom of Goyame
34	革法得	Ge Fa De	*Cafates*	革法得 (Ge Fa De) is the transliteration of Cafates. On the Africa-KWQ, 革法得 (Cafates) is depicted at the southwest side of 齊歷湖 (Lake of Cichlids; today's Lake Victoria). But on the Africa-Ortelius (1570) and the Africa-Plancius (1594), Cafates is a region located at the northeast or eastern side of a large unnamed lake (most likely today's Lake Victoria). The Africa-Mercator (1569) does not depict Cafates. The Africa-KWQ and the three major sixteenth century maps all place Lake Victoria south of the Equator; actually, Lake Victoria touches the Equator on its northern side.
• 35	泥羅河泉	Ni Luo He Quan	Source or spring of the Nile: Italian: *Sorgente del Nilo*	泥羅河 means the Nile, and 泉, "source or spring". 泥羅河泉 (Ni Luo He Quan) on the Africa-KWQ is designated as the source of today's White Nile, but incorrectly flows into the Indian Ocean. There have been many controversies about the remotest source of the Nile. Beginning in the mid-1800s, Lake Victoria was largely accepted as the source of the White Nile, since British Captain John Hanning Speke (1827–1864)[110] was the first European to see Lake Victoria. Speke renamed the lake after British Queen Victoria in 1858 and mapped it. But the ultimate source of the Nile is still being sought today. The Kagera River forms part of the upper headwaters of the Nile and carries water from its most distant source, but it is unknown which of the two feeder rivers to the Kagera River — Nyabarongo of Rwanda and Ruvubu of Burundi — is the longer. Hence, explorers continue to push farther to find the Nile's "true" source.[111] On the Africa-KWQ, the source of White Nile is depicted specifically as 泥羅河泉, hence, 齊歷湖 (Lake of Cichlids) would naturally correspond to today's Lake Victoria.

<table>
<tr>
<td>• 36</td>
<td>齊歷湖</td>
<td>Qi Li Hu</td>
<td>Lake of Cichlids</td>
<td>Cichlids (齊歷; Qi Li)[112] are fish from the family Cichlidae.[113] Nearly all African cichlids live in one of the three great lakes in East Africa: Lake Malawi, Lake Victoria and Lake Tanganyika.
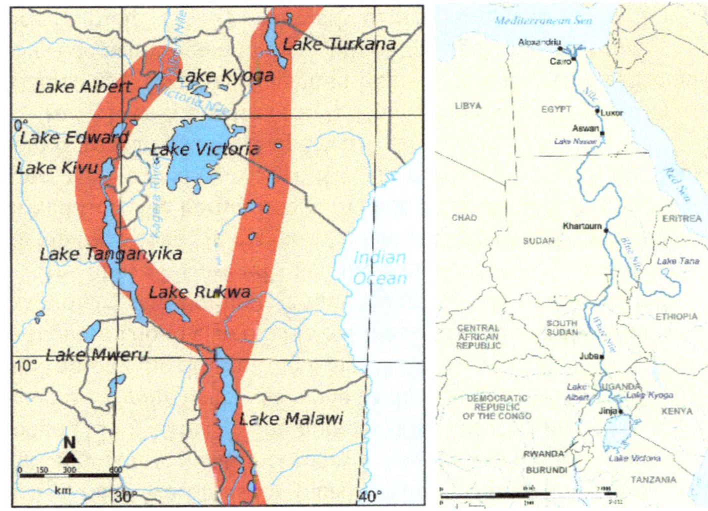

Fig. A4.4. Left panel: Lake Victoria and the Great Rift Valley (see endnote 114, public domain); right panel: today's Nile flowing from south to north through eastern Africa; it begins in the rivers that flow into Lake Victoria (located in today's Uganda, Tanzania and Kenya), and empties into the Mediterranean Sea (see endnote 115, public domain).

On the Africa-KWQ, the source of the White Nile is depicted as 齊歷湖 (Qi Li Hu) which is the transliteration of Lake of Cichlids (today's Lake Victoria). However, today's Congo/Kongo River (its past name was Zaire River),[116] is depicted on the Africa-KWQ as an unnamed river flowing through the entire Kingdom of Congo, and incorrectly connecting with the western and southern sides of 齊歷湖 (Lake of Cichlids). The river's sources are in the highlands and mountains of the East African Rift, as is the case for Lake</td>
</tr>
</table>

(Continued)

(*Continued*)

Item #	Name	Pinyin	Etymology	Analysis of history, politics, geography; comparison or comment
				Tanganyika and Lake Mweru. These lakes feed the Lualaba River, which then becomes the Congo River/Zaire River below Boyoma Falls.[117] The Chambeshi River (the longest tributary) in Zambia is generally taken as the source of the Congo/Kongo River. More detailed discussions are in the main text: Subsection (1) of Item 5 in Section 6.
• 37	假佛尔	Jia Fu Er	Cafri[118]	
38	請波羅	Qin Bo Luo	Zimbro	Kingdom/City-state of Zimbro
• 39	仙勞冷祖濱	Xian Lao Leng Zu Bin	*Lorenzo. Spiaggia di S.*[119]	"Spiaggia" means "beach". 仙勞冷祖濱 (Xian Lao Leng Zu Bin) is the transliteration of San Lorenzo Beach.
40	漢卧剌	Han Wo La	Angola or Ngola	漢卧剌 (Han Wo La) is the transliteration of Angola or Ngola. The first large political entity in today's Angola was the Kingdom of Kongo (1390–1914).[120] The founder (king) of the kingdom was called the *Ngola*, hence the name of the country.[121] "Angola" was listed among the titles of the King of Kongo in 1535. The Portuguese Diogo Cão (c. 1452–1486) explored in 1484 what today is Angola[122]; and Portuguese founded the colony of Angola in 1575. The above history tells that the Kingdom of Angola did not exist in the Zheng He era, nor the name "Manicongo" (which was derived from the Portuguese in the late fifteenth century). Both names were added by Matteo Ricci to make the Africa-KWQ look more European and more updated. The Africa-Mercator (1569) depicts Angola to the south of "Manicongo" (explained again in Item 42) and the Africa-Plancius (1594) depicts Angola to the south of "Congo", whereas the Africa-Ortelius (1570) incorrectly depicts Angola to the north of "Manicongo".
• 41	兀兒曷	Wu Er He	Urho[123]	

42	馬泥工哥	Ma Ni Gong Ge	Portuguese *Manicongo* was derived from Mwene-Kongo which was the title of the ruler of the Kingdom of Kongo.[124] The ruler of Kongo was the most powerful Mwene in the region and whom the Portuguese regarded as the king upon their arrival in 1483. This title was later used to denote the entire kingdom.[125]	馬泥工哥 (Ma Ni Gong Ge) is the transliteration of Manicongo. The Kingdom of Kongo (1390–1914)[126] was located on the western coast of central Africa and south of the Congo/Kongo River (formerly known as the Zaire River).  Fig. A4.5. The 1711 Kingdom of Kongo/Congo (see endnote 127, public domain), and the Kingdom of Angola to its south. The Kingdom of Kongo was ruled by a king called Mwene-Kongo, meaning ruler of Kongo, but the Portuguese colonialists could not pronounce the name, so it became Manicongo. The Portuguese explorer Diogo Cão was the first European to discover the unchartered Congo River/Zaire River in 1482 and visit the Kingdom of Kongo, when the Kingdom was already a highly developed state at the centre of an extensive trading network.[128] It is Matteo Ricci who altered the Chinese nomenclature of the Kingdom of Kongo into Manicongo. The reason is explained in detail in Item 40 (漢卧剌/Angola).
43	大拇	Da Mu	Kingdom/City-state of Damut	

(*Continued*)

(*Continued*)

Item #	Name	Pinyin	Etymology	Analysis of history, politics, geography; comparison or comment
44	大摩速突	Da Mo Su Tu	This may be the city of Damut (?)[129]	
• 45	岸辨叔旺	An Ban Shu Wang	Ambiancativa[130]	
46	磕辣	Ke La	Cotla[131]	
47	孰亞	Shu Ya	Azuga	
48	風入湖	Feng Ru Hu	Lake Fungi	
• 49	干只入悶	Gan Zhi Ru Men	Canzegemen[132]	
50	契陸	Qi Lu	Quiloa[133]	It is a coastal city in today's southeast Kenya along the Indian Ocean.
51	門巴察	Men Ba Cha	Mombasa	It is also a coastal city in today's southeast Kenya along the Indian Ocean.
52	默令德	Mo Ling De	Malinde	默令德 (Mo Ling De) is the transliteration of Melinde which is the old name of the port city Malindi (near Mombasa) before 1498. This old name is depicted on the Africa-KWQ, which lends further support that the map was drawn before the Portuguese arrived. This city was on the eastern coast of Kenya and part of the Indian Ocean trade networks. The expeditions made by Zheng He's fleets to Africa during their fifth, sixth and seventh voyages basically followed long-established Arabic–Chinese trade routes, evidenced by Song Dynasty porcelains and copper coins excavated at Swahili burial sites.[134] On the fourth voyage (1413–1415), in 1414, Zheng He met in Bengal (in modern-day India) with envoys from Melinde (now part of Kenya). These men had brought with them as tribute giraffes,[135] and they gave one of those giraffes to the Chinese, who took it home. In recent years, the People's Republic of China took interest in Admiral Zheng He's exploration and began sponsoring archaeological digs of sites in places like modern Malindi. The discoveries confirm that the date of the first international trade with East Africa is decades before the arrival there of Vasco da Gama.

Zone III: Eastern and northeastern Africa (north of the Equator)				
 Fig. A4.6. Extracted from Fig. 4.10, right panel, public domain.				
53	砑麻蝋	Ya Ma La	Amara[136]	The Kingdom of Amara was for the most part a plain without mountains and was very fertile.[137]

(*Continued*)

(Continued)

Item #	Name	Pinyin	Etymology	Analysis of history, politics, geography; comparison or comment
● 54	黑湖	Hei Hu	Lake Tana	Lake Tana is the largest lake in Ethiopia and the source of the Blue Nile.[138] Spanish priest Pedro Paez (1564–1622) was the first European to find the origins of the Blue Nile in 1618.[139] But the lake was already depicted as 黑湖, meaning "black lake", on the Africa-KWQ when it was published in 1602. This chapter has shown that the source map of the Africa-KWQ was drawn by the Chinese based on the information obtained by the Ming mariners in Africa in early 1433 during their seventh (and last) voyage to the Western Ocean.
55	亞毘 心域	Ya Pi Xin Yu	Abyssinia	In 1270, an unholy alliance between Muslim and Christian states occurred to found a Solomonic Dynasty/Ethiopia Empire or simply Ethiopia (1270–1974),[140] which historically spanned a geographical area in today's regions of Amhara and Tigray. This ancient Ethiopia was ruled by the Abyssinians, from whom Abyssinia gets its name.[141] 亞毘心域 (Ya Pi Xin Yu) is the transliteration of Abyssinia. But Abyssinia was just one state of Ethiopia, hence it is incorrect to present the name Abyssinia as synonymous with Ethiopia. Ethiopia was able to grow by military expansion to include other primarily Afro-Asiatic-speaking communities and that made it dominate the Horn of Africa.[142] This Imperial Ethiopia was not the Ethiopia of today. Towards the close of the fifteenth century, the Portuguese started their operations in Ethiopia.[143] Yet, for more than a century, scholars have been greatly confused regarding the name Abyssinia, which later became Ethiopia. The "historical" country of Abyssinia is in today's North Ethiopia and Southern Eritrea-Hamasien, Akaleguzai and Sera'e. The Abyssinians refer to their country as "Abyssinia: A Christian Island surrounded by enemy Islam and pagans" or Abyssinia-"Ethiopia" (Christian Ethiopia/Northern Ethiopia). Moreover, the languages of highland Ethiopia were different from those of lowland Ethiopia (in the east and south).

• 56	入曷謎的里	Ru He Mi Di Li	Bagamidri[144]	This was the Kingdom of Bagamidri. 入曷謎的里 (Ru He Mi Di Li) should be corrected as 八曷謎的里 (Ba He Mi Di Li). The Ethiopian *Empire/Ethiopia* comprised no less than 26 kingdoms and 14 provinces. The Kingdom of Bagamidri[145] was one of them: it was a large and fertile kingdom, watered with many rivers. The Africa-KWQ depicts 入曷謎的里 (Item 56) and 把曷末多羅 (Item 22). Both are interpreted as Bagamidri. But 把曷末多羅 (Item 22) is depicted at the west side of southern Africa, not in Ethiopia. This cannot be correct.
• 57	帝入蔔卯翁	Di Ru Le Mao Weng	Tigremahon[146]	Tigremahon was a very large kingdom lying between the River Marabo, the Nile, the Red Sea and the Kingdom of Angote. This kingdom is composed of over 17 provinces; the most northerly of these lies in today's Egypt.[147]
58	伐底 曷尔	Fa Di He Er	Fatigar	Fatigar[148] was a kingdom in Africa, which belonged formerly to the King of Ethiopia.
59	清入	Qing Ru	Zingi	Kingdom of Zingi
• 60	下黑地阿皮亞	Xia Hei Di A Pi Ya	*Ethiopia Inferiore*,[149] meaning lowland Ethiopia.	Here 下 means "lowland"; 黑地阿皮亞 (Hei Di A Pi Ya) is the transliteration of Ethiopia. 下黑地阿皮亞 (Xia Hei Di A Pi Ya) means "lowland Ethiopia" or *Ethiopia Inferiore*. Item 55 is closely related to Item 60. The Africa-KWQ depicts Abyssinia (in highland Ethiopia) to the northwest of 下黑地阿皮亞 (*Ethiopia Inferiore*/lowland Ethiopia). None of the three major sixteenth century European maps depicts this geographical item, suggesting their lack of knowledge of this major empire in East Africa.
61	馬加大作	Ma Jia Da Zuo	Mogadishu: the sixteenth century explorer Leo Africanus knew the city as *Magadazo* (alt. *Magadoxo*).[150]	馬加大作 (Ma Jia Da Zuo) is the transliteration of Mogadishu which has a long history. The city was in today's Somalia and located in the coastal Banadir region on the Indian Ocean. It served as the capital of the influential Sultanate of Mogadishu in the ninth to thirteenth centuries and was an important port connecting with traders from the Indian Ocean for millennia. Mogadishu enjoyed the height of its prosperity during the fourteenth and fifteenth centuries.

(*Continued*)

(*Continued*)

Item #	Name	Pinyin	Etymology	Analysis of history, politics, geography; comparison or comment
				Vasco da Gama passed by Mogadishu in 1498.[151]
• 62	多亞 剌	Duo Ya La	Doara	多亞剌 (Duo Ya La) is the transliteration of Doara which was a wadi (a valley or ravine, bounded by relatively steep banks, which in the rainy season becomes a watercourse) in the central part of today's Somalia.
63	訝德兒	Ya De Er	Adal or Adel	The Adal Sultanate (1415–1577)[152] or Kingdom of Adal or Awdal or Bar Sa'ad ad-din was a historic Islamic state of eastern Africa, in the region southwest of the Gulf of Aden. Adal is depicted on the Africa-KWQ, indicating that this map cannot be dated earlier than the year of 1415 and later than 1577. The year 1433 falls nicely within this date range.
64	瓦兒大付峰	Wa Er Da Fu Feng	Cape Guardafu; Italian: *Capo Guardafui*	Cape Guardafui is a headland in the autonomous Puntland region in Somalia.[153]
• 65	党各哩	Dang Ge Li	Dangali[154]	On the Africa-KWQ, 党各哩 (Item 65; the Kingdom of Dangali) is in eastern Africa near the border of today's Ethiopia and Somalia.
• 66	多拔西	Duo Ba Xi	Dobas	The Kingdom of Dobas is located to the southwest of Dangali.
67	波尔諾湖	Bo Er Nuo Hu	Lake Borno; the lake has the same name as the Kingdom of Borno.[155]	Readers can refer to Subsection 1 of Item 5 in Section 6 of this chapter's main text for a full discussion of 波尔諾湖 (Bo Er Nuo Hu; Item 67), which is the transliteration of Lake Borno.
• 68	黑地兀皮亞	Hei Di Wu Pi Ya	Ethiopia	After introducing 亞昆心域 (Item 55; Abyssinia in highland Ethiopia) and 下黑地阿皮亞 (Item 60; *Ethiopia Inferiore*), the Africa-KWQ depicts 黑地兀皮亞 (Ethiopia; 1270–1974)[156] as an empire with another portion of its territory near the Red Sea.

69	大非力	Da Fei Li	Dafila	
70	伯六	Bo Liu	Bello	
• 71	巴皮羅泥亞	Ba Pi Luo Ni Ya	Babylon in Egypt; *Babylon* is the Latin representation of Greek *Babylṓn* (Βαβυλών), derived from the native (Babylonian) *Bābilim*, meaning "gate of the god(s)".[157]	巴皮羅泥亞 (Ba Pi Luo Ni Ya) is the region of Babylon in Egypt. Babylon was originally the dominant city of Mesopotamia.[158] Around the turn of the fourth century, the Romans established a fortress town along the east bank of the Nile. This fortress, known as Babylon, was the nucleus of the Roman and then the Byzantine[159] (395–1453) city and is the oldest structure in the city of Cairo. In 1168, Egypt's capital was permanently moved to Cairo,[160] and Babylon advanced to become the second metropolis of Egypt in the fifteenth century after the city of Memphis; the meaning of the term Babylon of Egypt has been identified with or used as the symbol and synonym for Egypt itself.[161] The Africa-Mercator (1569) does not depict the name of Egypt, but the Africa-Ortelius (1570) and the Africa-Plancius (1594) use ancient Greek *Aegyptvs*, which has one of the longest histories of any country, tracing its heritage along the Nile Delta back to the period 6000–4000 B.C. The Africa-KWQ uses 巴皮羅泥亞 (Babylon in Egypt), tracing Egypt's history to the Roman era: it signifies the existence of Roman–China contacts in Antiquity.
• 72	薩係得	Sa Xi De	Sahid[162]	On the Africa-Ortelius (1570), there is Zibit near where Sahid should be.
73	泥羅河	Ni Luo He	River Nile. Neilos (Nile) was the River-God of Aigyptos (Egypt) in North Africa. Neilos was the Greek name for the Nile River. [63]	Much discussion has been given in Items 35 and 36. 泥羅河 (Ni Luo He) is the transliteration of the Greek name *Neilos* (today's Nile) which is a major north-flowing river in northeastern Africa and the longest river in Africa.[164] There is only one year-round river in Egypt, the Nile. Today, it has no non-seasonal tributaries for its entire length in Egypt. It has two further tributaries upstream, the Blue Nile and White Nile, which merge in central Sudan. In the Nile Delta, the river splits into several branches and lesser channels. The Nile flows through Egypt, Sudan, South Sudan, Ethiopia, Uganda, the Democratic Republic of the Congo, Kenya, Tanzania, Rwanda and Burundi.[165]

(*Continued*)

(*Continued*)

Item #	Name	Pinyin	Etymology	Analysis of history, politics, geography; comparison or comment
@3	中有七百州最大者末罗耶有城沿河名门菲此城为天下极大城行十日程地产宝石乌木		Zhong You Qi Bai Zhou Zui Da Zhe Mo Luo Ye You Cheng Yan He Ming Men Fei Ci Cheng Wei Tian Xia Ji Da Cheng Xing Shi Ri Cheng Di Chan Bao Shi Wu Mu	My translation and comments (between square brackets) are as follows: In this region, there are 700 states, among which the largest is Moloye [末罗耶] in which there is a city named Memphis [the capital of ancient Egypt] along a river. Memphis is a very large city in the world and can be reached by a ten-day journey. The place produces precious stones and ebony. Memphis was the capital of ancient Egypt (*Kemet* or *Kumat*) during the Old Kingdom and remained an important city throughout ancient Egyptian history.[166] The Africa-KWQ accurately describes the geographical environment of this city and its impressive size.
74	拨尔捺曷速	Bo Er Na He Su	Barnagasso[167]	Barnagasso (Bahr Negasti) was a province in Ethiopia.[168]
@4	天下惟此江至大以七口入海其國近年無雲雨故國人精於天文其江每年次泛溵地甚肥澤如糞其 田故國人種之五穀以一收百國稱富 饒。		Tian Xia Wei Ci Jiang Zhi Da Yi Qi Kou Ru Hai Qi Kou Jin Nian Wu Yun Yu Gu Guo Ren Jing Yu Tian Wen Qi Jiang Mei Nian Ci Fan Chang Di Shen Fei Ze Ru Fen Qi Tian Gu Guo Ren Zhong Zhi Wu Gu Yi Yi Shou Bai Guo Cheng Fu Rao	My translation and comment (between square brackets) are as follows: This is the largest river in the world. It flows into the sea through seven estuaries. The country has had no clouds or rain in recent years; hence their people become good at astronomy [to know how to make *meteorological* forecasts]. Every year this river flooded, leaving silt and plant debris like manure to enrich the soil and produce fertile lands; hence the grains they sow can be harvested with high productivity. It is a well-known rich country. The annotation correctly states that Nile is the largest (longest) river in the world (called the father of African rivers). It rises south of the Equator and flows northward through northeastern Africa to drain into the Mediterranean Sea. In addition to Egypt, the Nile runs through or along the border of ten other African countries, namely, Burundi, Tanzania, Rwanda, Eritrea, the Democratic Republic of the Congo, Kenya, Uganda, Sudan, Ethiopia and South Sudan. Its three main tributaries are the White Nile, the Blue Nile and the Atbara.[169] I have discussed the White and Blue Nile in detail in the main text.

				It is also well known that the Nile flooded every year between June and September. The Nile spreads out in Lower Egypt and forms the Nile Delta and drains into the Mediterranean Sea.[170] It is also well known that Egyptian astronomy[171] began in prehistoric times; the observation of stars was important in determining the annual flooding of the Nile. The annotation on the Africa-KWQ correctly describes the Nile's length and flooding, Egypt's climate and agriculture as well as the advanced Egyptian astronomy.
Zone IV: Central Africa (north of the Equator; Region I)				

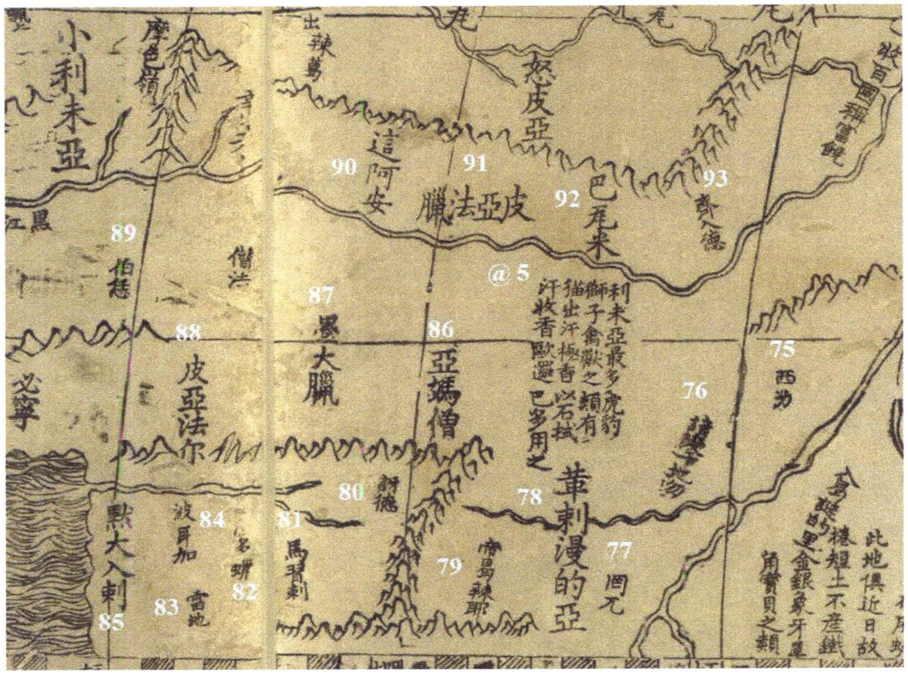

Fig. A4.7. Extracted from Fig. 4.10, right panel, public domain.

(*Continued*)

(*Continued*)

Item #	Name	Pinyin	Etymology	Analysis of history, politics, geography; comparison or comment
75	西力	Xi Li	Syre	Kingdom/City-state of Syre
76	諳變干地勿	An Bian Gan Di Wu	Ambiancantina	Region of Ambiancantina
77	罔兀	Wang Wu	Vangue	Kingdom/City-state of Vangue
● 78	革剌漫的亞	Ge La Man Di Ya	Calamat[172]	
● 79	帝曷辣耶	Di He La Ye	Tigrai[173]	
● 80	訝德	Ya De	*Iaté*[174]	
81	馬習剌	Ma Xi La	Maceira	Kingdom/City-state of Maceira
82	漫多蠟	Man Duo La	Medra	Kingdom of Medra[175]
● 83	雷地	Lei Di	*Terra dei lampi*[176]	
84	波耳加	Bo Er Jia	Borca	Kingdom/City-state of Borca
● 85	默大入剌	Mo Da Ru La	Media Guilan[177]	
@ 5	利未亞最多虎豹獅子禽獸之類有猫出汗極香以石拭汗收香歐邏巴多用之	Li Wei Ya Zui Duo Hu Bao Shi Zi Qin Shou Zhi Lei You Mao Chu Han Ji Xiang Yi Shi Shi Han Shou Xiang Ou Luo Ba Duo Yong Zhi		My translation and comment (between square brackets) are as follows: Africa [利未亞] has the most tigers, leopards, lions, and other beasts. There are cats that can sweat with a fragrant odor. The Europeans often wipe their sweat with stones to smell the fragrance. Africa is the paradise of animals. The list of animals goes far beyond what are mentioned in the above annotation. But none of the three major sixteenth century European maps makes similar annotation about African animals.

86	亞媽僧	Ya Ma Seng	Amazen	
87	墨大臘	Mo Ba La	Medra	There are 漫多蠟 (Medra; Item 82) and 墨大臘 (Medra; Item 87) on the Africa-KWQ. Similarly, on the Africa-Ortelius (1570), there is MEDRA which may depict the region, and Medra which may depict the kingdom/city-state.
88	皮亞法尔	Pi Ya Fa Er	Biafar, Birfara, Biafra or Biafares	Early modern maps of Africa from the fifteenth to the nineteenth centuries, drawn from accounts written by explorers and travellers, show references to Biafar, Biafara, Biafra and Biafares.[178]
• 89	伯恁	Bo Nin	Benin City (called "Bini" by locals)	Benin City was called *Bini* by locals. 伯恁 (Bo Nin) can be regarded as the transliteration of "Bini". Benin City was the capital and largest city of Edo State in southern Nigeria. Further discussion is given in Item 94 of 必寧.
• 90	這阿安	Zhe A An	Cheogane[179]	
91	皮亞法臘	Pi Ya Fa La	Biafra	This is a repetition of 皮亞法尔 of Item 88, but this place is at a different location.
• 92	巴瓦米	Ba Wa Mi	Bagamidu[180]	
93	齊入德	Qi Ru De	Zegzeg[181]	The location of 齊入德 (Qi Ru De) or Zegzeg on the Africa-KWQ is quite different from that on the Africa-Ortelius (1570).

(Continued)

(Continued)

Item #	Name	Pinyin	Etymology	Analysis of history, politics, geography; comparison or comment
Zone V: Western Africa (Region I)				

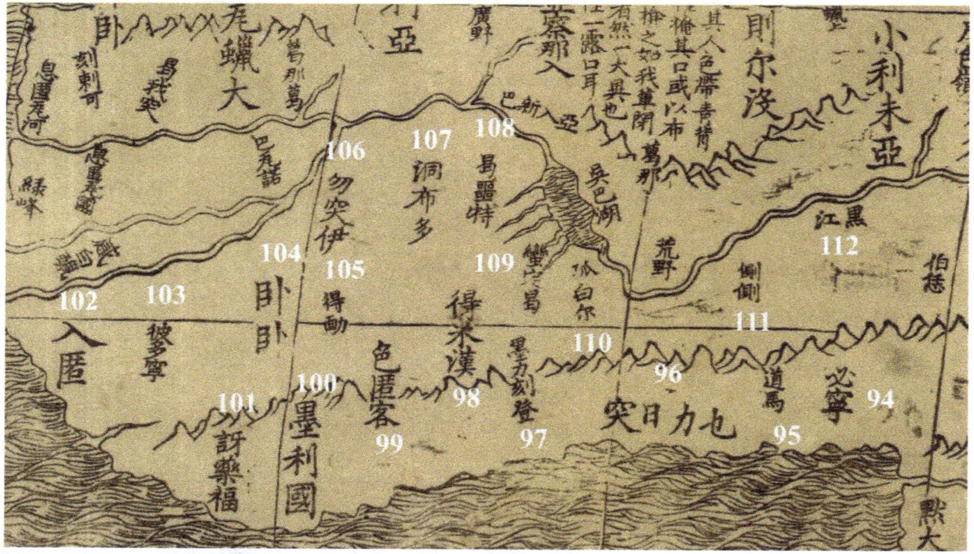

Fig. A4.8. Extracted from Fig. 4.10, right panel, public domain.

Item #	Name	Pinyin	Etymology	Analysis of history, politics, geography; comparison or comment
94	必寧	Bi Ning	Benin was a city-state. The Kingdom of Benin's capital was Edo, later known as Benin City (after the arrival of the Portuguese in the late fifteenth century).	必寧 (Bi Ning) is the transliteration of Benin (the Kingdom of Benin), also known as the Kingdom of Edo (1180–1897),[182] which was a kingdom in what is now southern Nigeria.[183] The Kingdom of Edo's capital was Edo, later known as Benin City, because the Portuguese who arrived in 1485 would refer to the place as Benin, the city centre as Benin City (called "Bini" by locals) and the empire as Benin Empire. The Kingdom of Benin was one of the oldest and most developed states in the coastal hinterland of western Africa. It has no historical relation to the modern nation-state of Benin, which was historically known as Dahomey (c. 1600–1904).[184]

				On the Africa-KWQ, 伯恁 (Bo Nin) in Item 89 can be regarded as the transliteration of the city *Bini*. Since the source map of the Africa-KWQ is of Chinese origin and was drawn in 1433 when the kingdom's name was still the Kingdom of Edo (before the arrival of the Portuguese), this tells that Matteo Ricci has changed Edo into 必寧 (Bi Ning; Benin) to make the Africa-KWQ look more European.
95	道馬	Dao Ma	Dauma or Davrna; Dahomey	道馬 (Dao Ma) is the transliteration of Dahomey.[185] According to the Fon oral tradition, Dahomey (also called Abomey, after its capital city) was the state of the Fon people. It was originally a dependency of the Kingdom of Allada which lasted from c. 1100 to 1724; a coastal West African kingdom in what is now southern Benin). Only in the seventeenth century did a Dahomey ruler called Wegbaja declare himself king and make Dahomey an independent state.[186]
• 96	乜力日突	Mi Li Ri Tu	Meleghete or Melegete	The Province of Meleghete was a well-known place.[187] Also, see Item 97. On the Africa-Ortelius (1570), the item also shows as Melegete and MELEGETE.
97	墨力剋登	Mo Li Ke Deng	Melegete or Melagueta[188]	This is the coast of Melegete (Melagueta) in Western Africa.[189]
98	得米漢	De Mi Han	Temian	Kingdom of Temian[190]
• 99	色匿客	Se Ni Ke	Scenico[191]	
100	墨利國	Mo Li Guo	Melli[192] or Mali Empire	墨利 (Mo Li) is the transliteration of "Melli" or "Mali" and 國 (Guo) means "kingdom". The Kingdom of Melli[193] was named after their primary village where their king resided. (Leo Africanus said that the capital city was called Melli.[194]) The village was very large and contained over six thousand families. It was "Westerly bounded with a mighty Forest, which runs sixty miles along the Banks of Niger, to the Verges of the Ocean".[195] This is the Mali Empire (c. 1230–1670) which reached its political peak in the fourteenth century.

(*Continued*)

(*Continued*)

Item #	Name	Pinyin	Etymology	Analysis of history, politics, geography; comparison or comment
				On the Africa-KWQ, 墨利國 (Kingdom of Melli) is by the coast of the Atlantic Ocean and its location is south of the Niger River, close to the Equator; there is a mountain range (presumably with forests) to its north. However, on the three major sixteenth century European maps, Melli is not near the coast of the Atlantic Ocean, nor near the Equator, and has no mountains to its north; their Melli corresponds better to the historical Mali Empire. The Mali Empire started to disintegrate from the early sixteenth century, while the Songhai Empire (c. 1464–1591)[196] gradually took centre stage in western Africa. The source map of the Africa-KWQ is clearly drawn before c. 1464, because the Songhai Empire is not on it. But why do the three major sixteenth century European maps also not depict the Songhai Empire? Because a series of plots and coups forced the empire into a period of decline and instability between 1539 and 1549. Although political stability was restored between 1549 and 1582/1583,[197] the news might not have passed to these European cartographers who had never set foot on the lands of Africa. Later, the Empire collapsed in 1591 due to a disastrous military defeat[198]; hence Songhai naturally is not depicted on the Africa-Plancius (1594). The Africa-KWQ's depiction of the geographical location of the Mali (Melli) Empire is too far to the south, hence, not accurate. This further shows that it is an older map of the early fifteenth century in comparison with the three major sixteenth century European maps.
101	訝樂福	Ya Le Fu	Gilolof	On the Africa-KWQ, 訝樂福 (Ya Le Fu) means Gilolof; it seems to have been written mistakenly as 訝藥福 (Ya Yao Fu).

102	入匿	Ru Ni	Guinea[199]	入匿 (Ru Ni) is the Chinese name given to the region of Guinea which was on the fringes of the major western African empires. The Ghana Empire (c. 300–c. early 1200s)[200] is believed to be the earliest of these, followed by the Sosso Empire in the twelfth century,[201] then the Islamic Mali Empire (c. 1230–1670).[202] The Mali Empire began to decline, and its power began to be supplanted by its vassal states in the early sixteenth century. The most successful of these was the Songhai Empire (c. 1464–1591).[203] The Songhai Empire is not depicted on the Africa-KWQ, signifying that the political landscape of this map reflects Africa before c. 1464. The English term Guinea comes from the Portuguese word *Guiné*, which emerged in the mid-fifteenth century to refer to the lands inhabited by the *Guineus*, a generic term for the black-skinned African peoples south of Senegal.[204] But 入匿 (Ru Ni) is not a transliteration of Guinea. This shows that 入匿 is of Chinese origin and had existed long before the Portuguese colonisation of Guinea in 1588.
103	彼多寧	Bi Buo Ning	Bitonin	Kingdom/City-state of Bitonin
104	卧卧	Wo Wo	Gago[205]	
105	得動	De Mian	Ternian	Ternian was a city.[206]
• 106	勿突伊	Wu Tu Yi	Vodei[207]	
107	洞布多	Dong Bu Duo	Tonbuto	
• 108	曷噩特	He E Te	Hongoté[208]	

(*Continued*)

(Continued)

Item #	Name	Pinyin	Etymology	Analysis of history, politics, geography; comparison or comment
109	蠻定曷	Man Ding He	Mandingo (Mandinka)	蠻定曷 (Man Ding He) is the transliteration of the Mandingo Kingdom. The ancestors of the Mandingo of today's region of Gambia and Senegal lived in Kangaba, which was a part of the ancient Mali Empire. They became independent in 1235 and gradually some of them moved westwards and settled near the big rivers Gambia, Senegal and Casamance. This move also gave them better trading possibilities near the Trans-Sahara-Route and allowed them to reign over their own land.[209] But in the middle of the seventeenth century, the Europeans' influence started. They were mainly interested in trading slaves, gold and ivory.[210]
110	孤白尔	Gu Bai Er	Guber or Gvber	Kingdom/City-state of Guber
111	側側	Ce Ce	Zegzeg[211]	側側 (Ce Ce) is the transliteration of Zegzeg. Zaria, formerly Zazzau or Zegzeg was a historic kingdom traditionally dated from the eleventh century.[212] The kingdom's name changed to Zaria at the end of the sixteenth century with its headquarters at Zaria City.[213] On the Africa-KWQ, there is 齊入德 (Item 93), which is also interpreted as Zegzeg. But this does not seem to be correct.
112	黑江	Hei Jiang	Niger River: the earliest European use of the name "Niger" for the river is by Leo Africanus. The name may be a direct translation by Leo Africanus from Arabic bilād as-sūdān (بلاد السودان), meaning "Land of the Blacks", into Latin (Niger "black").	黑 (Hei) means "black" and 江 (Jiang), "river". 黑 江 is the Niger River, which is the third-longest river in Africa, after the Nile and the Congo/Kongo River. It runs in a crescent shape through Mali, Niger, on the border with Benin and then through Nigeria, discharging through a massive delta, known as the Niger Delta (the Oil Rivers), into the Gulf of Guinea in the Atlantic Ocean (see Fig. 4.16). On the Africa-KWQ, 黑江 (the Niger River) appears to the west of 這阿安 (Zhe A An; Item 90; Cheogane). It continues westward, reaching 呉巴湖 (Wu Ba Hu; Item 142; Lake Guber) before discharging into the Atlantic Ocean near 入匿 (Ru Ni; Item 102; Guinea), 綠峰 (Lu Feng; Item 119; Cape Verde) and as 息匿瓦河 (Xi Ni Wa He; Item 120; Senegal River). The Africa-KWQ seems to have been influenced by the early Arabic geographers who believed the upper Senegal River and the upper Niger River were connected to each other and formed a single

				river flowing from east to west, which they called the "Western Nile". Although some of the headwaters of the Senegal River are near the Niger River in Mali and Guinea, but there is no "Western Nile".
Zone VI: Western Africa (Region II)				
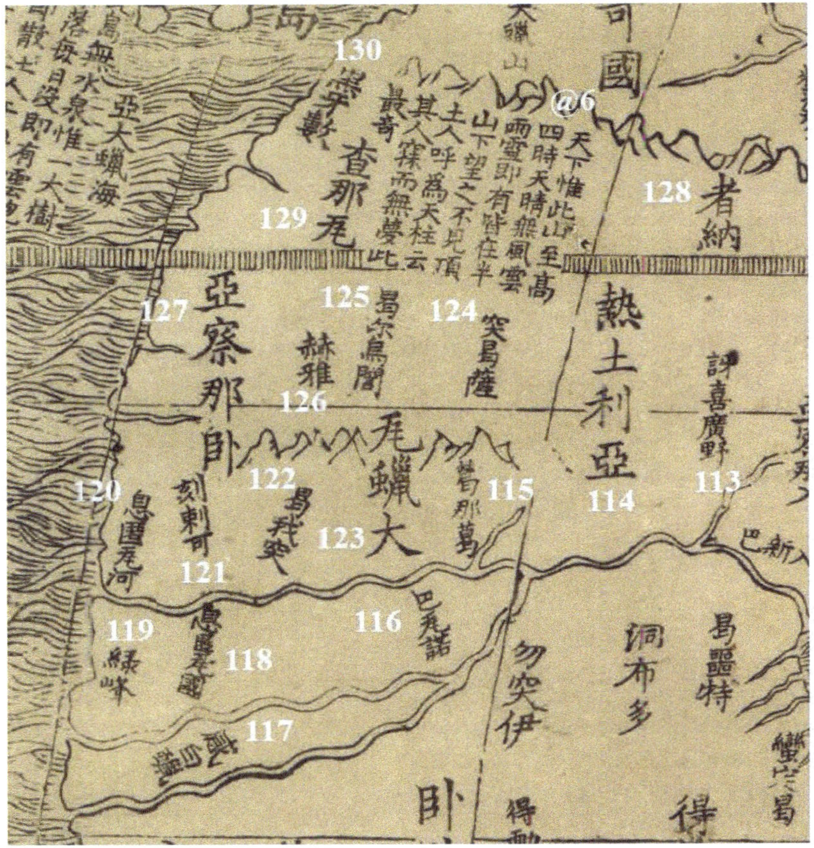 Fig. A4.9. Extracted from Fig. 4.10, right panel, public domain.				

(*Continued*)

(*Continued*)

Item #	Name	Pinyin	Etymology	Analysis of history, politics, geography; comparison or comment
• 113	訝喜黄野	Ya Xi Guang Ye	*Iasci, deserto*[214]	
• 114	熱土利亞	Re Tu Li Ya	Getulia[215]	熱土利亞 (Re Tu Li Ya) is the transliteration of Getulia. Gaetuli was the Romanised name of an ancient Berber tribe inhabiting Getulia.[216] The Getulia district covered the large desert region south of the Atlas Mountains, bordering the Sahara.
• 115	葛那葛	Ge Na Ge	Conaco[217]	
• 116	巴瓦諾	Ba Wa Nuo	Pauano[218]	
117	感白蠟	Gan Bai La	Gambra	Kingdom of Gambra
118	息匿瓦國	Xi Ni Wa Guo	Senegal: it is the African realm in upper Guinea, along the length of the Senegal River.[219] It owes its name to the Senegal River, which borders it to the east and north.	息匿瓦國 (Xi Ni Wa Guo) is the transliteration of Senegal. In the thirteenth and fourteenth centuries, the area of today's Senegal came under the influence of the empires to the east; the Jolof Empire of Senegal (1350–1549), a western African state that ruled parts of modern-day Senegal.[220] But the Jolof Empire later became the vassal of the Mali Empire (c. 1230–1670) and collapsed around 1549.[221] Before the arrival of European settlers, the history of the Saharan region is mainly characterised by the consolidation of settlements in large state entities — the Ghana Empire (c. 300–c. early 1200s), the Mali Empire (c. 1230–1670) and the Songhai Empire (c. 1464–1591). The cores of these great empires were located on the territory of the current Republic of Mali, so Senegal occupied a peripheral position. The first French settlement in modern-day Senegal dates to the fourteenth century,[222] followed by Dinis Dias exploring the mouth of the Senegal River in c. 1444 to reach the westernmost point of Africa he calls *Cabo Verde*[223] and the Dutch, the English and the French also reached the coast of western Africa in the mid-fifteenth century. They first settled along the coasts, on islands in the mouths of rivers and then a little further upstream.

				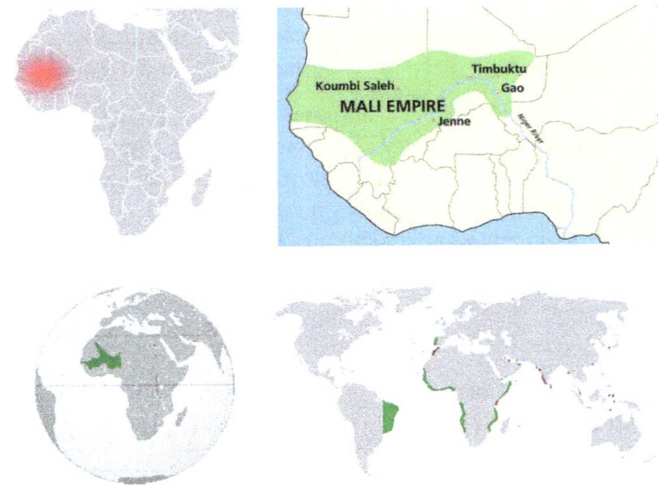 Fig. A4.10. The top left map shows the location of the Ghana Empire (see endnote 224, public domain); the top right map shows the extent of the Mali Empire at its height (see endnote 225, public domain); the bottom left map shows the territorial extent of the Songhai Empire c. 1500 (see endnote 226, public domain); and the bottom right map shows Portuguese colonies and posts in the sixteenth century (see endnote 227, public domain).
119	綠峰	Lu Feng	Cape Verde Peninsula	Cap-Vert, or the Cape Verde Peninsula, is a peninsula in Senegal and the westernmost point of the African continent.[228] Portuguese explorers called it Cabo Verde or "Green Cape", the same meaning as 綠峰. The Portuguese first sighted the cape in c. 1444.[229] The Cape Verde islands, 570 km/350 mi further west, are named after the cape, but were discovered a few years later after the name was given.

(*Continued*)

(Continued)

Item #	Name	Pinyin	Etymology	Analysis of history, politics, geography; comparison or comment
• 120	息匿瓦河	Xi Ni Wa He	Senegal River	As the Senegal River reached into the heart of the gold-producing Ghana Empire and later the Mali Empire,[230] Trans-Saharan traders gave the Senegal River its famous nickname "River of Gold". On the Africa-KWQ, the Senegal River is depicted as a tributary of the Niger River. But today's Senegal River is not.
• 121	刻剌可	Ke La Ke	Colaco[231]	
122	曷我突	He Wo Tu	Hoden	曷我突 (He Wo Tu) is the transliteration of Hoden.
123	瓦蠟大	Wa La Da	Gualata	Kingdom of Gualata[232]
124	突曷薩	Tu He Sa	Tegazza	Kingdom of Tegazza
125	曷尔烏閣	He Er Wu Ge	Cheneg (?)[233]	This geographical item is the same as Item 128. On the Africa-Ortelius (1570), Cheneg also appears twice.
126	赫雅	He Ya	Hayr	
127	亞察那卧	Ya Cha Na Wo	Azanaga[234]	Azanaga is on the coast of western Africa.
128	者納	Zhe Na	Cheneg	
129	查那瓦	Cha Na Wa	Azanaga (?)[235]	亞察那卧 (Item 127) and 查那瓦 (Item 129) both are interpreted as Azanaga. On the Africa-Ortelius, there are also two entries: Azanaga and AZANAGA.
@ 6	天下惟此山至高四時天晴無風雲雨雪即有皆在半山下望之不見頂土人呼為天柱云其人寐而無夢此最奇	Tian Xia Wei Ci Shan Zhi Gao Si Shi Tian Qing Wu Feng Yun Yu Xue Ji You Jie Zai Ban Shan Xia Wang Zhi Bu Jian Ding Tu Ren Hu Wei Tian Zhu Yun Qi Ren Mei Er Wu Meng Ci Zui Qi		My translation is as follows: This is the only very high mountain in the world. In all four seasons, the sky is clear and there is no wind, cloud, rain or snow. Even if there were, it would be all in the middle of the mountain and invisible from down below. The local people call it Celestial Pillar of Clouds. They have no dream during their sleep. This is the strangest thing. "They have no dream during their sleep" may mean that they sleep so soundly in such a peaceful environment that they do not remember their dreams after waking up.

• 130	黑牙數	Hei Ya Shu	Heasus[236]	
Zone VII: Central Africa (north of the Equator; Region II)				
131	瓦和瓦	Wa He Wa	Gaoga	瓦和瓦 (Wa He Wa) is the transliteration of Gaoga.[237] Gaoga is depicted on the Africa-Ortelius (1570) and the Africa-Plancius (1594).
132	嗷瓦	Ac Wa	Darga	嗷瓦 (Ao Wa) is the transliteration of Darga. Darga is depicted on the Africa-Plancius (1594).
133	怒皮亞	Nu Pi Ya	Nubia	怒皮亞 (Nu Pi Ya) is the transliteration of Nubia, which is an ancient region of southern Egypt and northern Sudan, including the Nile valley between Aswan and Khartoum and the surrounding area.[238] Following the collapse of the Kingdom of Kush[239] during the fourth century B.C., a political void in the region it had controlled was filled by several smaller emerging Nubian kingdoms. Later, the influx of Arabs and Nubians to Egypt and Sudan contributed to the suppression of the Nubian identity following the collapse of the last Nubian kingdom around 1504.
134	瓦約瓦	Wa Yue Wa	Gheogan[240]	Kingdom of Gheogan

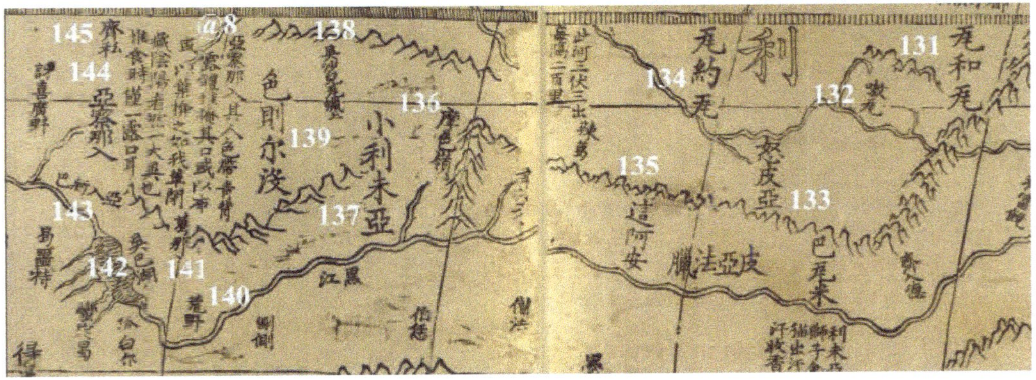

Fig. A4.11. Extracted from Fig. 4.10, right panel, public domain.

(*Continued*)

(Continued)

Item #	Name	Pinyin	Etymology	Analysis of history, politics, geography; comparison or comment
@7	此河三伏三出每隔二百里		Ci He San Fu San Chu Mei Ge Er Bai Li	My translation and comment (between square brackets) are as follows: This river disappears and reappears approximately every 200 *li* [1 *li* is about ½ km] distance apart. From the Africa-KWQ, there seems to have been only one disappearing stretch (200 *li* may be too short to draw on this map's scale); there seems to be a mountain range or hilly region, with two rivers descending west and east, respectively.
• 135	辣葛	La Ge		Lacca[241]
• 136	摩色嶺	Mo Se Ling		Muslin[242]
• 137	小利未亞	Xiao Li Wei Ya		*Libia minore*[243]
• 138	呉沙兒瓦蠟上	Wu Sha Er Wa La Shang		Usciarualasciam[244]
• 139	色則尔没	Se Ze Er Mo		Secelme[245]
• 140	荒野	Huan Ye		This geographical term cannot be identified.[246]
• 141	葛那	Ge Na		This geographical term cannot be identified.
142	呉巴湖	Wu Ba Hu	Lake Guber	呉巴湖 (Wu Ba Hu) is the transliteration of Lake Guber, which is today's Lake Débo.
• 143	亞入新巴	Ya Ru Xin Ba	Agisymba[247]; the origin of the name Agisymba is unknown.	亞入新巴 (Ya Ru Xin Ba) is the transliteration of Agisymba which was an unidentified country in Africa mentioned by Ptolemy in the middle of the second century A.D. According to Ptolemy, Julius Maternus (or Matiernus) undertook an expedition around 90 A.D., and arrived at the rivers Bahr Salamat and Bahr Aouk, near the current Central African Republic in a region then called Agisymba.[248]

				On the Africa-KWQ, 亞入新巴 (Agisymba) is shown close to 呉巴湖 (Lake Guber) and on its north side. 呉巴湖 (Lake Guber) is also known as the Great Lake or Lake Débo.[249]
144	亞察那入	Ya Cha Na Ru	Azanaque[250]	
@ 8	亞察那入其人色帶青背露體惟掩其口或以布或以葉揜之如我輩閉藏陰陽者然一大異也惟飡時僅一露口耳	Ya Cha Na Ren Qi Ren Se Dai Qing Bei Lu Ti Wei Yan Qi Kou Huo Yi Bu Huo Yi Ye Yan Zhi Ru Wo Bei Bi Cang Yin Yang Zhe Ran Yi Da Yi Ye Wei Shi Shi Jin Yi Lu Kou Er		My translation and comment (between square brackets) are shown as follows: The local people in Azanaque [亞察那入] have cyan-coloured skin and like to expose their back body nude. But they show a much different habit by covering their mouths with cloth or leaves, just like us who try to converge our internal energy to keep a balance of Yin and Yang in our body. They reveal their mouths and ears only when they are eating. In Ancient Chinese philosophy, Yin and Yang can be thought of as complementary, rather than opposing forces that interact to form a dynamic system in which the whole body is greater than the assembled parts. Everything has both Yin and Yang aspects.[251] The Yin and Yang concept is uniquely of Chinese origin. None of the European maps ever employed the Yin and Yang concept.
145	齊私	Qi Si	Zis	
Zone VIII: North Africa				
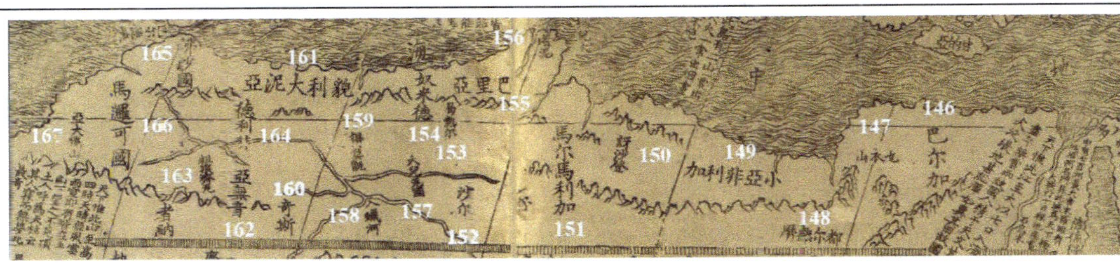Fig. A4.12. Extracted from Fig. 4.10, right panel, public domain.				
146	巴尔加	Ba Er Jia	Barcha	巴尔加 (Ba Er Jia) is the transliteration of Barcha which was the name of a region incorporating western Egypt and eastern Libya.

(*Continued*)

(Continued)

Item #	Name	Pinyin	Etymology	Analysis of history, politics, geography; comparison or comment
147	乜永山	Mie Yong Shan	*Meis mons*[252] means "My Mountain".	乜永 (Mie Yong) is the transliteration of Meis, 山 means "mountain".
148	都尔热曆	Dou Er Re Li	Bilidulgerid	In Africa, there is an immense tract of land between Barbary and the great desert named Bilidulgerid, "the land of dates", from the profusion of the trees growing there.[253] However, on the Africa-KWQ, only a portion 都尔热曆 (dulgerid) of this geographical term is written on the map and the front portion "Bili" is left out due to map damage.
• 149	小亞非利加	Xiao Ya Fei Li Jia	*Africa Minore*[254]	The Africa-KWQ names the north-eastern part of today's Libya as 小亞非利加 ("Africa Minor"; 小 means "small" or "minor"). In ancient days, Libya could refer to the country immediately west of Egypt, namely, Marmarica (*Libya Inferior*) and further west, Cyrenaica (*Libya Superior*).[255] A fifteenth century map shows Marmarica Lybia immediately to the west of Egypt.[256] 小亞非利加 (Africa Minor) seems to represent Marmarica/Lybya Inferior.
• 150	訝沙登	Ya Sha Deng	Iasciatem[257]	
• 151	馬尔馬利加	Ma Er Ma Li Jia	Marmarica	馬尔馬利加 (Ma Er Ma Li Jia) is the transliteration of Marmarica, which in ancient geography was a littoral area in ancient Libya, located between Cyrenaica and Aegyptus.[258] As discussed in Item 149, 馬尔馬利加 depicted on the Africa-KWQ should be Cyrenaica, also known as *Libya Superior*. This is a mistake made by the Africa-KWQ.
• 152	沙尔	Sha Er	Sar[259]	
• 153	曷热尔	He Re Er	Hoger[260]	

• 154	奴米德	Nu Mi De	Numidia[261]	奴米德 (Nu Mi De) is the transliteration of Numidia. In Antiquity, Algeria was known as the Kingdom of Numidia (202 B.C.–40 B.C.) and its people were called Numidians and Imazighen which means "Free men".[262] The kingdom was on the North African coast and comprised roughly the territory of northeast Algeria. It was one of the first major states in the history of Algeria and the Berbers.[263] It began as a sovereign state and later alternated between being a Roman province and a Roman client state.[264] After the fall of the Western Roman Empire in 476, Numidia became the Praetorian Prefecture of Africa (534–591) under the Eastern Roman (Byzantine) Empire, and then the Exarchate of Africa (591–698).[265] Muslim Arabs conquered Algeria in the early eighth century. But the Berber people in North Africa remained independent. After the fall of the Umayyad Caliphate, numerous local dynasties emerged.[266] French colonisation started in 1830 and lasted until 1962. Numidia is depicted on the Africa-KWQ, but not on any of the three major sixteenth century European maps, indicating that the Numidia known to the Chinese explorers was in the era before Christ, long before the Europeans' colonisation of the area.
155	巴尔巴里亞	Ba Er Ba Li Ya	Barbarie[267] or Barbaria[268]	巴尔巴里亞 (Ba Er Ba Li Ya) is the transliteration of Barbaria, which was the name used by the ancient Greeks[269] for a region along northeast Africa which contained several city-states. That region would dominate trading in the Red Sea and the Indian Ocean. On the Africa-KWQ, due to paper damage to the map, the first two Chinese characters 巴尔 are missing: only 巴里亞 is depicted.

(*Continued*)

(Continued)

Item #	Name	Pinyin	Etymology	Analysis of history, politics, geography; comparison or comment
156	堵泥素	Du Ni Su	Tunis; Latin: *Tunes*; the term *Tunis* can mean "camp at night", "camp", or "stop", or may have referred to "the last stop before Carthage" by people who were journeying to Carthage by land.[270]	堵泥素 (Du Ni Su) is the transliteration of Tunis, which was one of the Barbary States[271] and formed a considerable territory of the Ottoman Empire (1299–1922) in Northern Africa. Its history is nearly identical with the city of the same name (shown in what follows).[272] In the city's core lies its ancient Medina, a central market, and an UNESCO World Heritage Site. Between the seventh and eighth century, an Arab Muslim conquest occurred in the region. The first Islamic city was founded in Kairouan, in today's Tunisia, where the Great Mosque of Uqba was constructed in 670 A.D. This mosque is the oldest and most prestigious sanctuary in the Muslim West with the oldest standing minaret in the world.[273] In the early fifteenth century, Tunis was ruled by Arabic-speaking Muslims. In my English book titled *The Last Journey of the San Bao Eunuch, Admiral Zheng He*, I have devoted the entire Chapter 2 to describe and verify how one of Zheng He's translators, Ma Huan (马欢; c. 1380s–c. 1460), and six other senior officers were sent in December 1432 by Assistant Envoy Hong Bao to Tunisia for trade and pilgrimage in a Calicut merchant ship (accompanying the Calicut merchants). This took place when the Ming Treasure Fleets were conducting their seventh (and last) voyage.[274] They visited the Great Mosque of Uqba in Kairouan. Ma Huan later made a vivid description of the Mosque in his 1451 book titled *Ying Ya Sheng Lan* 《瀛涯胜览》 or *The Overall Survey of the Ocean's Shores*.[275] The first Ottoman conquest of Tunis took place in 1534. After some complications, the Ottomans permanently acquired the former Hafsid Tunisia, retaining it until the French conquest of Tunisia in 1881.[276]
• 157	大兒瓦國	Da Er Wa Guo	*Tarua, regno di*[277]	
158	入蠟河	Ru La He	River Ghir	Detailed analysis is given in Section 6 of the main text of this chapter.

159	得莫甑	De Mo Zheng	Tebesse[278]	Detailed analysis is given in Table 4.1.
160	奇斯	Qi Si	Zis	Item 145 齊私 is also Zis, a repetition.
• 161	貌利大泥亞	Mao Li Da Ni Ya	*Mauretania*	*Mauretania* is the Latin name for a region in the ancient Maghreb (also known as Northwest Africa or the Arab Maghreb). It stretched from today's central Algeria westwards to the Atlantic, covering northern Morocco, and southward to the Atlas Mountains.[279] In 25 B.C., the kings of Mauretania became Roman vassals until about 44 A.D., when the area was annexed to Rome and divided into two provinces.[280] *Mauretania* is not depicted on the three European maps, indicating that the Africa-KWQ still contains very old geographical information dating back to the Roman era.
• 162	亞無音	Ya Wu Yin	Abuinam[281]	
163	粧痕齊瓦	Cu Hen Qi Wa	Zuenziga	The sandy region of the desert of Zuenziga is most dangerous for travellers.[282]
• 164	德利非	De Li Fei	Trisidis (?)[283]	
165	佛沙國	Fo Sha Guo	Fessa	Kingdom of Fessa
166	馬邏可國	Ma Luo Ke Guo	Morocco	The first Moroccan state was established by Idris I in 788 A.D. The country reached its zenith in the eleventh and twelfth centuries under the Almoravid and Almohad Dynasties, when it controlled most of the Iberian Peninsula and the Maghreb.[284] Beginning in the fifteenth century, the Portuguese extended their area of control to include parts of Morocco. But Morocco was the only North African nation to escape occupation by the Ottoman Empire. The Alaouite Dynasty, which rules Morocco to this day, seized power in 1631.[285]
167	亞大蠟山	Ya Da La Shan	Atlas Mountains; Latin: *Atlas, monte*	The Atlas Mountains is a mountain range in the Maghreb. It separates the Mediterranean and Atlantic coastlines from the Sahara Desert.[286]

(*Continued*)

(Continued)

Item #	Name	Pinyin	Etymology	Analysis of history, politics, geography; comparison or comment
Zone IX: Islands, oceans and seas				

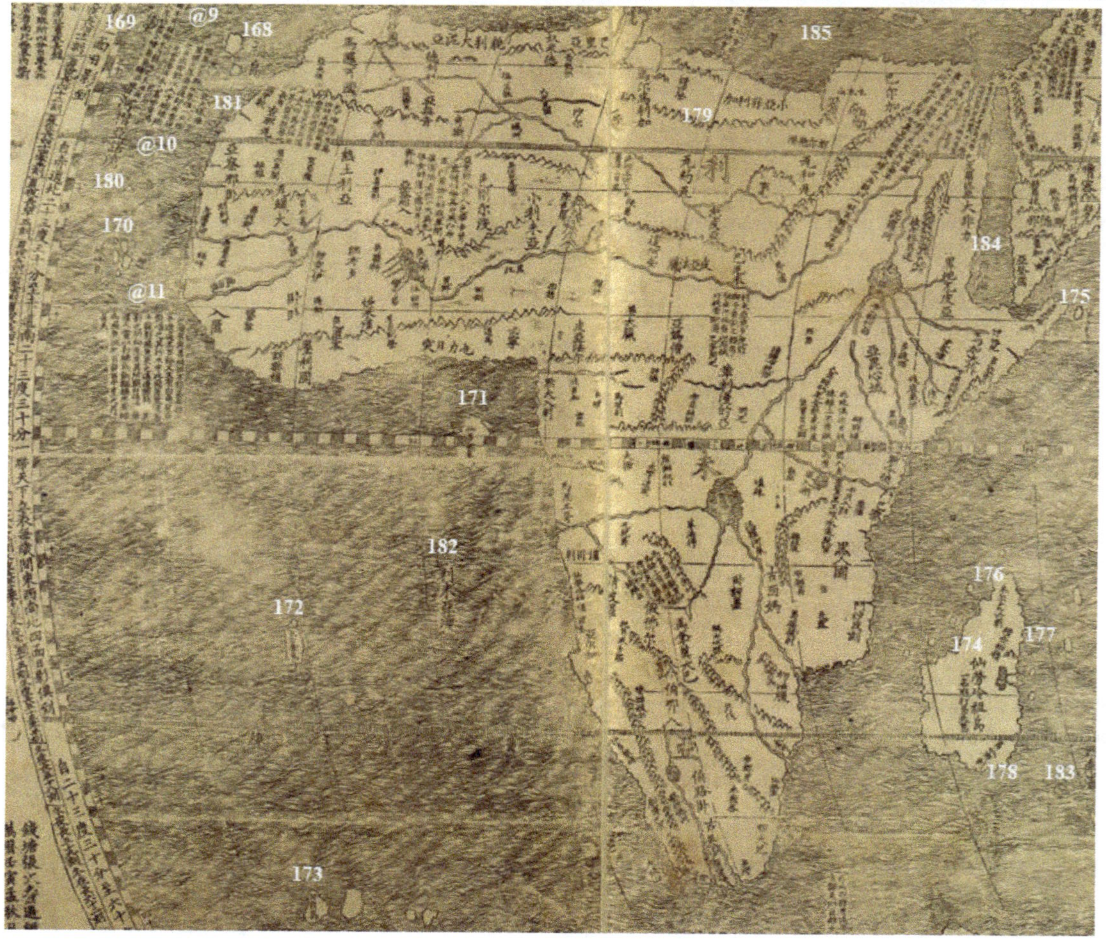

Fig. A4.13. Extracted from Fig. 4.10, right panel, public domain.

Islands				
168	福島	Fu Dao	Fuerteventura	Please refer to Table 4.1, Item 10 in the main text of this chapter for a detailed discussion.
169	木島	Mu Dao	Madeira or Madera; the island is geologically located on the African Tectonic Plate, but culturally, economically and politically it is European.[287]	Madeira is the larger island of an archipelago (it consists of only two islands, plus a small archipelago, in the Macaronesia region) situated in the North Atlantic Ocean. Madeira was claimed by Portuguese sailors in 1419 and settled after 1420. They called the larger island Madeira.[288] *Madeira* means "wood" in Portuguese; the Chinese name 木 means "tree or wood" and 島, "island"; 木島 means an island covered with many trees, because there used to be a subtropical rainforest that once covered the whole island. On the Africa-KWQ, 木島 (Mu Dao) is today's Madeira Island. On the Africa-Mercator (1569) and the Africa-Plancius, the island is depicted as "Madera", but its position corresponds to today's Madeira Island. The island is not depicted on the Africa-Ortelius (1570).
@9	木島去波爾杜瓦爾半月程樹木繁翳地肥美波爾杜瓦爾人至此焚之八年始盡今種葡萄釀酒絕佳		Mu Dao Qu Bo Er Du Wa Er Ban Yue Cheng Shu Mu Fan Yi Di Fei Mei Bo Er Du Wa Er Ren Zhi Ci Fen Zhi Ba Nian Shi Jin Jin Zhong Pu Tao Niang Jiu Jue Jia	My translation and comments (between square brackets) are as follows: Madeira [木島] is about half a month's journey from 波爾杜瓦爾 [see my explanation shown in what follows]. The island is covered with many trees and the land is fertile. It took the people who arrived here from 波爾杜瓦爾 eight years to burn out all the trees to grow grapes nowadays. The wine made from grapes tastes excellent. Madera (木島) is geologically located on the African Tectonic Plate, but the archipelago is culturally, economically and politically European. 波爾杜瓦爾 (Portucale, Portugale or Portugália) was the old name of Portugal before its full independence. I have shown that,[289] in 1179, Portugal's independence was recognised by Pope Alexander III. I have also shown that the European portion of KWQ reflects this political landscape of Europe in the period between 1157 and 1166, before Portugal's official independence was fully recognised.

(*Continued*)

(*Continued*)

Item #	Name	Pinyin	Etymology	Analysis of history, politics, geography; comparison or comment
10	鐵島此島無水泉惟一大樹葉不落每日沒即有雲抱之日出即散土人樹根跑一池雲降成水人畜皆資焉		Tie Dao Ci Dao Wu Shui Quan Wei Yi Da Shu Ye Bu Luo Mei Ri Mo Ji You Yun Bao Zhi Ri Chu Ji San Tu Ren Shu Gen Pao Yi Chi Yun Jiang Cheng Shui Ren Xu Jie Zi Yan	My translation is as follows: On the Iron Island, there is no spring water resource, but only a big tree with leaves that never fall. When the sun sets, there are always clouds around the tree, and when the sun rises the clouds are dispersed. The aboriginals dig a pool near the tree roots to collect water resulting from the condensation of the clouds. Both the men and their livestock benefit from this water. There is a small unnamed island located right next to annotation @10 in Fig. A4.13. This island should be the Iron Island (鐵島), which is the smallest and youngest of the El Hierro Canary Islands.[290] Legends have this: the island is named after the abundant iron ore on the island; but the ironmen claim it is instead named after their own unyielding nature. Jean de Bethencourt (1362–1425), a Norman-French explorer, conquered this small island in the period from 1402 to 1405.[291] The El Hierro Canary archipelago is the second-smallest one of the eight main islands of the Canary Islands, also known informally as the Canaries[292] (a Spanish archipelago in the Atlantic Ocean). At their closest point to the African mainland, they are 100 km/62 mi west of Morocco. They are the southernmost of the autonomous communities of Spain, and are in the African Tectonic Plate, but economically and politically European. None of the three major sixteenth century European maps depicts the Iron Island. This indicates that the Chinese had explored the island, whereas the three European cartographers did not know of its existence.
170	綠峰島	Lu Feng Dao	Cape Verde archipelago	Cape Verde archipelago consists of ten volcanic islands in the central Atlantic Ocean. The name *Cabo Verde* ("Green Cape"; the Cape Verde Peninsula) in Portuguese is the name that Portuguese navigators gave a cape on mainland Africa in c. 1444; they applied the same name to the Cape Verde archipelago they discovered a few years later. This indicates that the name "Cape Verde archipelago"

				was not directly a reference to the geographical condition of the archipelago itself. The Portuguese colonised the islands in 1456, thus establishing the first European settlement in the tropics.[293] But the Chinese explorers derived the name 綠峰島 based on their observation of the geographical condition of the archipelago itself for the following reasons: because of their proximity to the Sahara, most of the Cape Verde islands are dry, but on islands with high mountains and farther away from the African coast, the higher humidity provides a rainforest habitat. On the Africa-KWQ, 綠 means "green", 峰, "mountain or mountain peak" and 島, "island". The name 綠峰島 *precisely* reflects the impact of the orography on the humidity and the rainforest habitat on the island known to the Chinese explorers.
171	仙多默岛	Xian Duo Mo Dao	São Tomé Island: the name is Portuguese for "Saint Thomas". The island is named after St. Thomas the Apostle.	仙多默岛 (Xian Duo Mo Dao) is the transliteration of São Tomé Island, which is the largest island[294] of São Tomé and Príncipe (English: Saint Thomas and Prince). The island is located 2 km/1.25 mi north of the Equator, in the Gulf of Guinea, off the western equatorial coast of central Africa. The island was uninhabited until the Portuguese explorers reached there around 1470.[295] The Africa-KWQ, the Africa-Ortelius (1570) and the Africa-Plancius (1594) depict the São Tomé Island on both sides of the Equator, only the Africa-Mercator (1569) depicts it correctly north of the Equator.
172	仙衣力拿島	Xian Yi Li Na Dao	St. Helena: it was named after Saint Helena of Constantinople.	仙衣力拿島 (Xian Yi Li Na Dao) is the transliteration of Saint Helena Island, which is a remote volcanic tropical island in the Atlantic Ocean lying west of the coast of southwestern Africa. It is one of the most remote islands in the world. There are several stories about how the first European sighted or reached this uninhabited island in the early sixteenth century.[296] In the main text of this chapter, I have given arguments and evidence to show that the Africa-KWQ is based on Chinese geographical information obtained by the Ming mariners in early 1433. As I have written at the beginning of this Appendix, Matteo Ricci most likely has replaced the original Chinese name of the island and used the European island name to make the Africa-KWQ look more European.

(*Continued*)

(*Continued*)

Item #	Name	Pinyin	Etymology	Analysis of history, politics, geography; comparison or comment
● 173	勿自島	Wu Zi Dao	*Vuiza* (?), *isola di*[297]	
174	仙勞冷租島 / 麻打曷失葛	Xian Lao Leng Zu Dao/ Ma Da He Shi Ge	*Isla San Lorenzo*; Madagascar	麻打曷失葛 (Ma Da He Shi Ge) is the transliteration of Madagascar, which is the fourth-largest island in the world. In the Malagasy language, the island of Madagascar is called *Madagasikara*. The island's name Madagascar is not of local origin but rather was popularised in the Middle Ages by Europeans.[298] It was first recorded in the memoirs of thirteenth century Venetian explorer Marco Polo as a corrupted transliteration of the name Mogadishu, the Somali port with which Polo had confused the island.[299] No single Malagasy-language name predating *Madagasikara* appears to have been used by the local population to refer to the island.[300] On St. Laurence's Day in 1500, Portuguese explorer Diogo Dias landed on the island and named it *São Lourenço,* which can be transliterated as 仙勞冷租 (Xian Lao Leng Zu). On the Africa-KWQ, both island names of European origin appear; but Polo's name of the Island, Madagascar, already existed in the thirteenth century, hence, Zheng He's mariners in the fifteenth century would have had the opportunity to know this name. Matteo Ricci must have added the sixteenth century name, San Lorenzo Island, to the Africa-KWQ.
175	沙哥多剌島	Sha Ge Duo La Dao	Zocotora, Zacototera or Zacatora	
● 176	未尔曷六剌	Wei Er He Liu La	*Ueirholieula*[301]	
● 177	檀香樹嶺	Tan Xiang Shu Ling	*Sandalo odoro-so, picco del*[302]; meaning the mountain ridge where the sandal-wood trees grow.	檀香樹 (Tan Xiang Shu) means sandalwood tree. Sandalwood is a class of wood from trees in the genus Santalum. Northern Madagascar does export sandalwood.[303] 嶺 means "ridge". 檀香樹嶺 is the mountain ridge where sandalwood grows.

• 178	曷叭布剌	He Ba Bu La	*Hopapula*[304]	
• 179	利未亞	Li Wei Ya	On the Africa-KWQ, the Africa continent has a Chinese name 利未亞 (Li Wei Ya). This name is not connected with the name "Africa" given by the ancient Greeks and Romans[305] and depicted on the ancient European maps, or with Alkebulan (before the name Africa was created).[306] Libya is the Greek term for the northwestern part of the African continent on some European maps.	
Oceans and seas				
180	河摺亞諾滄	He Zhe Ya Nuo Cang	河摺亞諾 is the transliteration of the first Latin word of *Oceanus Atlanticus* or *Oceano Occidentale*; *Oceanus* (a water deity) is known as the gigantic river that encircled the world; 滄 means "ocean".	On KWQ, Matteo Ricci placed 大西洋 (literally Great Western Ocean) off the Strait of Gibraltar and placed 河摺亞諾滄 (meaning Ocean Sea) off the west coast of Africa, seemingly to indicate that 大西洋 is the smaller water area off the Strait of Gibraltar, whereas 河摺亞諾滄 (*Oceanus*) is the gigantic river that encircled the world in the Greek mythological literature: *Oceanus* was the primordial Titan god of the great, Earth-encircling River Okeanos, font of all of the Earth's fresh-water — rivers, wells, springs and rain-clouds. On all three sixteenth century European maps, *Oceanus* appears in the Southern Hemisphere.
181	亞大蠟海	Ya Da La Hai	Atlantic Ocean or *Oceanvs Atlanticvs* is the ocean's name, derived from Greek mythology; it means the "Sea of Atlas". The oldest known	The term "Atlantic" originally referred specifically to the Atlas Mountains in Morocco as well as the sea off the Strait of Gibraltar and the North African coast. It then expanded to the ocean beyond Gibraltar. During the Age of Discovery (early fifteenth century to early seventeenth century), the Atlantic was also known to English cartographers as the Great Western Ocean, which can be translated as 大西洋. Matteo Ricci was the first European to introduce 大西洋 (the Great Atlantic Ocean) on KWQ. Previously, only 西洋 was used by the Chinese.

(*Continued*)

(*Continued*)

Item #	Name	Pinyin	Etymology	Analysis of history, politics, geography; comparison or comment
			mention of this name is contained in *The Histories of Herodotus*[307] (in c. 450 B.C.). The Greeks thought that there is only one interconnected world ocean, split into several sub-oceans, for them the Atlantic was on the other side of the Pillars of Atlas.	On the Africa-KWQ, 亞大蠟海 (Ya Da La Hai) is off the northwestern coast of Africa and 大西洋 is off the western coast of the Iberian Peninsula, each covers only a portion of today's Atlantic Ocean. Off the northeastern coast of Africa and to the east of the Arabian Sea, the map also depicts an ocean named 小西洋 (Small Western Ocean). This 小西洋 (Small Western Ocean) used to be the 西洋 (Western Ocean) — the Indian Ocean — the Chinese were familiar with. On the Africa-Mercator (1569), the portions of today's Atlantic Ocean on either side of the Equator (the North and the South Atlantic Oceans) are depicted as *Oceanus Æthiopicus*. On the Africa-Ortelius (1570), the portion of the Atlantic Ocean to the north of the Equator is depicted as *Mar del Nort*, meaning the North Sea, and the portion to the south of the Equator is depicted as *Oceanvs Atlanticvs* (Atlantic Ocean). On the Africa-Plancius (1594), the portion of the Atlantic Ocean to the north of the Equator is depicted as *Mar del Nort* (written on the left side of the double hemisphere). But we know that today's North Sea is bounded by the Orkney Islands and the east coast of Great Britain to the west and the northern and central European mainland to the east and south.
• 182	利未亞海	Li Wei Ya Hai	South Atlantic Ocean	On the Africa-KWQ, 利未亞海 (Li Wei Ya Hai) is depicted as the ocean in the Southern Hemisphere, between the 仙衣力拿島 (St Helena Island; Item 172) and the southwestern coast of the African continent; it is today's South Atlantic Ocean. On the three major sixteenth century European maps, the Latin expression *Oceanus Æthiopicus* is the name given to the southern half of today's Atlantic Ocean. This name appeared on European maps from ancient times up to the turn of the nineteenth century. The Chinese nomenclature is unique and has no connection with the Latin expression.

@11	此海有魚善飛但不能高舉掠水平過遠至百餘丈又有白角兒魚能噬之其行水中比飛魚更速善于窺影飛魚畏之遠遁然能伺其影之所向先至其所開口待啖海濱人嘗以白練魚為餌飄搖水面紿為飛魚捕之百發百中烹之其味甚美		Ci Hai You Yu Shan Fei Dan Bu Neng Gao Ju Lue Shui Ping Guo Yuan Zhi Bai Yu Zhang You You Bai Jiao Er Yu Neng Shi Zhi Qi Xing Shui Zhong Bi Fei Yu Geng Su Shan Yu Kui Ying Fei Yu Wei Zhi Yuan Dun Ran Neng Si Qi Ying Zhi Suo Xiang Xian Zhi Qi Suo Kai Kou Dai Dan Hai Bin Ren Chang Yi Bai Lian Yu Wei Er Piao Yao Shui Mian Gei Wei Fei Yu Bu Zhi Bai Fa Bai Zheng Peng Zhi Qi Wei Shen Mei	My translation and comment (between square brackets) are shown as follows: This ocean has fish which are good at flying. But they cannot lift themselves too high above the water level to fly across a hundred *zhang* [1 *zhang* is about 3 *chi*, 1 *chi* is about 1/3 m, hence 1 zhang is about 1 m]. There is also a kind of fish named white-horned fish which can bite the flying fish. The white-horned fish move faster in water than the flying fish and are good at following the shadows of the flying fish. The flying fish are afraid of the white-horned fish and try to distance them. But the white-horned fish can follow the flying fish's shadow to know their moving direction and arrive at their destination beforehand. To swallow the flying fish, the white-horned fish only need to open their mouths and wait for the moment to act. The seaside men often use the bai lian fish as baits, floating them on the water surface, to catch the flying fish when they bite the baits. This method always works. After they are cooked, the fish taste delicious. Again, none of the three major sixteenth century European maps has an annotation so vividly describing the fish and the fishermen's activities in the Atlantic Ocean.
183	西南海	Xi Nan Hai	Indian Ocean	On the Africa-KWQ, 西南海 (Xi Nan Hai) is today's South Indian Ocean. Since Antiquity, Chinese already grasped the technique of measuring latitude on the Earth by observing the height of the Polaris star in the Northern Hemisphere; hence they knew the direction of north well.

(*Continued*)

(*Continued*)

Item #	Name	Pinyin	Etymology	Analysis of history, politics, geography; comparison or comment
				Chinese invented the compass for navigation in the Song Dynasty (960–1279). Their compass has four cardinal directions: clockwise, they are north (北), east (東), south (南) and west (西), each separated by 90 degrees. They are secondarily divided by four ordinal (inter-cardinal) directions: northeast, southeast, southwest and northwest, each located halfway between two cardinal directions. The Chinese also used these directions to name the oceans surrounding their territory and elsewhere in the world. 海 means "sea or ocean": 西南海 is the ocean to the southwest of China, which is the Indian Ocean. The Africa-KWQ has the following oceans or seas: 大東洋 (to the west of California Peninsula), 小東洋 (to the east of Japan), 大西洋 (to the west of Europe), 小西洋 (near the west coast of India), 北海 (near the Arctic region), 南海 (today's South Pacific Ocean), 西南海 (today's South Indian Ocean), 東紅海 (to the east of the California Peninsula) and 西紅海 (the Red Sea). 洋 means "ocean", and 海 means "sea". On the Africa-Mercator (1569), the Indian Ocean is not depicted; on the 1570 *Theatrum Orbis Terrarum* by Ortelius,[308] the Latin expression *Mar Di India* is used to denote the Indian Ocean at its proper location, but this is not done for the Africa-Ortelius (1570). On the Africa-Plancius (1594), *Oceanus Indicus* is displaced to where today's Arabian Sea is. But the Africa-KWQ depicts the Arabian Sea correctly to the northeast of the African continent and to the east of the Red Sea; and finally, to the south of the Arabian Sea lies 西南海 (the Indian Ocean).
184	西紅海	Xi Hong Hai	Red Sea or *Mar Rosso, olim Sinus Arabicus*	Detailed discussion is given in Table 4.1.
185	地中海	Di Zhong Hai	Mediterranean Sea	The Mediterranean Sea separates the European continent from the African continent.

Endnotes

[1]Library of Congress. “Kun yu wan guo quan tu.” www.loc.gov, Library of Congress, 2022, loc.gov/item/2010585650.
[2]Ronnie Po-Chia Hsia. *Matteo Ricci and the Catholic Mission to China, 1583–1610: A Short History with Documents*. Indianapolis, IN, USA: Hackett Publishing Company, Inc., 2016.
[3]“華夷圖.” 自由的百科全书, 百度百科, 2021, baike.baidu.hk/item/華夷圖/9816135.
[4]Charles H. Hapgood. “Chapter five: The ancient maps of the East and West.” *Maps of the Ancient Sea Kings: Evidence of Advanced Civilization in the Ice Age*. Kemptom, IL, USA: Adventures Unlimited Press, 1997, pp. 135–147.
[5]“File:Da-ming-hun-yi-tu.jpg.” *Wikipedia Commons*: *The Free Media Repository*, Wikimedia Foundation, 17 Oct. 2016, upload.wikimedia.org/wikipedia/commons/c/cd/Da-ming-hun-yi-tu.jpg.
[6]Sheng-Wei Wang. “Chapter nine: Secrets revealed by ancient maps.” *The Last Journey of the San Bao Eunuch, Admiral Zheng He*, Hong Kong, China: Proverse Hong Kong, 2019, p. 281.
[7]Jack Weatherford. *Genghis Khan and the Making of the Modern World*. New York, NY, USA: Crown, 2005.
[8]Yun Shi. *Yongle Emperor*. Wuhan, China: Changjiang Literature Press, 2017.
[9]Sheng-Wei Wang. *The Last Journey of the San Bao Eunuch, Admiral Zheng He*. Hong Kong, China: Proverse Hong Kong, 2019; Edward L. Dreyer. *Zheng He: China and the Oceans in the Early Ming Dynasty, 1405–1433*. New York, NY, USA: Pearson College Div, 2006.
[10]Zhao Zhong Nan. *Biography of Emperor Xuande*. Beijing, China: China Social Sciences Press, Ltd., 2008.
[11]Sheng-Wei Wang. *The Last Journey of the San Bao Eunuch, Admiral Zheng He*. *Op. cit*.; Edward L. Dreyer. *Zheng He: China and the Oceans in the Early Ming Dynasty, 1405–1433*. *Op. cit*.
[12]Dr. Sue Gronewald. “The Ming Voyages.” http://afe.easia.columbia.edu/, Asia for Educators, Columbia University, 2022, afe.easia.columbia.edu/special/china_1000ce_mingvoyages.htm.
[13]Sheng-Wei Wang. *The Last Journey of the San Bao Eunuch, Admiral Zheng He*. *Op. cit.*
[14]Swahili is a coastal area of the Indian Ocean in Southeast Africa inhabited by the Swahili people. It includes Sofala (Mozambique), Mombasa, Gede, Pate Island, Lamu, Malindi and Kilwa. In addition, several coastal islands are included in the Swahili coast such as Zanzibar, Comoros and Horn of Africa.
[15]Claudius Ptolemy, *et al*. *Geography of Claudius Ptolemy*. New York, NY, USA: Cosimo Classics, 2011.
[16]“File:PtolemyWorldMap.jpg.” *Wikimedia Commons*, 29 Sept. 2022, upload.wikimedia.org/wikipedia/commons/2/23/PtolemyWorldMap.jpg.
[17]Peter Russell. *Prince Henry the Navigator*. New Haven, CN, USA: Yale University Press, 2001.
[18]Jennifer Swanson. *Bartolomeu Dias: First European Sailor to Reach the Indian Ocean (Spotlight on Explorers and Colonization)*. New York, NY, USA: Rosen Young Adult, 2017.
[19]Nigel Cliff. *The Last Crusade: The Epic Voyages of Vasco da Gama*. New York, NY, USA: Harper Perennial, 2012.
[20]“File:DeVirgaDetail.jpg.” *Wikipedia Commons*, 12 Dec. 2014, upload.wikimedia.org/wikipedia/commons/7/7b/DeVirgaDetail.jpg.
[21]“File:Zhenghemap.jpg.” *Wikipedia Commons*, 6 Apr. 2021, upload.wikimedia.org/wikipedia/commons/b/bc/Zhenghemap.jpg.
[22]Liu Gang 刘钢. *Gu Di Tu Mi Ma: Zhong Guo Fa Xian Shi Jie De Mi Tuan Xuan Ji* 《古地图密码:中国发现世界的谜团玄机》 or *Secret Code of Ancient Maps: The Mystery of China's Discovery of the World*. Guangxi, China: Guangxi Normal University Press Group (广西师范大学出版社集团), 2009.

[23]Sheng-Wei Wang. "Chinese Explored Cape Breton Island Long before the Europeans." *Op. cit.*
[24]"File:FraMauroDetailedMapInverted.jpg." *Wikipedia Commons*, 10 Sept. 2016, upload.wikimedia.org/wikipedia/commons/8/8b/FraMauroDetailedMapInverted.jpg.
[25]"File:FraMauro1420Ship.png." *Wikipedia Commons*, 28 July 2019, upload.wikimedia.org/wikipedia/commons/7/70/FraMauro1420Ship.png.
[26]"File:Cantino planisphere (1502).jpg." *Wikipedia, The Free Encyclopaedia*, Wikimedia Foundation, 25 Aug. 2014, upload.wikimedia.org/wikipedia/commons/9/9c/Cantino_planisphere_(1502).jpg.
[27]"File:Caverio Map circa 1506.jpg." *Wikipedia Commons*, 16 Oct. 2020, upload.wikimedia.org/wikipedia/commons/d/d7/Caverio_Map_circa_1506.jpg.
[28]"File:Ruysch map.jpg." *Wikipedia Commons*, 15 May 2018, upload.wikimedia.org/wikipedia/commons/4/47/Ruysch_map.jpg.
[29]"File:Waldseemuller map 2.jpg." *Wikipedia Commons*, 21 June 2009, upload.wikimedia.org/wikipedia/commons/c/c0/Waldseemuller_map_2.jpg.
[30]"File:PietroCoppo.jpg." *Wikipedia Commons*, 18 Nov. 2020, upload.wikimedia.org/wikipedia/commons/b/b5/PietroCoppo.jpg.
[31]"File:Map Diego Ribero 1529.jpg." *Wikipedia Commons*, 28 May 2022, upload.wikimedia.org/wikipedia/commons/7/74/Map_Diego_Ribero_1529.jpg.
[32]"File:Mercator 1569 world map composite.jpg." *Wikimedia Commons*, 26 Nov. 2016, upload.wikimedia.org/wikipedia/commons/4/4b/Mercator_1569_world_map_composite.jpg.
[33]"File:Africae tabula nova.jpg." *Wikipedia Commons*, 28 Sept. 2021, upload.wikimedia.org/wikipedia/commons/d/d1/Africae_tabula_nova.jpg.
[34]"File:1594 double hemisphere world map by Petrus Plancius.jpg." *Wikipedia Commons*, 7 Oct. 2020, upload.wikimedia.org/wikipedia/commons/0/0d/1594_double_hemisphere_world_map_by_Petrus_Plancius.jpg.
[35]Frank McLynn. *Hearts of Darkness: The European Exploration of Africa*. London, UK: Sharpe Books, 2020.
[36]"File:Kunyu Wanguo Quantu by Matteo Ricci Plate 1-3.jpg." *Wikipedia Commons*, 23 Oct. 2020, upload.wikimedia.org/wikipedia/commons/b/b8/Kunyu_Wanguo_Quantu_by_Matteo_Ricci_Plate_1-3.jpg.
[37]"File:ZA-WC-kap-vom-flugzeug-1.jpg." *Wikipedia Commons*, 3 Feb. 2021, commons.wikimedia.org/wiki/File:ZA-WC-kap-vom-flugzeug-1.jpg.
[38]Jeannie Evers, *et al.*, eds. "cape." https://education.nationalgeographic.org/, National Geographic Society, 2022, education.nationalgeographic.org/resource/cape.
[39]Christian Larbi Ayisi. "Review: Reproductive biology of Cichlidae." *International Journal of Agricultural Sciences*, vol. 4, no. 8, Aug. 2014, pp. 248–251. www.internationalscholarsjournals.org.
[40]Merriam-Webster Inc. "cichlid." www.merriam-webster.com, Merriam-Webster, Inc., 2022, merriam-webster.com/dictionary/cichlid.
[41]Gomes Eanes de Zurara. *Chronica do descobrimento e conquisita de Guiné (Portuguese Edition)*, Neuilly sur Seine, France: Ulan Press, 2012, p. 1; *op. cit.*, p. 158.
[42]Huang Shijian (黄时鉴) and Gong Yingyan (龚缨晏). *Li Ma Dou Shi Jie Di Tu Yan Jiu* 《利玛窦世界地图研究》 or *Research on Matteo Ricci's World Map*. Shanghai, China: Shanghai Chinese Classics Publishing House (上海古籍出版社), 2004, p. 202.
[43]David Brawn. *Fuerteventura Tour & Trail Super-Durable Map*. London, UK: Discovery Walking Guides Ltd, 2018.
[44]George Glas. *The History of the Discovery and Conquest of the Canary Islands*. Schleswig-Holstein, Germany: Hansebooks, 2019.

[45]T.C. Bell, private email communication; Raoul McLaughlin. *Rome and the Distant East: Trade Routes to the Ancient Lands of Arabia, India and China*. London, UK: Continuum, 2010.
[46]Adrian Goldsworthy. *The Punic Wars*. London, UK: Cassell, 2001.
[47]Kevin Shillington. *History of Africa*. New York, NY, USA: Red Globe Press, 2018; Cheikh Anta Diop. Mercer Cook, ed. *The African Origin of Civilization: Myth or Reality*. New York, NY, USA: Lawrence Hill & Company, 1989. As found in the electronic version on the Internet at https://www.almendron.com/tribuna/wp-content/uploads/2019/10/african-origin-of-civilization-complete.pdf.
[48]"File:OrteliusWorldMap1570.jpg." *Wikipedia Commons*, 1 June 2022, commons.wikimedia.org/wiki/File:OrteliusWorldMap1570.jpg.
[49]Giacomo Gastaldi. "The Drawing of the Modern Geography of the Whole Africa." www.wdl.org, Library of Congress, 22 Mar. 22, 2016, wdl.org/en/item/6764/.
[50]*Ibid.*
[51]*Ibid.*
[52]Sheng-Wei Wang. "Chapter nine: Secret revealed by ancient maps." *The Last Journey of the San Bao Eunuch, Admiral Zheng He, op. cit*., p. 290.
[53]"File:CapeHopeDetail.png." *Wikipedia Commons*, 31 Oct. 2020, upload.wikimedia.org/wikipedia/commons/4/4e/CapeHopeDetail.png.
[54]Mao Yuanyi (茅元仪; 1594–1640). Nanjing Shi Di Fang Zhi Bian Zuan Wei Yuan Hui Ban Gong Shi (南京市地方志编纂委员会办公室) or Office of the Compilation Committee of Nanjing Atlas, ed. *Zheng He Hang Hai Tu*《郑和航海图》or *The Charts of Zheng He's Voyages*. Nanjing, China: Nanjing Publishing House (南京出版社), 2019.
[55]Sheng-Wei Wang. "Chapter nine: Secret revealed by ancient maps." *The Last Journey of the San Bao Eunuch, Admiral Zheng He, op. cit*., p. 278.
[56]Peter Forbath. *The River Congo: The Discovery, Exploration, and Exploitation of the World's Most Dramatic River*. Boston, MA, USA: Houghton Mifflin, 1991.
[57]"File:River Congo.svg." *Wikimedia Commons*, 1 Aug. 2022, upload.wikimedia.org/wikipedia/commons/thumb/1/13/River_Congo.svg/2502px-River_Congo.svg.png.
[58]Philip Koslow. *Kanem-Borno: 1,000 Years of Splendor (The Kingdoms of Africa)*. Broomall, PA, USA: Chelsea House Publishers, 1995; Amadou Ba. *Kanem Bornu 8th–19th Centuries: Record of Longevity of an Empire in the Heart of Africa. Kindle Edition*. Canada: AB Editions, 2021.
[59]*Ibid.*
[60]*Ibid.*
[61]*Ibid.*
[62]Kevin White and David J. Mattingly. "Ancient lakes of the Sahara." *American Scientist*, vol. 94, no. 1, 2006, pp. 58–65. doi:10.1511/2006.57.983.
[63]Sebastian Münster. "Evolution of the Map of Africa." https://library.princeton.edu/, Princeton Library, 2022, library.princeton.edu/visual_materials/maps/websites/africa/maps-continent/continent.html.
[64]MacGregor Laird. *Narrative of an Expedition into the Interior of Africa: By the River Niger, in the Steam-Vessels Quorra and Alburkah, in 1832, 1833 and 1834 (Vol. 1)*. London, UK: Alpha Edition, 2021.
[65]Sylvia Sikes. *Lake Chad Versus the Sahara Desert: A Great African Lake in Crisis*. New Berry, UK: Mirage, 2003.
[66]Gerald J. Rizzo. "The Inland Niger Delta: A Cartographic Beacon." www.afriterra.org, Afriterra Foundation, 2022, https://web.archive.org/web/20160303223241/http://www.afriterra.org/Research/InlandNigerDelta/InlandNigerDelta_files/v3_document.htm.
[67]"File:Map of River Niger.svg." *Wikipedia Commons*, 8 Sept. 2022, upload.wikimedia.org/wikipedia/commons/thumb/7/79/Map_of_River_Niger.svg/2560px-Map_of_River_Niger.svg.png.

[68]Chapter 3 of this book.
[69]John Stewart. *African States and Rulers: An Encyclopedia of Native, Colonial and Independent States and Rulers Past and Present*. Jefferson, NC, USA: McFarland Publishing, 1989.
[70]"File:Old Portuguese map of SA 529.JPG." *Wikipedia Commons*, 6 Sept. 2020, commons.wikimedia.org/wiki/File:Old_Portuguese_map_of_SA_529.JPG.
[71]Sheng-Wei Wang. "Chapter three: Luo Maodeng's Tianfang/Yun Chong was Mecca in Saudi Arabia." *The Last Journey of the San Bao Eunuch, Admiral Zheng He*, *op. cit*., pp. 91–138.
[72]*Ibid.*
[73]*Ibid.*
[74]Dr. Sue Gronewald. "The Ming Voyages." *Op. cit.*
[75]Siu-Leung Lee (李兆良). "Kunyu Wanguo Quantu Cheng Tu Duan Dai (坤輿萬國全圖成圖斷代)." *Kunyu Wanguo Quantu Jie Mi: Ming Dai Ce Hui Shi Jie*《坤輿萬國全圖解密:明代測繪世界》or *Deciphering the Kunyu Wanguo Quantu, A Chinese World Map — Ming Chinese Mapped the World before Columbus*, Taipei, Taiwan: Linking Publishing Company (聯經出版社), 2012, pp. 53–54.
[76]Chapter 3 of this book.
[77]Sheng-Wei Wang. "Chapter three: Luo Maodeng's Tianfang/Yun Chong was Mecca in Saudi Arabia." *The Last Journey of the San Bao Eunuch, Admiral Zheng He*, *op. cit*., pp. 91–138.
[78]*Ibid.*
[79]*Ibid.*
[80]Chapter 2 of this book.
[81]Information Office of the People's Government Fujian Province. *Zheng He's Voyages Down the Western Seas*. Fujian, China: Information Office of the People's Government Fujian Province, 1 Jan. 2005, p. 43; https://books.google.com.hk/books?id=QmpkR6l5MaMC&pg=PA43&lpg=PA43&dq=how+many+days+does+it+take+from+Calicut+to+Mogadishu+by+sea&source=bl&ots=IuCIJzD0Jv&sig=ACfU3U1FJr4a1Pl2tzsB7MNihrDuVm0VGA&hl=zh-TW&sa=X&ved=2ahUKEwi5y4DqiN7zAhXLKqYKHbo9B6IQ6AF6BAgQEAM#v=onepage&q=how%20many%20days%20does%20it%20take%20from%20Calicut%20to%20Mogadishu%20by%20sea&f=false.
[82]United States Hydrographic Office. *West Coast of Africa, from Cape Spartel to Cape Agulhas: Including the Islands in the Bight of Biafra, Ascension, St. Helena, Tristan Da Cunha, and Gough Islands*. Open Library: Arkose Press, 2015.
[83]Jennifer Swanson. *Bartolomeu Dias: First European Sailor to Reach the Indian Ocean (Spotlight on Explorers and Colonization)*. *Op. cit.*
[84]"File:ZA-WC-kap-vom-flugzeug-1.jpg." *Op. cit.*
[85]W. H. Ingrams. *Zanzibar: Its History and Its People*. Oxfordshire, UK: Psychology Press, 1967, p. 99.
[86]*Ibid.*
[87]"File:Spice Islands (Zanzibar highlighted).svg." *Wikipedia Commons*, 9 Oct. 2021, upload.wikimedia.org/wikipedia/commons/thumb/6/67/Spice_Islands_(Zanzibar_highlighted).svg/1662px-Spice_Islands_(Zanzibar_highlighted).svg.png.
[88]Huang Shijian (黄时鉴) and Gong Yingyan (龚缨晏). *Op. cit*, p. 188.
[89]Huang Shijian (黄时鉴) and Gong Yingyan (龚缨晏). *Op. cit*., p. 192.
[90]William Tronzo. *The Cultures of His Kingdom: Roger II and the Cappella Palatina in Palermo*, Princeton, NJ, USA: Princeton University Press, 1997, p. 96.
[91]Huang Shijian (黄时鉴) and Gong Yingyan (龚缨晏). *Op. cit*., p. 189.
[92]Huang Shijian (黄时鉴) and Gong Yingyan (龚缨晏). *Op. cit*., p. 205.

[93]John Stewart. *African States and Rulers: An Encyclopedia of Native, Colonial and Independent States and Rulers Past and Present. Op. cit.*
[94]*Ibid.*
[95]*Ibid.*
[96]Roger Crowley. *Conquerors: How Portugal Forged the First Global Empire*. New York, NY, USA: Random House, 2015, p. 24.
[97]Kat Eschner. "The Peculiar Story of Giraffes in 1400s China." www.smithsonianmag.com, Smithsonian, 21 June 2017, smithsonianmag.com/smart-news/peculiar-story-giraffes-medieval-china-180963737/.
[98]Huang Shijian (黄时鉴) and Gong Yingyan (龚缨晏). *Op. cit.*, p. 193.
[99]Huang Shijian (黄时鉴) and Gong Yingyan (龚缨晏). *Op. cit.*, p. 187.
[100]Huang Shijian (黄时鉴) and Gong Yingyan (龚缨晏). *Op. cit.*, p. 208.
[101]Hiob Ludolf. *A New History of Ethiopia: Being a Full and Accurate Description of the Kingdom of Abessinia*, Schleswig-Holstein, Germany: Hansebooks, 2016, p. 14.
[102]Huang Shijian (黄时鉴) and Gong Yingyan (龚缨晏). *Op. cit.*, p. 194.
[103]UNESCO World Heritage Convention. "Island of Mozambique." https://whc.unesco.org/, UNESCO World Heritage Convention, 2022, whc.unesco.org/en/list/599.
[104]Malyn Newitt. *A Short History of Mozambique*. Oxford, UK: Oxford University Press, 2017.
[105]Sanjay Subrahmanyam. *The Career and Legend of Vasco da Gama*. Cambridge, UK: Cambridge University Press, 1997.
[106]Malyn Newitt. *A Short History of Mozambique. Op. cit.*
[107]Huang Shijian (黄时鉴) and Gong Yingyan (龚缨晏). *Op. cit.*, p. 203.
[108]Huang Shijian (黄时鉴) and Gong Yingyan (龚缨晏). *Op. cit.*, p. 188.
[109]Huang Shijian (黄时鉴) and Gong Yingyan (龚缨晏). *Op. cit.*, p. 188.
[110]John Hanning Speke. *The Discovery of the* Source *of the Nile*. San Francisco, CA, USA: Blurb, 2021.
[111]NASA. "Quest for the Source of the Nile." https://earthobservatory.nasa.gov/, Earth Observatory-NASA, 2021, earthobservatory.nasa.gov/images/7236/quest-for-the-source-of-the-nile.
[112]M. K. Oliver and P. V. Loiselle. "A new genus and species of cichlid of the mbuna group (Pisces: Cichlidae) from Lake Malawi." *Revue de Zoologie et de Botanique Africaines*, vol. 85, 1972, pp. 309–320.
[113]Christian Larbi Ayisi. "Review: Reproductive biology of Cichlidae." *International Journal of Agricultural Sciences*. *Op. cit*. www.internationalscholarsjournals.org.
[114]"File:Rift en.svg." *Wikipedia Commons*, 8 Sept. 2020, upload.wikimedia.org/wikipedia/commons/thumb/0/0a/Rift_en.svg/1672px-Rift_en.svg.png.
[115]"File:River Nile map.svg." *Wikipedia Commons*, 15 Feb. 2022, upload.wikimedia.org/wikipedia/commons/thumb/2/25/River_Nile_map.svg/1118px-River_Nile_map.svg.png.
[116]Peter Forbath. *The River Congo: The Discovery, Exploration, and Exploitation of the World's Most Dramatic River. Op. cit.*
[117]Oscar Nideo, *et al*. "Mollusk communities of the central Congo River shaped by combined effects of barriers, environmental gradients, and species dispersal." *Journal of Limnology*, vol. 76, 2017. doi:10.4081/jlimnol.2017.1585.
[118]Huang Shijian (黄时鉴) and Gong Yingyan (龚缨晏). *Op. cit.*, p. 202.
[119]Huang Shijian (黄时鉴) and Gong Yingyan (龚缨晏). *Op. cit.*, p. 189.
[120]Mukenge Tshilemalema. *Culture and Customs of the Congo*, Santa Barbara, CA, USA: Greenwood Press, 2001, p. 18; Alisa LaGamma. *Kongo: Power and Majesty*. New York, NY, USA: Metropolitan Museum of Art, 2015.

[121]Mark R. Lipschutz and R. Kent Rasmussen. *Dictionary of African Historical Biography*. Berkeley, CA, USA: University of California Press, 1989, p. 150.
[122]Ernest George Ravenstein. *The Voyages of Diogo Cão and Bartholomeu Dias, 1482–1488*. Pretoria, South Africa: State Library of Pretoria, 1986.
[123]Huang Shijian (黄时鉴) and Gong Yingyan (龚缨晏). *Op. cit.*, p. 185.
[124]Ernest George Ravenstein. *The Voyages of Diogo Cão and Bartholomeu Dias, 1482–1488*. *Op. cit*; Beazley, Charles Raymond. "Cam (Cão), Diogo." In Hugh Chisholm, ed. *Encyclopædia Britannica, Vol. 5 (11th ed.)*. Cambridge, UK: Cambridge University Press, 1911, p. 79.
[125]*Ibid.*
[126]Anne Hilton. *The Kingdom of Kongo (Oxford Studies in African Affairs)*. Oxford, UK: Oxford University Press, 1985.
[127]"File:KingdomKongo1711.png." *Wikimedia Commons*, 13 Sept. 2020, commons.wikimedia.org/wiki/File:KingdomKongo1711.png.
[128]Ernest George Ravenstein. *The Voyages of Diogo Cão and Bartholomeu Dias, 1482–1488*. *Op. cit.*
[129]Huang Shijian (黄时鉴) and Gong Yingyan (龚缨晏). *Op. cit.*, p. 185.
[130]Huang Shijian (黄时鉴) and Gong Yingyan (龚缨晏). *Op. cit.*, p. 195.
[131]Huang Shijian (黄时鉴) and Gong Yingyan (龚缨晏). *Op. cit.*, p. 206.
[132]Huang Shijian (黄时鉴) and Gong Yingyan (龚缨晏). *Op. cit.*, p. 184.
[133]Hacke, William. "11) Magadoxa, Melinde, Mombaza, Quiloa." www.loc.gov, Library of Congress, 2021, loc.gov/resource/g7401pm.gct00372/?sp=22.
[134] Bing Zhao. "Chinese-style ceramics in East Africa from the 9th to 16th century: A case of changing value and symbols in the multi-partner global trade." https://journals.openedition.org/, OpenEdition Journals, 2022, journals.openedition.org/afriques/1836?lang=en.
[135]Kat Eschner. "The Peculiar Story of Giraffes in 1400s China." *Op. cit.*
[136]Leo Africanus. Robert Brown, ed., John Pory, trans. "A description of places undescribed by Leo." *The History and Description of Africa: And of the Notable Things Therein Contained, Volume I*, Cambridge, UK: Cambridge University Press, 2010, p. 40.
[137]*Ibid.*
[138]The European Space Agency. "Lake Tana, source of the Blue Nile." www.esa.int, The European Space Agency, 2022, esa.int/Applications/Observing_the_Earth/Lake_Tana_source_of_the_Blue_Nile.
[139]Isabel Boavida, *et al.*, eds. *Pedro Páez's History of Ethiopia, 1622/Volume I*. London, UK: Hakluyt Society, 2011; *op. cit.*, *Volume II*, 2019.
[140]A. H. M. Jones and Elizabeth Monroe. *The History of Abyssinia*. Whitefish, MT, USA: Kessinger Publishing, LLC, 2010; Harold G. Marcus. *A History of Ethiopia Updated Edition*. Berkeley, CA, USA: University of California Press, 2002.
[141]*Ibid.*
[142]*Ibid.*
[143]*Ibid.*
[144]Huang Shijian (黄时鉴) and Gong Yingyan (龚缨晏). *Op. cit.*, p. 184.
[145]J. P. Gent. "The history of Ethiopia: or the Kingdom of the Abessines." https://quod.lib.umich.edu/, Early English Books Online Text Creation Partnership, 2011, quod.lib.umich.edu/e/eebo2/A49450.0001.001/1:4.1.3?rgn=div3;view=fulltext.
[146]Leo Africanus. Robert Brown, ed., John Pory, trans. "A description of places undescribed by Leo." *The History and Description of Africa: And of the Notable Things Therein Contained, Vol. I*, *op. cit.*, p. 39.
[147]*Ibid.*

148*Ibid.*
149Huang Shijian (黄时鉴) and Gong Yingyan (龚缨晏). *Op. cit.*, p. 184.
150Leo Africanus. Robert Brown, ed., John Pory, trans. "A description of places undescribed by Leo." *The History and Description of Africa: And of the Notable Things Therein Contained, Vol. I, op. cit.*, p. 59.
151E. G. Ravenstein, ed. *A Journal of the First Voyage of Vasco da Gama, 1497–1499 (Hakluyt Society, First Series)*. London, UK: Hakluyt Society, 2010.
152Haggai Erlich. *The Cross and the River: Ethiopia, Egypt, and the Nile*. Boulder, CO, USA: Lynne Rienner Publishers, 2001.
153R. Strachey. "Meteorology of the Red Sea and Cape Guardafui." *Proceedings of the Royal Geographical Society and Monthly Record of Geography*, vol. 10, no. 11, Nov. 1888, pp. 704–708. https://www.jstor.org/stable/1801400.
154Huang Shijian (黄时鉴) and Gong Yingyan (龚缨晏). *Op. cit.*, p. 208.
155Leo Africanus. Robert Brown, ed., John Pory, trans. "John Leo his first book of the description of Africa." *The History and Description of Africa: And of the Notable Things Therein Contained, Vol. I, op. cit.*, p. 128.
156Henry Harrison Epps Jr. Ellery Truesdell, narr. *The Rise and Fall of the Ethiopian Empire: The Lost Tribe of Israel (Audible Audiobook)*. Southaven, MS, USA: Henry Epps Inc., 2013.
157Paul Kriwaczek. *Babylon: Mesopotamia and the Birth of Civilization*. Washington, DC, USA: Atlantic, 2012.
158*Ibid.*
159Jonathan Shepard, ed. *The Cambridge History of the Byzantine Empire c. 500–1492 Revised Edition*. Cambridge, UK: Cambridge University Press, 2019.
160Nezar AlSayyad. *Cairo: Histories of a City*. Cambridge, MA, USA: Belknap Press, 2013.
161A. J. Butler. *Babylon of Egypt: A Study in the History of Old Cairo*. Oxford, UK: Oxford University Press, 1914.
162Huang Shijian (黄时鉴) and Gong Yingyan (龚缨晏). *Op. cit.*, p. 207.
163Bruce Brande. *The River Nile*. Washington, DC, USA: National Geographic Society, 1966.
164*Ibid.*
165*Ibid.*
166Lynn Meskell. *Private Life in New Kingdom Egypt*. Princeton, NJ, USA: Princeton University Press, 2002, p. 34; Kathryn A. Bard, ed. *Encyclopedia of the Archaeology of Ancient Egypt*. Oxfordshire, UK: Routledge, 2014, p. 694.
167Huang Shijian (黄时鉴) and Gong Yingyan (龚缨晏). *Op. cit.*, p. 195.
168Christopher Clapham. "The European Mapping of Ethiopia, 1460–1856." *Journal of Ethiopian Studies*, vol. 40, no. 1/2, Festschrift Dedicated in Honour of Prof. Richard Pankhurst & Mrs. Rita Pankhurst (June–December 2007), pp. 293–307. https://www.jstor.org/stable/41988232.
169Photograph by David Boyer. "Nile River." https://education.nationalgeographic.org/, National Geographic, 2022, education.nationalgeographic.org/resource/nile river.
170Henri J. Dumont. *The Nile: Origin, Environments, Limnology and Human Use*. New York, NY, USA: Springer, 2016.
171Marshall Clagett. *Ancient Egyptian Science: A* Source *Book. Volume Two: Calendars, Clocks, and Astronomy*. Philadelphia, PA, USA: 2004.
172Huang Shijian (黄时鉴) and Gong Yingyan (龚缨晏). *Op. cit.*, p. 198.
173Huang Shijian (黄时鉴) and Gong Yingyan (龚缨晏). *Op. cit.*, p. 199.
174Huang Shijian (黄时鉴) and Gong Yingyan (龚缨晏). *Op. cit.*, p. 202.

[175]Leo Africanus. Robert Brown, ed., John Pory, trans. "John Leo his first book of the description of Africa." *The History and Description of Africa: And of the Notable Things Therein Contained, Vol. I, op. cit*., p. 128.
[176]Huang Shijian (黄时鉴) and Gong Yingyan (龚缨晏). *Op. cit*., p. 204.
[177]Huang Shijian (黄时鉴) and Gong Yingyan (龚缨晏). *Op. cit*., p. 207.
[178] "Map of Africa from 1669." catalog.afriterra.org. Afriterra Foundation. Archived from the original on 24 July 2011. Retrieved 14 March 2010; "Map of Africa from 1669." catalog.afriterra.org (zoomMap). Afriterra Foundation. Archived from the original on 24 July 2011. Retrieved 14 March 2010; "Map of North-West Africa, 1829." U. T. Libraries. TX, USA: University of Texas. Archived from the original on 8 August 2009. Retrieved 14 March 2010.
[179]Huang Shijian (黄时鉴) and Gong Yingyan (龚缨晏). *Op. cit*., p. 201.
[180]Huang Shijian (黄时鉴) and Gong Yingyan (龚缨晏). *Op. cit*., p. 187.
[181]Huang Shijian (黄时鉴) and Gong Yingyan (龚缨晏). *Op. cit*., p. 205.
[182]Dominique Malaquais. *The Kingdom of Benin*. London, UK: Franklin Watts, 1998.
[183]*Ibid*.
[184]Lombard Jacques. "The Kingdom of Dahomey." In: C. D. Forde and P. M. Kaberry, eds. *West African kingdoms in the Nineteenth Century*, Oxford, UK: Oxford University; Londres: International African Institute, 1967, pp. 70–92. https://horizon.documentation.ird.fr/exl-doc/pleins_textes/divers17-07/010056146.pdf.
[185]*Ibid.*
[186]*Ibid.*
[187]Leo Africanus. Robert Brown, ed., John Pory, trans. "His description of places undescribed by Leo." *The History and Description of Africa: And of the Notable Things Therein Contained, Vol. I, op. cit*., p. 78.
[188]John William Blake, ed. *Europeans in West Africa, 1540–1560: Volume I: Documents to Illustrate the Nature and Scope of Portuguese Enterprise in West Africa, the Abortive ... and Guinea (Hakluyt Society, Second Series).* London, UK: Hakluyt Society, 2017.
[189]*Ibid*.
[190]Leo Africanus. Robert Brown, ed., John Pory, trans. "His description of places undescribed by Leo." *The History and Description of Africa: And of the Notable Things Therein Contained, Vol. I, op. cit*., p. 78.
[191]Huang Shijian (黄时鉴) and Gong Yingyan (龚缨晏). *Op. cit*., p. 191.
[192]Leo Africanus. Robert Brown, ed., John Pory, trans. "John Leo his first book of the description of Africa." *The History and Description of Africa: And of the Notable Things Therein Contained, Vol. I, op. cit*., p. 128.
[193]John Ogilby. *Africa Being an Accurate Description of Africa*. Online Early English Books: Site Creation Partnership, 2021, p. 322. https://quod.lib.umich.edu/e/eebo2/A70735.0001.001/1:8.7.3?rgn=div3;view=fulltext.
[194]John O. Hunwick. "The mid-fourteenth century capital of Mali." *Journal of African History*, vol. 14, no. 2, 1973, pp. 195–206. doi:10.1017/s0021853700012512. JSTOR 180444.
[195]*Ibid*.
[196]John O. Hunwick. *Timbuktu and the Songhay Empire: Al-Saʿdi's Taʾrīkh Al-Sūdān Down to 1613, and Other Contemporary Documents*. Leiden, Netherlands: Brill, 2003.
[197]*Ibid*.
[198]*Ibid*.
[199]Leo Africanus. Robert Brown, ed., John Pory, trans. "John Leo his first book of the description of Africa." *The History and Description of Africa: And of the Notable Things Therein Contained, Vol. I, op. cit*., p. 128.
[200]David C. Conrad. *Empires of Medieval West Africa: Ghana, Mali, and Songhay (Great Empires of the Past)*. New York, NY, USA: Facts on File, 2005.

[201]Kevin Shillington. *History of Africa*, London, UK: Palgrave Macmillan, 2012, pp. 93, 101.
[202]Michael Gomez. *African Dominion: A New History of Empire in Early and Medieval West Africa*. Princeton, NJ, USA: Princeton University Press, 2019; Captivating History. *The Mali Empire: A Captivating Guide to One of the Largest Empires in West African History and the Legendary Mansa Musa*. Captivating History, 2022.
[203]John O. Hunwick. *Timbuktu and the Songhay Empire: Al-Saʿdi's Taʾrīkh Al-Sūdān Down to 1613, and Other Contemporary Documents*. *Op. cit.*; Roland Oliver and Anthony Atmore. *Medieval Africa, 1250–1800*. Cambridge, UK: Cambridge University Press, 2001; Captivating History. *The Songhai Empire: A Captivating Guide to One of the Largest States of Medieval West Africa*. Captivating History, 2022.
[204]Gomes Eanes de Zurara. *Chronica do descobrimento e conquisita de Guiné (Portuguese Edition), op. cit.*, p. 1; *op. cit.*, p. 158.
[205]Huang Shijian (黄时鉴) and Gong Yingyan (龚缨晏). *Op. cit.*, p. 198.
[206]Huang Shijian (黄时鉴) and Gong Yingyan (龚缨晏). *Op. cit.*, p. 202.
[207]Huang Shijian (黄时鉴) and Gong Yingyan (龚缨晏). *Op. cit.*, p. 187.
[208]Huang Shijian (黄时鉴) and Gong Yingyan (龚缨晏). *Op. cit.*, p. 199.
[209]Anonymous author. *A true history of the African chief Jingua and his comrades: with a description of the Kingdom of Mandingo, and of the manners and customs of the inhabitants, an account of King Sharka, of Gallinas: a sketch of the slave trade and horrors of the middle passage, with the proceedings on board the "long, low, black schooner", Amistad*. Hartford, UK: Published at Hartford, New York and Boston, for the booksellers, 1839. As found in the electronic version on the Internet at https://collections.library.yale.edu/catalog/2068873.
[210]*Ibid.*
[211]Leo Africanus. Robert Brown, ed., John Pory, trans. "John Leo his first book of the description of Africa." *The History and Description of Africa: And of the Notable Things Therein Contained, Vol. I, op. cit.*, p. 128.
[212] E. J. Arnett, "A Hausa Chronicle." *Journal of the Royal African Society*, vol. 9, 1910.
[213] *Ibid.*
[214]Huang Shijian (黄时鉴) and Gong Yingyan (龚缨晏). *Op. cit.*, p. 202.
[215]Huang Shijian (黄时鉴) and Gong Yingyan (龚缨晏). *Op. cit.*, p. 206.
[216]G. Camps (1999). "Gudâla/Guezula." *Encyclopédie berbère (in French)*, vol. 21, 1999, pp. 3223–3224. doi:10.4000/encyclopedieberbere.1788. Retrieved 18 Sept. 2020; J. Desanges. "Autolatae/Autololes/Autoteles." *Encyclopédie berbère (in French)*. vol. 8, 1990, pp. 1175–1176. doi:10.4000/encyclopedieberbere.154; Hector MacLean. "On the Kimmerian and Atlantean Races." *The Journal of the Anthropological Institute of Great Britain and Ireland*, vol. 1, 1872, xl–lxi. doi:10.2307/2841286. JSTOR 2841286.
[217]Huang Shijian (黄时鉴) and Gong Yingyan (龚缨晏). *Op. cit.*, p. 203.
[218]Huang Shijian (黄时鉴) and Gong Yingyan (龚缨晏). *Op. cit.*, p. 187.
[219]Aucourt, Louis chevalier de. Naomi J. Andrews, trans. "Senegal, Kingdom of." In Scholarly Publishing Office — University of Michigan Library. *The Encyclopedia of Diderot & d'Alembert Collaborative Translation Project*. Ann Arbor, MI, USA: Michigan Publishing, University of Michigan Library, 2007.
[220]Eunice A. Charles. *Precolonial Senegal: The Jolof Kingdom, 1800–1890*. Boston, MA, USA: African Studies Centre, Boston University, 1977, p. 3; Anthony Ham. *West Africa*. Fort Mill, SC, USA: Lonely Planet, 2009, p. 670.
[221]*Ibid.*
[222]Ferdinand Édouard Buisson. *Dictionnaire de pédagogie et d'instruction primaire*, 1887, p. 442; as found in the electronic version on the Internet at https://gallica.bnf.fr/ark:/12148/bpt6k24232h/f1301.item.texteImage.

[223]Duncan Castlereagh. *Encyclopaedia of Discovery and Exploration — The Great Age of Exploration*. London, UK: Aldus, 1971.
[224]"File:Empire ghana.png." *Wikipedia Commons*, 11 Sept. 2020, commons.wikimedia.org/wiki/File:Empire_ghana.png.
[225]"File:MALI empire map.PNG." *Wikipedia Commons*, 20 Feb. 2021, upload.wikimedia.org/wikipedia/commons/8/8f/MALI_empire_map.PNG.
[226]"File:Songhai Empire (orthographic projection).svg." *Wikipedia Commons*, 12 June 2020, upload.wikimedia.org/wikipedia/commons/thumb/f/f4/Songhai_Empire_(orthographic_projection).svg/2048px-Songhai_Empire_(orthographic_projection).svg.png.
[227]"File:Portuguese Empire map.jpg." *Wikipedia Commons*, 5 Sept. 2020, upload.wikimedia.org/wikipedia/commons/8/8e/Portuguese_Empire_map.jpg.
[228]National Geospatial-Intelligence Agency, Bethesda, MD, USA. "Cape Verde Peninsula: Senegal." https://geographic.org/, Geographical Names, 2022, geographic.org/geographic_names/name.php?uni=-3127495&fid=5740&c=senegal.
[229]Duncan Castlereagh. *Encyclopaedia of Discovery and Exploration — The Great Age of Exploration. Op. cit.*
[230]Department of the Arts of Africa, Oceania, and the Americas. "The Trans-Saharan Gold Trade (7th-14th Century Century)." www.metmuseum.org, The Metropolitan Museum of Art, Oct. 2000. metmuseum.org/toah/hd/gold/hd_gold.htm.
[231]Huang Shijian (黄时鉴) and Gong Yingyan (龚缨晏). *Op. cit.*, p. 196.
[232]Leo Africanus. Robert Brown, ed., John Pory, trans. "John Leo his first book of the description of Africa." *The History and Description of Africa: And of the Notable Things Therein Contained, Vol. I, op. cit.*, p. 128.
[233]Huang Shijian (黄时鉴) and Gong Yingyan (龚缨晏). *Op. cit.*, p. 199.
[234]Huang Shijian (黄时鉴) and Gong Yingyan (龚缨晏). *Op. cit.*, p. 195.
[235]Huang Shijian (黄时鉴) and Gong Yingyan (龚缨晏). *Op. cit.*, p. 198.
[236]Huang Shijian (黄时鉴) and Gong Yingyan (龚缨晏). *Op. cit.*, p. 203.
[237]Leo Africanus. Robert Brown, ed., John Pory, trans. "John Leo his first book of the description of Africa." *The History and Description of Africa: And of the Notable Things Therein Contained, Vol. I, op. cit.*, p. 124.
[238]Janice Kamrin and Adela Oppenheim. "The Land of Nubia." www.metmuseum.org, The Metropolitan Museum of Art, 2022, metmuseum.org/about-the-met/collection-areas/egyptian-art/temple-of-dendur-50/nubia; Dietrich Raue. *Handbook of Ancient Nubia*. Berlin, Germany: Walter de Gruyter GmbH, 2019.
[239]The British Museum. "Ancient Nubia and the Kingdom of Kush, an introduction." www.khanacademy.org, The British Museum, 2022, khanacademy.org/humanities/ancient-art-civilizations/egypt-art/x7e914f5b:kingdom-of-kush/a/ancient-nubia-and-the-kingdom-of-kush-an-introduction.
[240]Huang Shijian (黄时鉴) and Gong Yingyan (龚缨晏). *Op. cit.*, p. 186.
[241]Huang Shijian (黄时鉴) and Gong Yingyan (龚缨晏). *Op. cit.*, p. 205.
[242]Huang Shijian (黄时鉴) and Gong Yingyan (龚缨晏). *Op. cit.*, p. 207.
[243]Huang Shijian (黄时鉴) and Gong Yingyan (龚缨晏). *Op. cit.*, p. 186.
[244]Huang Shijian (黄时鉴) and Gong Yingyan (龚缨晏). *Op. cit.*, p. 192.
[245]Huang Shijian (黄时鉴) and Gong Yingyan (龚缨晏). *Op. cit.*, p. 191.
[246]Huang Shijian (黄时鉴) and Gong Yingyan (龚缨晏). *Op. cit.*, p. 198.
[247]Huang Shijian (黄时鉴) and Gong Yingyan (龚缨晏). *Op. cit.*, p. 194.
[248]Dierk Lange. "The Mune as the Ark of the Covenant between Duguwa and Sefuwa (in ancient Kanem)." *Borno Museum Society Newsletter*, vol. 66–67, 2006, pp. 15–25. http://dierklange.com/pdf/

contents/12Mune-Symbol_as_the_Ark.pdf; the article has a map (p. 6) of the ancient Central Sahara and proposes to identify Agisymba of 100 A.D. with the early Kanem state.
[249]Gerald J. Rizzo. "The Inland Niger Delta: A Cartographic Beacon." *Op. cit.*
[250]Huang Shijian (黄时鉴) and Gong Yingyan (龚缨晏). *Op. cit.*, p. 195.
[251]Robin R. Wang. *Yinyang: The Way of Heaven and Earth in Chinese Thought and Culture*. Cambridge, UK: Cambridge University Press, 2012.
[252]Huang Shijian (黄时鉴) and Gong Yingyan (龚缨晏). *Op. cit.*, p. 184.
[253]Munn & Co. "Dates and the Date Palm." *Scientific American*, vol. XXXVI, no. 8, 24 Feb. 1877. https://chestofbooks.com/crafts/scientific-american/XXXVI-8/Dates-And-The-Date-Palm.html.
[254]Huang Shijian (黄时鉴) and Gong Yingyan (龚缨晏). *Op. cit.*, p. 186.
[255]"File:Cyrenaica Marmarica.jpg." *Op. cit.*
[256]University of Minnesota. "15th Century, Marmarica." https://umedia.lib.umn.edu/, University of Minnesota, 2021, umedia.lib.umn.edu/item/p16022coll251:2824.
[257]Huang Shijian (黄时鉴) and Gong Yingyan (龚缨晏). *Op. cit.*, p. 202.
[258]"File:Cyrenaica Marmarica.jpg." *Wikipedia Commons*, 13 June 2021, commons.wikimedia.org/wiki/File:Cyrenaica_Marmarica.jpg.
[259]Huang Shijian (黄时鉴) and Gong Yingyan (龚缨晏). *Op. cit.*, p. 193.
[260]Huang Shijian (黄时鉴) and Gong Yingyan (龚缨晏). *Op. cit.*, p. 198.
[261]Leo Africanus. Robert Brown, ed., John Pory, trans. "John Leo his first book of the description of Africa." *The History and Description of Africa: And of the Notable Things Therein Contained, Vol. I, op. cit.*, pp. 126–127.
[262]M. Ferroukhi and Bob Idjennaden. *The Kingdom of Numidia (The Forgotten Civilisations of Africa Book 1). Kindle Edition.* 2012; James McDougall. *A History of Algeria.* Cambridge, UK: Cambridge University Press, 2017; SB. "Does Algeria have any former names?" www.answers.com, Answer, LLC., 2021, answers.com/Q/Does_Algeria_have_any_former_names.
[263]*Ibid.*
[264]*Ibid.*
[265]"File:Exarchate-Africa.png." *Wikipedia Commons*, 18 Oct. 2020, commons.wikimedia.org/wiki/File:Exarchate-Africa.png.
[266]James McDougall. *A History of Algeria*. Cambridge, UK: Cambridge University Press, 2017; "Algeria." www.historyfiles.co.uk, Kessler Associates, 1999–2021, historyfiles.co.uk/KingListsAfrica/AfricaAlgeria.htm.
[267]Leo Africanus. Robert Brown, ed., John Pory, trans. "John Leo his first book of the description of Africa." *The History and Description of Africa: And of the Notable Things Therein Contained, Vol. I, op. cit.*, pp. 126 and 129.
[268]Huang Shijian (黄时鉴) and Gong Yingyan (龚缨晏). *Op. cit.*, p. 188.
[269]Lionel Casson. "Barbaria." In: Glen W. Bowersock, *et al.*, eds. *Late Antiquity: A Guide to the Postclassical World*, Cambridge, MA, USA: Belknap Press of Harvard University Press, 1999, p. 334.
[270]Peter M. Rossi and Wayne Edward White. *Articles on the Middle East, 1947–1971: A Cumulation of the Bibliographies from the Middle East Journal*, Ann Arbor, MI, USA: Pierian Press, University of Michigan, 1980, p. 132; Paul Sebag. *Tunis: Histoire d'une ville. Histoire et Perspectives Méditerranéennes*. Paris, France: L'Harmattan, 1998.
[271]The Barbary States were a collection of North African states, many of which practiced state-supported piracy to exact tribute from weaker Atlantic powers.

[272]Leo Africanus. Robert Brown, ed., John Pory, trans. "John Leo his first book of the description of Africa." *The History and Description of Africa: And of the Notable Things Therein Contained, Vol. I, op. cit.*, p. 126.
[273]UNESCO. "Kairouan." https://whc.unesco.org/, UNESCO World Heritage Convention, 2022, whc.unesco.org/en/list/499.
[274]Sheng-Wei Wang. "Chapter two: Ma Huan's Tianfang/Mo-jia (天方/默伽) was ancient Tunisia in North Africa." *The Last Journey of the San Bao Eunuch, Admiral Zheng He, op. cit.*, pp. 53–90.
[275]Ma Huan. J. V. G. Mills, trans. *Ying-yai Sheng-lan: The Overall Survey of the Ocean's Shores (1433)*. London, UK: Hakluyt Society, 1970.
[276]Eamonn Gearon. *The Sahara: A Cultural History*. Oxford, UK: Oxford University Press, 2011, p. 117.
[277]Huang Shijian (黄时鉴) and Gong Yingyan (龚缨晏). *Op. cit.*, p. 185.
[278]Huang Shijian (黄时鉴) and Gong Yingyan (龚缨晏). *Op. cit.*, p. 202.
[279]Phillip C. Naylor. *Historical Dictionary of Algeria*, Lanham, MD, USA: Rowman & Littlefield Publishers, 2015, p. 376.
[280]Duncan Fishwick. "The annexation of Mauretania." *Historia: Zeitschrift für Alte Geschichte Bd*, vol. 20, no. 4, 1971, pp. 467–487. JSTOR, http://www.jstor.org/stable/4435213. Accessed 25 Oct. 2022.
[281]Huang Shijian (黄时鉴) and Gong Yingyan (龚缨晏). *Op. cit.*, p. 194.
[282]John Ogilby. "An accurate description of Africa, Lybia, or Zaaea." https://quod.lib.umich.edu/, Early English Books (eebotcp), 2022, quod.lib.umich.edu/e/eebo2/A70735.0001.001/1:8.6.3?rgn=div3;view=fulltext.
[283]Huang Shijian (黄时鉴) and Gong Yingyan (龚缨晏). *Op. cit.*, p. 206.
[284]John G. Hall. *North Africa*. New York, NY, USA: Infobase Publishing, 2002; C. R. Pennell. *Morocco: From Empire to Independence*. London, UK: Oneworld Publications, 2009.
[285]C. R. Pennell. *Morocco: From Empire to Independence. Op. cit.*
[286]Leo Africanus. Robert Brown, ed., John Pory, trans. "John Leo his first book of the description of Africa." *The History and Description of Africa: And of the Notable Things Therein Contained, Vol. I, op. cit.*, p. 169.
[287]Document 12002E299. "EUR-Lex — 12002E299 — EN." https://eur-lex.europa.eu/, An official website of the European Union, 2017, eur-lex.europa.eu/legal-content/EN/ALL/?uri=CELEX%3A12002E299.
[288]Gomes Eanes de Zurara. Charles Raymond Beazley and Edgar Prestage, trans. *The Chronicle of the Discovery and Conquest of Guinea (Cambridge Library Collection — Hakluyt First Series)*. Cambridge, UK: Cambridge University Press, 2010.
[289]Chapter 3 of this book.
[290]"File:Map of the Canary Islands.svg." *Wikipedia Commons*, 24 Oct. 2021, commons.wikimedia.org/wiki/File:Map_of_the_Canary_Islands.svg.
[291]Jean de Béthencourt. Richard Henry Major, ed. *The Canarian, or, Book of the Conquest and Conversion of the Canarians in the Year 1402*. London, UK: Hakluyt Society, 2010.
[292]Valentin R. Troll and Juan C. Carracedo. *The Geology of the Canary Islands*. Amsterdam, Netherlands: Elsevier, 2016.
[293]Richard A. Lobban. *Cape Verde: Crioulo Colony to Independent Nation*. Oxfordshire, UK: Routledge, 1998.
[294]"File:Obo National Park.png." *Wikipedia Commons*, 14 Sept. 2020, commons.wikimedia.org/wiki/File:Obo_National_Park.png.
[295]Albertino Francisco and Nujoma Agostinho. *Exorcising Devils from the Throne: São Tomé and Príncipe in the Chaos of Democratization*, New York, NY, USA: Algora Publishing, 2011, p. 24.

[296]Ian Bruce. "St Helena Day." *The Journal of the Friends of St Helena*, no. 44, 2015, pp. 32–46; A. E. Chamot. M. A. Aldanov, trans. *Saint Helena Little Island.* New York, NY, USA: Knopf, 1924.
[297]Huang Shijian (黄时鉴) and Gong Yingyan (龚缨晏). *Op. cit.*, p. 187.
[298]William Edward Cousins. *Madagascar of To-Day: A Sketch of the Island, with Chapters on Its Past History and Present Prospects*. London, UK: The Religious Tract Society, 1895, pp. 11–12.
[299]Adrian Room. *Placenames of the World: Origins and Meanings of the Names for 6,600 Countries, Cities, Territories, Natural Features, and Historic Sites*, Jefferson, NC, USA: McFarland, 2006, p. 230.
[300]*Ibid.*
[301]Huang Shijian (黄时鉴) and Gong Yingyan (龚缨晏). *Op. cit.*, p. 188.
[302]Huang Shijian (黄时鉴) and Gong Yingyan (龚缨晏). *Op. cit.*, p. 207.
[303]Samuel F. Sanchez. "The State and Raw Commodities in Madagascar: A Historical Approach to Madagascar's State-regulated External Trade in the 19th and 20th Centuries." *Afrique Contemporaine*, vol. 251, no. 3, 2014, pp. 157–166.
[304]Huang Shijian (黄时鉴) and Gong Yingyan (龚缨晏). *Op. cit.*, p. 199.
[305]Rukewve Ochuko. "What Is Africa's Original Name?" https://guardian.ng/, Guardian, 2022, guardian.ng/life/what-is-africas-original-name/.
[306]*Ibid.*
[307]Herodotus. John M. Marincola, ed. Aubrey de Sélincourt, trans. *The Histories*. London, UK: Penguin Classics, 2003.
[308]"File:OrteliusWorldMap1570.jpg." *Op. cit.*

Chapter 5

Did Zheng He's Mariners Explore the Americas Before Columbus? An Analysis Offers Essential New Insights

Abstract

Did Ming mariners explore the Americas?

This chapter analyses the Americas depicted on the world map Kunyu Wanguo Quantu《坤輿萬國全圖》or the Complete Geographical Map of All the Kingdoms of the World, published by Matteo Ricci in 1602 in China. The work thoroughly examines the region's 17 original annotations, and the history and coordinates of 241 geographical items: one unnamed river which I conclude to be the Mississippi River, 17 named rivers, three named lakes and 220 other geographical items of chiefdoms/kingdoms, regions, cities/towns/villages, mountains, islands, bays, gulfs, seas and oceans. These geographical items are compared with their counterparts on the three major sixteenth century European maps and today's map. The comparison involves identifying more than one thousand geographical coordinates based on different Primary Meridians used by these maps. The results show that (1) the Americas of Kunyu Wanguo Quantu are of Chinese origin; Matteo Ricci merely edited the Chinese source map to make it look more European; and (2) the Ming Treasure Fleets led by Admiral Zheng He (郑和) to the Western Ocean have explored the Americas.

Additional findings are: (1) the Americas of Kunyu Wanguo Quantu reflect the political landscape of the region between 1400 and 1428, decades before Christopher Columbus set foot in the Bahamas; (2) after taking into account other Chinese historical records, this political era can be narrowed to between the 1420s and 1428, a period overlapping with the sixth voyage of Zheng He's Treasure Fleets but preceding their seventh (and last) voyage, implying the geographical data contained in the Americas on Kunyu Wanguo Quantu were obtained during their sixth voyage; (3) the map shows correctly and uniquely the southernmost tip of the South American mainland (Cape Froward), the southwestern tip of Alaska (Cape Peirce; also the westernmost tip of the American continent), and the northernmost part of the all-encompassing World Ocean (Arctic Ocean); (4) the map shows correctly that Cape Breton Island in the east coast of the Americas was separated into two islands by an ancient Chinese-built canal; and (5) this work lends support to the claim that Chinese might have migrated to the Americas more than 4,000 years ago as written in *Shan Hai Jing* 《山海經》 or *Classic of Mountains and Seas*.

Keywords: Alaska, Americas, Arctic Ocean, Cape Breton Island, Kunyu Wanguo Quantu, Matteo Ricci, *Shan Hai Jing*, Treasure Fleets, Zheng He

1. Introduction

The important world map — Kunyu Wanguo Quantu 《坤輿萬國全圖》, abbreviated here as KWQ[1] — written with Chinese characters and having latitudinal and longitudinal lines, has been regarded as being drawn in 1602 by Matteo Ricci — an Italian Jesuit priest and one of the founding figures of the Jesuit China missions — and his Chinese collaborators, based on the European maps which Ricci brought with him to China in 1582.[2]

In the previous four chapters, I analysed and provided evidence to show that: (1) on KWQ, the map of Cape Breton Island (in today's southeastern Canada), and the map of Australia, New Zealand, Land of Fire and Antarctica embedded inside Mo Wa La Ni Jia (墨瓦蠟泥加) or *Terra Australis* of KWQ, are based on geographical information obtained in the 1420s by the Ming (明代; 1388–1644) mariners; the date overlaps with their sixth voyage to the Western Ocean; (2) the map of Europe on KWQ reflects the political landscape of Europe between 1157 and 1166, overlapping with the era of the Southern Song Dynasty (南宋; 1127–1279); and (3) the map of Africa on KWQ reflects the political landscape of Africa in 1433, and is based on geographical information obtained in 1433 by the Ming mariners during their seventh voyage to the Western Ocean. I have provided strong evidence to show that in all these cases, Matteo Ricci merely edited some of the geographical names on the Chinese source maps (no longer in existence) and brought in the *Terra Australis Incognita* (Unknown South Land), a hypothetical continent, to make the resulting KWQ look more European and to match the era of his stay in China.

The goal of this chapter is to examine the American portion of KWQ — abbreviated as America-KWQ — to find its true origin. Also, N. America-KWQ and S. America-KWQ will be used as abbreviations for the North American and South American portions of America-KWQ, respectively. The three major sixteenth century European maps to be compared with the America-KWQ are: (1) the Americas extracted from Gerardus Mercator's world map of 1569, abbreviated as America-Mercator (1569); (2) the 1570 *Americae Sive Novi Orbis, Nova Descriptio* by Abraham Ortelius, abbreviated as America-Ortelius (1570); and (3) the Americas extracted from the 1594 double hemisphere world map by Petrus Plancius, abbreviated as America-Plancius (1594). I will use abbreviations like N. America-Mercator (1569), S. America-Mercator (1569), N. America-Ortelius (1570), S. America-Ortelius (1570), N. America-Plancius (1594) and S. America-Plancius (1594) to denote North and South America on each cartographer's map of the Americas.

2. Europeans Were Greatly Interested in the New World

According to the Sagas of Icelanders, Norse sailors (often called Vikings) from Iceland first settled in Greenland in the 980s (they themselves claim starting in the ninth century). In other parts of

North America, L'Anse aux Meadows on the northernmost tip of Newfoundland, and a second site in southwestern Newfoundland, are the only known sites of a Norse village. But the Viking voyages did not become common knowledge in the Old World, and Europeans remained unaware of the existence of the Americas as a whole, until the first decades following the year 1492 when Christopher Columbus (1451–1506) landed on the Bahamas.

But how did Columbus get the idea of sailing west?

He was inspired by the Florentine mathematician, astronomer and cosmographer Paolo dal Pozzo Toscanelli (1397–1482)[3] who proposed in 1474 to the Portuguese court to sail west as a shortcut to reach the fabled Spice Islands in the east. We do not know what kind of map was in Toscanelli's letter that was sent to the court. Although he underestimated the Earth's circumference,[4] and Columbus underestimated the Earth's diameter and overestimated the width of Asia, Columbus took Toscanelli's map with him on his first transatlantic voyage in 1492 and landed in the Bahamas.

The Vikings did not draw maps. Christopher Columbus, likewise, never left behind any known maps of his various explorations. It was the Castilian navigator and cartographer Juan de la Cosa (c. 1450–1510), who accompanied him on up to three expeditions to the American continent, who drew in 1500 the first known European map. The map shows the new continent in greenish colour on bull leather (dimension 93 cm × 183 cm)[5] (see Fig. 5.1, left panel).

The map aroused the Europeans' enthusiasm in exploring and conquering this new land for wealth and territories. The combined North American[6] and South American[7] continents stretch c. 14,000 km/8,700 mi[8] from north to south, nearly touching both the Arctic and Antarctic Circles, and boast an area of over c. 16.43 million square miles, which is only a little shy of 30 percent of the total land area of the Earth (about 57.31 million square miles).

The Europeans started exploring and conquering the Americas in earnest from the sixteenth century, as evidenced by the second map (see Fig. 5.1, right panel), published in 1507 by German cartographer and humanist scholar Martin Waldseemüller (c. 1470–1520). The map gives the name "America" for the first time to this new continent, in honour of Amerigo Vespucci (1451–1512) who participated in at least two voyages between 1497 and 1504 during the Age of Discovery.[9]

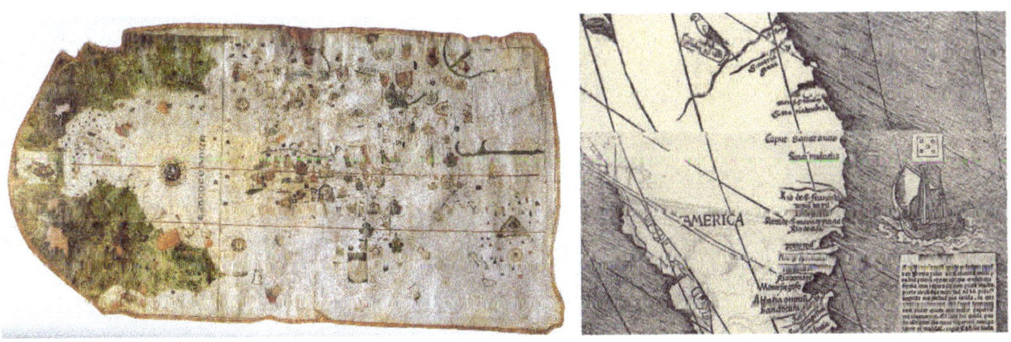

Fig. 5.1. The left panel shows the new continent in greenish colour on the 1500 Juan de la Cosa map (see endnote 5, public domain); the right panel is extracted from Martin Waldseemüller's 1507 world map (see endnote 10, public domain), which depicts the name "America" for the first time.

3. Europeans Rushed to the Americas During the Sixteenth Century, but a Vast Area Remained Unexplored

Early on, scholars believed that prior to the arrival of the Europeans, the Americas were inhabited by the first migrants who walked, over many generations, up to the northeast extreme of Asia and across a land bridge, since submerged by ocean, and then reached Alaska.[11] They settled in their new lands and became the indigenous peoples of the Americas. However, this is just one of the several theories[12] and there is no consensus. One of the other theories includes suggestions[13] and scientific evidence[14] showing that these migrants crossed the North Pacific Ocean by boat.

Figure 5.2 shows a map of some major European explorations of North America and a portion of South America from the late fifteenth century to 1602.

This figure shows that before 1602 when Matteo Ricci published KWQ in China, a vast area of North America had not been reached by Europeans, yet many geographical items unexplored are already depicted on the three major sixteenth century European maps. This "vast unexplored area" should cover North America roughly north of 37°N (the northern latitude of Arkansas reached by the Spanish explorer and conquistador Hernando de Soto) and west of 77°W

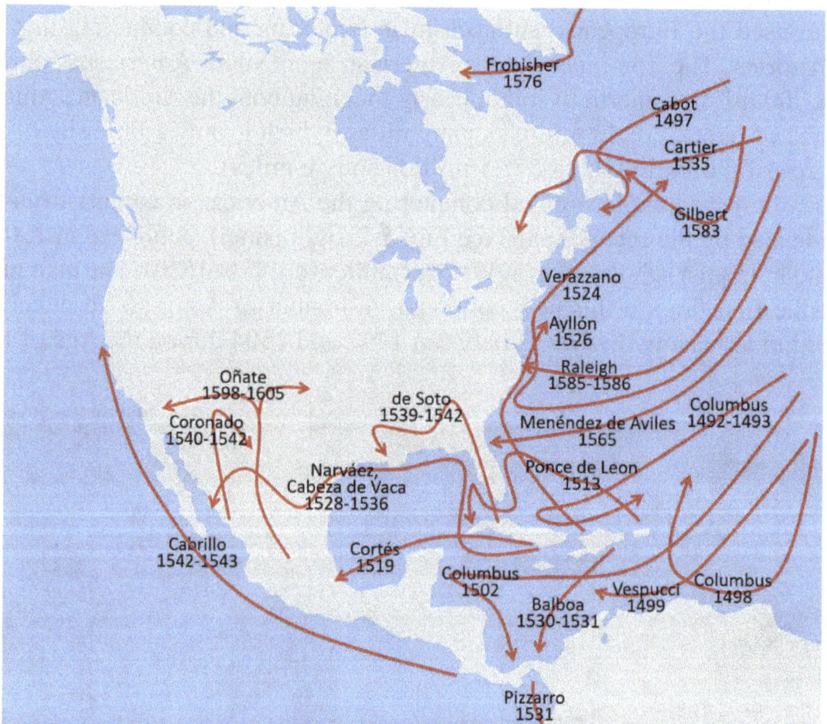

Fig. 5.2. Some major European explorations of North America and a small part of South America from the late fifteenth century to 1602 (map drawn by Michel A. Van Hove).

(Hudson Bay East: c. 59°N, c. 77°W; Hudson Bay was first reached by Europeans in 1611[15] after KWQ was published).

Where and from whom did the European cartographers get their geographical information? It seems an impossible task to imagine depicting so many unknown place names and locations if they were not obtained by "copying". Hence, to resolve the issue, in Section 4 of this chapter, I categorise the geographical items on the America-KWQ into the following non-overlapping groups and compare them to their counterparts on the three prominent sixteenth century European maps selected in this chapter to find clues.

These non-overlapping groups are: (1) geographical items showing up on the America-KWQ, but not on any of the three European maps; (2) geographical items of Chinese origin, but not included in the first group (to avoid double counting); the geographical information in groups (1) and (2) must have been obtained from the Chinese sources; (3) geographical items on both the America-KWQ and any of the three European maps, with written historical records showing that they were explored by Europeans before 1602; and (4) geographical items depicted on both the America-KWQ and any of the three Europeans maps, but without written historical records and without modern English names.

For geographical items in group (4), it is possible that some of these places appear in repositories that are difficult to access. However, it is discovered that most of these places fall into the vast region unexplored by Europeans as shown in Fig. 5.2, while exhaustive searches on the Internet cannot find them, so they most likely were not explored by the Europeans. The reason is as follows: the pre-colonial indigenous peoples did not have writing, their written history only started from the days of the European exploration or colonisation; no written history implies "not explored or colonised by the Europeans", hence, these geographical items should not appear on a European map.

However, the appearance of geographical items in group (4) on any of the three European maps means that their cartographers had two possibilities to acquire the information: (1) they heard about these places from non-official sources (through legends, accounts of fishermen, traders or their own imagination, etc.); or (2) they copied these geographical data from a non-European map source — the only likely source was the Chinese maps, because in the pre-colonial era only China had the capability, historical track record and motivation to engage in such marine activities to gather geographical data about the world. In the Appendix, I make remarks on these geographical items which fall into group (4) but give no further comment on the data sources of the European cartographers.

I also want to point out that because geographical items in group (4) were most likely obtained through "copying", because of cartographers' lack of on-hand exploration experiences, large errors in coordinates often show up, as opposed to those geographical items in group (3). This will become very clear as I proceed with my analysis and present the detailed comparisons in the Appendix.

Another point to make is that before the European colonisation, the Americas did not have nation-states of the sort that we know today (with some possible exceptions, such as the Aztec Empire),[16] but they did have nations of a sort or chiefdoms for peoples who shared common cultural traits, especially language. However, the Aztec Empire (1428–1521) is not depicted on the

America-KWQ and the three European maps to be analysed. This information reveals the political landscape of the Americas of these maps and can give guidance to the determination of the era revealed by the America-KWQ.

4. The America-KWQ Is of Chinese Origin, Not a Copy or Adapted Copy of the Three Major Sixteenth Century European Maps

The four maps to be analysed in great depth in Fig. 5.3 are: (1) the America-Mercator (1569); (2) the America-Ortelius (1570); (3) the America-Plancius (1594) and (4) the America-KWQ.

My analysis is as follows:

(1) The America-KWQ depicts a better shape of the American continent and gives smaller longitudinal errors in general than do the three European maps. European cartographers made progress in leaps and bounds in their mapmaking because of their navigators, explorers and conquistadors sailing to the Americas, colonizing and opening trade routes. However, there are serious problems in their maps with longitudinal measurements because their Americas are grossly expanded in the longitudinal direction at most latitudes.

For example, for North America, from the east to the west coast, the maximum longitudinal width of the North American continent is c. 160°, c. 145°, c. 160° and c. 122°, respectively, for the N. America-Mercator (1569), the N. America-Ortelius (1570), the N. America-Plancius (1594) and the N. America-KWQ. On a modern globe, that width is only about 110°. This tells us that the N. America-KWQ is more realistic in longitudinal shape and often gives smaller longitudinal errors of geographical items than the three European maps.

For South America, from the east to the west coast, the maximum longitudinal width is c. 60°, c. 60°, c. 62° and c. 58°, respectively, for the S. America-Mercator (1569), the S. America-Ortelius (1570), the S. America-Plancius (1594) and the S. America-KWQ. The real width is about 48°. This tells us that the S. America-KWQ will also give smaller longitudinal errors than the three European maps, but their differences will be smaller in comparison with the North American maps.

Moreover, for North America, since the Europeans had not explored the vast area to the north of c. 37°N and to the west of 77°W, the drawing of that area on the three European maps is fanciful. For example, the shapes of the western coast of today's Alaska are poorly represented on these European maps and the geographical items depicted within this area must have been copied from non-European sources such as Chinese maps. Detailed comparisons of the relevant geographical items are made in the Appendix.

For South America, the three European maps present a rounded-shape South America in comparison with today's narrower triangular-shaped South America. The S. America-KWQ, on the other hand, gives a closer representation of the triangular-shaped South America.

Overall, the N. America-KWQ depicts a better representation of the western coast of Alaska and the S. America-KWQ depicts a more realistic triangular-shaped South America. Chinese people have known these two regions since Antiquity: *Shan Hai Jing* 《山海經》 or *Classic of Mountains and Seas* (also known as *Shan Hai Jing*) records that Chinese migrated to the Americas

Fig. 5.3. The top left panel shows the America-Mercator (1569; see endnote 17, public domain); the top right panel shows the America-Ortelius (1570; see endnote 18, public domain; Credit: Library of Congress Geography and Map Division); the bottom left panel shows the America-Plancius (1594; see endnote 19, public domain) and the bottom right panel shows the America-KWQ (see endnote 20, public domain).

more than 4,000 years ago. Detailed explanations of the relevant geographical items are made in the Appendix.

Moreover, when comparing the America-KWQ with the three European maps, we also notice that the latitudes of the Northern and Southern Hemispheres of the former do not terminate at 90°N and 90°S, respectively, but a little more than 90°. The reason has been explained in Section 3 of Chapter 1 of this book. In brief: ancient Chinese not only knew that the Earth is round, but some of them also knew that it takes 365.25 days for the Earth to revolve around the sun; hence, the Chinese set a complete circular rotation as 365.25° (365.25 degrees). Since the Earth is round, in the above passage, the two poles of the Earth are therefore a little more than 182.5° apart (365.25°/2 = 182.625°). The Equator is the celestial body's belt, being a little more than 91° from each Pole (182.625°/2 = 91.3125°). Hence, there is a small but fundamental difference between the latitudinal coordinates on the America-KWQ and those on the three European maps, albeit the differences may be around 1.5 percent. Nevertheless, for the longitudinal coordinates, all four maps use 360° to go around the Earth at any given latitude.

Since the four maps use different primary meridians, for direct comparisons, I have converted all the longitudinal readings on these maps to their corresponding readings using the modern Greenwich Meridian. The formulas to use are as follows:

To convert longitude "x" from the America-KWQ, use $y = [360 - (x - 17.5)]$°W

To convert longitude "x" from the America-Mercator (1569), use $y = [360 - (x - 17.5)]$°W

To convert longitude "x" from the America-Ortelius (1570), use $y = [360 - (x - 23)]$°W

To convert longitude "x" from the America-Plancius (1594), use $y = [360 - (x - 23)]$°W

The basis of these conversions is as follows: the westernmost point on the African mainland is Pointe des Almadies on Cape Verde Peninsula (the westernmost point of Senegal and the African mainland) at 14.7°N, 17.5°W;[21] and Boa Vista, Cape Verde (a desert-like island that belongs to the Cape Verde Islands) is located at 16.1°N, and c. 23°W (the more precise reading is 22.8°W).

A quick glance at the three European maps in Fig. 5.3 and their comparison with Fig. 5.2 tells us immediately that there are many geographical names depicted in the vast unexplored region of North America on these three European maps. You may ask: from whom did these European cartographers get their geographical data? How much does the America-KWQ differ from them? These questions have prompted me to make a detailed and in-depth analysis of these four maps to resolve the issues.

(2) There are 17 annotations on the America-KWQ which are absent on the three European maps, the only exception being the annotation about the "Country of the Tall People". These 17 annotations are realistic descriptions (see Items @1 to @17 in the Appendix) of the Americas, including geography, natural environment (lake, islands, rivers) and habits (sleeping habits and cannibalism), life styles, living conditions, clothing, transportation methods, wars, animals, birds, plants and produce on land and sea. Since these annotations are not present on the European maps (except in one case), and since Matteo Ricci was working on KWQ in China, the information can only come from Chinese sources, and was obtained by the Chinese explorers who had visited these places since Antiquity including Zheng He's mariners.

For example, concerning the Country of the Tall People (Item 160 in the Appendix), in 1520 Ferdinand Magellan (1480–1521) reached Patagonia/Country of the Tall People, and did discover that the Patagons (Tehuelche) tended to be taller than Europeans of the time.[22] In fact, according to the Chinese record, there is already an account of the Country of the Tall People in a chapter titled "*Da Huang Dong Jing* 《大荒東經》" or "Classic of the Great Wilderness: East", of the Chinese Classic *Shan Hai Jing* 《山海經》 or *Classic of Mountains and Seas* (also known as *Shan Hai Jing*).[23] The book is a Chinese classic text and a compilation of mythic geography and beasts, and may have existed since as early as the fourth century B.C.[24] The era it describes can be traced back to the Xia Dynasty (c. 2070 B.C.–1600 B.C.) more than 4,000 years ago. The content of the book involves detailed descriptions of locations in the cardinal directions of the Mountains, Regions beyond Seas, Regions within Seas, and Wilderness inside and outside of China. In the Appendix of this chapter, Annotation @12 gives a detailed description of this "Country of the Tall People" and mentions that the men and women there coloured their faces as decoration. This special custom was not annotated in any of the three European maps.

(3) Among the 241 geographical items depicted on the America-KWQ, 97 + 1 (the "1" is due to the unnamed Mississippi River) of them (about 41 percent) are not depicted in any of the three European maps; and among the 97 items, 22 are of Chinese origin, eleven are of European origin (added by Matteo Ricci to the source map of America-KWQ), the remaining 64 are derived from aboriginal languages; hence, Chinese must have explored at least 87 (64 + 22 + 1; the "1" is for the unnamed Mississippi River) of these places before the Europeans arrived. In my analysis, the America-KWQ is categorised into six zones: Zone I: North America (Eastern Region); Zone II: North America (Western Region); Zone III: North America (Central America Region, Islands and Oceans/Seas); Zone IV: South America (Northern Region); Zone V: South America (Western Region); and Zone VI: South America (Eastern Region).

In the Appendix, Items denoted by "•" are geographical items which do not appear on the America-Mercator (1569), or the America-Ortelius (1570), or the America-Plancius (1594) maps. The 97 terms are shown in Table 5.1. For brevity, all the references related to these geographical items are cited in the Appendix where each item is discussed in detail.

(4) There are twelve more geographical items on the America-KWQ with a Chinese origin. These terms have no connection with names given by the Europeans after they arrived at the Americas or with the aboriginal names, and this means that the Chinese must have explored these twelve places. Table 5.2 lists these geographical items and gives strong evidence of Chinese presence in the Americas before the Europeans arrived. Each item is followed by a second identification such as "Item ≈ 1" or "Item 18", to indicate its listing position "Item ≈ 1" or "Item 18" in the Appendix. To avoid repetitions, all the references related to these geographical items are cited in the Appendix. In Table 5.1, there are already 22 geographical items of Chinese origin, but to avoid double counting, they are not included again in Table 5.2.

From Table 5.1, we already know that the Chinese explorers visited at least 86 (here, the 11 items of European origin and added by Matteo Ricci are not included) places in the Americas. With the inclusion of these additional 12 places in Table 5.2, the total number of places the Chinese explored grows to at least 98.

Table 5.1. Ninety-seven geographical items, including 22 of Chinese origin, depicted on the America-KWQ, but not on any of the three European maps.

Zone	**Category**	**Number of geographical items not on any of the European maps; number of items of Chinese origin or added by Matteo Ricci is inside brackets.**	**Geographical items not depicted on any of the three European maps; these are places which I conclude to have been explored by the Chinese before Europeans arrived. There are a few items which are added by Matteo Ricci. Items in "bold" characters are of Chinese origin.**
I	North America (Eastern Region)	11 (6, 0)	勒革氏國 (Lagus; Item 2), 革利國 (Coree Chiefdom; Item 3), 訝巴剌亦尔 (Iapalair; Item 4), **眾仙河** (Alabama River; ≈ 2), 何多亞蠟瓦 (Hotuoiolaua; Item 10), **甘峰** (Chon, promontorio di; Item 18), **香峰** (Hian, promontorio; Item 19), 泥德 (Fiste; Item 20), **冰海** (Arctic Ocean; Item 21), **如里漢島** (Rulihan Islabd/part of Cape Breton Island; Item 22), **鬼島** (Ghost Island/ part of Cape Breton Island; Item 23)
II	North America (Western Region)	24 (9, 3)	沙蠟丁諾 (Scialatimno; Item 27), **林濱** (Linpin; Item 28), 亞如的私的 (Aiotuxtit; Item 30), 新瓦力茶 (Nueva Galicia; Item 40), 火吒蠟瓦 (Huocialaua; Item 42), 七城國 (Cibola; Item 47), 亞哈庫 (Auacal; Item 52), 多兒瓦瓦 (Tuoruaua; Item 55), 哥入河 (Cogib, *fiume*; ≈ 3), 亞外嫣 (Iauaema; Item 57), 北亞墨利加 (North America; Item 58), 亞沙 (Ascia; Item 61), 沙兒倍 (Sciarpé; Item 62), 多籠 (Tuolom; Item 63), 白吳尔 (Peur; Item 70), **平地坡** (*Pianura, Pendio delta*; Item 72), **夜叉國** (A kingdom on the Wrangel Island; Item 74), **流鬼** (Demoni vagabonki; Item 75), **冰海** (Arctic Ocean; Item 76), **北海** (North Sea; Item 77), **別山** (Los Boleanes or Pie, monte; Item 79), **大東洋** (North Pacific Ocean; Item 80), 止會 (Cehoei; Item 85), **曚山** (*Mom*, monte; Item 88) *Note*: 新瓦力茶 (Nueva Galicia; Item 40), 七城國 (Cibola; Item 47) and 北亞墨利加 (North America; Item 58) are of European origin; these three items were added by Matteo Ricci to the source map of the America-KWQ.
III	North America (Central America, islands, and oceans/ seas)	11 (2, 2)	新以西把你海 (*Nuova Hispania, mare di*; Item 98), 白赫瑪 (Bahama; Item 99), 路格禺 (Lucayoneque; Item 100), 瓦投/止瓦投 (Ciguateo: Item 101), 曷弥亞那 (*Camercane, isole*; Item 105), 曷勿洗勿 (Higüey; Item 109), **仙如漢島** (*Isla de Puerto Rico*; Item 110), **對島** (Virgin Islands; Item 111), 河摺亞諾滄 (Pacific Ocean; Item 116), 鄂度蠟 (Honduras; Item 117), 瓦的馬革 (Uatimaco; Item 118) *Note*: 新以西把你海 (*Nuova Hispania, mare di*; Item 98), 河摺亞諾滄 (*Oceanus Atlanticus*; Item 116) are of European origin; these two items were added by Matteo Ricci to the source map of America-KWQ.

IV	South America (Northern Region)	6 (2, 2)	金加西蠟 (*Castilla de Oro*; Item 123), 新唵大魯西亞 (*Nuova Andalusia*; Item 128), 亞那牙 (Guaiana or Guyana; Item 129), 富令那國 (*Fulemna, regno*; Item 131), **漁人地** (Pescheria; Item 133), **青珠島** (Margarita Island; Item 134) *Note*: 金加西蠟 (*Castilla de Oro*; Item 123) and 新唵大魯西亞 (*Nuova Andalusia*; Item 128) are of European origin; these two items were added by Matteo Ricci to the source map of America-KWQ.
V	South America (Western Region)	19 (1, 3)	打勒那 (Takena; Item 135), 盤峩 (Pango; Item 137), 止巴泥瓦 (Cepaniua; Item 138), 角蠟巴沙 (Chiola pascia; Item 139), 馬亞柯 (Maiaco; Item 140), 亞馬鑽國 (Amazonas; Item 145), 哥吒麻 (Cociama; Item 147), 故私哥國 (City State of Cusco or Cuzco; Item 148), 乾櫶國 (*Chience, regno di*; Item 149), 亞彼那加 (Iapeinachia; Item 154), 徑得力 (Chinteli; Item 157), 巴羅謎哪 (Palomina; Item 158), 喻末近曷聶 (Chiomochinhonié; Item 161), 加默真瓦泥 (Camechencani or Chiamécenuani; Item 162), **苦峰** (*Cu, promontorio*; Item 163), 祁勒國 (Chico; Item 167), 仙如里亞諾 (*San Giuliano*; Item 168), 早沙家 (Zaoshajia; Item 174), 椰子島 (Cocos Island; Item 175) *Note*: 亞馬鑽國 (Amazonas; Item 145), 仙如里亞諾 (*San Giuliano*; Item 168), 椰子島 (Cocos Island; Item 175) are of European origin; these three items were added by Matteo Ricci to the source map of America-KWQ.
VI	South America (Eastern Region)	26 (2, 1)	其其山巴 (Chichichampa; Item 178), 邁哥 (Mayco; Item 179), **潮水峯** (*Flusso, capo del.*; Item 180), 可狼氏 (*Colante*; Item 182), 沙那耶 (Sciananes; Item 183), 都柄稽詣 (Tupimchii; Item 189), 帕齊那國 (*Pésina, regno*; Item 190), 半島 (Peninsula; Item 191), 峨勿大葛特 (Ngoutacote; Item 195), 馬加大突 (Machiatatu; Item 197), 麻尔碍曷突 (Morghaehotu; Item 198), 達坡那 (Tapuona; Item 199), 打巴砑勒 (Tapaialé; Item 200), 革那非所河 (*Cona feisuo, fiume*; ≈ 15), 毗麻剌 (*Pimala*; Item 206), 巴利孤打 (Paricuta; Item 208), 麻剌加岳湖 (*Maracayo, lago*; ¤ 2), 察尔加私 (Chaerjiasi; Item 211), 億大盤窪 (Itapanua; Item 212), 皮諾利那 (Pinolina; Item 213), 渴察革馬 (Cociacoma; Item 215), 題別里 (Tipilli; Item 216), 巴烏的 (Pauti; Item 218), 多巴欲 (Tuopaiü; Item 219), **金魚湖** (Goldfish Lagoon; ¤ 3), 南亞墨利加 (South America; Item 220) *Note*: 南亞墨利加 (South America; Item 220) is of European origin; this item was added by Matteo Ricci to the source map of America-KWQ.

Table 5.2. Additional twelve geographical items of Chinese origin on the America-KWQ.

#	**Geographical items of Chinese origin**
1 狗河 Item ≈ 1	The name 狗河 is of Chinese origin. It is today's St. Mary's River in Nova Scotia, Canada. More details are given in the Appendix.
2 松樹林 Item 64	The name 松樹林 is of Chinese origin. Washington State, Oregon and California are homes to pine forests.
3 雪山 Item 65	The name 雪山 is of Chinese origin. It is either Mount Hood about 80 km/50 mi east-southeast of Portland, Oregon, or Mount Rainier about 95 km/59 mi south-southeast of Seattle in the U.S. state of Washington.
4 祁未蠟 Item 66	The name 祁未蠟 is the transliteration of "Quivira". The meaning of this name can be found in the Dictionary of Chinese Buddhist Terms: "Kuvera" is the "God of Riches".
5 美灣 Item 69	美灣 is of Chinese origin. It is today's Bristol Bay which is the easternmost arm of the Bering Sea.
6 亞泥俺國 Item 71	In the late thirteenth century, Marco Polo wrote that Anian was in the northeast of Asia. Later, Hendon M. Harris, Jr. believed that Polo's Anian is in fact a mispronunciation of what is shown as Da Han (it is the pinyin of 大汉, meaning "Great Guy" or "Big Guy"; it is in today's Alaska) on an old Asian world map which he associated with an ancient Chinese geography: the *Classic of Mountains and Seas*, also known as *Shan Hai Jing* (versions of the text may have existed since as early as the fourth century B.C.).
7 亞泥俺峽 Item 78	The strait probably took its name from Ania, a Chinese province mentioned in a 1559 edition of Marco Polo's book. The Strait of Anián was a semi-mythical strait for the Europeans, which first appeared on a map issued by the Italian cartographer Giacomo Gastaldi (1500–1566) in 1562. The strait was believed by early modern cartographers to mark the boundary between North America and Asia and to permit access to a Northwest Passage from the Arctic Ocean to the Pacific. The first European to discover the true strait is the Danish-born Russian navigator Vitus Bering in 1728 and it became known as the Bering Strait (66.5°N, 169°W).
8 十字山尾 Item 91	十字山尾 (c. 23°N, 120°W) is a name of Chinese origin. It correlates closely with Cape San Lucas (22.9°N, 109.9°W), which is the extreme southern tip of the Baja California peninsula, Mexico. The rocky headland forms the southern extremity of the Sierra de San Lazaro and includes the western shore of San Lucas Bay. The isolated town of San Lucas lies 3 km/2 mi north of the cape.
9 東紅海 Item 93	The name 東紅海 is of Chinese origin. On the N. America-KWQ, the water separating Baja California from the Mexico mainland, now called "Gulf of California", is depicted as 東紅海 ("Eastern Red Sea").
10 巴大溫即長人國 Item 160	In the Chinese record, there is a 長人國 (Country of the Tall People) mentioned in the chapter titled "*Da Huang Dong Jing* 《大荒東經》" or "Classic of the Great Wilderness: East", of the Chinese classic *Shan Hai Jing* 《山海經》" or *Classic of Mountains and Seas*. These tall people live far beyond the East China Sea, in the great wilderness where there is a mountain named Dayan. It was where the sun and the moon rise from. This refers to today's Patagonia (巴大溫) region. Patagonia is the European name Matteo Ricci added to the source map of the America-KWQ.
11 第一角 Item 169	The name 第一角 (53.5°S, 79.5°W) is of Chinese origin. It is today's Cape Froward (53.9°S, 71.3°W), which is the southernmost point of the South American mainland. It is in the Magallanes region of Chile, along the north shore of the Magellan Strait, being the southern tip of the Brunswick Peninsula.
12 孛露海 Item 176	The name 孛露海 is of Chinese origin. It is today's eastern side of the North Pacific Ocean.

(5) There are 79 geographical items depicted on both the America-KWQ and any of the three European maps, which have written historical records showing that the Europeans have explored or colonised these places before 1602; but the coordinates of these geographical items on the America-KWQ show differences from those on the European maps, indicating that these data were obtained independently by the Chinese. Adding this number will make the places being explored by the Chinese to at least 177 (98 + 79). Matteo Ricci has used European names for some of these places, the rest of the names are derived from aboriginal languages. Thus far, we know that the Chinese have explored at least 177 places out of the 241 places (over 73 percent) depicted on the America-KWQ. The 79 places are shown in Table 5.3.

Table 5.3. Seventy-nine geographical items depicted on both the America-KWQ and any of the three European maps, with historical records showing that they were explored by Europeans before 1602 (but they were also explored independently by the Chinese).

Zone	**Category**	**Number of geographical items on the America-KWQ and on any of the three European maps**	**Geographical items depicted on the America-KWQ and on any of the three European maps. There are written historical records to show Europeans' exploration or colonisation of these places before 1602. The coordinates of these geographical items on the America-KWQ show differences from those on the Europeans maps, indicating these places were independently explored by the Chinese.**
I	North America (Eastern Region)	14	諾龍伯尓瓦 (Norumbega; Item 1), 得尓勿羅洛/譯云花地 (*Flora, terra*; Item 5), 亞伯尓耕國 (Apalchen; Item 6), 哥沙國/哥妙國 (Coosa; Item 8), 摩可沙國 (Mocosa; Item 9), 翁尼答 (Honguedo; Item 11), 新拂郎察 (New France; Item 12), 可尓得勒亚尓地 (Newfounland/*Terra de Corterealis*; Item 13), 得尓洛勿洛多/譯云耕農地 (Labrador; Item 14), 亦利多的蘭地 (Estotilant; Item 15), 巴革老地 (*Terra de baccalaos/* Newfoundland; Item 16), 飛私得島 (Greenland or Frisland; Item 17), 默羅它海 (*Gulfo Merosro*; Item 24), 地北極界 (Arctic Circle: Item 25)
II	North America (Western Region)	22	大入尓國 (Tagil; Item 26), 新以西把尼亞 (*Hispania Noua*; Item 32), 墨是可 (Mexico City; Item 34), 沙里思可國 (Jalisco; Item 37), 固列 (Cullacán; Item 38), 墨珠亞甘 (Michoacán; Item 39), 突爾利其祁默奇國 (Terlichichimechi; Item 45), 何察剌瓦 (Hochgelaga; Item 49), 祖瓦蠟 (Kingdom of Ceuola; Item 51), 沙瓦乃國 (Saguenai; Item 56), 加拿大國 (Canada; Item 60), 孟多齊峰 (Cape Mendocino; Item 68), 得光得白 (Tecoantepec; Item 82), 第瓦施 (Tlguex; Item 84), 弟克屬 (Tlguex; Item 86), 亞沙河 (*Axa flu/* Colorado River; ≈ 5), 多朵德亞河 (*Totonteac fl*; ≈ 6), 帝靜河 (*Tiguas rio*; ≈ 7), 多朵德亞國 (Totonteac; Item 87), 小安農降 (*C. del Ingano*; Item 89), 角利弗尓聶 (Baja California Sur; Item 90), 角利弗尓海 (the lower portion of the Gulf of California; Item 92)

(*Continued*)

Table 5.3. (*Continued*)

Zone	Category	Number of geographical items on the America-KWQ and on any of the three European maps	Geographical items depicted on the America-KWQ and on any of the three European maps. There are written historical records to show Europeans' exploration or colonisation of these places before 1602. The coordinates of these geographical items on the America-KWQ show differences from those on the Europeans maps, indicating these places were independently explored by the Chinese.
III	North America (Central America, islands, and oceans/ seas)	12	哇的麻剌 (Guatemala; Item 95), 宇革堂 (Iucatan; Item 97), 哇那罕 (Guanahaní; Item 103), 瑪牙瓦那 (Mayaguana; Item 104), 古巴島 (Cuba Island; Item 106), 牙賣加 (Jamaica; Item 107), 小以西巴你亞 (*Hispaniola*; Item 108), 鶴島 (Corvo; Item 113), 第三起島 (Terceira Island; Item 114), 怕霧打島 (Bermuda; Item 115), 諂柯 (Naco; Item 119), 里漢 (León; Item 120)
IV	South America (Northern Region)	7	坡巴牙那 (Popayán; Item 121), 祁臘正瓦斯 (Quillacinga; Item 122), 祈多 (Quito; Item 124), 苔尼戀 (Darién; Item 126), 小耨勿茶 (Little Venice; Item 127), 烏水河 (Black River; ≈ 8), 渴里白那 (*Caribana*; Item 132)
V	South America (Western Region)	11	巴吳亞那 (Paguana; Item 136), 孛露 (Peru; Item 141), 利禡 (Lima; Item 143), 意里加 (Arica; Item 151), 北度西山 (Potosi; Item 152), 大剌巴箇 (*Tarapacá*; Item 153), 智里 (Chile; Item 155), 亞大加馬 (Atacama Plateau; Item 159), 無氐澳 (Bottomless Bay; Item 165); 墨瓦蠟泥峽 (Magellan Strait; Item 170), 仙色巴松 (*San Sebastián Bay*; Item 171)
VI	South America (Eastern Region)	13	沙濱 (Arenagarda; Item 177), 肇加國 (Charcas; Item 186), 吳路漢河 (Uruguay River; ≈ 11), 伯剌那大江 (*El gran rio de Parana*; ≈ 12), 黑江 (Black River; ≈ 13), 銀河 (*Río de la Plata*; ≈ 14), 亞古齊亞 (Acutia; Item 194), 巴兒難伯可 (*Pernambuco, R.*; Item 202), 伯西兒 (Brazil; Item 203), 阿瑪加那 (*Amazones*; Item 204), 馬良温河 (*Maragnon fl.*; ≈ 16), 霸剌那 (Paraná; Item 214), 何勒利西那河 (*Oregliana fl.*; ≈ 17)

(6) Finally, there are the remaining 52 geographical items which are depicted on the America-KWQ as well as on any of the three European maps but they have no written historical records or modern English names. For these places, there are two possibilities: (1) European hunters, traders, miners and fishermen may have explored them without leaving historical records known to me, but these places were known to the European cartographers and/or the European cartographers imagined some of these place names; or (2) the Europeans had not explored these places, but European cartographers copied these place names and locations from Chinese maps which they had access to. For example, about 38 percent (20/53) of these European unexplored places are located in the vast Region II — roughly north of 37°N (the northern latitudes of Arkansas reached by the Spanish explorer and conquistador Hernando de Soto) and west of 77°W

(Hudson Bay East: c. 59°N, c. 77°W) — as shown in Fig. 5.2; it seems very plausible that the European cartographers copied these place names and locations from the Chinese maps that they had access to. Quite often their coordinates of these places are less accurate and show larger differences from those depicted on the America-KWQ.

Adding 52 more places to the 177 places that Chinese have explored (from Table 1.3), we know that at least *229 places out of the 241 places (over 95 percent)* depicted on the America-KWQ have been explored by the Chinese. Table 5.4 lists these 52 geographical items:

After detailed examination of over 1,000 coordinates on the four maps, I found that the America-KWQ often gives better latitudinal and longitudinal readings of the geographical items listed in Tables 5.1–5.4 than those on the three European maps when compared with a modern map. This is detailed item by item in the Appendix.

Table 5.4. Fifty-two geographical items depicted on both the America-KWQ and any of the three Europeans maps, but with no written historical records or modern English names.

Zone	**Category**	**Number of geographical items on the America-KWQ and on any of the three European maps**	**Geographical items depicted on both the America-KWQ and any of the three European maps; but there are no records to show Europeans' exploration of these places before 1602 when KWQ was published, indicating that these geographical data were obtained by the Chinese.**
I	North America (Eastern Region)	1	亞勿加尔國 (Avacal; Item 7)
II	North America (Western Region)	20	革哥私 (Gacos; Item 29), 多勿國 (Toua; Item 31), 古數沙 (Costosta; Item 33), 多皮蠟 (Topira; Item 35), 知加土蠟 (Cacatula; Item 36), 知鬧 (Chinao; Item 41), 亞私大達闌 (Astatlan; Item 43), 媽剌大 (Marata; Item 44), 加巴斯祁國 (Capaschi; Item 46), 祁蠟 (Chilaga; Item 48), 蘇亞蠟山 (*Suala mons*; Item 50), 多兒美 (Tolm; Item 53), 諸著 (Ciucio; Item 54), 哥泥白斯湖 (Lake Conibaz; ¤1), 何皮六河 (Obilo River/ Mackenzie River; ¤4), 加那瓦 (Canado; Item 59), 都茶那 (Tuchano; Item 67), 水潮峰 (Cape Peirce; Item 73), 亞奴皮亞大 (*La Anubiada*; Item 81), 支古訝 (Cicuie; Item 83)
III	North America (Central America, islands, and oceans/seas)	4	馬金色 (Maxa; Item 94), 大哇識个 (Tousco; Item 96), 瓦尼瑪 (Guanima/Cat Island; Item 102), 色氏測島 (Sept Cites Archipelago; Item 112)
IV	South America (Northern Region)	2	加利巴那 (*Calimana, la*; Item 125), 容皮 (Iumbi; Item 130)

(*Continued*)

Table 5.4. (*Continued*)

Zone	Category	Number of geographical items on the America-KWQ and on any of the three European maps	Geographical items depicted on both the America-KWQ and any of the three European maps; but there are no records to show Europeans' exploration of these places before 1602 when KWQ was published, indicating that these geographical data were obtained by the Chinese.
V	South America (Western Region)	10	亞牙勿里 (Aiauri; Item 142), 馬加利 (Hacari; Item 144), 氐薩那國 (*Tisnada*; Item 146), 茄盧麻達 (Giurumada; Item 150), 巴可 (Paca; Item 156), 三角 (*C. de 3 puntas*; Item 164), 何剌佛江 (*Orufes p.*; ≈ 9), 萬室女 (*Verginik le diecimila*; Item 166), 仙瑪里亞峯 (*S. Maria. C. de*; Item 172), 金特突 (*Quintete*; Item 173)
VI	South America (Eastern Region)	15	白峯 (*Peña Blanca*; Item 181), 如路馬大 (Giurumata; Item 184), 瑪八作 (Mapazo; Item 185), 止加 (Chinca; Item 187), 寧瓦大 (Ningatas; Item 188), 白氐河 (Peti River; ≈ 10), 巴臘牙 (Plaia; Item 192), 摩勒祓甕 (Morpion; Item 193), 多兀彼那 (Tuoupeina; Item 196), 阿林荅 (*Alinde, C. de*; Item 201), 椰林峯 (*Palmar, capo di*; Item 205), 白峯 (*Blanco, cabo*; Item 207), 弟私那大 (Tisnada; Item 209), 喝祃哇 (Aiauiri; Item 210), 皮可蠟 (Picora; Item 217)

Fig. 5.4. Cape Froward and the Strait of Magellan in the Magallanes region (see endnote 25, public domain).

Thus far in Section 4, I have already found *layers of evidence showing that the America-KWQ is not a copy or adapted copy of the three European maps but is of Chinese origin.* More evidence follows.

(7) The S. America-KWQ uniquely and correctly depicts the southernmost tip — Cape Froward (第一角) — of the South American mainland, the south-western tip — Cape Peirce

(水潮峰) — of Alaska, and the Arctic Ocean (冰海) which includes the North Pole region; moreover, the map also shows correctly that Cape Breton Island in the east coast of the Americas was separated into two islands by an ancient Chinese-built canal. The name 第一角 (53.5°S, 79.5°W; Item 169) on the Americas-KWQ is of Chinese origin; 第一 means "No. 1" and 角 "cape". It is today's Cape Froward (53.9°S, 71.3°W), the southernmost tip of the South American mainland (see Fig. 5.4). The latitude 53.5°S is very close to that of today's Cape Froward of 53.9°S, but the longitude 79.5°W is c. 8.2° more to the west, showing a larger error, which is not uncommon on the America-KWQ.

Cape Froward is in the Magallanes region of Chile, along the north shore of the Magellan Strait, being the southern tip of the Brunswick Peninsula. The earliest European map to show this southernmost tip is the 1520 Map of the Magellan Strait, in which *Cap de Seado* is on the north shore of the strait and is the southernmost tip of the South American mainland.[26] In January 1587, the English corsair Thomas Cavendish (1560–1592)[27] named the place Cape Froward after the region's rough climate (strong rains and winds). Hence, we can conclude that on the S. America-KWQ, 第一角 is today's Cape Froward, the southernmost tip of the South American mainland.

However, none of the three European maps correctly depicts the southernmost tip of the South American mainland (see Item 169 in the Appendix for detailed explanations), as can be seen in Fig. 5.5. Only the bottom right panel of Fig. 5.5 shows that the S. America-KWQ depicts 第一角 on the northern shore of the Magellan Strait and right at the southernmost location of the South American mainland.

Next, on the N. America-KWQ, 水潮峰 (59°N, 164.5°W) is depicted as the south-western tip of today's Alaska and the westernmost tip of the American continent, facing the Anian Strait (today's Bering Strait). Cape Peirce (58.6°N, 161.8°W) is almost where 水潮峰 is located. 水 means "water", 潮, "wave or tide" and 峰, "peak"; 水潮峰 means a peak or a headland with certain height sticking into the sea and washed with high tides. Alaska's Cape Peirce fits well with 水潮峰 by definition, since (1) it sticks into the sea and has a height of 81.99 m/269 ft; and (2) it is near the Bristol Bay (57.3°N, 159.8°W; this corresponds to 美灣 on the N. America-KWQ) where the upper reaches of Bristol Bay experience some of the highest tides in the world: one such area near Dillingham (59°N, 158.5°W) and another near Naknek (58.7°N, 157°W) have tidal extremes of more than 10 m/30 ft. Moreover, the extreme number of shoals, sandbars and shallows makes navigation troublesome, especially during the area's frequently strong winds. As the shallowest part of the Bering Sea, Bristol Bay is one of the most dangerous regions for large vessels.

水潮峰 is not Anchorage (61.2°N, 149.9°W) — on a strip of coastal lowland and extending up the lower alpine slopes of the Chugach Mountains in south-central Alaska and near the Gulf of Alaska — not only because the longitudinal position of Anchorage is too far away from that of 水潮峰 (59°N, 164.5°W) and its latitudinal position is a couple of degrees north of 水潮峰, but also because it is not a "峰", "peak". Although it is near Turnagain Arm (61°N, 149.8°W), which is a fjord that has some of the world's highest tides, Anchorage does not appear to be the westernmost tip of the American continent.

Lastly, for brevity, I refer readers to the Appendix about the Arctic Ocean (冰海; Items 21 and 76), and Chapter 1 of this book about why Cape Breton Island is depicted as two islands on the N. America-KWQ.

Fig. 5.5. In the top left panel, the S. America-Mercator (1569) depicts *C. de Seado* on the southern shore of the Magellan Strait and at the western side of the strait (public domain); in the top right panel, the S. America-Ortelius (1570) depicts *C. de Seado* on the northern shore of the Magellan Strait and at the western side of the strait (public domain); in the bottom left panel, the S. America-Plancius (1594) depicts *C. de Seado* on the southern shore of the Magellan Strait and at the eastern side of the strait (public domain); and in the bottom right panel, the S. America-KWQ depicts 第一角 (Item 169) on the northern shore of the Magellan Strait and at the southernmost location of the South American mainland (public domain). All figures here are extracted from those in Fig. 5.3.

(8) On the N. America-KWQ, Labrador and Newfoundland are incorrectly represented. As a European cartographer, Matteo Ricci would have known this part of North America well. Recently, scientists using a new dating technique, which analysed tree rings of three pieces of wood cut from the Norse settlement at L'Anse aux Meadows, have provided evidence that Vikings settled in North America in 1021.[28] Since the Norse visits to Vinland, the Italian navigator and explorer John Cabot (c. 1450–c. 1500) was the earliest-known European exploring in 1497 the coast of North America (either Cape Breton Island or Newfoundland was his first landing site).[29] Later in the 1530s, Jacques Cartier (1491–1557)[30] made three voyages to map the Gulf of St. Lawrence and the shores of the St. Lawrence River by sailing through the Strait of Belle Isle (the waterway that separates the Labrador Peninsula from the island of Newfoundland).

Notice that in Fig. 5.6, Mercator and Ortelius depict Newfoundland not only as an island but also partly as a peninsula called *Terra Corterealis* to the north and northwest of *Terra de Baccalaos*.

Fig. 5.6. The top left panel shows the East and North coasts of Canada on the N. America-Mercator (1569; public domain); a dashed arrow points to the Cape Breton Island labelled as *C. de Breton*, a solid-line arrow points at Newfoundland labelled as *Terra de Baccalaos* and a second solid-line arrow points to *Terra Corterealis,* which is a peninsula and also considered part of Newfoundland at the time; similar arrows are applied to the top right panel for the N. America-Ortelius (1570; public domain), and the lower left panel for the N. America-Plancius (1594; public domain). But the characters *Terra Corterealis* no longer show up on Plancius's map, which gives then the best representation of this portion of Canada to a modern map. The lower right panel shows N. America-KWQ on which the Cape Breton Island is separated into two islands (public domain) and Newfoundland becomes two peninsulas (public domain), which is incorrect. All figures here are extracted from Fig. 5.3.

Plancius, however, corrected this. (Later, the European cartographers recognised that *Terra Corterealis* was in fact part of today's Labrador and "*Terra Corterealis*" disappeared from their maps.) However, here all three European cartographers correctly depict Labrador as *Estotilant* facing the Atlantic Ocean and placed it to the north of *Terra de Baccalaos*. A more detailed explanation can be found in Items 13–16 in the Appendix.

But the bottom right panel in Fig. 5.6 shows that the N. America-KWQ depicts both 可尔得勒亚尔地 (*Terra de Corterealis*; Item 13) and 巴革老地 (*Terra de Baccalaos*; Item 16) as a portion of Newfoundland, and each as a peninsula lying to the south and north of 默羅它海 (*Melotuo, mare*; today's Lake Melville-Hamilton Inlet-Groswater Bay; Item 24),[31] respectively; whereas both 得尔洛勿洛多 (*Lavrador*; Item 14) and 亦利多的蘭地 (*Estotilant*; Item 15) lying inland represent mistakenly a portion of today's Labrador and to the west of 可尔得勒亚尔地 and 巴革老地, respectively. There is no reason for Matteo Ricci not to have known that Newfoundland is an island and Labrador should face the Atlantic Ocean if he were truly the cartographer of the N. America-KWQ.

(9) The N. America-KWQ uniquely depicts the Mississippi River and its tributaries. The confluence of the Mississippi and Ohio Rivers is at today's Cairo, Illinois (37°N, 89.2°W). The Mississippi River meets the Illinois River at Pere Marquette State Park, which is near Grafton (39°N, 90.4°W). In Fig. 5.7, on the N. America-KWQ, there is an unnamed river which has two

Fig. 5.7. The left panel shows an unnamed river on the N. America-KWQ (extracted from Fig. 5.3; bottom right panel; public domain) with its two tributaries; it flows into today's Gulf of Mexico. The confluence of the Mississippi and Ohio Rivers is at today's Cairo, Illinois, near position #1 and Mississippi River meets the Illinois River at Pere Marquette State Park, which is near Grafton at position #2. The right panel shows the course, watershed and major tributaries of the Mississippi River (see endnote 33, public domain).

major tributaries meeting this unnamed river at location #1 (36°N, 103.5°W) and location #2 (37°N, 104°W), respectively. This unnamed river flows generally south to the Mississippi River Delta located in Plaquemines Parish, Louisiana, south of the city of Venice, then into 新以西把你海 (*Nuova Hispania, mare di*; Item 98) or today's Gulf of Mexico.

This unnamed river can be identified as the Mississippi River, location #1 (36°N, 103.5°W) as Cairo (37°N, 89.2°W), and location #2 (37°N, 104°W) as Grafton (39°N, 90.4°W). We can see that the latitude readings on the N. America-KWQ are reasonably good, but the longitudinal errors are much larger, which is not uncommon on the America-KWQ. However, the Missouri River and the upper stream of the Mississippi River are not depicted on the map, indicating that they were not explored. The upper stream of the Mississippi River would be discovered in 1433–1434 by Zheng He's mariners during their seventh voyage. This is the finding I obtained after extracting the entire navigational routes, timelines and evidence from Luo Maodeng's 1597 masterwork titled *San Bao Tai Jian Xi Yang Gi* 《三宝太监西洋记》 or *An Account of the Western World Voyage of the San Bao Eunuch*. Later, I published a book titled *The Last Journey of the San Bao Eunuch, Admiral Zheng He*[32] to explain Zheng He mariners' entire journey of their seventh (and last) voyage which brought them to Cahokia, today the UNESCO World Heritage Site in the Central Mississippi valley.

On the European side, it was on May 17, 1673, that Father Jacques Marquette (1637–1675) and fur trader Louis Joliet (1645–after May 1700) set out on a four-month voyage which carried them thousands of miles through the heart of North America to explore the path of the Mississippi River. They are the first documented European discoverers of the Missouri River when they passed its violent mouth on their voyage down the Mississippi.[34] But that was more than 70 years after the publication of the KWQ and almost 240 years after Zheng He's mariners made the trans-Atlantic voyage to enter the Gulf of St. Lawrence, then the St. Lawrence River, and then discovered four of the five Great Lakes before sailing into the Illinois River and Mississippi Rivers to reach the American heartland in 1434 (along this water route, they had to pass through two short land portages).

None of the three European maps depicts the Mississippi River, even though Hernando de Soto (c. 1500–1542) was the first European documented as having crossed the Mississippi River in 1541.[35]

5. The America-KWQ Reflects the Americas in 1420–1428

The determination of the above era is based on the political landscape of the Americas revealed by the America-KWQ as follows:

(1) 哥妙國/哥沙國 (Coosa, Comos or Cossa, 32°N, 100°W; Item 8) is depicted on the N. America-KWQ. This was the Coosa Chiefdom (c. 1400–c. 1600),[36] in today's Georgia where Indigenous Americans lived. It was one of the most powerful chiefdoms in the southeast at the time. The Indigenous Americans lived on the Coosa Valley where the Coosa River (34.3°N, 85.2°W) flows through, before Hernando de Soto (c. 1500–1542) and his men became the first Europeans to visit the valley in 1540.[37] The Coosa River is a tributary of the Alabama River in the U.S. states of

Alabama and Georgia. Because 哥妙國/歌沙國 (Comos, Cossa or Coosa) is depicted on the N. America-KWQ, the political era revealed by the map should be after c. 1400, but before c. 1600. On the N. America-Ortelius (1570), Comos is depicted at 33.5°N, 95°W. More details are given in Item 8 of the Appendix.

(2) On the S. America-KWQ, 故私哥國 (City State of Cusco/Cuzco; 1197–1438; 20°S, 72.5°W; Item 148) was a small kingdom based in the city of Cusco (13.5°S, 72°W) in southeastern Peru (see Fig. 5.8, left and right panels) near the Urubamba Valley of the Andes Mountain range. The S. America-KWQ gives poor latitudinal reading, but very good longitudinal reading for Cusco/Cuzco.

Cusco/Cuzco was founded by the Incas,[40] and through warfare or peaceful assimilation, it began to grow and was succeeded by the Inca Empire (1438–1533/1572).[41] The city was the capital of the Inca Empire from the thirteenth century until the sixteenth century Spanish conquest. Since the Inca Empire is not depicted on the N. America-KWQ, but Cusco/Cuzco is still depicted as 故私哥國 (City State of Cusco/Cuzco), this means that the political era revealed by the America-KWQ is before 1438.

Combining (1) and (2), we can say that for the time being, the political era revealed by the America-KWQ is between c. 1400 and 1438.

Neither the City State of Cusco/Cuzco nor the Inca Empire is depicted on any of the three sixteenth century European maps, although Cusco/Cuzco as a city should still be there.

(3) On the America-KWQ, the Aztec Empire (1428–1521)[42] is not depicted. This means that the political era revealed by the America-KWQ can further be narrowed to between c. 1400 and 1428. This date is consistent with the following historical fact: the Inca emperor began to conquer 肇加國 (Charcas; Item 186) from c. 1430, but Charcas is depicted on the S. America-KWQ as an existing country on the map, not the Inca Empire. This means that the political era revealed by the

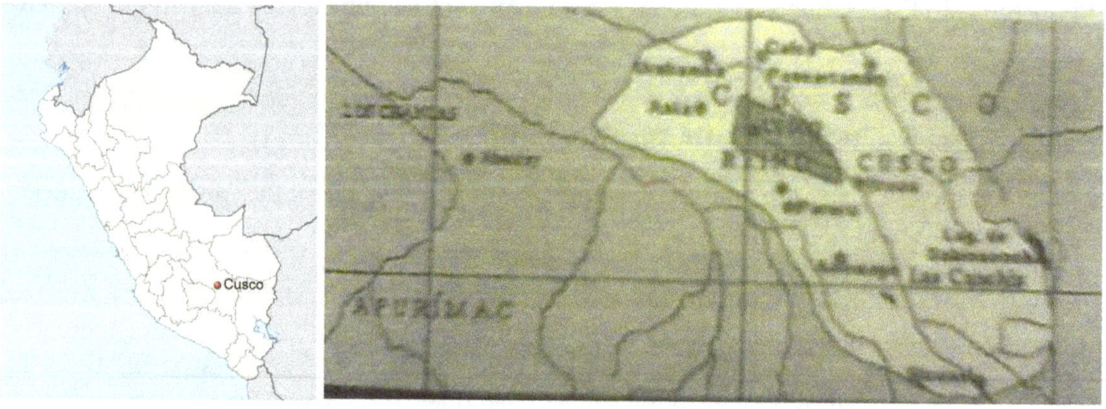

Fig. 5.8. The left panel shows the location of Cusco (13.5°S, 72°W) within Peru (see endnote 38, public domain); and the right panel shows the map of the City State of Cusco (1187–1438; see endnote 39, public domain).

America-KWQ should be before c. 1430; and the political era of c. 1400–1428 is before c. 1430. This era also rules out the possibility of the source map of the America-KWQ being based on the geographical information gathered by Zheng He's mariners during their seventh voyage in the early1430s.

(4) Combining the above findings of (1)–(3), the political era revealed by the America-KWQ is already narrowed to between c. 1400 and 1428. Since Zheng He set sail on his first voyage in 1405, and the first four voyages did not go beyond the Indian Ocean, the political era revealed by the America-KWQ can be further narrowed to between 1415 (the end of the fourth voyage) and 1428.

For the fifth voyage, there has been a controversial map (with questions about its authenticity) abbreviated as the 1418 Chinese World Map or the 1418 Zheng He Map (a copy of the original 1418 map),[43] in which a modern-shape America is depicted, suggesting that the Ming sailors reached this foreign land during their fifth voyage (1417–1419). But even if someday the dust settles and this map is shown to be authentic, the map contains far less geographical information in comparison with the very detailed America-KWQ. Hence, we can safely rule out that the rich and detailed geographical information on the America-KWQ was produced by Zheng He's fifth voyage. Then, we are left with the sixth voyage, which officially started from 1421 and ended in 1422 (the year when Zheng He returned to China to report to Emperor Yongle at his new capital Beijing). But some of Zheng He's squadrons did not return to China until a few years later, otherwise from whom was the rich and detailed geographical information on the America-KWQ obtained?

After combining all the findings in (1)–(4), we can finally narrow the political era revealed by the America-KWQ to 1420–1428.

6. Why Are the Great Lakes Missing from the America-KWQ?

In my book titled *The Last Journey of the San Bao Eunuch, Admiral Zheng He*,[44] I have extracted the complete navigational routes, timelines and evidence from the 1597 masterwork by historian Luo Maodeng (罗懋登) titled *San Bao Tai Jian Xi Yang Gi*《三宝太监西洋记》or *An Account of the Western World Voyage of the San Bao Eunuch*. The result shows that Zheng He's mariners made a trans-Atlantic voyage and their ships accidentally landed at the Labrador coast in the autumn of 1433. From there, they sailed into the Gulf of St. Lawrence and St. Lawrence River and discovered four of the five Great Lakes. This happened during their seventh voyage and a little over 10 years after their sixth voyage.

The above explains why the Great Lakes are not depicted on the America-KWQ, because it is a map which reveals the political era of the Americas between the 1420s and 1428, instead of the 1430s.

Now we may have some idea about where the Chinese mariners entered the Mississippi River during their sixth voyage, since the river is depicted in the left panel of Fig. 5.7. They most likely entered the Mississippi River from the Gulf of Mexico, which was named in Chinese as 新以西把你海 (*Nuova Hispania, mare di*; Item 98), instead of entering from the Labrador coastal area as they did during their seventh voyage. The north-eastern water route was then the most efficient route to go from the North Atlantic Ocean to the North American heartland: it requires sailing into the St. Lawrence River, four of the five Great Lakes, the Illinois and Mississippi Rivers and passing two short land portages connecting different sections of the water route.[45] The missing of the Great Lakes, plus the detailed geographical depiction of Florida, Cuba and the many islands in the neighbourhood, gives strong evidence that the Chinese sailors most likely explored the Caribbean Sea during their sixth voyage before entering the Mississippi River.

7. Were the Pericúes Chinese?

The original indigenous inhabitants of the southernmost portion of Baja California Sur including Cape San Lucas (十字山尾; Item 91) were the mysterious and lost tribe of the Pericúes since the eighteenth century.[46] The archaeological record for Pericú territory can be traced as far back as the early Holocene,[47] about 10,000 years ago, and perhaps into the late Pleistocene[48] which lasted from about 2,850,000 to 11,700 years ago. They were living in their tribal lands when the first Spanish settlers arrived in the region in the 1530s.[49] The Spanish found them markedly different from other local tribes — noting that they had a very different language and fast speech.[50] This led the Spanish settlers to name the tribe Pericos (parakeets), hence the modern-day name, the Pericúes.

The Pericúes were taller and stronger than the average mainland Mexican inhabitants, and the Pericú society possessed sophisticated maritime technology, making use of wooden rafts and double-bladed paddles.[51]

Genetic studies of the skulls of the Pericúes suggest that their ancestors were the ancient populations of southern Asia, Australia and the South Pacific Rim or remnants of some of the New World's earliest colonisers.[52] Were the Pericúes ancient Chinese?

8. Conclusions

Historians generally think that the extent of travel by the Ming Treasure Fleets did not exceed the east coast of Africa. But this viewpoint has faced challenges including my findings presented in my 2019 book titled *The Last Journey of the San Bao Eunuch, Admiral Zheng He*, which suggests that Zheng He's mariners made a trans-Atlantic voyage via southern Africa to arrive at the North American coast in the autumn of 1433.[53]

According to my research findings in Chapters 1–4 of this book, the Chinese mastered marine navigation and drew maps of Cape Breton Island in Canada, of Australia, of New Zealand, of Land of Fire, of Antarctica, of Europe and of Africa on the world map — Kunyu Wanguo Quantu《坤輿萬國全圖》, abbreviated here as KWQ — at different times long before the Europeans reached these places. The KWQ is considered by many the first modern world map and has been regarded as an important map because of its rarity, importance and exoticism. Admiral Zheng He set sail to the Western Ocean in the early fifteenth century: his seven voyages (1405 to mid-1430s) remain unmatched in world history. But the Ming rulers halted these epic expeditions soon after the seventh voyage ended to focus on domestic needs. They either burned the great fleet in the harbours or allowed the ships to rot. Most of the valuable navigational records and maps vanished in fires or were lost in history. However, some of these maps seem to still have existed in China when Matteo Ricci was making KWQ, while others seem to have passed to European hands for them to draw maps of places they had not explored.

In Section 1 of this chapter, I state that my goal for this chapter is to conduct an in-depth investigation of the Americas on KWQ, abbreviated as America-KWQ, as I did for other parts of KWQ, to find out when these regions were explored from the political landscapes they reveal to the world. Although Matteo Ricci claimed to be the cartographer of the map, serious doubts have been raised by modern-day researchers about the truth of his claim.

Section 2 shows that Europeans had a vast interest in the Americas after Cristopher Columbus landed in the Bahamas in 1492.

Section 3 shows that the Europeans rushed to the New World during the sixteenth century. Despite the fact that a vast area remained unexplored, many geographical items in the Americas were drawn into their maps. From whom did they get the information, if not from the Chinese?

Section 4 provides strong evidence to show that the America-KWQ is a Chinese-based map, not a copy of the three sixteenth century European maps discussed in this chapter. The key facts are: (1) the America-KWQ gives a better representation of the shape of the Americas and generally gives smaller longitudinal errors than those on the three European maps; (2) there are 17 original annotations on the America-KWQ, which are absent on the three Europeans maps, the only exception being about the Country of the Tall People, known to the Chinese since the fourth century B.C; (3) among the 241 geographical items depicted on the America-KWQ, 98 of them (about 41 percent; including the unnamed Mississippi River) are not depicted on any of the three European maps, implying that the Chinese had explored them prior to the Europeans; and among these 22 are of Chinese origin; (4) there are additional 12 geographical items on the America-KWQ of Chinese origin, revealing that the true cartographer of the map must be Chinese; (5) there are 79 geographical items on both the America-KWQ and any of the three European maps which are backed by recorded history of European exploration before 1602; some of the names of these places on the America-KWQ have been edited by Matteo Ricci to look European; but these places were explored independently by the Chinese before the Europeans arrived, because the coordinates of these geographical items are almost always better than those of the three European maps, implying these data were independently obtained by the Chinese explorers; (6) there are 52 geographical items on both the America-KWQ and any of the three European maps, but none of them is backed by recorded history; this implies that the Europeans might not have explored these places and the European cartographers copied the name and locations from the Chinese maps which they had access to onto their maps, often resulting in serious mistakes; (7) the S. America-KWQ uniquely and correctly depicts the southernmost tip Cape Froward (第一角; Item 169) of the South American mainland, the southwestern tip Cape Peirce (水潮峰; Item 73) of Alaska (also the westernmost tip of the American continent), and the Arctic Ocean (冰海; Items 21, 76) which includes the North Pole region; (8) on the N. America-KWQ, Labrador and Newfoundland are incorrectly represented, and Newfoundland (island) is missing; but such a mistake is absent from the three European maps, implying the old age of the N. America-KWQ in representing this region; however, the map uniquely and correctly depicts Cape Breton Island in today's Canada as two islands due to an ancient canal built or rebuilt by Zheng He's mariners, which separated the island into two islands, during their sixth voyage to the Western Ocean in the 1420s; and (9) the N. America-KWQ uniquely and correctly depicts the Mississippi River and its two major tributaries.

Section 5 shows that the depiction of 哥沙國/哥妙國 (the Coosa Chiefdom) and 故私哥國 (City State of Cusco or Cuzco) on the America-KWQ has determined the political era revealed by the America-KWQ to be between c. 1400 and 1438. The absence of the Aztec Empire further narrows this era to c. 1400–1428. Based on other historical records and analysis, the date can be finally narrowed to 1420s–1428.

Section 6 shows that the Great Lakes were not discovered by Zheng He's mariners during their sixth voyage in the 1420s, hence, not showing on the America-KWQ, but the discovery was made during their seventh (and last) voyage. My analyses thereby indicate that they most likely entered the Mississippi River from the Gulf of Mexico during their sixth voyage.

Lastly, Section 7 raises the question: Were the Pericúes Chinese?

Did Ming mariners explore the Americas? Did Mercator, Ortelius and Plancius copy the Chinese maps? *The answers should be quite clear after reading this chapter and this book.*

Acknowledgements

The author is grateful to Professor Emeritus Michel A. Van Hove for his critical reading and valuable comments, as well as technical assistance in preparing the figures in this chapter in published form. Special thanks go to M. Hayes-Shuptar (华铭叮) for the academic and editorial assistance with the endnotes.

Appendix

This appendix analyses all the geographical items and annotations on the N. America-KWQ and S. America-KWQ. These geographical items and annotations are categorised into six different zones. On the America-KWQ, Chinese traditional characters are used, except for a handful of cases where one or two Chinese simplified characters appear. The author uses modern Chinese simplified characters in writing the main text and the Appendix of this chapter. The non-Chinese-based geographical items are either aboriginal names or European-based names transliterated by Matteo Ricci into Chinese. Items preceded by "•" are geographical items which do not appear on the America-Mercator (1569), or the America-Ortelius (1570) or the America-Plancius (1594). The sign "@" means "annotation", "≈" denotes "river" and "¤" denotes "lake", and all the geographical items and annotations are numbered in sequence in this Appendix. The Chinese "pinyin" of each item is given right after it is introduced.

N. America-KWQ

Item #	Name	Pinyin	Etymology	Analysis of history, politics, geography; comparison or comment

Zone I: North America (Eastern Region)

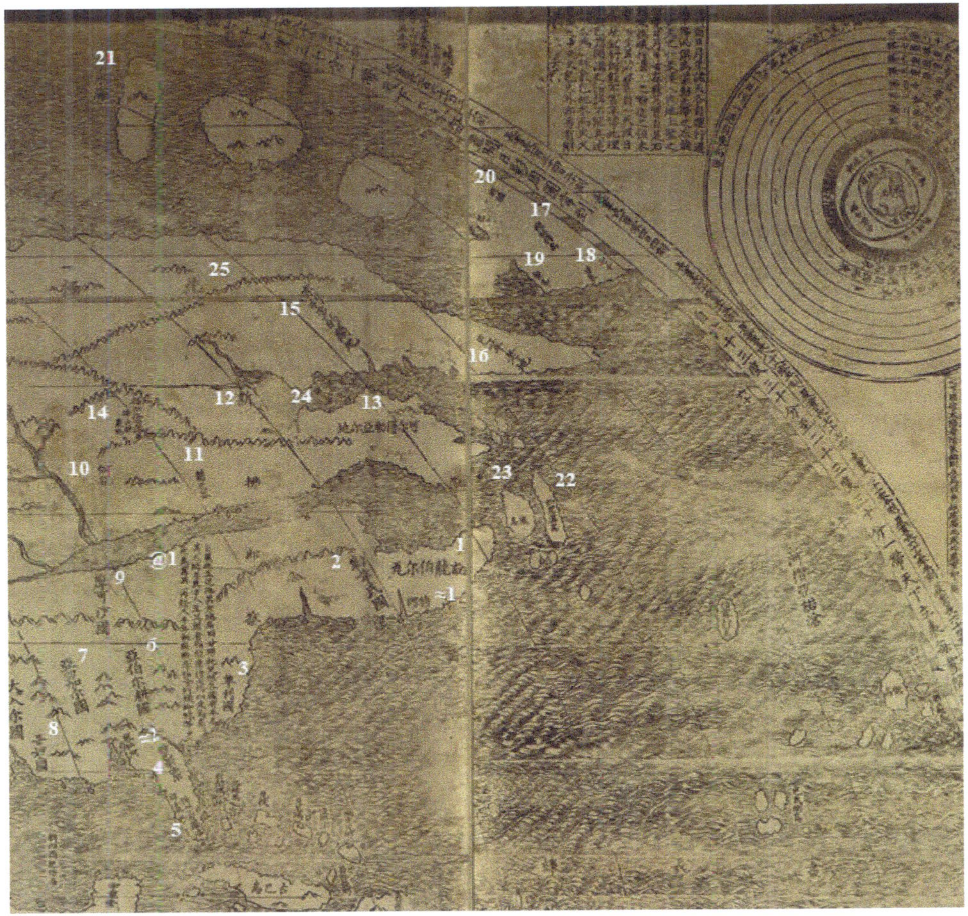

Fig. A5.1. Extracted from Fig. 5.3, bottom right panel, public domain.

(*Continued*)

(*Continued*)

Item #	Name	Pinyin	Etymology	Analysis of history, politics, geography; comparison or comment
1	諾龍伯尔瓦	Nuo Long Bo Er Wa	Norumbega was originally spelled *Oranbega* in Giovanni da Verrazzano's 1529 map of America,[54] and this name may be derived from one of the Algonquian languages spoken in New England.[55]	Norumbega or Nurembega was a legendary settlement in northeastern North America which showed up on many early maps from the 1500s until French colonisation of the region in 1604.[56] 諾龍伯尔瓦 is the transliteration of either the old spelling of Oranbega, or the later spelling of Norumbega or Nurembega; none of them has a European origin. The Norumbega shown on the N. America-KWQ is the place before the French colonisation of the Acadia Peninsula shown in Fig. A5.2. The entire French Colony of Acadia (1604–1713) also contains the Continental Acadia, hence includes parts of today's Maritime provinces (New Brunswick, Nova Scotia and Prince Edward Island), the Gaspé Peninsula and Maine to the Kennebec River.[57] The land that comprises what is now Nova Scotia was inhabited by the Mi'kmaq people at the time of the European exploration. Their territory was the first portion of North America that the Europeans exploited at length for resource extraction. As early as 1520, European fishermen began to set up camps ashore.[58] Fig. A5.2 also shows that New England lies further south of the Acadia Peninsula where the old place name *Oranbega* originated.

<table>
<tr><td></td><td></td><td></td><td></td><td>

Fig. A5.2. Acadia in 1754 (see endnote 59, public domain).

On the N. America-Ortelius (1570) the region is depicted as Norvmbega; but on the N. America-Mercator (1569) and the N. America-Plancius (1594), as Norombega. Since the N. America-KWQ and the three European maps were drawn before the French colony was founded in 1604, the French name Acadie naturally does not appear on any of them.</td></tr>
<tr><td>≈
1</td><td>狗河</td><td>Gou He</td><td>St. Mary's River</td><td>The name 狗河 is of Chinese origin; 狗 means "dog" and 河, "river". On the N. America-KWQ, 狗河 (43.5°N, 66.5°W) is today's St. Mary's River (45.1°N, 62°W).
On the N. America-Ortelius (1570), the coordinates of the river are 44°N, 71°W and depicted as R. grande. The longitudinal reading is too westward. St. Mary's River is one of Nova Scotia's longest rivers. In the seventeenth century, explorer Samuel de Champlain (1567–1736) named the river Rivère Isle Verte, meaning "Green Island River".[60] The current name is from Fort Sainte Marie, a French-built fort in 1654.[61]</td></tr>
</table>

(*Continued*)

(Continued)

Item #	Name	Pinyin	Etymology	Analysis of history, politics, geography; comparison or comment
• 2	勒革氏國	Le Ge Shi Guo	Lagus[62]	On the N. America-KWQ, the location of 勒革氏國 (44.5°N, 67.5°W) is in today's New Brunswick (46.7°N, 66.1°W) bordered by Quebec to the north, Nova Scotia to the east, the Gulf of St. Lawrence to the northeast, the Bay of Fundy to the southeast and the U.S. state of Maine to the west. New Brunswick was first inhabited by First Nations including the Mi'kmaq, Maliseet/Wəlastəkwiyik and Passamaquoddy.[63] The first documented European visits of New Brunswick were by Jacques Cartier (1491–1557) in 1534.[64] Prior to European arrival, Indigenous tribes did not leave a written record, hence we do not know the history of Lagus before that. On the N. America-KWQ, neither today's Bay of Fundy nor Prince Edward Island are depicted, hence, the Nova Scotia Peninsula is not well represented. A similar problem also exists on all the three sixteenth century European maps discussed in this chapter. Lagus is not even depicted on these three maps.
• 3	革利國	Ge Li Guo	Coree Chiefdom	The Coree (also Connamox, Cores, Corennines, Connamocksocks, Coranine Indians, Neuse River Indians) were a very small Indigenous American tribe, who once occupied a coastal area south of the Neuse River in southeastern North Carolina in the area now covered by Carteret and Craven counties.[65] The coordinates of 革利國 (36°N, 83°W) on the N. America-KWQ correlates closely with today's North Carolina (35.8°N, 80.8°W), which is a state in the southeastern region of the United States. The Coree were not described by English colonists until 1701,[66] by which time their population had already been reduced to as few as 125 members, likely due to epidemics of infectious disease and warfare. Hence, not surprisingly, none of the three European maps depicts the Coree Chiefdom.

@ 1	自農地至花地其方總名曰甘那托兒然各國有本名其人醇善異方人至其國者雅能厚待大約以皮為裘以魚為業其山內餘人平年相殺戰奪惟食蛇蟻蜘蛛等蟲		Zi Nong Di Zhi Hua Di Qi Fang Zong Ming Yue Gan Na Tuo Er Ran Ge Guo You Ben Ming Qi Ren Chun Shan Yi Fang Ren Zhi Qi Guo Zhe Ya Neng Hou Dai Da Yue Yi Pi Wei Qiu Yi Yu Wei Ye Qí Shan Nei Yu Ren Ping Nian Xiang Sha Zhang Duo Wei Shi She Yi Zhi Zhu Deng Chong	My translation and comments (between square brackets) are as follows: From Labrador [得尔洛勿洛多/譯云耕農地] to Florida [得尔勿羅洛/譯云花地], the entire area is named Gan Na Tuo Er [甘那托兒] where, however, each kingdom has its own name. Their people are very kind. When foreigners visit their countries, their people and leaders receive the visitors with generosity and respect. Most of time, they wear leather as clothing, and their livelihood depends on fishing. For those who live in the mountains, they often fought battles and killed each other in ordinary years. They eat snakes, ants, spiders and other insects. This annotation describes the countries, people and lifestyles of the North American Indians (from Labrador to Florida) and their hospitality to the foreigners. However, the tribes which lived in the mountains often engaged in wars. The Chinese term "平年" usually refers to the years without leap days in the Gregorian calendar. Here, it is better to interpret them as ordinary years with or without leap days.
• 4	訝巴剌亦尔	Ya Ba La Yi Er	Iapalair[67]	The latitude of 訝巴剌亦尔 (Iapalair; 31°N, c. 90°W) on the N. America-KWQ correlates closely with today's U.S. state of Georgia (33.3°N, 83.4°W). An unnamed river lying to the northeast of 訝巴剌亦尔 correlates well with the Savannah River (length: 484 km/301 mi[68]; 32.1°N, 81.1°W; see Fig. A5.3), which is a major river in southeastern U. S., forming most of the border between the states of South Carolina and Georgia. The Altamaha River (220 km/137 mi; 31.3°N, 81.3°W) is a shorter river and has a bending flowing path at lower latitude (see Fig. A5.3), hence, it is a less suitable candidate for the unnamed river depicted near 訝巴剌亦尔.

(*Continued*)

(*Continued*)

Item #	Name	Pinyin	Etymology	Analysis of history, politics, geography; comparison or comment
				The area of Georgia was inhabited by Indigenous American tribes for thousands of years. A modest Spanish presence was established in the late sixteenth century.[69] None of the three sixteenth century European maps depicts Iapalair. The N. America-KWQ is apparently a map before the late sixteenth century.  Fig. A5.3. Map of the Province of Georgia (see endnote 70, public domain).
5	得尔勿羅洛/譯云花地	De Er Wu Luo Luo/Yi Yun Hua Di	Flora, terra[71]; "flora" means "flower" or "plants", "terra" means "land".	Florida was named for the day (2 April, 1513) on which Spanish explorer Juan Ponce de León (1474–1521) became the first known European to make landfall, calling the region *La Florida* for its lush greenery and the Easter season.[72] Various Indigenous American groups have inhabited Florida for at least 14,000 years. Florida has no written history until the arrival of Europeans; the Spanish explorer Juan Ponce de León (1474–1521) in 1513 made the first textual records.[73]

				得尔勿羅洛 is the transliteration of Flora, terra. On the N. America-KWQ, 得尔勿羅洛 (27°N, 90°W) corresponds to roughly today's Florida (28°N, 82°W). Florida borders the Gulf of Mexico to the west, Alabama to the northwest, Georgia to the north, the Bahamas and Atlantic Ocean to the east, and the Straits of Florida and Cuba to the south. On both the N. America-Mercator (1569) and the N. America-Ortelius (1570), *La Florida* is used to depict today's Florida; but the N. America-Plancius (1594) simply uses Florida.
6	亞伯尔耕國	Ya Bo Er Geng Guo	Apalchen	亞伯尔耕國 is in the Apalchen Village of the tribe (the Apalachee) of North American Indians who spoke a Muskogean language (they had no written language) and inhabited the area in northwestern Florida between the Aucilla and Apalachicola rivers above Apalachee Bay.[74] A Spanish expedition led by Pánfilo de Narváez (1470 or 1478–1528) in 1528[75] encountered the indigenous Apalachee people but did not reach their main village near today's Tallahassee (30.4°N, 84.3°W), Florida. These explorers transcribed the village's name as Apalchen or Apalachen.[76] Later, the Spanish altered the name to Apalachee and applied it to the coastal region bordering today's Apalachee Bay and spread inland to the north, as well as to the tribe which lived in it.[77] Hernando de Soto (c. 1500–1542) and his expedition occupied the Apalachee town of Anhaica (now Tallahassee) in the winter of 1538–1539.[78] On the N. America-KWQ, 亞伯尔耕國 (37.5°N, 90.5°W) is related to the geographical location where the Apalachee tribe historically lived between the Aucilla River and Ochlockonee River, but not very close to the head of Apalachee Bay (in the north-eastern Gulf of Mexico), before the Spanish arrived. The N. America-Ortelius (1570) depicts Apalchen further north and close to the Atlantic Ocean, indicating the later Spanish development related to this geographical region.

(*Continued*)

(*Continued*)

Item #	Name	Pinyin	Etymology	Analysis of history, politics, geography; comparison or comment
● ≈ 2	眾仙河	Zhong Xian He	Alabama River	眾仙河 is a name of Chinese origin. 眾 means "many", 仙, "saints or god" and 河, "river". 眾仙河 (32°N, 92.5°W) literally means "River of All Gods". It is today's Alabama River (31.8°N, 87.1°W), which flows into Mobile Bay of the Gulf of Mexico, see Fig. A5.4.  Fig. A5.4. The Mobile, Alabama and Coosa rivers are basically a single river, the name of which changes at the confluences of major tributaries (see endnote 79, public domain). The Alabama River was named by early European explorers after the Indian tribe that lived in the territory and first appeared in 1540 spelled as "Alibamu", "Alibamo" and even "Limamu" in the journals of the Spanish explorer Hernando De Soto (c. 1500–1542).

				Documented by Europeans first in 1701, the Alabama, Coosa and Tallapoosa rivers were central to the homeland of the Creek Indians before their removal by United States forces to the Indian Territory in the 1830s.[80]
7	亞勿加尔國	Ya Wu Jia Er Guo	Avacal or Auacal[81]	On the N. America-KWQ, 亞勿加尔國 (37°N, 95°W) may be in today's Kentucky, Illinois or Indiana. The uncertainty is due to the longitudinal reading shown on the N. America-KWQ, which tends to depict the place more westward by 5–10 or more degrees. There is no written historical record or a modern English name for this chiefdom. This means the Europeans had not explored the place during the sixteenth century. On the N. America-Mercator (1569), the kingdom is depicted as Auacal and on the N. America-Plancius (1594) Avana: res. Avacal; both are north of 40°N. The kingdom is depicted on the N. America-Ortelius (1570) as Avacal, just south of 40°N.
8	哥沙國/哥妙國	Ge Sha Guo/ Ge Miao Guo	Coosa, Cossa[82] or Comos	歌妙國 (Comos; 32°N, 100°W) is depicted on the N. America-KWQ. It should be interpreted as 歌沙國 (Coosa or Cossa).[83] Both Comos and Cossa are depicted not far apart on *Americae Sive Qvartae Orbis Partis Nova Et Exactissima Descriptio* (1562) by Diego Gutiérrez and both are close to an unnamed river which may be the Coosa River.[84] The Coosa Chiefdom (c. 1400–c. 1600)[85] was one of the most powerful in the southeast at the time. The Indigenous Americans lived in the Coosa Valley where the Coosa River (34.3°N, 85.2°W) flows through before Hernando de Soto (c. 1500–1542) and his men became the first Europeans to visit it in 1540.[86] The Coosa River is a tributary of the Alabama River in the U.S. states of Alabama and Georgia, see Fig. A5.4.

(Continued)

(*Continued*)

Item #	Name	Pinyin	Etymology	Analysis of history, politics, geography; comparison or comment
				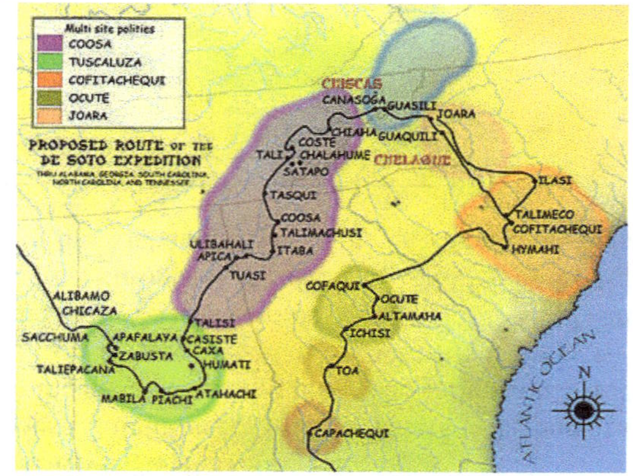 Fig. A5.5. The map shows the de Soto expedition route through Georgia, South Carolina, North Carolina, Tennessee and Alabama, based on the Charles M. Hudson map of 1997 (see endnote 87, public domain). The Coosa Chiefdom is inside the purple-coloured area. It seems that the N. America-KWQ depicts 歌沙國 erroneously as a coastal chiefdom facing the Gulf of Mexico, whereas Fig. A5.5 depicts Coosa inland. If its longitudinal error on the N. America-KWQ can be reduced, the Coosa Chiefdom should be located along the Coosa River at the northeastern end of the Alabama River (眾仙河; ≈ 2). Despite this geographical error, the depiction of the Coosa Chiefdom reveals the era of the Americas after c. 1400 and before c. 1600. On the 1570 World Map by Ortelius,[88] Comos is depicted to the west of the upper stream of an unnamed river which may be the Coosa River; but on the N. America-Ortelius (1570), Comos is depicted to the north of the upper stream of a river (but it has a different name; Cossa is also depicted, right next to Comos). Ortelius seemed confused about the locations of Comos and Cossa. Neither the N. America-Mercator (1569) nor the N. America-Plancius (1594) depicts Comos, Cossa or Coosa.

9	摩可沙國	Mo Ke Sha Guo	Mocosa[89] or Monongahela	Monongahela (or transliterally Mocosa) culture was an Iroquoian Indigenous American cultural manifestation of Late Woodland peoples from 1050 to 1635 in today's western Pennsylvania, western Maryland, eastern Ohio and West Virginia.[90] On the N. America-KWQ, 摩可沙國 (Mocosa; 42.5°N, 91°W) is depicted more westward than its actual geographical location to the southeast of Lake Erie (41.4°N, 83.5°W to 42.9°N, 78.8°W) shown in Fig. A5.6. On the N. America-Mercator (1569), Mocosa is depicted at 41.5°N, 75°W; and on the N. America-Ortelius (1570), at 42.7°N, 83°W. Both give better longitudinal readings than that on the N. America-KWQ. Mocosa is not depicted on the N. America-Plancius (1594). 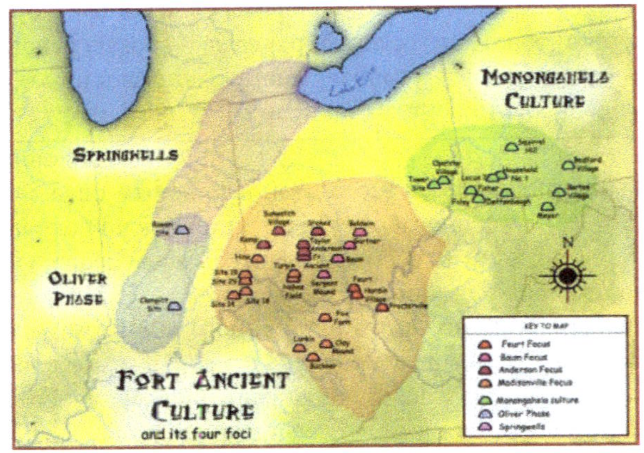 Fig. A5.6. Monongahela cultural region, with some of its major sites and neighbours in 1050–1635 (see endnote 91, public domain).

(*Continued*)

(*Continued*)

Item #	Name	Pinyin	Etymology	Analysis of history, politics, geography; comparison or comment
• 10	何多亞蠟瓦	He Duo Ya La Wa	Hotuoiolaua[92]	On the N. America-KWQ, 何多亞蠟瓦 (Hotuoiolaua; 51.5°N, 84.5°W) was a region in today's western Quebec (53°N, 70°W), Canada. Again, the N. America-KWQ depicts the geographical location more westward than its actual position. Indigenous peoples in Quebec total eleven distinct ethnic groups. Between 1534 and 1763, Quebec was called *Canada* and was the most developed colony in New France (northeastern North America).[93] Jacques Cartier (1491–1557) became the first European explorer to discover and map Quebec when he landed in Gaspé in 1534.[94] But the place is not depicted on any of the three European maps. There is no written historical record or a modern English name for Hotuoiolaua.
11	蓊㞋答	Weng E Da	Honguedo: the name first appeared in the reports of Jacques Cartier of 1535–1536; but the origin of the name is uncertain.	蓊㞋答 (52.5°N, 76.5°W) is the transliteration of Honguedo. French-Breton maritime explorer Jacques Cartier (1491–1557) in 1534 reached Gaspé — as called by the St. Lawrence Iroquoians Honguedo — in today's Quebec. On the N. America-KWQ, 蓊㞋答 is depicted at a distance away from the Honguedo Strait (French: *Détroit d'Honguedo*), which is a strait in eastern Quebec, Canada, between Anticosti Island and the Gaspé Peninsula. It is one of the two outlets of the St. Lawrence River into its estuary, the Gulf of St. Lawrence. The other is the Jacques Cartier Strait on the north side of Anticosti Island. In the sixteenth century, the Honguedo Strait was known as the Saint-Pierre Strait as shown on the N. America-Mercator (1569). On the N. America-Ortelius (1570), Hanguedo is placed near a tributary, *Saguenai fl.*, of the St. Lawrence River. On the N. America-Plancius (1594), Hanguedo is placed near an unnamed tributary of the St. Lawrence River. All three maps depict Honguedo at quite a distance away from the Honguedo Strait.

				 Fig. A5.7. The location of the Honguedo Strait (*détroit d'Honguedo*) is between Anticosti Island and the Gaspé peninsula (see endnote 95, public domain).
12	新拂郎察	Xin Fu Lang Cha	New France: French: *Nouvelle-France*	新拂郎察 is the Chinese translation of New France (French: *Nouvelle-France*) which was the area colonised by France in North America. It began with the exploration of the Gulf of St. Lawrence by Jacques Cartier in 1534 and ended with the cession of New France to Great Britain and Spain in 1763 under the Treaty of Paris (1763).[96] New France, at its peak, controlled an area that extended from Newfoundland to the Canadian Prairies and from Hudson Bay to the Gulf of Mexico, including all the Great Lakes of North America.[97] Notice that on the N. America-KWQ published in 1602, the Great Lakes are not depicted, because only around 1615 did Étienne Brulé (c. 1592–c. 1633) become the first European to discover the Great Lakes;[98] and only in the 1430s did Zheng He's mariners reach the Gulf of St. Lawrence and St. Lawrence River to explore the N. America inland and discover at least four of the five Great Lakes; however, the geographical data on the America-KWQ were obtained during their sixth voyage in the 1420s when they entered the American continent from the Gulf of Mexico as I explained in Section 6 of the main text of this chapter.

(*Continued*)

(*Continued*)

Item #	Name	Pinyin	Etymology	Analysis of history, politics, geography; comparison or comment
13	可尔得勒亚尔地	Ke Er De Le Ya Er Di	*Terra de Cortereali*s or part of Newfounland	"Terra Cortereali*s*" began appearing on European maps in one form or another in the very early sixteenth century,[99] referred to as Newfoundland (it was not thought of as an island then), because two Europeans (Gaspar Corte-Real and his brother Miguel Corte-Real)[100] reached northeast America in 1501–1502. 可尔得勒亚尔 is the transliteration of *Cortereali*s, and 地/*Terra* means "land". Portuguese explorer Gaspar Corte-Real (1450–1501) who, alongside his father and brother, participated in various exploratory voyages which are said to have been some of the first by Europeans to reach Newfoundland and possibly other parts of eastern Canada.[101] However, Italian navigator John Cabot (Zuan Chabotto/Giovanni Caboto; c. 1450–c. 1500) was the first European on record since the Norse settlers to set foot in 1497 on Newfoundland (an island).[102] This is the part of North America the Europeans should know well. Newfoundland is an island separated from the Labrador Peninsula by the Strait of Belle Isle, and from Cape Breton Island by the Cabot Strait. It blocks the mouth of the St. Lawrence River, creating the Gulf of St. Lawrence, the world's largest estuary. But on the N. America-KWQ, 可尔得勒亚尔地 (*Terra de Cortereali*s was then interpreted as part of Newfoundland) is depicted as a peninsula separated by a large bay, depicted as 默羅它海 (Item 24), from 巴革老地 (*Terra de baccalaos*; Item 16; it was interpreted then as the other part of Newfoundland), also a peninsula. On the N. America-Mercator (1569) and the N. America-Ortelius (1570), Newfoundland seems to not only contain the *Terra Cortereali*s peninsula but also several additional islands blocking the Gulf of St. Lawrence including *C. de baccalaos*. The N. America-Plancius (1594) does not depict *Terra Cortereali*s anymore, but only a single island named *I. dos Bacalhaos* (Portuguese: *Terra de baccalaos*) as today's Newfoundland. Clearly, the depiction of N. America-Plancius (1594) resembles most closely today's Newfoundland as shown in the bottom panel of Fig. A5.8.

				If Matteo Ricci were the true cartographer of the N. America-KWQ, how could he make such a mistake in contrast to the three major European maps? *This evidence lends strong support to the conclusion that the America-KWQ is not a direct or adapted copy of the three European maps.*
14	得尔洛勿洛多/譯云耕農地	De Er Luo Wu Luo Duo/ Yi Yun Geng Nong Di	Labrador; Portuguese *Lavrador* means "farmer".	Labrador is named after Portuguese explorer João Fernandes Lavrador (1453–1501), who sailed along the coasts of today's Labrador Peninsula in 1498–1499.[103] He was a farmer from the Azores Islands. The N. America-KWQ erroneously places 得尔洛勿洛多 (Labrador) inland and to the west of 可尔得勒亚尔地 (Newfoundland). The *Speculum Orbis Terrae* ("Mirror of the World"), an atlas published by Cornelis de Jode (1568–1600) in Antwerp in 1593, does depict *Ter. d Laborador*, but as an island in the opening of the Gulf of St. Lawrence.[104] The N. America-Plancius (1594) depicts *Terra de Labrador* to the north of *G de Morostro*. More discussions are presented in Sub-section (8) of Section 4 in the main text.
15	亦利多的蘭地	Yi Li Duo De Lan Di	*Estotilant*	亦利多的蘭地 is the transliteration of *Estotilant,* which is a region that appeared on the Zeno map in the top panel of Fig. A5.8, which corresponds to today's Labrador. The name "Estotilande" is also written on a map of New France from 1660.[105] On the N. America-KWQ, 亦利多的蘭地/*Estotilant*/Labrador is incorrectly depicted as an inland region to the west of 巴革老地 (*Terra de Baccalaos*; it was then interpreted as part of Newfoundland). On the three European maps, this land region, depicted as Estotilant or Estotiland, is now known to be Labrador, correctly facing the North Atlantic Ocean. The error of the N. America-KWQ is that it does not have today's Newfoundland (island) to begin with!

(Continued)

(*Continued*)

Item #	Name	Pinyin	Etymology	Analysis of history, politics, geography; comparison or comment
				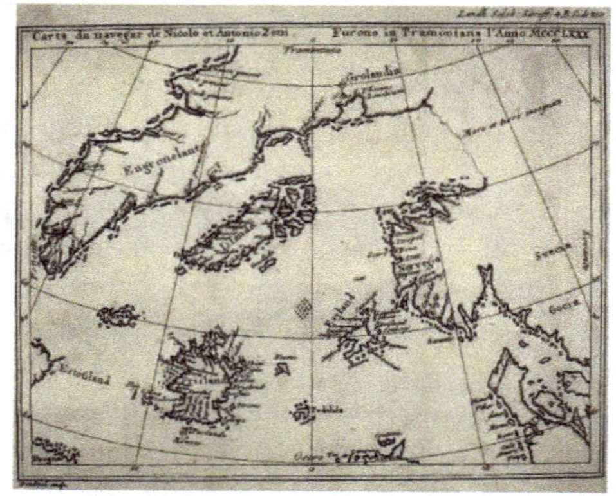  Fig. A5.8. The top panel shows "Estotilant" appearing on the 1558 Zeno map (see endnote 106, public domain); it is located on the western side of the Atlantic Ocean, in the approximate location of what is now known to be Labrador. The Zeno map is based on an exploration made by the Zeno brothers, Nicolò (c. 1326–c. 1402) and Antonio (died c. 1403), of the North Atlantic and Arctic waters in the 1390s. In the lower panel, the upper, yellow-coloured region is Labrador, and the lower portion, forming a large island, is Newfoundland (see endnote 107, public domain); together they make today's Labrador and Newfoundland Province.

16	巴革老地	Ba Ge Lao Di	*Terra de baccalaos; baccalao* in Spanish means "cod".	巴革老地 is the transliteration of *Terra de baccalaos*. On the N. America-KWQ, 巴革老地 is depicted as a peninsula and part of Newfoundland (the other part is 可尔得勒亚尔地 *Terra Corterealis* discussed in Item 13). The N. America-KWQ made a serious error in depicting today's Newfoundland as composed of two peninsulas separated by a large bay. But on the N. America-Mercator (1569), *Terra de baccalaos* is explicitly written where today's Newfoundland is and *Terra Corterealis* becomes part of today's Labrador, the other part is *Estotilant* (facing the Atlantic Ocean) to the north of *Golfam de Merofro*. On the N. America-Ortelius (1570), *C. de bona vista baccalaos* is written next to several small islands which block the Gulf of St. Lawrence and *Terra Corterealis* becomes part of today's Labrador, the other part is *Estotilant* (facing the Atlantic Ocean) to the north of an unnamed bay. On the N. America-Plancius (1594), *I. dos Balcahaos* is today's Newfoundland and *Estotilant* is today's Labrador facing the Atlantic Ocean, *Terra Corterealis* is not depicted anymore. There is *G. de Merostro,* which is in *Estotilant.* This is the closest representation of today's Labrador and Newfoundland.
17	飛私得島	Fei Si De Dao	Greenland or Frisland	Frisland, also called Frischlant, Friesland, Frislanda, Frislandia or Fixland, was a phantom island (in European minds) that appeared on virtually all the maps of the North Atlantic from the 1560s through the 1660s due to the confusion between an imaginary island and the actual southern part of Greenland. The N. America-Ortelius (1570) depicts only Frisland (62°N, 25.5°W), whereas the N. America-Mercator (1569) depicts both Frisland (62°N, 18.5°W) and Greenland (75°N, 17.5°W), and the N. America-Plancius (1594) also depicts both Frisland (62°N, 33°W) and Greenland (71°N, 35°W), showing the confusion of these cartographers.

(Continued)

(*Continued*)

Item #	Name	Pinyin	Etymology	Analysis of history, politics, geography; comparison or comment
				On the N. America-KWQ, only 飛私得島 (71°N, 30.5°W) is depicted and it is today's Greenland (71.7°N, 42.5°W). The latitudinal reading is good, but the longitudinal reading is more eastward. Greenland has been politically and culturally associated with Europe, beginning in 986,[108] even though physio-graphically it is a part of the continent of North America. Most of its residents are Inuit, whose ancestors migrated from Alaska by the thirteenth century,[109] while Norsemen settled the uninhabited southern part of Greenland beginning in the tenth century and named the island as Greenland.[110] Greenland was formally under the Norwegian crown from 1261.[111] In 1499 the Portuguese briefly explored and claimed the island.[112] Greenland was under Danish rule from 1776 to 1979, when Denmark granted it home rule. It is very interesting and correct that the N. America-KWQ depicts 飛私得島 as today's Greenland without drawing a separate "Frisland Island".
• 18	甘峰	Gan Feng	*Chon, promontorio di*[113]	The name is of Chinese origin. There is no written historical record or a modern English name for this mountain.
• 19	香峰	Xiang Feng	*Hian, promontorio*[114]	The name is of Chinese origin. There is no written historical record or a modern English name for this mountain.
• 20	泥德	Ni De	Fiste[115]	There is no written historical record or a modern English name for this place.
• 21	冰海	Bing Hai	Arctic Ocean	The name is of Chinese origin. The Arctic Ocean includes the North Pole region in the middle of the Northern Hemisphere and extends south to about 60°N. It is mostly covered by sea ice throughout the year. Hence, the N. America-KWQ depicts it as 冰海, literally meaning "sea of ice". The Arctic Ocean is surrounded by Eurasia and North America, and the borders follow topographic features: the Bering Strait on the Pacific side and the Greenland Scotland Ridge on the Atlantic side.

				For much of European history, the north polar regions remained largely unexplored and their geography, conjectural. Early cartographers were even unsure whether to draw the region around the North Pole as land in Johannes Ruysch's map of 1507[116] or Gerardus Mercator's map of 1595,[117] or as water in Martin Waldseemüller's world map of 1507.[118] 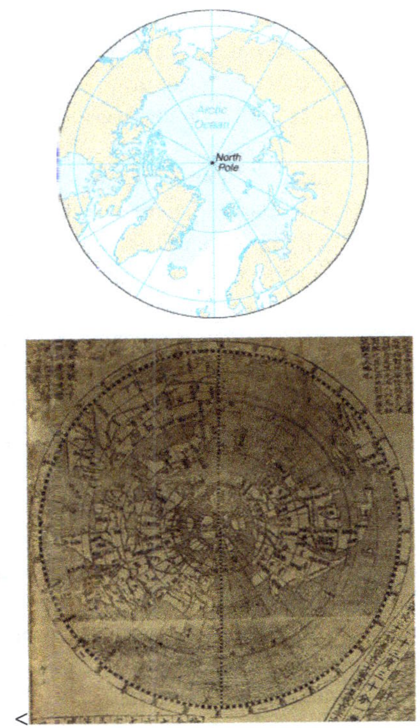 Fig. A5.9. The top panel shows the Arctic Ocean on a modern map, with the Americas to the left, including Hudson Bay (some of which is south of 57°N latitude, off the map: see endnote 119, public domain); the bottom panel shows the north polar regions extracted from KWQ (see endnote 120, public domain), with the Americas to the right; the drawing is very realistic.

(Continued)

(Continued)

Item #	Name	Pinyin	Etymology	Analysis of history, politics, geography; comparison or comment
				Only in 1648 did Russian explorer Semyon Ivanovich Dezhnyov (c. 1605–1673)[121] open the Northeast Passage — the now famous Bering Strait between America and Asia — in fact, the Northeast Passage is a broad term for any route lying north of the Eurasian continent and stretching from the waters north of the Norwegian Sea to the Bering Strait; the Northwest Passage connects the Atlantic and Pacific Oceans via the Arctic Ocean. Finally, Americans Robert Peary (1586–1920) and Matthew Henson (1866–1955), along with several Inuits, reached the North Pole on 6 April, 1909, by traversing across the sea ice on dog sleds.[122]
• 22	如里漢島	Ru Li Han Dao	Rulihan Island	The name is of Chinese origin. Both islands, 如里漢島 (Rulihan Island; Item 22) and 鬼島 (Ghost Island; Item 23), together form Cape Breton Island — the entire mother island. In Chapter 1, I have shown that the Island is depicted as two separate islands on KWQ, because there existed an ancient canal, built or rebuilt in the 1420s by Zheng He's mariners, across today's village of St. Peter's. The canal connected Bras d'Or Lake and the Gulf of St. Lawrence to facilitate a direct sailing into the St. Lawrence River. The canal separated Cape Breton Island into two separate islands. The date was long before the Europeans arrived. On the N. America-Mercator (1569), there is a single island depicted as *C de Breton*. Similarly, on the N. America-Ortelius (1570), there is a single island depicted as *C. de Bertã*; and on the N. America-Plancius (1594), there is also a single island depicted as *C: de Bertã*. None of these European cartographers knew that there was a Chinese-built canal connecting the Bras d'Or Lake with the Atlantic Ocean. The canal was still functional when the French military arrived there in the seventeenth century.[123]
• 23	鬼島	Gui Dao	Ghost Island	The name is of Chinese origin. Please see Item 22 for a detailed explanation.

24	默羅它海	Mo Luo Ta Hai	*Gulfo Merosro* or *Melotuo*, mare	On the N. America-KWQ, 默羅它海 is the *Gulfo Merosro* or *Melotuo, mare,* which is today's Lake Melville/Hamilton Inlet/Groswater Bay system, see Fig. A5.10. Groswater Bay is in south-central Labrador, Canada. Its Hamilton Inlet and Melville Lake extensions stretch 140 km/87 mi inland.[124] In 1586, an Inuit attack on the expedition of John Davis (c. 1550–1605) killed two and wounded others.[125] Hamilton Inlet got its name in honour of Charles Hamilton (1767–1849), commodore-governor of Newfoundland in the early 1800s and former namesake of the inlet's affluent, the Hamilton River (now the Churchill).[126] On the N. America-Mercator (1569), there is *Golfam de Merofro*; on the N. America-Ortelius (1570), there is an unnamed gulf; and on the N. America-Plancius (1594), there is *G. de Merostro.* The sea or gulf also appears on the 1593 Americae Pars Borealis Florida, Baccalaos, Canada, Corterealis.[127] 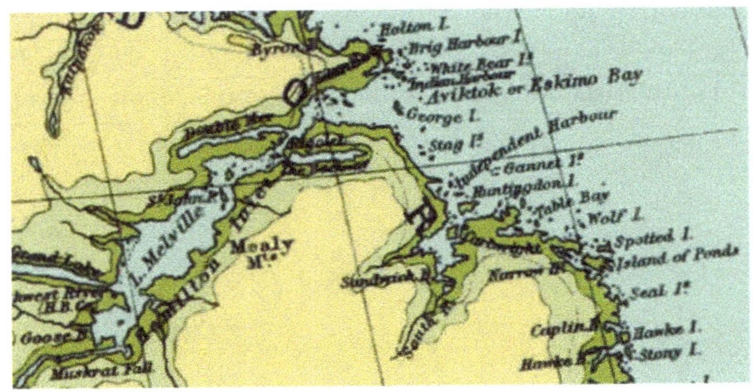 Fig. A5.10. This map depicts the Lake Melville/Hamilton Inlet/Groswater Bay system, showing Hamilton Inlet as inclusive of L. Melville.[128]

(Continued)

(Continued)

Item #	Name	Pinyin	Etymology	Analysis of history, politics, geography; comparison or comment
25	地北極界	Di Bei Ji Jie	Arctic Circle: it is the southern-most latitude in the Northern Hemisphere at which the centre of the sun can remain continuously above or below the horizon for twenty-four hours.[129]	地 means "Earth", 北, "north", 極, "pole" and 界, "boundary". 地北極界 is today's Arctic Circle, which marks the northernmost region at which the centre of the noon sun is just visible on the December solstice and the southernmost point at which the centre of the midnight sun is just visible on the June solstice.[130] It is one of the two polar circles and the most northerly of the five major circles of latitude as shown on maps of Earth. The position of the Arctic Circle is not fixed and currently runs at 66.6°N. The region north of this circle is known as the Arctic, and the zone just to the south is called the Northern Temperate Zone. The term "Arctic" comes from the Greek word "arktos", which means "bear". Greek seafarers called the Arctic region, into which they had presumably already ventured for the first time around 325 B.C., "land under the constellation of the Great Bear".[131] Both the N. America-Mercator (1569) and the N. America-Plancius (1594) depict Circulus Arcticus, meaning Arctic Circle, but not the N. America-Ortelius (1570).

Zone II: North America (Western Region)

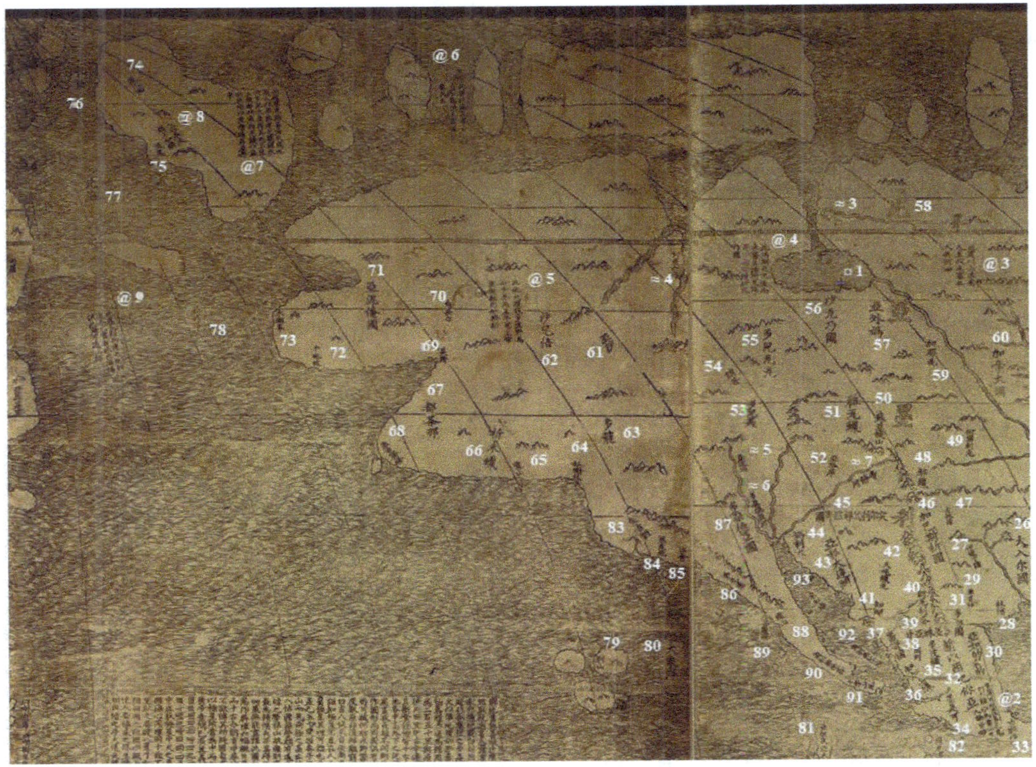

Fig. A5.11. Extracted from Fig. 5.3, bottom right panel, public domain.

26	大入尔國	Da Ru Er Guo	Tagil	On the N. America-KWQ, 大入尔國 (35°N, 100°W) is in today's state of Mississippi (33°N, 90°W). Again, the N. America-KWQ gives a more westward longitudinal reading. There is no written historical record or a modern English name for Tagil. But Tagil appears on the three European maps.

(*Continued*)

(*Continued*)

Item #	Name	Pinyin	Etymology	Analysis of history, politics, geography; comparison or comment
				The unnamed large river near 大入尔國 is the Mississippi River as discussed in Subsection (9) of Section 4 in the main text. The Missouri and Arkansas Rivers are not depicted on the N. America-KWQ, implying that they were not explored at the time. The Mississippi River either borders or passes through the states of Minnesota, Wisconsin, Iowa, Illinois, Missouri, Kentucky, Tennessee, Arkansas, Mississippi and Louisiana.[132] The river is divided into the Upper Mississippi, from its source at Lake Itasca in northwestern Minnesota to its confluence with the Ohio River at Cairo, Illinois; and the Lower Mississippi, from Cairo to its mouth near New Orleans, Louisiana, where it flows into the Gulf of Mexico. However, on the N. America-KWQ, the Upper Mississippi does not pass-through Minnesota, Wisconsin and Iowa, implying that the Upper Mississippi was not mapped. The Illinois River was important among Indigenous Americans and early French traders as the principal water route connecting the Great Lakes with the Mississippi. Notice that the Great Lakes are not depicted either, implying they were not discovered yet. Indigenous Americans have lived along the Mississippi River and its tributaries for thousands of years. Most were hunter-gatherers, but some were Mound Builders (such as the Cahokians).[133] Named by Algonkian-speaking Indians, Mississippi can be translated as "Father of Waters" (meaning Great River).[134] On 8 May 1541, Spanish conquistador Hernando de Soto (c. 1500–1542) was the first European to reach the Mississippi River, south of today's Memphis, Tennessee.[135] He then led his 400 ragged troops across the great river at night and headed into today's Arkansas in search for gold and silver in the American wilderness, but to no avail.

				In 1673, the expedition of Jacques Marquette (1637–1675) and Louis Jolliet (1645–1700)[136] succeeded in mapping the Mississippi within 700 km/435 mi of the Gulf of Mexico. But this great river was already depicted on the N. America-KWQ in 1602 and explored in the 1420s by Zheng He's mariners! None of the three sixteenth century European maps depicts the Mississippi River. The 1584 Geronimo de Chaves map[137] does depict a river named *Rio del Espiritu Santo,* which correlates with the Mississippi River.
• 27	沙蠟丁諸/沙蠟丁諾	Sha La Ding Nuo	Scialatimno	There is no written historical record or a modern English name for 沙蠟丁諸/沙蠟丁諾 (35.2°N, 105°W). The place should be in today's Arkansas or Missouri, after considering the longitudinal error in the westward direction on the N. America-KWQ.
• 28	林濱	Lin Bin	Linpin[138]	The name 林濱 (30.1°N, 104°W) is of Chinese origin. It correlates closely with today's Houston (29.7°N, 95.4°W). The Houston area was the home of the Karankawa and the Atakapa indigenous peoples for at least 2,000 years before the first known settlers arrived. These tribes are almost non-existent today. The Houston area remained largely uninhabited until European settlement in the 1830s.
29	車哥私	Ge Ge Si	Gacos	On the N. America-KWQ, 革哥私 (31°N, 106.5°W) is in today's Texas. The first European to see Texas was Alonso Álvarez de Pineda (1494–1520)[139] in 1520, and he was the first European to have created the first map of the northern Gulf Coast. Between 1528 and 1535, four survivors of the Narváez expedition spent six and a half years in Texas as slaves and traders among various indigenous groups. Among them, Spanish explorer Álvar Núñez Cabeza de Vaca (c. 1490–c. 1559)[140] was the first European to explore the interior of Texas. But there is no written historical record or a modern English name for this place, Gacos.

(Continued)

(*Continued*)

Item #	Name	Pinyin	Etymology	Analysis of history, politics, geography; comparison or comment
• 30	亞如的私的	Ya Ru De Si De	Aiotuxtit	On the N. America-KWQ, 亞如的私的 (27°N, 107.5°W) should be in today's Texas. It does not appear on the N. America-Ortelius (1570), but it is depicted on the 1587 Map of America by Ortelius.[141]
31	多勿國	Duo Wu guo	Toua[142]	On the N. America-KWQ, 多勿國 (29°N, 108°W) should be in today's Texas. On the N. America-Mercator (1569), Toua is depicted at 28°N, 106°W; and on the N. America-Ortelius (1570) Tova is depicted at 29.5°N, 111°W. The N. America-Plancius (1594) does not depict the place. But there is no written historical record or a modern English name for the place.
32	新以西把尼亞	Xin Yi Xi Ba Ni Ya	*Hispania Noua* or *Hifpania Noua*: in Latin, the name means "New Hispania" (Hispania was the ancient name of Spain) or New Spain.	新 means "new" and 以西把尼亞 is the transliteration of "Hispania". The term "Hispania" is a link to colonies that were formerly part of the Spanish Empire (1492–1968),[143] including the Americas, parts of the Asia-Pacific region and Africa. *Hispania Nov(u)a* is depicted on the three European maps, because Spanish explorers mapped the North American coastline north of Florida up to Newfoundland, Labrador, and Greenland by 1501, and Juan Ponce de León (1474–1521)[144] made the first major effort to establish a permanent settlement in North America in 1521.

				 Fig. A5.12. Giacomo Gastaldi's 1548 map of New Spain, *Nueva Hispania Tabula Nova* (in the centre of this figure; see endnote 145, public domain). After the 1521 Spanish conquest of the Aztec Empire, conqueror Hernán Cortés (1485–1547)[146] named the territory New Spain and established the new capital of Mexico City on the site of Tenochtitlan, the capital of the Mexica (Aztec) Empire. New Spain or Kingdom of New Spain (1521–1821)[147] was an integral territorial entity of the Spanish Empire during the Spanish colonisation of the Americas. At its peak, its jurisdiction comprised a vast area that included what are now Mexico, much of the Southwestern U.S. and California in North America, Central America, northern parts of South America and several Pacific Ocean archipelagos.[148]

(*Continued*)

(*Continued*)

Item #	Name	Pinyin	Etymology	Analysis of history, politics, geography; comparison or comment
33	古數沙	Gu Shu Sha	Costosta[149]	On the N. America-KWQ, 古數沙 (18°N, 106.5°W) should be in today's southern Mexico. The longitude is more westward than its actual position, as is often the case for the longitude readings on the N. America-KWQ. Costosta is also depicted on the N. America-Ortelius (1570), but there is no written historical record or a modern English name for the place.
34	墨是可	Mo Shi Ke	Mexico City	On the N. America-KWQ, 墨是可 (21°N, 112.5°W) should be today's Mexico City (19.4°N, 99.1°W). The city of Mexico-Tenochtitlan (1325–1521) was founded by the Mexica people in 1325.[150] After the 1521 Spanish conquest of the Aztec empire, New Spain was established, and the site of the Tenochtitlan became the new capital Mexico City (a great deal of Tenochtitlan was destroyed in the fighting). The old Mexica city is now referred to as Tenochtitlan.
@ 2	墨是可地產各色鳥羽人輯以為畫山水人物皆妙		Mo Shi Ke Di Chan Ge Se Niao Yu Ren Ji Yi Wei Hua Shan Shui Ren Wu Jie Miao	My translation and comment (between square brackets) are as follows: Mexico is a land producing various bird feathers. People collect them [as pens] to draw wonderful paintings of landscapes and human figures. Mexico is home to about fifty hummingbird species, and about a dozen with ranges extending into the Mexico City area. They are small birds.[151] Mexico also has many other kinds of birds which may live in Mexico City.[152]
35	多皮蠟	Duo Pi La	Topira	On the N. America-KWQ, 多皮蠟 (25°N, 112.5°W) should be in today's Mexico. There is no written history or a modern English name for the place.

36	知加土蠟	Zhi Jia Tu La	Cacatula or Zacatula[153]	On the N. America-KWQ, the latitude of 知加土蠟 (23.5°N, 114.5°W) is almost the same as the Tropic of Cancer (the most northerly circle of latitude on Earth at which the Sun can be directly overhead; the circle is currently at 23.4°N). 知加土蠟 should be in today's Mexico and facing the Pacific Ocean. On the N. America-Ortelius, Cacatula (19.8°N, 125.5°W) is more westward (a larger longitudinal error) and at lower latitude. There is no written history or a modern English name for the place.
37	沙里思可國	Sha Li Si Ke Guo	Xalisco[154] or Jalisco (the name comes from the Nahuatl Xalixko)[155]	On the N. America-KWQ, 沙里思可國 (26°N, 115.5°W) is Jalisco (20.6°N, 103.7°W), which is the former name of Xalisco. It was a city state and, with its surrounding municipality of the same name, was in the present Mexican state of Nayarit. From the tenth to the sixteenth centuries, many nomadic tribes lived in Jalisco's central valley. Spanish explorers arrived in Jalisco soon after the conquistador Hernán Cortés (1485–1547) conquered the Aztec capital of Tenochtitlan (now Mexico City) in 1521.[156] The N. America-Mercator (1569) depicts Xalisco at 23°N, 126.5°W; the N. America-Ortelius (1570) depicts it at 26°N, 128°W; and the N. America-Plancius (1594) depicts it at 22°N, 115°W. Only the N. America-Plancius (1594) has made improvements in both the latitude and longitude readings of Jalisco/Xalisco.
38	固列	Gu Lie	Culiacán or Culuacan[157]	On the N. America-KWQ, 固列 (25°N, 113.5°W) is Culiacán (24.8°N, 107.4°W) — officially Culiacán Rosales — which is a city in northwestern Mexico. The city was co-founded in 1531 by the Spanish conquerors Lázaro de Cebreros[158] and Nuño Beltrán de Guzmán[159] under the name of *Villa de San Miguel*. On the N. America-Plancius (1594), the city is depicted as *Culiacan S. Michael* at 24.5°N, 119°W. The latitude reading is as good as, but the longitudinal reading is worse than, those on the N. America-KWQ.

(*Continued*)

(*Continued*)

Item #	Name	Pinyin	Etymology	Analysis of history, politics, geography; comparison or comment
39	墨珠亞甘	Mo Zhu Ya Gan	Michoacán	Today's Michoacán (19.2°N, 101.9°W) is in western Mexico (being one of the thirty-two states which comprise the Federal Entities of Mexico) and has a stretch of coastline on the Pacific Ocean to the southwest.[160] But on the N. America-KWQ, 墨珠亞甘 (27°N, 112.5°W) is inland and at a higher latitude. The Spaniards' first years in Michoacán were 1522–1527. The first Spaniard to the area was Cristóbal de Olid (1487–1524).[161] On the N. America-Mercator (1569), Mecñuacan is depicted at 21.5°N, 113.5°W; and on the N. America-Ortelius (1570), Mechvacan is depicted at 19.8°N, 118.5°W. We can see that their latitude readings are better than that on the N. America-KWQ, but longitudinal readings are still more westward than those on the N. America-KWQ.
• 40	新瓦力茶	Xin Wa Li Cha	*Nueva Galicia*	New Kingdom of Galicia (*Nuevo Reino de Galicia*; Galician: *Reino de Nova Galicia*) or simply New Galicia (*Nueva Galicia* or *Nova Galicia*) was an autonomous kingdom of the Viceroyalty of New Spain (1521–1821).[162] Spanish exploration of the area began in 1531 with the expedition of Nuño Beltrán de Guzmán (c. 1490–1558).[163] The area was named in 1531 "Reino de Nueva Galicia" after Galicia in Spain. Nueva Galicia's territory became today's Mexican states of Aguascalientes, Guanajuato, Colima, Jalisco, Nayarit and Zacatecas. The name must have been added by Matteo Ricci onto the N. America-KWQ, because it is not of Chinese origin and did not exist in the fifteenth century when the source map of the N. America-KWQ was drawn.
41	知鬧	Zhi Nao	Chinao[164]	On the N. America-KWQ, 知鬧 (30°N, 115.5°W) is in the west coast region of today's Mexico. However, on the N. America-Mercator (1569), Chinao is depicted at 23°N, 110.5°W, being far away from the coast as an inland city. Similarly, on the N. America-Ortelius (1570), it is at 23°N, 117°W, also being far away from the coast as an inland city. There is no written history or a modern English name for the place.

• 42	火吒蠟瓦	Huo Zha La Wa	Huocialaua[165]	On the N. America-KWQ, 火吒蠟瓦 (33°N, 114°W) may be in today's Texas or New Mexico. Due to the longitudinal error of the map, it is not easy to pinpoint its actual location on a modern map. There is no written history or a modern English name for the place.
43	亞私大達闌	Ya Si Da Da Lan	Astatlan	On the N. America-KWQ, 亞私大達闌 (35°N, 118.5°W) is Astatlan, which was a town in the coastal region of western Mexico, facing the Gulf of California. There is no written history or a modern English name for the place.
44	媽剌大	Ma La Da	Marata	On the N. America-KW, 媽剌大 (36°N, 121.5°W) is a city/kingdom located to the east of the Baja California peninsula. On the N. America-Ortelius (1570), Marata is depicted at 37.5°N, 135°W; and on the N. America-Plancius (1594), at 33.5°N, 120.5°W. Although older texts have very little to support the existence of places like Marata, etc., it does not mean that such places were made up.[166] There is no historical record showing that the Europeans explored the place.
45	突爾利其祁默奇國	Tu Er Li Qi Qi Mo Qi Guo	Terlichichimechi	On the N. America-KWQ, 突爾利其祁默奇國 (39°N, 114.5°W) is today's Texas (31°N, 100°W). The latitude is too much to the north. On the N. America-Mercator (1569) and the N. America-Ortelius (1570), the land that will become Texas is labelled "Terlichichimechi" in reference to both a land and a people that were thought to occupy this area. In 1519, the explorer Alonso Álvarez de Piñeda (1494–1520)[167] became the first European to map the Texas Gulf Coast. However, it would be another nine years before any Spaniards explored the Texas interior.
46	加巴斯祁國	Jia Ba Si Qi Guo	Capaschi[168]	On the N. America-KWQ, 加巴斯祁國 (37.5°N, 100°W) is Capaschi, which was in today's Texas or Oklahoma. There is no written history or a modern English name for the place.

(Continued)

(*Continued*)

Item #	Name	Pinyin	Etymology	Analysis of history, politics, geography; comparison or comment
• 47	七城國	Qi Cheng Guo	Seven Cities of Gold, also known as the Seven Cities of Cibola; Italian: *Regno delle sette cittaã*	In the fifteenth century, the Age of Discovery began in Europe. The maritime empires' world exploration, and their discovery of the ocean route to the East earned them great wealth. Hence, in the sixteenth century when local legends spoke of the Seven Cities of Cibola — the Seven Cities of Gold — now known as the Zuni Indian villages (Zuni, North American Indian tribe of what is now west-central New Mexico, on the Arizona border), they inevitably spurred adventurous conquistadors to launch expeditions in search of the elusive cities.[169] Besides "Cibola", names often associated with similar lost cities of gold also include: El Dorado, Paititi, City of the Caesars, Lake Parime at Manoa, Antilia and Quivira. However, several European explorers later discovered that the stories were unfounded and that there were no treasures there, only adobe towns. In the early twenty-first century, the Canadian architect Paul Chiasson discovered that the Cape Breton Island in Canada was also an Island of Seven Cities,[170] and it did have gold mines.[171] In alphabetical order, the seven cities are Cape Dauphin, Chéticamp, Ingonish, Louisbourg, St. Ann's, St. Peter's and the area *now* known as *Sydney.*[172] On top of Cape Dauphin, an enormous Chinese ruin was discovered.[173] This is also further discussed in Chapter 1 of this book. On the N. America-KWQ, 七城國 (39°N, 105.5°W) is depicted to the east of 突爾利其祁默奇國 (Item 45) in today's Texas, whereas the actual location of Cibola/Zuni Indian villages was in west-central New Mexico, on the Arizona border. Hence, they should be to the west of Texas. This shows that Matteo Ricci did not know where to place the Seven Cities of Gold/Cibola correctly and how many such places existed.

48	祁蠟	Qi La	Chilaga[174]	On the N. America-KWQ, 祁蠟 (42.5°N, 106°W) correlates with today's Chicago and the name Chilaga may have been an Indian word for Chicago (41.9°N, 87.7°W) in the 1500s or even earlier.[175] Jean Baptiste Point du Sable (c. 1750–1818)[176] travelled to North America in his twenties and settled on the shores of Lake Michigan, an area that would eventually develop into the city of Chicago. But Lake Michigan is not depicted on the N. America-KWQ, because Zheng He's mariners discovered the Great Lakes (at least four of them) later in the 1430s, during their seventh voyage, as I have explained in Section 6 of this chapter. The date is much earlier than the seventeenth century when the European explorers "discovered" these lakes. On the N. America-Mercator (1569), Chilaga is depicted at 46°N, 84.5°W; on the N. America-Ortelius (1570), at 47°N, 90.5°W; and on the N. America-Plancius (1594), at 49°N, 98°W. We can see that the coordinates of Chilaga on the N. America-KWQ give the best latitude value, but the worst longitude value in comparison with the three European maps, and the modern map. The Europeans had not explored Chilaga before 1602.
49	何察剌瓦	He Cha La Wa	Hochgelaga: it may be a French corruption of an Iroquoian term.[177]	On the N. America-KWQ, 何察剌瓦 (46°N, 98.5°W) is Hochelaga (45.6°N, 73.5°W) which was a sixteenth century fortified village by St. Lawrence Iroquoian (they were an Indigenous people who existed from the fourteenth century to about 1580). The village was on or near Mount Royal in today's Montreal, Quebec, Canada. Jacques Cartier (1491–1557) arrived there by boat on 2 October 1535 and visited the village the next day.[178] The longitudinal error on the N. America-KWQ is more than 20° westward. On the N. America-Ortelius (1570), Hochgelaga is depicted at 48.5°N, 83°W; and on the N. America-Plancius (1594), at 52°N, 85°W. We can see that the coordinates of Hochgelaga on the N. America-KWQ give the best latitude value, but the worst longitude value in comparison with the three European maps and the modern map.

(Continued)

(Continued)

Item #	Name	Pinyin	Etymology	Analysis of history, politics, geography; comparison or comment
50	蘇亞蠟山	Su Ya La Shan	*Suala mons*[179]	On the N. America-KWQ, 蘇亞蠟山 (47.5°N, 107.5°W) should still be in today's Quebec or Ontario, Canada. It is hard to determine, because the N. America-KWQ often depicts the longitudinal reading more westward from the place's actual longitude. There is no written history or a modern English name for the mountain.
51	祖瓦蠟	Zu Wa La	Kingdom of Ceuola[180] or Cibola	On the N. America-KWQ, 祖瓦剌 (48.5°N, 109°W) is the legendary Ceuola/Cibola. But the latitude is too far away from the real location of Cibola (Háwikuh; 34.9°N, 109°W; now in New Mexico) discovered in 1539 by Father Frier Marco de Nica (c. 1495–1558).[181] Later explorers went there but discovered that there were in fact no treasures as the friar had described. On the N. America-Ortelius (1570), the latitude of Ceuola/Cibola (36.5°N, 143°W) correlates with Cibola/Háwikuh in New Mexico, but the longitude is too westward.
• 52	亞哈庫	Ya Ha Ku	Auacal[182]	On the N. America-KWQ, 亞哈庫 (44°N, 114°W) is regarded as the transliteration of Auacal, which also represents 亞勿加尔國 (Item 7) depicted as a different chiefdom not far away from the Atlantic Ocean. However, 亞哈庫 is located not far from the Gulf of California. On the three European maps, Auacal corresponds to 亞勿加尔國 (Item 7) and only appears once. There is no written history or a modern English name for the place.
53	多兒美	Duo Er Mei	Tolm[183]	On the N. America-KWQ, 多兒美 (49°N, 119.5°W) should be a place in today's Montana (47°N, 110°W). The N. America-KWQ again marks the place more westward. Mountains are depicted near 多兒美, because the name Montana derives from Spanish and describes mountain or mountainous region.[184]

				Various indigenous peoples lived in the territory of today's state of Montana for thousands of years. The oldest dated human burial site in North America discovered in 1968 is near Wilsall, Montana. The human remains of a male infant were found at the Anzick site along with Clovis culture artifacts (roughly 13,200 to 12,900 years ago[185]; this is the first genetic evidence that the Clovis people were descended from Asians). On the N. America-Mercator (1569), Tolm is depicted at 45°N, 154.5°W, close to the North Pacific coast; and on the N. America-Ortelius (1570), at 45°N, 161°W, close to the North Pacific coast and further to the west. On the N. America-Plancius (1594), Tolm is depicted at 49°N, 168°W, also close to the North Pacific coast and even further to the west. On the 1768 Vaugondy Map of California and Alaska,[186] Tolm is depicted as closer to the North Pacific coast than it is on the N. America-KWQ. The first non-Indigenous American explorers known to have set foot in Montana were the members of the expedition led by American explorer Meriwether Lewis (1774–1809) and William Clark (1770–1838) from 1804 to 1806.[187] There is no written history or a modern English name for the place.
54	諸著	Zhu Zhu	Ciucio[188]	On the N. America-KWQ, 諸著 (53°N, 118.5°W) should be in the neighbourhood of today's province Manitoba or Saskatchewan, Canada. But on the N. America-Ortelius (1570), there is a closely related geographical term "Cicuic" (40.5°N, 167°W) near the coastal area facing the North Pacific Ocean. Similarly, on the N. America-Plancius (1594), there is *C. Blanco Ciouic* (36°N, 169°W) right on the North Pacific coast. There is no written history or a modern English name for the place.
• 55	多兒瓦瓦	Duo Er Wa Wa	Tuoruaua[189]	On the N. America-KWQ, 多兒瓦瓦 (54.5°N, 113.5°W) should be in today's Manitoba (56.4°N, 98.7°W), Canada. We continue to see this size of longitudinal error on the N. America-KWQ.

(*Continued*)

(*Continued*)

Item #	Name	Pinyin	Etymology	Analysis of history, politics, geography; comparison or comment
56	沙瓦乃國	Sha Wa Nai Guo	Saguenai was today's city of Saguenay established in 2002; the name is possibly derived from the Innu word.	On the N. America-KWQ, 沙瓦乃 (58°N, 104.5°W) is the transliteration of "Saguenai", which was a city-state; 國 means "kingdom or chiefdom". The city is now called Saguenay (48.4°N, 71°W) in Quebec, Canada. It is on the Saguenay River, about 126 km/78 mi upriver and northwest of Tadoussac, which is a village in Quebec, Canada, at the confluence of the Saguenay and St. Lawrence rivers.[190] In 1542, Jacques Cartier (1491–1557), in search of the rich and mythical country "Kingdom of Saguenay",[191] supposedly reached today's Quebec, Canada, and founded the Charlesbourg-Royal settlement (46.7°N, 71.3°W). His crew initially thought they had found large amounts of diamonds and gold in the area. The treasures were shipped back to France but turned out to be quartz crystals.[192] On the N. America-Mercator (1569), the coordinates of Saguenai (56.5°N, 84.5°W) are close to those on the N. America-Ortelius (1570), depicted as Sagvenai ("v" is the same as "u"; 56.5°N, 87°W) and on the N. America-Plancius, depicted as Saguenai (56.5°N, 86°W). The latitude readings of these four maps are comparable, but too northward in comparison with the modern map. However, the N. America-KWQ shows a very poor longitudinal reading compared to the three European maps. The N. America-KWQ depicts this city-state close to Hudson Bay. When people got off from their boats, they could reach the Saguenay City State by land as described by Annotation @3 shown in the following.
¤ 1	哥泥白斯湖	Ge Ni Bai Si Hu	Lake Conibaz: today's Hudson Bay	On the N. America-KWQ, 哥泥白斯湖 (63.5°N, 101°W) is today's Hudson Bay (60°N, 86°W). But its location is depicted about 16° in longitude west of its actual location. The N. America-Mercator (1569) does not depict this lake. The N. America-Ortelius (1570) only depicts a partial Conibaz Lake on the top edge of the map starting from c. 58°N, 108°W, being further west than that on the N. America-KWQ.

				The N. America-Plancius (1594) depicts *Lago de Conibaz* at c. 58°N, 117°W, even further west than that on the N. America-Ortelius (1570); and in the lake, there is an Indian village named Conibaz. The lake is connected to the north with the Arctic Sea. Near that estuary, two unnamed rivers merge to become River Obilo (Item ≈ 4; today's Mackenzie River), which also flows into the Arctic Sea; this map is incorrect as we know today. But why draw an Indian village in the lake? This came with a strange fiction concerning the "Lake Conibaz" which appears on several of the maps of the late sixteenth century.[193] The lake was often represented with a large island and its waters upon which was placed an Indian village called Conibaz. This lake is usually placed far to the west and sometimes discharges into a northern ocean by Conibaz River. The originator of this geographical anomaly was Andre Thevet (1516–1590) who had talked with Jacques Cartier himself and seems to have gotten his idea for Lake Conibaz from Cartier's description of the Indian method of collecting wampum shells, which Cartier says they called "Cornibotz". Thevet, from this description and from Cartier's accounts of large interior Lakes, evolved his "Lac De Conibaz", which was introduced upon his map of 1575.[194] Thevet's Lake Conibaz was copied by Cornelius de Judaeis on his map in *Speculum Orbis Terrarum*, 1593; by Petrus Plancius in his *Orbis Terrarum*, 1594; by Cornelius Wytfliet in *Descriptiones Ptolemaicae augmentum*, 1597; by Giacomo Gastaldi, in his woodcut wall map of 1561, who finally gave Conibas in 1597 the form we know today for Hudson Bay (extending inland via a narrowish channel and opening into a large inland bay; also showing Tolm Regnum),[195] by Mathias Quadus in *Novi Orbis pars borealis*, 1600; and probably by others. Although we can see no Lake Conibaz on the N. America-Mercator (1569), the lake is depicted on the 1606 Mercator Hondius Map of the Arctic (First Map of the North Pole).[196]

(*Continued*)

(*Continued*)

Item #	Name	Pinyin	Etymology	Analysis of history, politics, geography; comparison or comment
				Henry Hudson (c. 1565–disappeared 23 June 1611) is the first European to discover Hudson Bay, named after him, in 1611, during his search for the Northwest Passage. But Lake Conibaz already fully appears on the N. America-Plancius (1594). Did Plancius copy the lake from the Chinese map? In particular, the longitudinal position of this bay is better represented (101°W) on the N. America-KWQ than that (117°W) on the N. America-Plancius (1594), in comparison with its actual value of 85°W. Besides, Hudson Bay has no connection with the Mackenzie River. This is correctly depicted on the N. America-KWQ. European settlement of Canada began in 1605 (Port Royal, Nova Scotia) and 1608 (Quebec City). Canada-based Europeans reached the Arctic Ocean in 1789 and the Pacific in 1793, both expeditions led by Alexander Mackenzie (c. 1764–1820).[197] All these explorations happened after the publication of the N. America-KWQ in 1602.
@ 3	此洪湖之水淡而未審其涯所至依是下舟可達沙瓦乃國		Ci Hong Hu Zhi Shui Tan Er Wei Shen Qi Ya Suo Zhi Yi Shi Xia Zhou Ke Da Sha Wa Nai Guo	My translation is as follows: The water of this vast lake is fresh, but the extent of the lake has not been examined. If you get off the boat and walk from the lakeside, you can reach the Saguenai City State. Hudson Bay is a large body of saltwater in northeastern Canada.[198] The bay, however, is fed by countless rivers. They pour torrents of fresh water into it. In addition, the bay is ice-covered for much of the year, the rate of evaporation is low.[199] Hence, the bay water is salty, but a lot less salty than the waters of the oceans. If you follow the bay through Hudson Strait, you come to the Atlantic Ocean. If you follow it north around Baffin Island, you come to Baffin Bay.[200] Sooner or later the channels between the islands north of Hudson Bay bring you to the chilly Arctic Ocean.[201]

● ≈ 3	哥入河	Ge Ru He	*Cogib, fiume*[202]	In today's Quebec, the major rivers, from east to west, are the La Grande, Eastmain and Rupert Rivers. But none of them has a latitude larger than the Arctic Circle, which currently runs at 66.6°N. There is also no "Cogib, fiume" in Western records. The only land to the east of Hudson Bay which contains a part of the area north of the Arctic Circle is Baffin Island. Yet, on this island there is no river flowing from east to west into Hudson Bay. This leads me to think that 哥入河 may correlate with the Hudson Strait (62.4°N, 78.1°W), which relates to Hudson Bay, and its latitude is only about 4° below the Arctic Circle. On the N. America-KWQ, to the north and the east of 哥入河 are mountains. This is also consistent with the existence of the Baffin Mountains (which are part of the Arctic Cordillera) along the northeastern coast of Baffin Island and Bylot Island in Nunavut, Canada. The strait lies between the land of northern Quebec and the land and mountains of Baffin Island; hence it was mistaken as a big river (about 800 km/500 mi long and 65–240 km/40–150 mi wide). On the N. America-KWQ, 哥入河 is a long river which spans about 10° in longitude. Since the Earth's circumference is 40,075 km/24,901 mi, this gives an estimated length of the river as 1,113 km/692 mi, a reasonable approximation of the strait's length. It also implies that the explorers entered this strait from the Hudson Bay side but did not sail through the entire strait to reach the Labrador Sea before realising that it was not a river. On the Map from Magin 1611 BHoU-p6, Cogib flu is depicted as the funnel portion of Hudson Bay.[203] This is incorrect.
≈ 4	何皮六河	He Pi Liu He	Obilo River; today's Mackenzie River	On the N. America-KWQ, 何皮六河 (62.5°N, 114.5°W) corresponds to today's Mackenzie River (62.1°N, 122.8°W). To the west of 何皮六河 (Obilo River/Mackenzie River) is the Liard River (61.7°N, 121.2°W), which is a major tributary of the Mackenzie River and drains into it at Fort Simpson, Northwest Territories in Canada. Mackenzie River flows into the Arctic Ocean. The N. America-KWQ depicts a second river next to 何皮六河 (Mackenzie River) near the river's estuary. It should not be the case.

(*Continued*)

(*Continued*)

Item #	Name	Pinyin	Etymology	Analysis of history, politics, geography; comparison or comment
				Scottish explorer Alexander Mackenzie (c. 1761–1820) travelled the river in the hope it would lead to the Pacific Ocean, but instead reached its mouth on the Arctic Ocean on 14 July 1789.[204] Mackenzie River is not depicted on the N. America-Mercator (1569) but appears on his 1595 map of the North Pole.[205] The N. America-Ortelius (1570) does not show this river either, while the N. America-Plancius (1594) depicts this river without a name. On the N. America-KWQ, the longitudinal separation between the Mackenzie River and the Hudson Bay (see Item ¤1) is 13.5°. The actual longitudinal separation between them is around 36.8° (122.8 – 86°). The N. America-KWQ shrinks their separation by more than 20°. Henry Hudson (c. 1565–1611)[206] was the first European to discover the bay in 1611 during his search for the Northwest Passage. As for mapping Hudson Bay more accurately, its east coast was mapped by him two years later in 1613; the south coast was traced in 1631, and the explorer Luke Foxe (1586–c. 1635)[207] lent his name to Foxe Channel in the same year[208]; the west coast was not mapped until the early 1820s.[209]  Fig. A5.13. Mackenzie River (thin blue line) and Liard River (thick blue line; see endnote 210, public domain); the black line west of the Mackenzie is the border between two Canadian provinces (Yukon and Northwest Territories).

				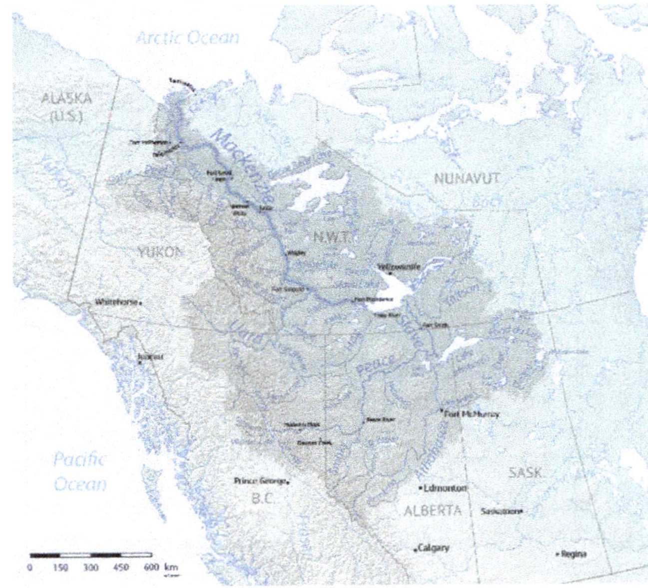 Fig. A5.14. Map of the Mackenzie River system in Canada (see endnote 211, public domain); Hudson Bay is at the right edge of the figure.
@ 4	此處以上未有人至故未審其人物何如		Ci Chu Yi Shang Wei You Ren Zhi Gu Wei Shen Qi Ren Wu He Ru	My translation is as follows: The land to the north of here has not been reached by humans. Hence, the place and people there have not been surveyed and studied. On the N. America-KWQ, the above "land to the north of here" is to the north of 地北極界 (Arctic Circle; Item 25) or the Arctic border (the Arctic border is an imaginary line at latitude 66.6°N; the Arctic is the area within the Arctic Circle drawn by this imaginary line), which is very cold. In those days, the area had not been reached by humans.

(*Continued*)

(*Continued*)

Item #	Name	Pinyin	Etymology	Analysis of history, politics, geography; comparison or comment
• 57	亞外媽	Ya Wai Ma	Iauaema[212]	On the N. America-KWQ, 亞外媽 (58°N, 100°W) should be in today's Quebec or Ontario. There is no written history or a modern English name for the place.
58	北亞墨利加	Bei Ya Mo Li Jia	North America	The 1507 Waldseemüller world map or *Universalis Cosmographia* ("Universal Cosmography") is known as the first map to use the name "America". As explained in *Cosmographiae Introductio*, the name was bestowed in honour of the Italian Amerigo Vespucci (1454–1521),[213] because he set forth the then-revolutionary concept that the lands that Christopher Columbus sailed to in 1492 were part of a separate continent.[214] On the America-KWQ, the names 北亞墨利加 (North America; Item 58) and 南亞墨利加 (South America; Item 220) are separately placed on the continent of today's North America and South America.
59	加那瓦	Jia Na Wa	Canado[215]	On the N. America-KWQ, 加那瓦 (55°N, 97.5°W) should be in today's St. Lawrence River region of Quebec, Canada. There is no written history or a modern English name for the place.
60	加拿大國	Jia Na Da Guo	Canada: the name was known before Europeans arrived; it came from the St. Lawrence Iroquoian word *kanata*, meaning "village" or "settlement".[216]	On the N. America-KWQ, 加拿大國 (52.5°N, 93.5°W) is today's region of Quebec City (46.8°N, 71.3°W) in Canada. Here, the longitudinal error is large. On the N. America-Mercator (1569), Canada is depicted at 55°N, 73.5°W; on the N. America-Ortelius (1570), at 55°N, 78.5°W; and on the N. America-Plancius (1594), at 53.5°N, 65.5°W. These coordinates show that the three European maps give better longitudinal readings, but the N. America-KWQ gives better latitude. Indigenous peoples have continuously inhabited what is now Canada for thousands of years.

				In 1535, Indigenous inhabitants of today's Quebec City region used the word Canada to direct French explorer Jacques Cartier to the village of Stadacona. Cartier later used the word *Canada* to refer not only to that village but to the entire area subject to Donnacona (the chief at Stadacona).[217] By 1545, European books and maps had begun referring to this small region along the St. Lawrence River as *Canada*. Between 1534 and 1763, Quebec was called *Canada* and was the most developed colony in New France (northeastern North America).[218] Jacques Cartier (1491–1557) became the first European explorer to discover and map Quebec when he landed in Gaspé on 24 July 1534.[219]
● 61	亞沙	Ya Sha	Ascia[220]	On the N. America-KWQ, 亞沙 (57°N, 128.5°W) should be in today's British Columbia (53.7°N, 127.7°W), Canada.
● 62	沙兒倍	Sha Er Bei	Sciarpé[221]	On the N. America-KWQ, 沙兒倍 (58.5°N, 134°W) may be in today's British Columbia (53.7°N, 127.7°W) or Yukon (63°N, 135°W) in Canada.
● 63	多籠	Duo Long	Tuolom[222]	On the N. America-KWQ, 多籠 (49°N, 134°W) may be in today's British Columbia (53.7°N, 127.7°W), Canada or in Washington State (47.8°N, 120.7°W), USA. Recall that the N. America-KWQ often places a geographical term more westward than its actual position.
64	松樹林	Song Shu Lin	*Baya de las Pinaz*[223]	松樹林 is of Chinese origin, because it is mentioned in the ancient Chinese book *Shan Hai Jing*. On the N. America-KWQ, 松樹林 (45°N, 140.5°W) can be in the state of Washington (47.8°N, 120.7°W), Oregon (44°N, 120.5°W) or along California's coastline (e.g. the coordinates of Eureka in Northern California are 40.8°N, 124.2°W). We have noticed that the N. America-KWQ often gives a more westward longitudinal reading. The main cause of this error may be due to the American continent on this map has been expanded in the east–west direction compared to the real width of the continent. Nevertheless, the following records show that Washington State, Oregon and California are homes to pine forests.

(*Continued*)

(*Continued*)

Item #	Name	Pinyin	Etymology	Analysis of history, politics, geography; comparison or comment
				In Washington State, native pine trees include the Western White Pine (Pinus monticola), Whitebark Pine (Pinus albicaulis), Ponderosa Pine (Pinus ponderosa) and the Lodgepole Pine (Pinus contorta).[224] In Oregon, there is a great variety of types of pine trees growing throughout the state, from the Pacific Ocean to the Rocky Mountains.[225] In fact, forests cover over 30.5 million acres of Oregon, almost half of the state.[226] Also, in much of California, Yellow Pine forests grow which usually include the Jeffrey Pine (Pinus jeffreyi) or Ponderosa or Yellow Pine Pinus Ponderosa but can sometimes have other middle-elevation pines like the Coulter Pine Pinus coulteri.[227] On the *Maris Pacifici* by Abraham Ortelius,[228] *Baia de las Pinas* is depicted. But *Baia* means “bay” and *Pinas* may be related to pine fruits, not pine forest. On the N. America-Plancius (1594), *Los Penales* (35°N, 158°W) is depicted as pine forests, but it is in Southern California and its longitudinal position is too far west (placing it in the North Pacific Ocean on a modern map).
65	雪山	Xue Shan	Snow Mountain, it may be Mount Hood or Mount Rainier	On the N. America-KWQ, 雪山 (44°N, 147°W; 雪 means “snow” and 山 “mountain”) may be Mount Hood (45.5°N, 121.9°W)[229] about 80 km/50 mi east-southeast of Portland, Oregon, or Mount Rainier (46.9°N, 121.8°W),[230] about 95 km/59 mi south-southeast of Seattle in the U.S. state of Washington. Again, the N. America-KWQ gives the mountain a more westerly location. Mount Hood (45.5°N, 121.9°W) is Oregon’s highest mountain, and one of the loftiest mountains in the U. S., based on its elevation (3,429 m/11,249 ft),[231] and it offers the only year-round lift-served skiing in North America.

				Mount Rainier (46.9°N, 121.8°W) is also a good candidate for 雪山. The mountain is almost directly north of Mount Hood and their latitudes are almost the same. Mount Rainier is the highest mountain (4,392 m/14,410 ft)[232] in the U.S. state of Washington, and the most topographically prominent mountain in the contiguous United States; it is the tallest in the Cascade Volcanic Arc. 雪山 is not the Sierra Nevada (36.6°N, 118.3°W) which lies primarily in central and eastern California, while the Carson Range spur lies primarily in Nevada. 雪山 is farther to the west of 松樹林 (*Baya de las Pinaz*; Item 64) and near the North Pacific Ocean, whereas the Sierra Nevada lies to the east of the pine forests in Washington State, Oregon and California, and is not close to the North Pacific Ocean.  Fig. A5.15. The left panel shows Mount Hood (see endnote 233, public domain) and the right panel shows Mount Rainier (see endnote 234, public domain). On the N. America-Mercator (1569), there is *Sierra neuada* (42°N, 169.5°W) near the Pacific coast. On the N. America-Ortelius (1570), there is also *Sierra neuada* (42°N, 175.5°W) near the Pacific coast. On the N. America-Plancius (1594), too, there is *Sierra neuada* (45°N, 171.5°W). All their longitudinal positions are too far west, because these three European maps grossly overestimated the width of the North American continent. All three maps erroneously depict Sierra Nevada roughly where 雪山 is (albeit further westward).

(*Continued*)

(*Continued*)

Item #	Name	Pinyin	Etymology	Analysis of history, politics, geography; comparison or comment
				On the N. America-Plancius (1594), there is *C. Blanco* (meaning "White Mountain" or "Snow Mountain (30°N, 160°W)". Since its latitude is so low (like Baja California and Mexico), it is unlikely to see snow covering this mountain throughout the year. It cannot be correct. There are also *C. Blanco* at three other locations. It seems that Plancius did not know where to place *C. Blanco.* The first white men who "discovered" Mount Hood, on 29 October 1792, were British Navy Lt. William E. Broughton (1762–1821) and his crew (representing King George III). They saw it from the Columbia River near the mouth of the Willamette River.[235] In the same year, Captain George Vancouver (1757–1798)[236] of the British Royal Navy observed Mount Rainier while surveying the Pacific coast in 1792 and decided to name the mountain after his friend, Rear Admiral Peter Rainier.[237]
66	祁未蠟	Qi Wei La	Quivira	The origin of the word "Quivira" is uncertain. But the meaning of this name can be found in the Dictionary of Chinese Buddhist Terms: "Kuvera" was the "God of Riches".[238] Hence, it can be considered as of Chinese origin. In the sixteenth century, Quivira was a mythical nation. Spanish conquistador Francisco Vásquez de Coronado (1510–1554) heard about it. The nation was related to the Seven Cities of Gold that he never found. The location of Quivira is believed by most authorities to be in central Kansas. The indigenous city of Etzanoa in Kansas,[239] which flourished between 1450 and 1700, was thought to be part of Quivira. On some early sixteenth and seventeenth century maps of North America, a large region including what is now Kansas, Oklahoma, southeastern Colorado, northeastern New Mexico and the Texas Panhandle was called "Quivira".[240]

				However, on the N. America-KWQ, 祁未蠟 (47.5°N, 147.5°W) should correlate with today's Alaska, although its latitude value on this map is a little low to be on the mainland of Alaska. This shows the inaccuracy of the N. America-KWQ in depicting Southern Alaska. On the three sixteenth century European maps, Qvivira or Quiuira is depicted between 40°–50°N and 180°–190°W; it is further west than on the N. America-KWQ. However, it is well known that Europeans had not explored Alaska in the sixteenth century. In 1728, the Danish explorer Vitus Jonassen Bering (1681–1741) documented the Bering Strait between Asia and North America. In 1741, a Russian expedition led by Bering along with George Wilhelm Steller (1709–1746)[241] made the first "discovery" of Alaska, landing near today's Kayak Island (59.9°N, 144.5°W). Since the meaning of "Quivira" is related to Chinese Buddhism, it seems very reasonable to think that the Quivirans were originally Chinese who migrated to Alaska and later moved to the American Central Plain.
67	都荼那	Du Cha Na	Tuchano	On the N. America-KWQ, 都荼那 (49.5°N, 153°W) should be in today's Alaska. The low latitude again shows the inaccuracy of the N. America-KWQ in depicting Southern Alaska. On the N. America-Mercator (1569), Tuchano is depicted at 45°N, 182.5°W; on the N. America-Ortelius (1570), at 45°N, 186°W; and on the N. America-Plancius (1594), at 48°N, 183°W. All these longitudinal positions are too far to the west. There is no written history or a modern English name for the place.

(*Continued*)

(*Continued*)

Item #	Name	Pinyin	Etymology	Analysis of history, politics, geography; comparison or comment
68	孟多齊峰	Meng Duo Qi Feng	Cape Mendocino: Spanish: *Cabo Mendocino*, meaning "Cape of Mendoza"	On the N. America-KWQ, 孟多齊峰 (46°N, 159.5°W) seems to correlate with today's Alaska. But the real location of Cape Mendocino (40.4° N, 124.4°W) is roughly 320 km/200 mi north of San Francisco (37.8°N, 122.4°W); it is the westernmost point on the coast of California. On the N. America-KWQ, such a large error is caused by adjoining the Alaska Peninsula south of 美灣 (Bristol Bay) with the west coast of Canada and California to form a fictional broad peninsula between 40°–55°N and 140°–160.5°W, as shown in Fig. A5.11. Cape Mendocino is not depicted on the N. America-Mercator (1569) and the N. America-Ortelius (1570). But it is depicted on the 1589 Ortelius as *Maris Pacifici*[242] at 47.5°N, 188°W, which is too far west of today's Bering Sea. On the N. America-Plancius (1594), *C. Mendocino* (44°N, 196°W) is depicted at a worse location near today's Japan in the North Pacific Ocean. The above information shows that all these cartographers must have imagined the location of *C. Mendocino* and they were not sure about the general shape of the west coast of North America between latitudes c. 40°N and c. 60°N. Cape Mendocino has been a landmark since the sixteenth century, when Manila galleons followed the prevailing westerlies across the Pacific to the Cape, then followed the coast south to Acapulco, Mexico.[243] It is a great surprise that not only has the N. America-KWQ misplaced this cape, but the three European maps have done so, too.

				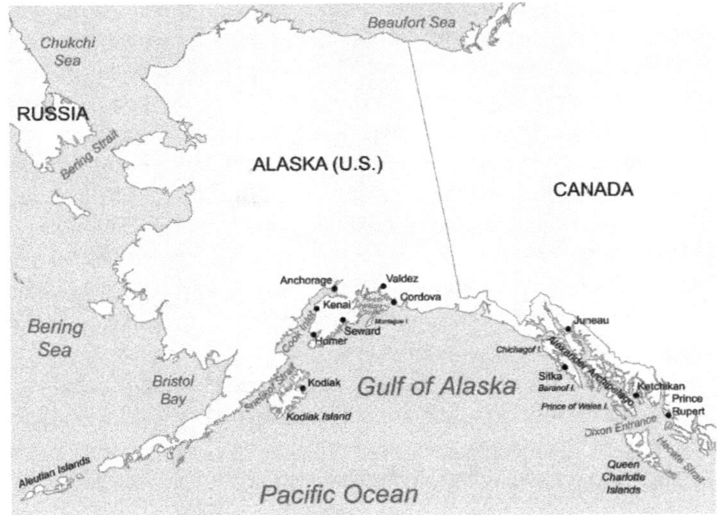 Fig. A5.16. This is a map of Alaska (see endnote 244, public domain). The Alaska Peninsula (also called Aleut Peninsula or Aleutian Peninsula) extends about 800 km/497 mi to the southwest from the mainland of Alaska and ends in the Aleutian Islands. The peninsula separates the Pacific Ocean from Bristol Bay, an arm of the Bering Sea.[245]
69	美灣	Mei War	*Baya hermosa*[246]; it is today's Bristol Bay.	The name 美灣 is of Chinese origin. *Baya hermosa* is the translation of 美灣. On the N. America KWQ, 美灣 (55°N, 147°W) is a bay geographically correlated with today's Bristol Bay (57–59°N, 157–162°W), which is the easternmost arm of the Bering Sea. On the modern-day map, both Naknek (58.7°N, 157°W) and Dillingham (59°N, 158.5°W) have tidal extremes more than 10 m/30 ft, are close to the location of 美灣 (57–59°N, 157–162°W).

(*Continued*)

(*Continued*)

Item #	Name	Pinyin	Etymology	Analysis of history, politics, geography; comparison or comment
				Neither the N. America-Mercator (1569) nor the N. America (1570) depicts *B. hermosa*. But on the N. America-Plancius (1594), *B. hermosa* is depicted at four locations. It seems that Plancius was not sure about where to place *B. hermosa*. In addition, he drew the shape of Alaska very poorly. The first European to discover Bristol Bay was Captain James Cook (1728–1779)[247] during his voyage through the area in 1778. He named the area "in honour of the Admiral Earl of Bristol" in England.
● 70	白吴尔	Bai Wu Er	Peur[248]	On the N. America-KWQ, 白吴尔 (60°N, 142.5°W) may be on the northern border between today's Canada and Alaska. But the westernmost point of Canada is Boundary Peak 187 (60.3°N, 141°W), Yukon. Hence, 白吴尔 (Peur) should be in today's Alaska.
@ 5	此地大壙故多生野馬山牛羊而其牛背上皆有肉鞍形如駱駝		Ci Di Da Kuang Gu Duo Sheng Ye Ma Shan Niu Yang Er Qi Niu Bei Shang Jie You Rou An Xing Ru Luo Tuo	My translation is as follows: This place is in the immense wilderness. Hence, there are many wild horses, mountain cattle and goats. These cattle have humps on their backs like camels. There are over 1,000 breeds of cattle recognised worldwide.[249] Today, cattle having "humps" on their back are mostly from India and Bangladesh. Could these mountain cattle be American bison? But they do have a pronounced shoulder hump. The existence of wild horses is a great surprise to those who believe in a horseless North America, because horses which originally evolved in North America became extinct on the continent approximately 8,000–12,000 years ago. In 1493, Spanish horses, representing *E. caballus*, were brought back to North America, first to the Virgin Islands. From early Spanish imports to Mexico and Florida, horses moved north.[250] However, originally these were not wild horses. The Chinese record *Shan Hai Jing* 《山海经》 or *the Classic of Mountains and Seas/Shan Hai Jing* mentions certain things that were "good for horses" in Fu Sang (the Americas).[251] It would be quite logical that these early Chinese brought horses with them to explore the new land in Alaska.

				This annotation is written on the map near the geographical items 白吳尔 (Peur; Item 70) and 亞泥俺國 (Anian Kingdom; Item 71), both were in today's Alaska. In *Xin Tang Shu* 《新唐書》 or *The New Records of the Tang Dynasty*, it is recorded that Da Han was rich in sheep and horses.[252] Da Han was today's Alaska.[253]
71	亞泥俺國	Ya Ni an Guo	Anian regnum[254] (kingdom) was today's Alaska	In the late thirteenth century, Marco Polo wrote that Anian was in the northeast of Asia. Hendon M. Harris, Jr. in the 1970s believed that Marco Polo's Anian is in fact a mispronunciation of what is shown as Da Han (it is the pinyin of 大汉, meaning "Great Guy" or "Big Guy", in today's Alaska) on an old Asian world map which he associated with an ancient Chinese geography: the *Classic of Mountains/Shan Hai Jing*[255] (versions of the text may have existed since as early as the fourth century B.C.).[256] 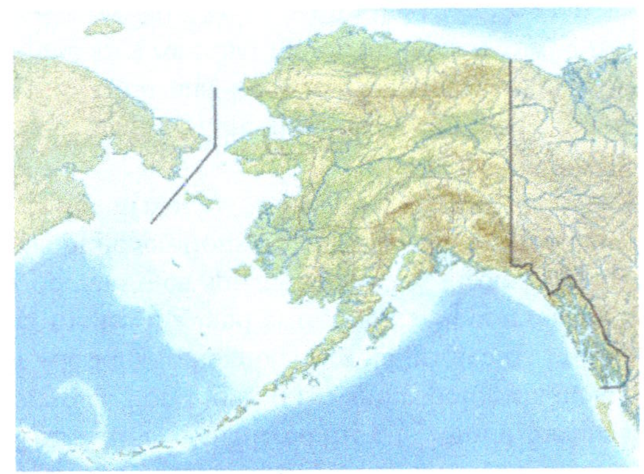 Fig. A5.17. The Bering Strait (with the left black line) separates Alaska from Siberia, and the right black line shows Alaska's boundary with Canada (see endnote 257, public domain).

(*Continued*)

(*Continued*)

Item #	Name	Pinyin	Etymology	Analysis of history, politics, geography; comparison or comment
				Alaska was occupied by various indigenous peoples for thousands of years before the Europeans arrived. The European discovery of Alaska came in 1741, when a Russian expedition led by Danish navigator Vitus Bering (1681–1741) sighted the Alaskan mainland. The state is considered the entry point for the settlement of North America. For a long time, anthropologists have believed that today's Alaska's indigenous people originated in Asia,[258] having crossed over the Bering land bridge from Siberia during the Last Glacial Period, and then spread southward throughout the Americas over subsequent generations.[259] But evidence from recent research also supports the arrival by sea to Alaska's coast 17,000 years ago.[260] Anian is depicted on the three European maps, but as explained in Item 66, the Europeans' knowledge about Alaska started from 1728. It is quite apparent that the cartographers of the three European maps copied "Anian" from the Chinese map and made much larger errors than the N. America-KWQ, as follows: On the N. America-KWQ, 亞泥俺國 (60°N, 152.5°W) is today's Alaska (66.2°N, 153.4°W). On the N. America-Mercator (1569), Anian is depicted at 65°N, 187.5°W; on the N. America-Ortelius (1570), at 61.5°N, 180°W; and on the N. America-Plancius (1594), at 58.5°N, 186°W. All their longitudinal readings are too far westward from the actual value of 153.4°W.
• 72	平地坡	Ping Di Po	*Pianura*, *Pendio delta*[261]	The name is of Chinese origin. "Pianura" is a large, flat area of land with very few trees on it and "Pendio" means sloping. On the N. America-KWQ, 平地坡 (56°N, 161.5°W) should be in today's Alaska, facing Bristol Bay. The exact location cannot be identified on a modern map. But it is not the Kenai Fjords National Park (60°N, 150°W), which is full of glaciers and mountains.[262]

73	水潮峰	Shui Chao Feng	Cape Peirce[263]	"Cape" is a headland sticking into the sea. On the N. America-KWQ, 水潮峰 (59°N, 164.5°W) is depicted on the south-western coast of today's Alaska facing the Anian Strait (today's Bering Strait). Its location is not far from 美灣 (55°N, 147.5°W) depicted on the N. America-KWQ — today's Bristol Bay (57.3°N, 159.8°W). The upper reaches of Bristol Bay (57.3°N, 159.8°W) experience some of the highest tides in the world. One such reach, Nushagak Bay at the mouth of the Nushagak River, an inlet of Bristol Bay, near Dillingham (59°N, 158.5°W), and another near Naknek (58.7°N, 157°W) in Kvichak Bay, have tidal extremes more than 10 m/30 ft, ranking them and the area as eighth highest in the world. Moreover, the extreme number of shoals, sandbars and shallows makes navigation troublesome, especially during the area's frequently strong winds. As the shallowest part of the Bering Sea, *Bristol Bay is one of the most dangerous regions for large vessels.* Alaska's Cape Peirce (58.6°N, 161.8°W) with its 81.99 m/269 ft in height[264] correlates very well with 水潮峰 (59°N, 164.5°W). Cape Peirce faces the Bristol Bay, which is one of the most dangerous regions for large vessels. 水 means "water", 潮, "wave or tide" and 水潮峰 means a cape washed with high tides. Cape Peirce was named in 1869 by U.S. Captain Banjamin Peirce as "M(ys) Peschera", meaning "cave point", and there is the "Calm Point" nearby named by Captain Cook, RN, "from our having calm weather when off it".[265] 水潮峰 is not Anchorage (61.2°N, 149.9°W) proposed by Siu-Leung Lee,[266] because Anchorage is in south-central Alaska, not in south-western Alaska. Anchorage is near the Gulf of Alaska, not Bristol Bay and the Bering Sea. It is too far to the east.

(Continued)

(*Continued*)

Item #	Name	Pinyin	Etymology	Analysis of history, politics, geography; comparison or comment
				On the N. America-Plancius (1594), *C: de Corrientes* (meaning Cape of Currents) is depicted at two locations (42.5°N, 195°W and 60°N, 198°W). Neither of them is a better representation of Cape Peirce (58.6°N, 161.8°W) than 水潮峰 (59°N, 164.5°W). Also, neither the N. America-Mercator (1569) nor the N. America-Ortelius (1570) depicts *C: de Corrientes*. However, the N. America-Ortelius (1589) does depict *Grandes Corrientes* (meaning Big Currents; 56°N, 183°W),[267] but it is too far to the west, because the shape of its western Alaska is poor.
@ 6	此係北海眾書云有多島但未載其數其名		Ci Xi Bei Hai Zhong Shu Yuan You Duo Dao Dan Wei Zai Qi Shu Qi Ming	My translation is as follows: This is the North Sea. Many books record that there are many islands in this sea, but they do not list the number and names of these islands. Situated in the northern extremity of North America and covering about 1,424,500 km^2/550,002 mi^2, there are 36,563 islands in the Arctic Ocean owned by many countries.[268] This annotation gives an account of the geological landscapes of the North Sea or today's Arctic Ocean.
• 74	夜叉國	Ye Cha Guo	A kingdom on the Wrangel Island	夜叉國 is of Chinese origin. On the N. America-KWQ, 夜叉國 (83°N, 156.5°W) may be on today's Wrangel Island (71.2°N, 179.4°W), which is about the same size as Yellowstone National Park and located in the Chukchi Sea (see Fig. A5.18). However, the island size on the N. America-KWQ seems to be much bigger than the actual size of Wrangel Island.

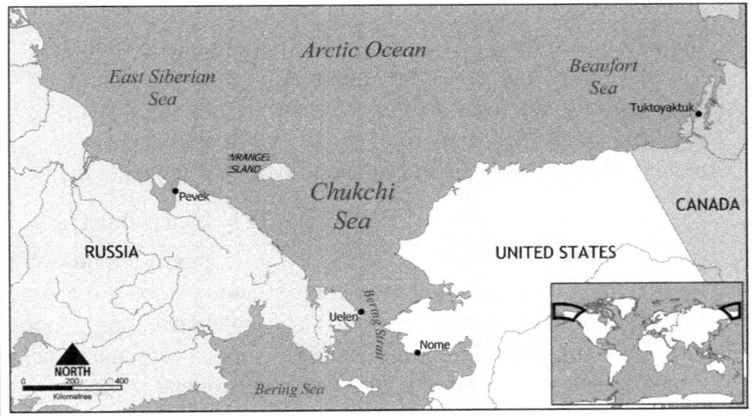

Fig. A5.18. Location of Wrangel Island (Herald Island to its east is too small to be seen; see endnote 269, public domain).

Despite the mountainous terrain and the high latitude, Wrangel Island is not glaciated. The small Herald Island (71.2°N, 175.5°W), located 60 km/37 mi to the east, rises in sheer cliffs, making it quite inaccessible by ship, hence, it is less of a candidate for 夜叉國. Eskimos established camps on the southern side of the Wrangel Island for marine hunters. Ferdinand P. Wrangel (1797–1870), after whom the island was later named, is the first European to determine its location from accounts of Siberian indigenous people. However, he did not land there during his mapping of the Siberian coast in the early 1820s.[270]

Wrangel Island was a way station on a trade route linking the Inuit settlement at Point Hope (68.4°N, 166.8°W) in Alaska with the north Siberian coast; the coast was colonised in late prehistoric and early historic times by Inuit settlers from North America. By 1300, the Paleo-Eskimos[271] had been completely replaced by the Thule people (the real Eskimo). By the time Europeans arrived at Wrangel Island, there was no aboriginal population.

(*Continued*)

(*Continued*)

Item #	Name	Pinyin	Etymology	Analysis of history, politics, geography; comparison or comment
• 75	流鬼	Liu Gui	Demoni vagabonki[272]	流鬼 (76°N, 164.5°W) is a name of Chinese origin. Demoni vagabonki is a literal translation of 流鬼. It was on today's Wrangel Island. Chertov Ovrag[273] is an archaeological site on Wrangel Island. In 1975, evidence was found at this site of human occupation (carbon dating c. 1700 B.C.). This is the westernmost site of Paleo-Eskimo habitation. Could Chertov Ovrag be the site of 流鬼?
@ 7	人穴居皮服不知騎		Ren Xue Ju Pi Fu Bu Zhi Qi	This annotation is written right next to 流鬼 (Demoni vagabonki; Item 75). My translation is as follows: The inhabitants live in caves, wear leather clothes, but do not ride horses. Eskimos live in the Arctic region. In the past they lived in ice caves. Living in caves keeps them warm, but also shows that their lifestyle was primitive. Wrangel Island was once home to wild Asiatic horses and Pleistocene bison, and wooly mammoths inhabited the island until about 3,000 years ago.[274] The above annotation only says that the people did not ride horses. This does not imply that there was no horse on the island
@ 8	此處寒凍極甚海水成冰國人以車馬度之鑿開冰穴多取大魚因其地不生五穀即以魚肉充飢以魚油點燈以魚骨舟車		Ci Chu Han Dong Ji Shen Hai Shui Cheng Bing Guo Ren Yi Che Ma Du Zhi Zao Kai Bing Xue Duo Qu	My translation is as follows: This place is extremely cold and frosty. The sea water becomes icy, and people use carts pulled by horses to ride over it. They chisel holes straight into the ice to catch many big fish. Because the land does not grow grains, fish meat is used to satiate hunger, fish oil to light up lamps, and fish bones for making carts.

			Da Yu Yin Qi Di Bu Sheng Wu Gu Ji Yi Yu Rou Chong Ji Yi Yu You Dian Deng Yi Yu Gu Zhou Che	On the N. America-KWQ, this is the area north of 70°N. Evidence of prehistoric human occupation was uncovered in 1975 at the Chertov Ovrag site on Wrangel Island, Russian Arctic (see Item 75). The annotation mentions "horse" again in the extremely cold region of North America.
• 76	冰海	Bing Hai	Arctic Ocean	The name is of Chinese origin. 冰海 (Arctic Ocean) is already discussed in Item 21.
• 77	北海	Bei Hai	North Sea or today's southern portion of the Arctic Ocean	Chinese invented the compass for navigation in the Northern Song Dynasty (960–1279). Their compasses have four cardinal directions: clockwise, they are north (北), east (東), south (南) and west (西), each separated by 90 degrees. They are secondarily divided into four ordinal (inter-cardinal) directions: northeast, southeast, south-west and northwest, each located halfway between two cardinal directions. The Chinese then used these main directions to name the oceans surrounding their territory, and elsewhere in the world. 海 means "sea or ocean": 北海 is located at the north of the Northern Hemisphere and north of China, and is the southern portion of today's Arctic Sea.
78	亞泥俺峽	Ya Ni An Xia	Strait of Anián	The strait probably took its name from Ania, a Chinese province mentioned in a 1559 edition of Marco Polo's book. The Strait of Anián was a semi-mythical strait for the Europeans, which first appeared on a map issued by the Italian cartographer Giacomo Gastaldi (1500–1566) in 1562.[275] The strait was believed by early modern cartographers to mark the boundary between North America and Asia and to permit access to a Northwest Passage from the Arctic Ocean to the Pacific. The first European to document the true strait is the Danish-born Russian navigator Vitus Jonassen Bering (1681–1741) in 1728.[276] In 1741, a Russian expedition led by Bering along with George Wilhelm Steller (1709–1746) made the first "discovery" of Alaska,[277] landing near today's Kayak Island (59.9°N, 144.4°W). The strait became known as the Bering Strait (66.5°N, 169°W).

(*Continued*)

(*Continued*)

Item #	Name	Pinyin	Etymology	Analysis of history, politics, geography; comparison or comment
				On the N. America-KWQ, 亞泥俺峽 (Strait of Anián; 60°N, 175°W) is depicted at a lower latitude where the Bering Sea (58°N, 178°W) is located. The true geographical location of Bering Strait/Strait of Anián should be at 66°N, 169°W. On both the N. America-Mercator (1569) and the N. America-Plancius (1594), there is *El Streto de Anian* at 68°N, 195°W and 60°N, 203°W, respectively, being too westward in its longitudinal position. The N. America-Ortelius (1570) does not depict the Strait of Anián.
• 79	别山	Bie Shan	*Los Boleanes* or *Pie, monte.*[278]	The name 别山 (27°N, 144.5°W) is of Chinese origin. It does not have a modern English name, and it is not depicted on the three European maps. However, on the N. America-Mercator (1569) and the N. America-Ortelius (1570), there is *Los bolcanes* meaning "The Volcanos", at 21°N, 167°W and 27°N, 180°W, respectively; and on the N. America-Plancius, there is *Los Monges* at 21°N, 155–159°W.
• 80	大東洋	Da Dong Yang	Big East Ocean; part of today's North Pacific Ocean	Item 77 explains that Chinese used four cardinal directions to name the oceans surrounding their territory and elsewhere in the world. 洋 means "sea or ocean": 東洋 is the big ocean to the east of China; it separates into 小東洋 (the Small East Ocean) to the east of Japan and 大東洋 (the Big East Ocean) to the far east of the Small East Ocean, specifically, it lies immediately to the west of the Baja California.
81	亞奴皮亞大	Ya Nu Pi Ya Da	*La Anubiada* or Anubiada[279]	On the N. America-KWQ, 亞奴皮亞大 (18°N, 124.5°W) is an archipelago. It is also depicted on the N. America-Ortelius (1570) at 17°N, 142°W, but not depicted on the N. America-Mercator (1569) and the N. America-Plancius (1594). There is no written history or a modern English name for the place.

82	得光得白	De Guang De Bai	Tecoantepec[280] is the Nahuatl name.[281]	On the N. America-KWQ, 得光得白 (Tecoantepec; c. 17°N, c. 115.5°W) is depicted as an archipelago. But Tecoantepec (16.3°N, 95.2°W) is in fact a piece of land connected with Mexico to form the Gulf of Tecoantepec (a large widemouthed inlet of the Pacific Ocean, forming the southern shore of the Isthmus of Tehuantepec in southeastern Mexico).  Fig. A5.19. Map showing the relief of the Tecoantepec isthmus (see endnote 282, public domain). The Tecoantepec route represents the shortest distance between the Gulf of Mexico and the Pacific Ocean. Before the opening of the Panama Canal, it was a major overland transport route. The route discovered by Hernán Cortés (c. 1485–1547) in 1526 was named the Tehuantepec Passage.[283] On all the three European maps, Tecoantepec (c. 17°N, c. 108°–115°W) is correctly depicted on today's Mexico mainland.
83	支古訝	Zhi Gu Ya	Cicuie[284]	On the N. America-KWQ, 支古訝 (39°N, 137.5°W) was a mythical city for the sixteenth century Europeans. Notice that the longitudinal position of 支古訝 is only about 3° to the east of 松樹林 (45°N, 140.5°W; *Baya de las Pinaz*; Item 64) and its latitudinal position is about 6° south of 松樹林. Since 松樹林 is in Washington State, Oregon or North California, 支古訝 would likely be in today's Idaho or Nevada.

(Continued)

(*Continued*)

Item #	Name	Pinyin	Etymology	Analysis of history, politics, geography; comparison or comment
				On the N. America-Mercator (1569), Cicuie is depicted at 38°N, 154°W; on the N. America-Ortelius (1570), at c. 40°N, 168°W; and on the N. America-Plancius (1594), at 36.5°N, 169°W. All are further west than that on the N. America-KWQ. This city cannot be found on modern maps.
84	第瓦施	Di Wa Shi	Tiguex[285]	On the N. America-KWQ, 第瓦施 (37°N, 136°W) was also a mythical city for the sixteenth century Europeans. Since historical records show that the famous Tiquex War, the first-named war between Europeans and Indigenous Americans in the United States, took place in New Mexico in the winter of 1540–1541,[286] 第瓦施 (Tiguex) was in today's New Mexico. But the N. America-KWQ depicts it too far westward, facing the North Pacific Ocean. On the N. America-Mercator (1569), Tiguex is depicted at 37.5°N, 159°W; on the N. America-Ortelius (1570) at 37°N, 160°W; and on the N. America-Plancius (1594) at 36.5°N, 163°W. All are further west than Tiguex on the N. America-KWQ.
• 85	止會	Zhi Hui	Cehoei[287]	On the N. America-KWQ, 止會 (36°N, 133.5°W) should be in today's New Mexico, because it is very close to 第瓦施 (37°N, 135°W; Item 83). But the N. America-KWQ depicts it too far westward, facing the North Pacific Ocean.
86	弟克屬	Di Ke Shu	Tiguex[288]	On the N. America-KWQ, there are 第瓦施 (37°N, 136°W; Item 84) and 弟克屬 (33°N, 130°W; Item 86). Both are interpreted by Huang Shijian and Gong Yingyan as Tiguex. The location of the Tiguex War was in the Rio Grande valley (33.7°N, 107°W)[289] in the vicinity of Albuquerque and Bernalillo, New Mexico. Hence, 弟克屬 (33°N, 130°W; Item 86) seems to correlate better with the actual location of Tiguex than 第瓦施 (37°N, 136°W; Item 84). On the N. America-Mercator (1569), Tiguex is depicted at 37°N, 157.5°W, on the N. America-Ortelius (1570), at 38°N, 163°W, and on the N. America-Plancius (1594), at 37.3°N, 163°W. Comparing with the coordinates of the Rio Grande valley (33.7°N, 107°W), all four maps give their longitudinal values too westward, but the error of N. America-KWQ is the smallest.

≈ 5	亞沙河	Ya Sha He	*Axa flu.* or today's Colorado River	On the N. America-KWQ, there are three rivers flowing into 東紅海 (Item 93) and 角利弗尔海 (Item 92); together they are today's Gulf of California. 亞沙河 (Axa River) is today's Colorado River. In 1540, when Hernando de Alarcón (c. 1500–1542) sailed up the Colorado River to visit the indigenous people, the river was known as *Rio de Tizon* or *Rio de Buena Guia*. He also sailed into the Gulf of California and was credited with being the first European to have determined that Baja California was a peninsula and not an island.[290] On the N. America-KWQ, the river's estuary at c. 36.5°N, c. 123.5°W opens today's Gulf of California[291] in today's Mexico. That estuary on the modern map is located at 31.7°N, 114.7°W. 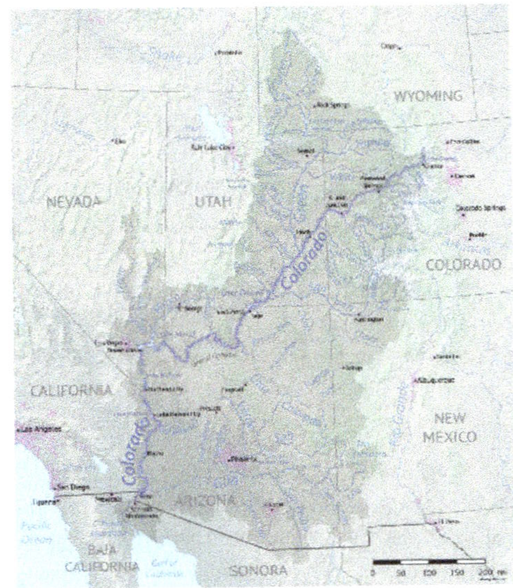 Fig. A5.20. Map of the Colorado River basin showing the Colorado, Little Colorado and Gila Rivers (see endnote 292, public domain)

(*Continued*)

(*Continued*)

Item #	Name	Pinyin	Etymology	Analysis of history, politics, geography; comparison or comment
				The Colorado River is a major river of North America, rising in the Rocky Mountains of Colorado, and flowing generally west and south for 2,330 km/1,450 mi into the Gulf of California in northwestern Mexico.[293] On the N. America-Mercator (1569), the N. America-Ortelius (1570) and the N. America-Plancius (1594), the Axa River is also depicted, and the river's estuary is at 37°N, 144°W; 36°N, 152°W and 37°N, 131°W, respectively. But the longitudinal errors are much larger than that on the N. America-KWQ.
≈ 6	多朵德亞河	Duo Duo De Ya He	*Totonteac fl*[294]	多朵德亞 is the transliteration of the indigenous name Totonteac. 河 means "river". It was today's Little Colorado River.[295] The Little Colorado River is a tributary of the Colorado River in the U.S. state of Arizona, but the N. America-KWQ erroneously depicts it as an independent river. On the N. America-Mercator (1569) and the N. America-Ortelius (1570), the Totonteac River is depicted correctly as a tributary of the Axa River (today's Colorado River), whereas on the N. America-Placius (1594), this tributary has a different name: Chucho. Europeans encountered this river in the sixteenth century.[296]
≈ 7	帝靜河	Di Jing He	*Tiguas rio*[297]	The name 帝靜河 denotes today's Gila River. The river was named because Tigua (Tiwa) Indians were a group of Pueblo peoples of northern New Mexico. The Gila River is a long (1,044 km/649 mi) tributary of the Colorado River, flowing through Arizona and New Mexico in the United States; it lies mainly within the U. S., but also extends into northern Sonora, Mexico. On the N. America-Mercator (1569), there is an unnamed river which may be correlated with the Gila River, but on this map, it is depicted as a very long river joining the St. Lawrence River in Canada. There exists no such connection.

				On the N. America-Ortelius (1570), *Tiguas rio* is correctly depicted, whereas on the N. America-Plancius (1594), there is an unnamed river which may be correlated with the Gila River. But it is very short. Europeans encountered this river in the sixteenth century.[298]
87	多朵德亞國	Duo Duo De Ya Guo	Totonteac[299]	多朵德亞國 (37.5°N, 126°W) is the transliteration of the Totonteac Regnum. It was in today's Arizona (34.0°N, 111.1°W). On the N. America-KWQ, this kingdom's longitudinal location is too westward. The country also appears on the three European maps. In the sixteenth century, the Spaniards in New Spain (now Mexico) began to hear rumours of "Seven Cities of Gold" named "Cibola", located across the desert, hundreds of miles to the north. According to legend, the seven cities of gold could be found throughout the pueblos of the New Mexico Territory. Besides "Cibola", some said that names associated with similar lost cities of gold also include El Dorado, Paititi, City of the Caesars, Lake Parime at Manoa, Antilia, and Quivira (Item 66). Other cities were also suggested. While there have always been mentions of a seventh city, no evidence of a site has been found. Totonteac, situated in the Totonteac Regnum, was thought to be the Seventh City of Cibola for a while. However, when conquistador Francisco Vázquez de Coronado (1510–1554) finally arrived at Cíbola in 1540, he discovered no treasures — only adobe towns.[300]
● 88	矇山	Meng Shan	*Mom*, monte[301]	The name 矇山 (29°N, 125.5°W) is of Chinese origin. On the N. America-KWQ, it is depicted on today's Baja (Lower) California peninsula. The Baja California peninsula is in northwestern Mexico.[302] It separates the Pacific Ocean from the Gulf of California. The peninsula extends from Mexicali, Baja California, in the north to Cabo San Lucas, Baja California Sur ("sur" meaning "south").

(*Continued*)

(*Continued*)

Item #	Name	Pinyin	Etymology	Analysis of history, politics, geography; comparison or comment
				The first Spaniard in Baja California Sur is believed to have been Fortún Ximénez (who died in 1533), arriving in 1533.[303] The peninsula is separated from mainland Mexico by the Gulf of California and the Colorado River (see Fig. A5.20). In 1539, Spanish explorer Francisco de Ulloa (who died in 1540)[304] proved that Baja California was a peninsula rather than an island. Nevertheless, the idea of the island persisted for well over a century and was included in many European maps. The Spaniards gave the name *Las Californias* to a region of North America spanning the United States and Mexico, including today's U.S. state of California and the Mexican states of Baja California and Baja California Sur. The peninsula is divided at latitude 28°N into the state of Baja California in the north, and the state of Baja California Sur in the south. On the N. America-KWQ, 曚山 (29°N, 125.5°W) is in Baja California, because it is depicted north of this dividing line. 曚山 correlates closely with the Sierra de San Pedro Mártir,[305] a mountain range that runs north–south along the middle part of northwestern Baja California, Mexico. The meaning of its Spanish name means "mountains of Saint Peter the Martyr". The area was first explored by Europeans in 1701, including Eusebio Francisco Kino (1645–1711).[306] Picacho del Diablo (Devil4s Peak; c. 31.0°N, 115.5°W) is the highest peak in Baja California with its summit reaching 3,100 m/10,171 ft.[307] The N. America-KWQ depicts a mountain chain along the Baja California peninsula,[308] whereas none of the sixteenth century European maps show any mountain there, presumably because they had not explored this mountain chain.
89	亦安農降	Yi An Nong Jiang	*C. del Ingano* or *Cabo del Engaño*	On the N. America-KWQ, 亦安農降 (28°N, 128.5°W) is the transliteration of *Cabo del Engaño*. It is today's Cape Deceit.

				On the N. America-Mercator (1569), there is *C. del Engano* (27°N, 143°W); on the N. America-Ortelius (1570), there is *C. del ingano* (26°N, 148°W); and on the N. America-Plancius (1594), there is *C. del Engaño* (28°N, 150°W). All the longitudinal readings are too far westward than that on the N. America-KWQ. In 1542, Juan Rodríguez Cabrillo (1499–1543) led an expedition to explore the west coast of today's Mexico. It was the first Spanish expedition to pass beyond *Cabo del Engaño*.[309]
90	角利弗尔聶	Jiao Li Fu Er Nie	Today's Baja California Sur: the name "California" likely derived from the mythical island of California in the fictional story of Queen Calafia, as recorded in a c. 1508 work *The Exploits of Esplandian* by Garcia Ordonez de Montalvo.[310]	On the N. America-KWQ, 角利弗尔聶 (24°N, 123.5°W) corresponds to today's Baja California Sur (25.5°N, 111.6°W). The longitudinal error is around 12° westward. Fortún Ximénez Bertandoña (who died in 1533) is the first European known to have landed in Baja California.[311] Later in 1536, Hernán Cortés (1485–1547)[312] and his men ventured to Baja California and explored the Pacific coast of Mexico. The Gulf of California was originally named the *Sea of Cortés* by its first European discoverer Francisco de Ulloa (who died in 1540) in 1539.[313] Juan Rodríguez Cabrillo (1499–1543)[314] is the first European to sight the region that is today's California in 1542. But not until 1542 did Spaniards sail north to Alta California[315] and land as far north as modern Santa Barbara (34.4°N, 119.7°W); Sebastián Vizcaíno (1548–1624)[316] explored and mapped the coast of California in 1602 for New Spain, putting ashore in Monterey; Gaspar de Portolá (1716–1786)[317] is credited to be the first European to have sighted San Francisco Bay in 1769; Juan de Ayala (1745–1797) and the crew of his ship are the first Europeans known to have entered San Francisco Bay in 1775.[318] For hundreds of years, there persisted a European misconception that California was an island (although in 1539, Spanish explorer Francisco de Ulloa proved that Baja California was a peninsula rather than an island; see detail in item 88), and many maps were made depicting it as such. Only by the eighteenth century did enough information about California reach the outside world to dispel that myth.

(Continued)

(Continued)

Item #	Name	Pinyin	Etymology	Analysis of history, politics, geography; comparison or comment
				The Spaniards gave the name *Las Californias* to the peninsula of Baja California and to Alta California, the region that became today's state of California. Baja California Sur was known as the *El Territorio Sur de Baja California* ("South Territory of Lower California"). It occupies the southern half of the Baja California peninsula, south of 28°N, plus the uninhabited Rocas Alijos[319] in the Pacific Ocean. On both the N. America-Mercator (1569) and the N. America-Ortelius (1570), the entire Baja California peninsula is unnamed; however, on the N. America-Ortelius (1587),[320] the entire peninsula is depicted as *California* just as it is on the N. America-Plancius (1594).
91	十字山尾	Shi Zi Shan Wei	Cape San Lucas; Spanish: *Cabo San Lucas*	十字山尾 (c. 23°N, 120°W) is a name of Chinese origin. Its latitude is very close to the Tropic of Cancer, currently at 23.4°N. 十字山尾 correlates closely with today's Cape San Lucas (22.9°N, 109.9°W). The error of roughly 10° westward has been observed for many geographical items on the N. America-KWQ. Cape San Lucas is the extreme southern tip of the Baja California peninsula, Mexico. The rocky headland forms the southern extremity of the Sierra de San Lazaro and includes the western shore of San Lucas Bay. The isolated town of San Lucas lies 3 km/2 mi north of the cape. On the N. America-KWQ, there is a group of mountain ranges spanning from 30°N to c. 40°N on the Baja California peninsula. On a modern map, the Peninsular Ranges (also called the Lower California province) are a group of mountain ranges that stretch 1,500 km/930 mi from Southern California to the southern tip of the Baja California peninsula; they are part of the North American Coast Ranges, which run along the Pacific Coast from Alaska to Mexico.[321] The Sierra de la Laguna High Point, at 2,090 m/6,857 ft in elevation, is the highest point of the range, located in Baja

				California Sur state. If you take an aerial view, you can see that the mountain ridges seem to form many crosses, hence, the name of 十字山 (meaning "mountain ridges shaped like crosses") and 尾 means "tail" or "end". On the N. America-KWQ, this southernmost range of the Peninsular Ranges System is not fully depicted. Only Cabo San Lucas is located at the end of the Sierra de la Laguna (23.4°N, 110°W) — a mountain range at the southern end of the Baja California peninsula in Mexico and the southernmost range of the Peninsular Ranges System. On the N. America-Ortelius (1570), the southern tip of the Baja California peninsula is depicted as *C. Califormia* (which should be corrected as *C. California*; 22°N, 141.5°W), but no mountain range is depicted on the Baja California peninsula. The longitudinal error is enormous (31.6° westward). Neither the N. America-Mercator (1569) nor the N. America-Plancius (1594) gives a name to the southern tip of Baja California Sur. None of these three European maps give a meaningful name to this southern tip to match its geological environment.

(Continued)

(*Continued*)

Item #	Name	Pinyin	Etymology	Analysis of history, politics, geography; comparison or comment
				 Fig. A5.21. The top panel shows the location of Cabo San Lucas on Baja California Sur, Mexico (see endnote 322, public domain); and the lower panel is an aerial photo of Cabo San Lucas (see endnote 323, public domain).

92	角利弗尔海	Jiao Li Fu Er Hai	The lower portion of the Gulf of California	On the N. America-KWQ, 角利弗尔海 is the lower portion of today's Gulf of California, and the name is the transliteration of the Gulf of California. In 1535, Hernán Cortés (1485–1547)[324] navigated into the Gulf of California, which he called the *Sea of Cortés*. He landed in what is now the bay of La Paz, which he named the Santa Cruz Port and Valley. On the N. America-Ortelius (1570), the entire Gulf is depicted as *Mar Vermeio* (meaning "Red Sea"). The truth is that the *Rio Colorado* (Red River) was named in the early seventeenth century by the Spanish after the red-coloured sediment that would flow in the river. From whom did Ortelius get the information that the Colorado River used to look red? On the N. America-Mercator (1569) and the N. America-Plancius (1594), the Gulf is unnamed, and there is no indication of the water colour of the Gulf being red. However, the next Item will show that the Chinese knew that only the water of the upper portion of the Gulf was red.
93	東紅海	Dong Hong Hai	Upper portion of the Gulf of California; *Mar Vermeio*	The name 東紅海 is of Chinese origin. On the N. America-KWQ, the water separating Upper Baja California from the Mexican mainland is depicted as 東紅海 ("Eastern Red Sea"). It is to the east of China, to distinguish it from the Red Sea, which is a seawater inlet of the Indian Ocean in the Middle East, between Egypt and Saudi Arabia. On the Africa-KWQ, this Red Sea is depicted as 西紅海 ("Western Red Sea"), because it is to the west of China. Notice that 東紅海 only denotes the upper portion of the Gulf of California, the lower portion is depicted as 角利弗尔海 (Item 92). The reason is that the red-coloured water of the Colorado River would have been diluted by the sea water and could not travel to the lower portion of the Gulf. The Gulf is about 1,126 km/700 mi long and 48 to 241 km/30 to 50 mi in width.[325] On the N. America-Ortelius (1570), the gulf is depicted as *Mar Vermeio*, whereas on the N. America-Mercator (1569) and the N. America-Plancius (1594), it is unnamed.

(*Continued*)

(*Continued*)

Item #	Name	Pinyin	Etymology	Analysis of history, politics, geography; comparison or comment
Zone III: North America (contains Central America and the Caribbean)				

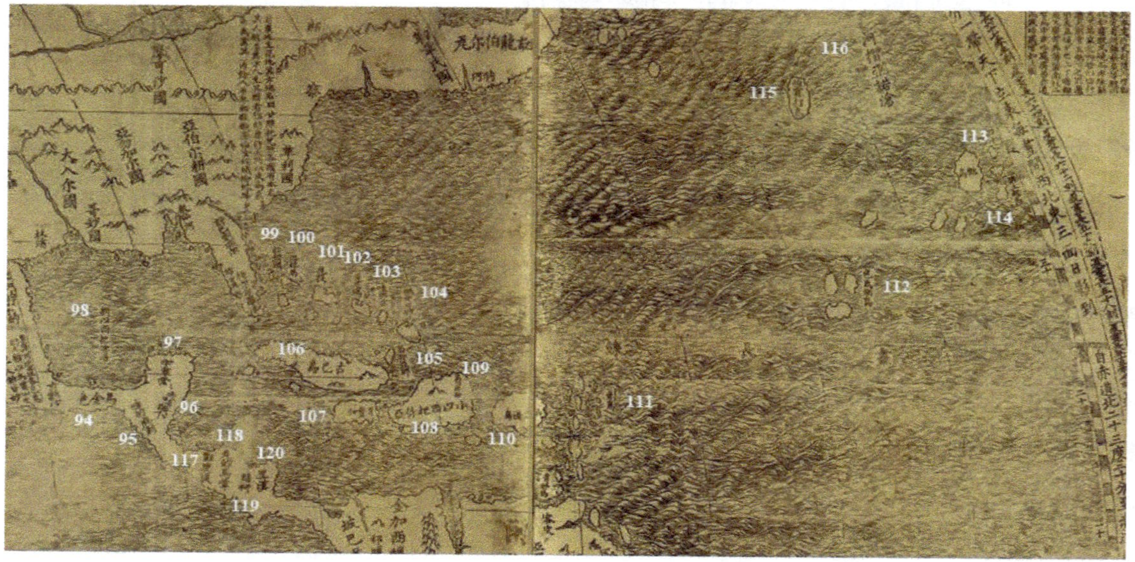

Fig. A5.22. Extracted from Fig. 5.3, bottom right panel, public domain.

Item #	Name	Pinyin	Etymology	Analysis of history, politics, geography; comparison or comment
94	馬金色	Ma Jin Se	Maxa	On the N. America-KWQ, 馬金色 (18°N, 103.5°W) was an ancient city/town. On the N. America-KWQ, it is depicted between 古數沙 (Costosta; 18°N, 106.5°W; Item 33) in today's Mexico (23.6°N, 102.6°W) and 哇的麻剌 (Guatimala; 16°N, 100.5°W; Item 95) in today's Guatemala (14.6°N, 90.5°W). Considering the correct longitudinal values of today's Mexico and Guatemala, we can see that the true longitude of 馬金色 should be between 90.5°W and 102.6°W and more likely in today's Mexico.

Maxa had a Mayan civilization. The Mayan peoples had the most highly developed writing system in the pre-Columbian Americas, as well as arts, architecture, mathematics, a calendar and an astronomical system.[326] However, from the late eighth through the end of the ninth century, the Mayan civilization declined for unknown reasons and by c. 900, the Mayan civilization had collapsed. Only in the lowlands of the Yucatan Peninsula (Mayan Lowland) did a few Mayan cities continue to flourish in the Post-Classic Period (900–1500).[327] The first European expedition to make landfall at the Yucatán Peninsula was led by Francisco Hernández de Córdoba (c. 1467–1517) in 1517.

Fig. A5.23. The top panel covers Mexico and Central America (see endnote 328, public domain); the lower panel shows the major Maya cities in this area (see endnote 329, public domain). Nomenclature: the southern part of the *North American* continent is composed of two regions: *Central America* and the Caribbean.

(Continued)

(*Continued*)

Item #	Name	Pinyin	Etymology	Analysis of history, politics, geography; comparison or comment
				Maxa is depicted on the N. America-Ortelius (1570) at 16°N, 103°W, but not on the N. America-Mercator (1569) or the N. America-Plancius (1594). There is no written history or a modern English name for the place.
95	哇的麻剌	Wa De Ma La	The name "Guatemala" comes from the Nahuatl word *Cuauhtēmallān* (nah-wiki), or "place of many trees", a derivative of the K'iche' Mayan word for "many trees".[330]	The core of the Mayan civilization, which extended across Mesoamerica (that comprises the modern-day countries of northern Costa Rica, Nicaragua, Honduras, El Salvador, Guatemala, Belize and central to southern Mexico), was historically based in the territory of modern Guatemala. In the sixteenth century, most of this area was conquered by the Spanish and claimed as part of the viceroyalty of New Spain. "Guatemala" was the name that the Tlaxcaltecan warriors, who accompanied Pedro de Alvarado (c. 1485–1541)[331] during the Spanish Conquest (beginning in 1519), gave to this territory. Since the Spanish knew this place well, the coordinates of Guatemala on the three European maps are comparable to those on the N. America-KWQ. On the N. America-KWQ, 哇的麻剌 (16°N, 100.5°W) is today's Guatemala (14.6°N, 90.5°W). The N. America-Mercator (1569) depicts Guatemala at 14.6°N, 95°W; the N. America-Mercator (1569), at 14.5°N, 99.5°W; and the N. America-Mercator (1569), at 14°N, 101°W. All their longitudinal readings are about 5–10° more westward than the modern-day value.

96	大哇識个	Da Wa Shi Ge	Tousco[332]	On the N. America-KWQ, 大哇識个 (17.5°N, 99.5°W) might be in today's Belize (17.2°N, 88.5°W) or Guatemala (14.6°N, 90.5°W). Tousco is not depicted on the N. America-Mercator (1569) or the N. America-Plancius (1594), but it is depicted on the N. America-Ortelius (1570) at 17°N, 99°W. These coordinates are comparable to those on the N. America-KWQ. There is no written history or a modern English name for the place.
97	宇革堂	Yu Ge Tang	Iucatan or Yucatán; the name is also assigned to the peninsula and came either from early explorations of the conquistadors from Europe or from a local plant name.[333]	On the N. America-KWQ, 宇革堂 (20.4°N, 98°W) is today's Iucatan/Yucatán Peninsula (20.4°N, 89.1°W). Again, longitudinal error on the N. America-KWQ persists. Iucatan/Yucatán is bordered by the states of Campeche to the southwest and Quintana Roo to the southeast, with the Gulf of Mexico off its north coast. In 1517, Francisco Hernández de Córdoba (c. 1467–1517)[334] was the first European to discover the Yucatán Peninsula. But the first encounter with the Yucatec Maya may have occurred in 1502, when during their fourth voyage Christopher Columbus and his crew came across a large trading canoe off Honduras.[335] The N. America-Mercator (1569) depicts Iucatan at 19°N, 89.5°W; the N. America-Ortelius (1570), at 19°N, 95°W; and the N. America-Plancius (1570) depicts the peninsula without giving it a name. In 1531, the Spanish moved their base of operations to Campeche, where they repulsed a significant Maya attack.[336] In late 1546, the eastern Maya were defeated in a single battle, which marked the final conquest of the northern portion of the Yucatán Peninsula. Hence, the European cartographers had access to the coordinates of Iucatan/Yucatán.

(*Continued*)

(*Continued*)

Item #	Name	Pinyin	Etymology	Analysis of history, politics, geography; comparison or comment
● 98	新以西把你海	Xin Yi Xi Ba Ni Hai	*Nuova Hispania, mare di*; it was today's Gulf of Mexico.	以西把你 is the transliteration of *Hispania,* which was the Roman name for the Iberian Peninsula and its provinces. 新 means "*Nuova* (New)" and 海, "mare (sea)". 新以西把你海 represented the new sea controlled by New Spain or Kingdom of New Spain (1521–1821), which was established during the Spanish colonisation of the Americas.[337] The first alleged European exploration of the Gulf of Mexico was by Amerigo Vespucci (1451–1512) in 1497.[338] Both the N. America-Mercator (1569) and the N. America-Plancius (1594) depict *Golf Mexicano*, but the N. America-Ortelius (1570) depicts the gulf without naming it.
● 99	白赫瑪	Bai He Ma	Bahama: the name is derived from *Bahama* (the Lucayan name, meaning "Large Upper Middle [Land]"),[339] used by the indigenous Taíno people for the island of Grand Bahama (the northernmost of the islands of the Bahamas).[340]	On the N. America-KWQ, 白赫瑪 (29°N, 86.5°W) is today's Grand Bahama Island (26.7°N, 78.3°W), which is the northernmost island of the Bahama Archipelago/Lucayan Archipelago (25°N, 78°W; named for the indigenous Lucayan people). The Lucayan Archipelago is part of the West Indies but is not located in the Caribbean Sea. The first inhabitants of the Bahamas were the Taino (Arawak) who are also known as Lucayan.[341] They originated from both Hispaniola (today's Dominican Republic) and Cuba, and migrated by canoe into the Bahamas, settling the entire archipelago by the twelfth century. Christopher Columbus's first landfall on the Bahamas was on an island he named San Salvador (known to the Lucayans as *Guanahani*).[342] But precisely which island is a matter of scholarly debate. The N. America-Mercator (1569) depicts *Lucayo* (meaning today's Lucayan Archipelago) at 27.5°N, 74.5°W and the N. America-Ortelius (1570) at 26.5°N, 80.5°W. Both are in reasonable agreement with the coordinates of the Lucayan Archipelago (25°N, 78°W). But the N. America-Plancius (1594) does not depict this Archipelago.

				Since none of these three maps depicts the Grand Bahama Island, this shows that the N. America-KWQ is not a direct or adapted copy of the three European maps.  Fig. A5.24. A map showing the Bahamas/Commonwealth of the Bahamas (see endnote 343, public domain).
• 100	路格禺	Lu Ge Yu	Lucayoneque[344]; today's Great Abaco Island	On the N. America-KWQ, 路格禺 (28.5°N, 85.5°W) denotes today's Lucayoneque Island (26.4°N, 77.1°W; Great Abaco Island or Abaco Island; meaning "The People's Distant Waters Land"),[345] which is an island of the Lucayan Archipelago, see Fig. A5.24. On the N. America-Mercator (1569) and the N. America-Ortelius (1570), Lucayo is used to denote the Lucayan Archipelago, not a single island; the N. America-Plancius (1594) does not depict the Lucayan Archipelago. None of the three maps depicts specifically the Lucayoneque Island.

(*Continued*)

(*Continued*)

Item #	Name	Pinyin	Etymology	Analysis of history, politics, geography; comparison or comment
• 101	瓦投/止瓦投	Wa Tou/ Zhi Wa Tou	Ciguateo[346]	On the N. America-KWQ, 瓦投 (27.5°N, 76.1°W) should be corrected to 止瓦投, which is the transliteration of Ciguateo (today's Eleuthera Island; 25.1°N, 76.1°W), meaning "Distant Rocky Place", see Fig. A5.24.
102	瓦尼瑪	Wa Ni Ma	Guanima; today's Cat Island	On the N. America-KWQ, 瓦尼瑪 (26°N, 80.5°W) is Guanima (today's Cat Island, meaning "Middle Waters Land"; 24.5°N, 75.5°W).[347] The first white settlers were Loyalists fleeing the American Revolution, who arrived in 1783. The island may have been named after Arthur Catt, a pirate, or the name may refer to its one-time large population of feral cats. Guanima is not depicted on the N. America-Mercator (1569) or the N. America-Ortelius (1570), but it is depicted on the N. America-Plancius (1594) at 26°N, 80.5°W. How did Plancius know of Guanima when the island was first explored by the Europeans no sooner than the eighteenth century?
103	哇那罕	Wa Na Han	Guanahaní	On the N. America-KWQ, 哇那罕 (25.5°N, 79.5°W) is Guanahaní (meaning "Small Upper Waters Land"; 24.1°N, 74.5°W). Christopher Columbus changed it from its native Taíno name to San Salvador. Guanahaní has traditionally been identified with Watlings Island, which was officially renamed San Salvador Island in 1925 as a result, but modern scholars are divided on the accuracy of this identification. Guanahaní could also be today's Little Inagua (21.5°N, 73.0°W),[348] which is to the south and east of San Salvador. However, San Salvador's coordinates (24.1°N, 74.5°W) are closer to those of 哇那罕 (25.5°N, 79.5°W) on the N. America-KWQ, I am inclined to identify 哇那罕 (25.5°N, 79.5°W) as San Salvador (24.1°N, 74.5°W). Guanahaní is depicted on the N. America-Mercator (1569) at 26.5°N, 72.5°W (closer to the position of San Salvador), but not on the N. America-Ortelius (1570) or the N. America-Plancius (1594).

104	瑪牙瓦那	Ma Ya Wa Na	Mayaguana; it was the Lucayan name and is still used today (meaning "Lesser Midwestern Land").[349]	On the N. America-KWQ, 瑪牙瓦那 (25°N, 78°W) is Mayaguana (22.4°N, 72.9°W), which is the easternmost island and district of the Bahamas, see Fig. A5.24. Mayaguana was inhabited by Lucayans prior to the arrival of the Spanish following 1492. Since I am inclined to identify 哇那罕 (25.5°N, 79.5°W) as San Salvador, then on the N. America-KWQ, 瑪牙瓦那 (25°N, 78°W) should be to the southeast of 哇那罕 (25.5°N, 79.5°W). This is the case. The N. America-Mercator (1569) depicts Mayaguana at 23.5°N, 69.5°W, to the north and east of Little Inaguaí (21.5°N, 73°W). This is correct. Mayagnana is not depicted on the other two European maps.
• 105	曷弥亞那	He Mi Ya Na	*Camercane, isole*[350]	On the N. America-KWQ, 曷弥亞那 (21.5°N, 79.5°W) is one among the many Camercane Islands. The name Camercane is no longer in use. 曷弥亞那 correlates well with the Great Inagua Island (21.1°N, 73.3°W)[351] lying to the northeast of Cuba. Today's West Indies are a subregion of North America, surrounded by the North Atlantic Ocean and the Caribbean Sea that includes 13 independent island countries and 18 dependencies, and other territories in three major archipelagos: the Greater Antilles, the Lesser Antilles and the Lucayan Archipelago.[352] In 1652, Maurile de Saint-Michel wrote *Voyage des isles Camercanes* (*Voyage to the Camercanes islands*) describing his trip to the Camercane Islands in the West Indies.[353] *Camercane, isole* does not appear on any of the three European maps.
106	古巴島	Gu Ba Dao	Cuba Island; "*Cuba*" comes from the Taíno language, however, its exact derivation is unknown.	On the N. America-KWQ, 古巴島 (20.5°N, 85.5°W) is today's Cuba (21°N, 85°W). Christopher Columbus (1451–1506) was the first European to arrive at Cuba in 1492, when the island was already inhabited by three different groups of indigenous peoples. In 1506, Hernán Cortés (1485–1547)[354] took part in the conquest of Hispaniola and Cuba.

(*Continued*)

(*Continued*)

Item #	Name	Pinyin	Etymology	Analysis of history, politics, geography; comparison or comment
				On the N. America-Mercator (1569), the entire Cuban island is unnamed. On the N. America-Ortelius (1570), the island is named *Cvba* ("v" is the same as "u"), and on the N. America-Plancius (1594), *Cuba*. The N. America-Mercator (1569) depicts Cuba at 22°N, 77.5°W; the N. America-Ortelius (1570), at 22°N, 83°W; and the N. America-Plancius (1594), at 22°N, 82°W. The N. America-KWQ shows better longitudinal value (85.5°W) in comparison with today's Cuba (85°) than those on the three European maps.
107	牙賣加	Ya Mai Jia	Jamaica or *Iamaica*; the original inhabitants of Jamaica named the island Xaymaca, which meant "Land of Wood and Water" or "Land of Springs".[355]	On the N. America-KWQ, 牙賣加 (17.5°N, 83.5°W) is today's Jamaica Island (18.1°N, 77.3°W). The island, which is part of the Greater Antilles, lies c. 140 km/87 mi south of Cuba, and c. 191 km/119 mi west of Hispaniola. It is in the West Indies. It is the third-largest island of the Greater Antilles and the Caribbean (after Cuba and Hispaniola).[356] The original inhabitants of Jamaica came from South America and are believed to be the Arawaks, also called Tainos.[357] They were a mild and simple people by nature and led quiet and peaceful lives until they were destroyed by the Spaniards some years after Christopher Columbus discovered the island in 1494.[358] The N. America-Mercator (1569) depicts Jamaica at 18°N, 80.5°W; the N. America-Ortelius (1570), at 17.5°N, 81°W; and the N. America-Plancius (1594), at 18°N, 81°W. The N. America-KWQ shows a slightly larger longitudinal error in depicting Jamaica than those on the three European maps.

108	小以西巴你亞	Xiao Yi Xi Ba Ni Ya	Hispaniola: the island was called by various names by its indigenous people because the Taino had no written language.	On the N. America-KWQ, 小以西巴你亞 (17.5°N, 78.5°W) is today's island of Hispaniola (17°N, 71°W), which is in the Caribbean that is part of the Greater Antilles. The primary indigenous people on the island of Hispaniola were the Taíno people. Hispaniola is the region's second largest in area, after the main island of Cuba. Christopher Columbus was the first European to land at Hispaniola in 1492. He named the island *La Isla Española* ("the Spanish Island"), later Latinised to *Hispaniola*.[359] The N. America-Mercator (1569) depicts Hispaniola at 18.5°N, 68.5°W; the N. America-Ortelius (1570), at 18°N, 74°W; and the N. America-Plancius (1594), at 18°N, 73°W. The N. America-KWQ shows a slightly larger longitudinal error in depicting Hispaniola than those on the three European maps.
• 109	曷勿洗勿	He Wu Xi Wu	Higüey	On the N. America-KWQ, 曷勿洗勿 (20°N, 75.5°W) is a city on today's island of Hispaniola (17°N, 71°W). Higüey was also the name of a former indigenous chiefdom in Hispaniola's easternmost end when Christopher Columbus arrived in 1492. Guey or Huiou is the sun in the Arawakan language of the island's indigenous Taíno people. The name Higüey might mean land where the sun is born, because it is located at the island's eastern end, which receives the rays of the sun first. None of the three sixteenth century European maps depicts Higüey, only the N. America-Ortelius (1570) depicts Quises on the east coast of Hispaniola. But the pronunciation of Quises is not close to that of Higüey.
• 110	仙如漢島	Xian Ru Han Dao	*Isla de Puerto Rico*; Puerto Rico is Spanish for "rich port".	The name 仙如漢島 (17°N, 72.5°W) is of Chinese origin. On the N. America-KWQ, the island's coordinates correlate closely with today's Puerto Rico Island (18.3°N, 66.6°W) in Fig. A5.25.

(*Continued*)

(*Continued*)

Item #	Name	Pinyin	Etymology	Analysis of history, politics, geography; comparison or comment
			The island is also popularly known in Spanish as *la isla del encanto*, meaning "the island of enchantment".[360]	The N. America-Mercator (1569) depicts an unnamed island to the east of *Hispaniola* at 17°N, 61.5°W; on the N. America-Ortelius (1570), it is at 17.5°N, 65.5°W; and on the N. America-Plancius (1594), at 16.5°N, 61°W.  Fig. A5.25. Map of today's Caribbean Sea (see endnote 361, public domain).
• 111	對島	Dui Dao	Virgin Islands; Spanish: *Islas Vírgenes*	The name 對島 (c. 18.5°N, 65.5°W) is of Chinese origin. The coordinates of the island correlate closely with today's Virgin Islands (18.2°N, 64.8°W), which is an archipelago in the Caribbean Sea. They are geologically the easternmost part of the Greater Antilles.[362] The Virgin Islands are now separated into Spanish Virgin Islands (green), United States Virgin Islands (red) and British Virgin Islands (black) as shown in Fig. A5.26. The archipelago is separated from the Lesser Antilles by the Anegada Passage and from the main island of Puerto Rico by the Virgin Passage.

				On the N. America-Mercator (1569) and the N. America-Ortelius (1570), there are many small islands, having different names, to the east of *Hispaniola*; and on the N. America-Plancius (1594), there are also many unnamed small islands to the east of *Hispaniola*. 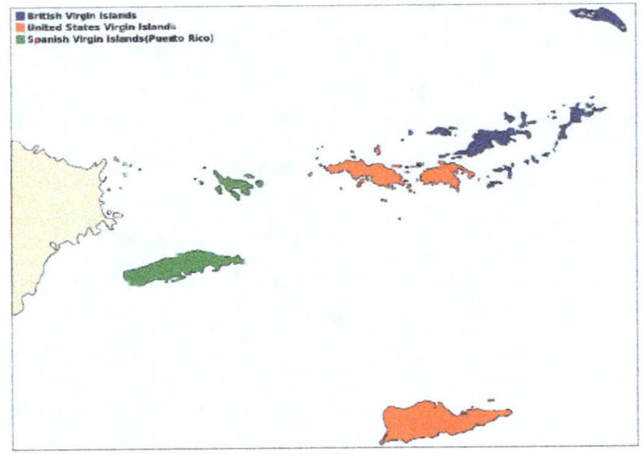 Fig. A5.26. Map of today's Caribbean Sea (see endnote 363, public domain); to the far left is the Puerto Rico Island.
112	色氏測島	Se Di Ce Dao	Sept Cites Archipelago	On the N. America-KWQ, 色氏測島 (27.5°N, 43°W) is the Sept Cites Archipelago. There is no written history or a modern English name for the place. It may be an imagined island group. It is also depicted on the N. America-Mercator (1569) and the N. America-Ortelius (1570) at similar geographical locations.
113	鶴島	He Dao	Coruo[364] or *Corvo*. Corvo is the Italian, Portuguese and Galician word for *crow*, and here it refers to Corvo Island.	On the N. America-KWQ, 鶴島 (36°N, 31.5°W) is today's Corvo Island (39.7°N, 31.1°W), which literally means the *Island of the Crow*, although the Chinese character 鶴 means "crane". It is the smallest and the northernmost island of the Azores archipelago.[365] Corvo does not have a marina, just an anchorage at the southern end of the island which is deep and rocky. The Azores lived in the European explorers' imagination since ancient days,[366] resulting in maps, such as the Genovese Atlas Medici from 1351.[367]

(*Continued*)

(*Continued*)

Item #	Name	Pinyin	Etymology	Analysis of history, politics, geography; comparison or comment
				The first European to discover Corvo and Flores Islands at the same time was the Portuguese explorer Diogo de Teive (fl. 1451–1472) around 1452, and it was initially called *Ilhas Floreira* (or literally, the *Flowered Islands*). The name referred to the two islands of Corvo and Flores, which also appeared on the *Aragonese Mapa Catalão* of 1375.[368] The N. America-Mercator (1569) depicts Corvo at 39°N, 26°W; on the N. America-Ortelius (1570), it is at 39°N, 30°W; and on the N. America-Plancius (1594), at 40.5°N, 34°W. These data are comparable to those on the N. America-KWQ.
114	第三起島	Di San Qi Dao	Portuguese: *Terceira*; the name *Terceira* means "third" in Portuguese, (as in "the third island" or "third to be discovered"), which was used to describe the island.	On the N. America-KWQ, 第三起島 (36°N, 28.5°W) is today's Terceira Island (38.7°N, 27.2°W), which is an island of the central Azores in the North Atlantic Ocean. It is one of the larger islands of the archipelago. In the Chinese name, 第三起 also means "the third" and 島, "island". Archaeological findings show that Terceira has a history of settlement that may date back 2,000 years, before the Portuguese arrived. Although there is uncertainty in the date and the discoverer associated with the islands of the Azores, there is an indication that Terceira may have been discovered by Vicente de Lagos in 1445.[369] The N. America-Mercator (1569) does not depict Terceira; the N. America-Ortelius (1570) depicts it at 37.5°N, 25.5°W; and the N. America-Plancius (1594), at 38°N, 28°W. These data are comparable with those on the N. America-KWQ.

115	怕霧打島	Pa Wu Da Dao	Bermuda: the islands making up Bermuda are named after Juan de Bermúdez (unknown date of birth–1570), an early sixteenth century Spanish sea captain and the first European explorer of the archipelago (in 1505).[370]	On the N. America-KWQ, 怕霧打島 (42°N, 41.5°W) is the transliteration of Bermuda (32.3°N, 64.8°W). The coordinates of 怕霧打島 (42°N, 41.5°W) show large errors in comparison with today's Bermuda (32.3°N, 64.8°W). Bermuda had no indigenous population when the Europeans arrived, nor during the initial British settlement a century later. It was mentioned in *Legatio Babylonica*, published in 1511 by historian Pedro Mártir de Anglería (1457–1526), and was included on Spanish charts of that year. Both Spanish and Portuguese ships used the islands as a replenishment spot to take on fresh meat and water. With its frequent storm-wracked conditions and dangerous reefs, the archipelago became known as the "Isle of Devils".[371] Neither Spain nor Portugal attempted to settle it. The N. America-Mercator (1569) depicts *La Bermuda* at 33°N, 56.5°W; the N. America-Ortelius (1570), at 32°N, 62°W; and the N. America-Plancius (1594), at 33.5°N, 69.5°W. On the three sixteenth century European maps, the coordinates of *La Bermuda* are better presented than those on the N. America-KWQ.
● 116	河摺亞諾滄	He Zhe Ya Nuo Cang	河摺亞諾 is the transliteration of the first Latin word of *Oceanus Atlanticus* or *Oceano Occidentale*; *Oceanus* (a water deity)[372] is known as the gigantic river that encircled the world; 滄 means "ocean" or "wave".	Matteo Ricci placed 大西洋 (literally Great Western Ocean) off the Strait of Gibraltar and 河摺亞諾滄 (meaning Ocean) off the west coast of Africa as well as to the east of North America. He seems to indicate that 大西洋/Great Western Ocean is the smaller water area off the Strait of Gibraltar, whereas *Oceanus* — the son of Uranus (Heaven) and Gaia (Earth) — is the gigantic river that encircled the world in the Greek mythological literature.[373] This geographical item has also been discussed in detail in the Appendix (Item 180) of Chapter 4 of this book. On the N. America-Mercator (1569), *Oceanvs Atlanticvs* denotes the ocean to the west of Africa and Europe and to the east of North America.

(*Continued*)

(Continued)

Item #	Name	Pinyin	Etymology	Analysis of history, politics, geography; comparison or comment
				On the N. America-Ortelius (1570), *Mare Atlanticvm* denotes the ocean to the east of North America, but on his 1570 World Map[374] and on the N. America-Plancius (1594), *Mar der Nort* is used to denote the ocean to the west of Europe and Africa and to the east of North America.
• 117	鄷度蠟	Feng Du La	Honduras: the name means "depths" in Spanish, referring to the bay of Trujillo as an anchorage, or to Christopher Columbus's quote that "*Gracias a Dios que hemos salido de esas Honduras*" ("Thank God we have departed from those depths").[375]	On the N. America-KWQ, 鄷度蠟 (13°N, 95.5°W) is today's Honduras (15.2°N, 86.2°W). The map still shows a more westward longitudinal reading than the actual value. Honduras was inhabited by many indigenous peoples when the Spanish arrived in the sixteenth century. The most advanced of these were the Maya. On his fourth and final voyage to the New World in 1502, Christopher Columbus landed near the modern town of Trujillo, near Guaimoreto Lagoon, becoming the first European to visit the Bay Islands on the coast of Honduras.[376] It was not until the end of the sixteenth century that *Honduras* was used for the whole province. Hence, none of the three European maps denotes "Honduras".  Fig. A5.27. Map of Central America on the North American continent (see endnote 377, public domain).

• 118	瓦的馬革	Wa De Ma Ge	Uatimaco[378]	On the N. America-KWQ, 瓦的馬革 (12.8°N, 93.5°W) should be in today's Honduras (15.2°N, 86.2°W). Recall that the N. America-KWQ often gives more westward longitudinal readings. There is no written history or a modern English name for the place.
119	詣柯	Yi Ke	Naco[379]	On the N. America-KWQ, 詣柯 (11.5°N, 92.5°W) is the *Valle de Naco* (15.4°N, 88.2°W) in northwestern Honduras. Naco was one of the late pre-Hispanic centres in the region.[380] Pedro de Alvarado (c. 1485–1541)[381] — Spanish conquistador and Governor of Guatemala — is considered the conquistador of much of Central America, including Guatemala, Honduras and El Salvador. Naco (*Valle de Naco*) is depicted on the N. America-Mercator (1569) at 12.5°N, 89.5°W; and on the N. America-Ortelius (1570) and the N. America-Plancius (1594) at 12.5°N, 95°W. The three maps give comparable coordinates of *Valle de Naco* to those on the N. America-KWQ.
120	里漢	Li Han	León[382]	On the N. America-KWQ, 里漢 (11.7°N, 91.5°W) is today's León (12.4°N, 86.9°W), which is a city in Nicaragua. In 1522, Spanish explorer Gil Gonzalez de Avila (1480–1520) named Nicaragua after a local Indian chief, Nicarao.[383] León is the second largest city in Nicaragua, after Managua. It was founded by the Spanish and named as *Santiago de los Caballeros de León* in 1524 by Francisco Hernández de Córdoba (c. 1467–1517) who completed the conquest of Nicaragua.[384] León is depicted on the N. America-Mercator (1569) at 11.5°N, 89°W; and on the N. America-Ortelius (1570) at 11°N, 94.5°W. The two maps give comparable coordinates of León to those on the N. America-KWQ. León is not depicted on the N. America-Plancius (1594).

(*Continued*)

(*Continued*)

Item #	Name	Pinyin	Etymology	Analysis of history, politics, geography; comparison or comment
Zone IV: South America (Northern Region)				
 Fig. A5.28. Extracted from Fig. 5.3, bottom right panel, public domain.				
121	坡巴牙那	Po Ba Ya Na	Popayán; the word *Popayán* comes from an indigenous language.	On the S. America-KWQ, 坡巴牙那 (6.5°N, 84.5°W) is today's Popayán (2.4°N, 76.6°W), which is a city located in southwestern Colombia (4.6°N, 74.3°W).

				 Fig. A5.29. Map of South America (see endnote 385, public domain). No records exist of the pre-Hispanic history of the indigenous village of Popayán. In 1537, the Spanish conquistador Sebastián de Belalcázar (1479/1480–1551) arrived in Popayán.[386] Popayán is depicted on the S. America-Ortelius (1570) at 2°N, 83°W, but not on the other two European maps. These coordinates fit better with Popayán (2.4°N, 76.6°W) than 坡巴牙那 (6°N, 85°W) on the S. America-KWQ.

(Continued)

(*Continued*)

Item #	Name	Pinyin	Etymology	Analysis of history, politics, geography; comparison or comment
122	祁臘正瓦斯	Qi La Zheng Wa Si	Quillacinga or Quillasinga	On the S. America-KWQ, 祁臘正瓦斯 (5°N, 83.5°W) is today's Quillacinga (2.4°N, 76.6°W). The Quillacingas (a people of southwestern Colombia and northern Ecuador) lived in the Atriz Valley. Quillacinga or Quillasinga is an indigenous village in the Colombian department of Nariño (1.3°N, 77.4°W). The first European conquistador who entered Nariño was Andagoya Pascual (1495–1548) in 1522, who travelled from the Colombian Pacific coast and then tried to organise the expedition that culminated in the conquest of Peru.[387] The place is depicted as *Los Quillacinga* (0.5°N, 81.5°W) on the S. America-Ortelius (1570). The coordinates are slightly better readings than those on the S. America-KWQ. The place does not show up on the other two European maps.
• 123	金加西蠟	Jin Jia Xi La	*Castilla de Oro*	Both 金 and "oro" mean "gold" and Castilla means "castle". *Castilla de Oro* or *del Oro* was the name given by the Spanish settlers at the beginning of the sixteenth century to the Central American territories from the Gulf of Urabá, near today's Colombian–Panamanian border, to the Belén River.[388] The name *Castilla de Oro* was made official in 1513 by King Ferdinand II of Aragon, then regent of the Crown of Castile.[389] *Castilla de Oro*, mapped by Amerigo Vespucci in 1550, is in today's Columbia and Venezuela as shown in Fig. A5.29, and correlates well with 金加西蠟 on the S. America-KWQ. However, none of the three sixteenth century European maps discussed in this chapter depicts *Castilla de Oro*.

				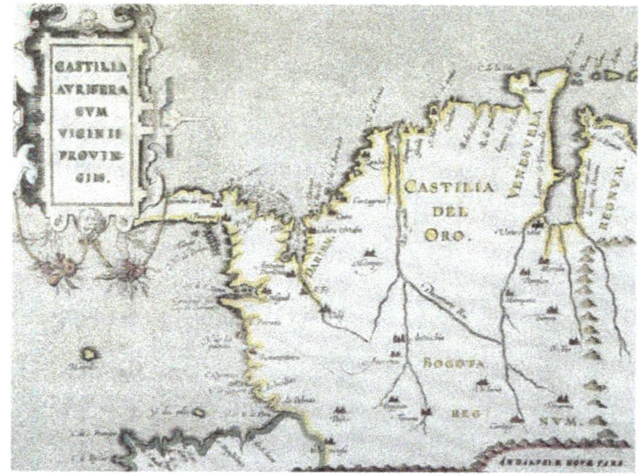 Fig. A5.30. *Castilla del Oro* mapped in 1550 by Amerigo Vespucci (see endnote 390, public domain).
124	祈多	Qi Duc	*Quito*[391]	On the S. America-KWQ, 祈多 (1°N, 86.5°W) is today's Quito (0.2°S, 78.5°W). Quito is the capital of Ecuador (Spanish for "Equator"), the closest capital city to the Equator. In 1534, Spanish soldiers, led by Sebastián de Belalcázar (1479/1480–1551), defeated the Inca in Quito, and named the town *Villa de San Francisco de Quito*.[392] Quito is depicted on the S. America-Oetelius (1570) at 0.5°N, 86°W and the S. America Plancius (1594) at 0°N, 94.5°W. All longitudinal readings are too far westward.
125	加利巴那	Jia Li Ba Na	*Calimana, la*[393]	On the S. America-KWQ, 加利巴那 (2°N, 84°W) is *Calimana,* which should be in today's Ecuador. On the S. America-Ortelius (1570), *Calimana* is depicted at 8°N, 70°W. There is no written history or a modern English name for the place.

(*Continued*)

(*Continued*)

Item #	Name	Pinyin	Etymology	Analysis of history, politics, geography; comparison or comment
126	荅尼戀	Da Ni Lian	Darién[394]	On the S. America-KWQ, 荅尼戀 (4.0°N, 80°W) is depicted in the region of the Darién Gap (7.9°N, 77.5°W),[395] which is a very dangerous part of the world's dense rainforest and jungle-covered mountains that separates North and South America. It covers Panama's Darién Province and the northern portion of Colombia's Chocó Department.[396] The Guna people lived in what is now Northern Colombia and the Darién Province of Panama at the time of the Spanish conquest. Vasco Núñez de Balboa (c. 1475–1519)[397] and Alonso de Ojeda (c. 1466–c. 1515)[398] explored the coast of Colombia in 1500 and 1501, and made contact with the Gunas. Balboa heard of the "South Sea" from locals while sailing along the Caribbean coast. On 25 September 1513, he saw the Pacific.[399] On the S. America-Ortelius (1570), Darién is depicted at 7.5°N, 81°W, but on the S. America-KWQ, it is depicted at a lower latitude of 4.0°N.

				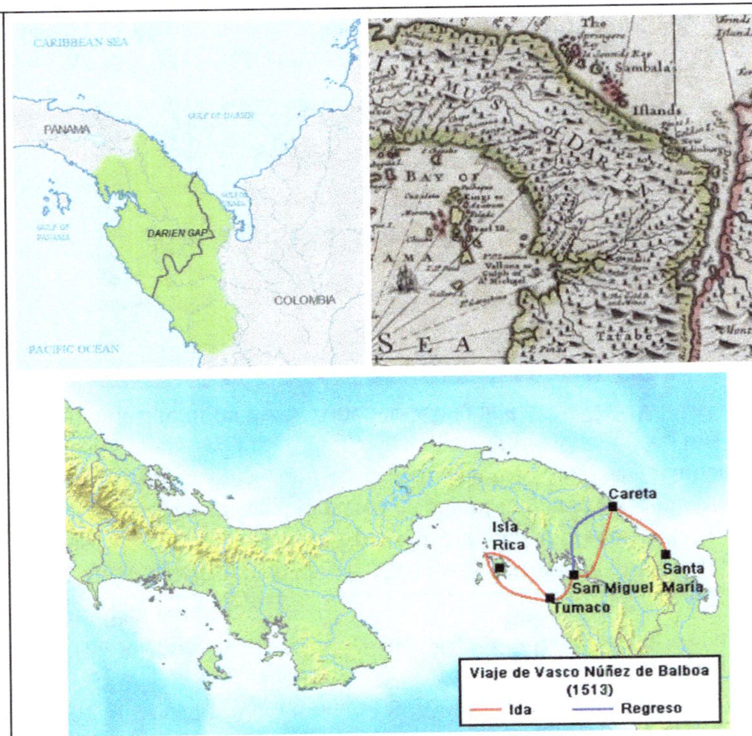 Fig. A5.31. Top left panel: Darién Gap at the Colombia–Panama border (see endnote 400, public domain); top right panel: a map of the Isthmus of Darién, the Bay of Panama, the Gulf of Vallona or St. Michael, with its Islands and countries adjacent, in 1699 (see endnote 401, public domain); bottom panel: Vasco Núñez de Balboa's travel route to the South Sea, 1513 (see endnote 402, public domain).

(*Continued*)

(*Continued*)

Item #	Name	Pinyin	Etymology	Analysis of history, politics, geography; comparison or comment
127	小耨勿茶	Xiao Nu Wu Cha	Little Venice or Venezuela[403]	On the S. America-KWQ, 小耨勿茶 (6.0°N, 76.5°W) is today's Venezuela (6.4°N, 66.6°W). In 1499, an expedition led by Alonso de Ojeda (c. 1466–c. 1515) explored the Venezuelan coast. The stilt houses in Lake Maracaibo reminded one of the navigators of the Italian city of Venice, so he named the region *Veneziola*, or "Little Venice".[404] *Venezuela* is the Spanish version of *Veneziola*.[405] Spain colonised the territory of Venezuela in 1522.[406] On all the three sixteenth century European maps, the area is depicted as *Benezuela*. On the S. America-Mercator (1569), *Benezuela* is depicted at 8°N, 67°W; on the S. America-Ortelius (1570), at 7°N, 72°W; and on the S. America-Plancius (1594), at 8°N, 71°W. The S. America-KWQ gives the best latitudinal but the worst longitudinal readings.
• 128	新唵大魯西亞	Xin An Da Lu Xi Ya	*Nuova Andalusia*[407]	New Andalusia Province (1537–1864) was a province of the Spanish Empire, and later of Greater Colombia and Venezuela. New Andalusia is not depicted on any of the three sixteenth century European maps, but it is depicted on the 1562 Diego Gutierrez Amazonas map (see Fig. A5.32).[408] Hence, Matteo Ricci added this geographical term on the S. America-KWQ at 4.5°N, 73.5°W. 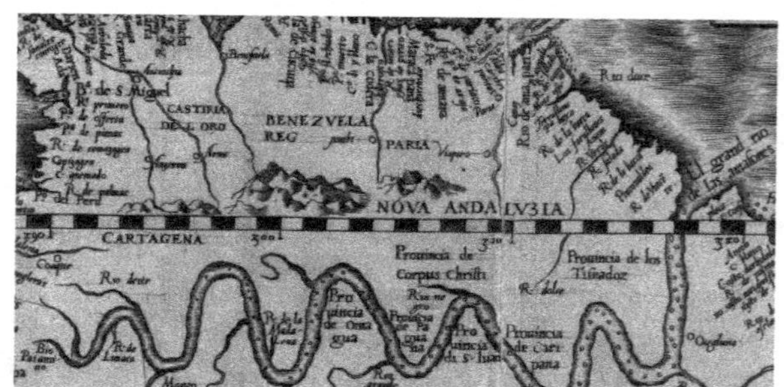 Fig. A5.32. The 1562 Diego Gutierrez Amazonas map depicts *Nuova Andalusia* (see endnote 409, public domain) near the Equator.

● 129	亞那牙	Ya Na Ya	Guyana; "Guyana" comes from an indigenous Amerindian language and means "land of many waters".	On the S. America-KWQ, 亞那牙 (6°N, 71°W) is today's Guyana (Guaiana; 4.9°N, 58.9°W), which is a country on the northern mainland of South America (see Fig. A5.29). The Spaniards explored the area in 1499.[410] Christopher Columbus was the first European to sight Guyana during his third voyage in 1498, and the Dutch were the first Europeans to establish colonies starting from 1581.[411] But Guyana is not depicted on any of the three European maps.
130	容皮	Rong Pi	Iumbi[412]	On the S. America-KWQ, 容皮 (9°N, 70°W) may be in today's Suriname (5.8°N, 55.2°W). On the S. America-Mercator (1569), Iumbi is depicted at 7°N, 61°W; and on the S. America-Mercator (1570), at 7°N, 66°W. Both are near a river which flows into the Caribbean Sea. There is no written history or a modern English name for the place.
● 131	富令那國	Fu Ling Na Guo	*Fulemna, regno*	On the S. America-KWQ, 富令那國 (6.5°N, 65°W) may be today's French Guyana (4.9°N, 52.3°W) or Brazil (14.2°S, 51.9°W). There are no written history or a modern English names for this chiefdom.
≈ 8	烏水河	Wu Shui He	*Rio Negro* or Black River	Today's 烏水河 is the *Rio Negro* ("Black River"; c. 2°N, c. 70°W) or Guainía as it is known in its upper part. The Black River is a major tributary of the Amazon River and the largest blackwater river (river with a slow-moving channel flowing through forested swamps or wetlands)[413] in the world. It does not flow into the Atlantic Ocean. But, on the S. America-KWQ, the name 烏水河 (*Rio Negro*/Black River) is mistakenly used to denote the Branco River ("White River"; 1.4°S, 61.9°W), the principal tributary of the *Rio Negro* from the north. The opposite is true.

(*Continued*)

(*Continued*)

<table>
<tr><th>Item #</th><th>Name</th><th>Pinyin</th><th>Etymology</th><th>Analysis of history, politics, geography; comparison or comment</th></tr>
<tr><td></td><td></td><td></td><td></td><td>The Rio Negro is correctly depicted on the three European maps, because the Spanish explorer Francisco de Orellana (1511–1546) came upon it in 1541 and named it.[414] Although the Portuguese arrived in Brazil in 1500 under Pedro Álvares Cabral (c. 1467 or 1468–c. 1520),[415] they did not discover the Branco River valley until two hundred years later. Nevertheless, the Branco River is depicted as an unnamed tributary on the three European maps. How did they know to draw this tributary?

Fig. A5.33. Map showing the Rio Negro and the Branco River in the Amazon Basin (see endnote 416, public domain).</td></tr>
<tr><td>132</td><td>渴里白那</td><td>Ke Li Bai Na</td><td>Caribana[417]</td><td>On the S. America-KWQ, 渴里白那 (2.5°N, 65.5°W) is in today's Brazil (14.2°S, 51.9°W).
Brazil was inhabited by numerous tribal nations prior to the landing in 1500 of explorer Pedro Álvares Cabral (c. 1467 or 1468–c. 1520), who claimed the area for the Portuguese Empire.
The three sixteenth century European maps discussed in this chapter show Caribana as a huge area.</td></tr>
</table>

● 133	漁人地	Yu Ren Di	Pescheria,[418] meaning a place rich in fishing resources	漁 means "fish", 人, "people" and 地, "ground". 漁人地 means "fishermen's ground". On the S. America-KWQ, 漁人地 (1°N, 58.5°W) correlates closely with the northern coast (near the Equator) of Brazil facing the North Atlantic Ocean, see Fig. A5.29. The ocean has some of the world's richest fishing resources, mostly for shrimps and large catfishes,[419] especially in the waters covering the shelves. Researchers report that biological production is high to the North of Brazil, because of the continental runoff from the Amazon River.[420]
● 134	青珠島	Qing Zhu Dao	Margarita Island	From Fig. A5.25, we know that Margarita Island is in the Caribbean Sea between latitudes 10.9°N and 11.2°N and longitudes 63.8°W and 64.4°W. The island is the largest island in the Venezuelan state of Nueva Esparta, situated off the northeastern coast of the country. The island is split into two peninsulas joined by an isthmus. The coast of the island is abundant in pearls; hence, the island's nick name is "Pearl of the Caribbean". Pearls are found in a variety of overtone colours, with some of these colours being very pricey and highly sought after. 青 means "green", 珠 means "beads". On the S. America-KWQ, 青珠島 (12°N, 71.5°W) suits the location of Margarita Island (11°N, 64°W) and Trinidad Island (10.5°N, 61.3°W). Since the map generally gives longitudinal readings that are more westward than the actual values, Margarita Island may be a more likely location, although Trinidad is worth consideration. Trinidad is in Lesser Antilles.[421] Trinidad and Tobago form a two-island nation of the West Indies. Trinidad and Tobago are closest to Barbados, Guyana, Venezuela and Grenada. They are often referred to as the southernmost islands in the West Indies[422] in North America. None of the three sixteenth century European maps depicts Margarita Island, but the 1562 Map of America by Diego Gutiérrez shows both Margarita Island and Trinidad Island.[423] Christopher Columbus was the first European to land on both islands on his third voyage in 1498.

(*Continued*)

(*Continued*)

Item #	Name	Pinyin	Etymology	Analysis of history, politics, geography; comparison or comment
Zone V: South America (Western Region)				

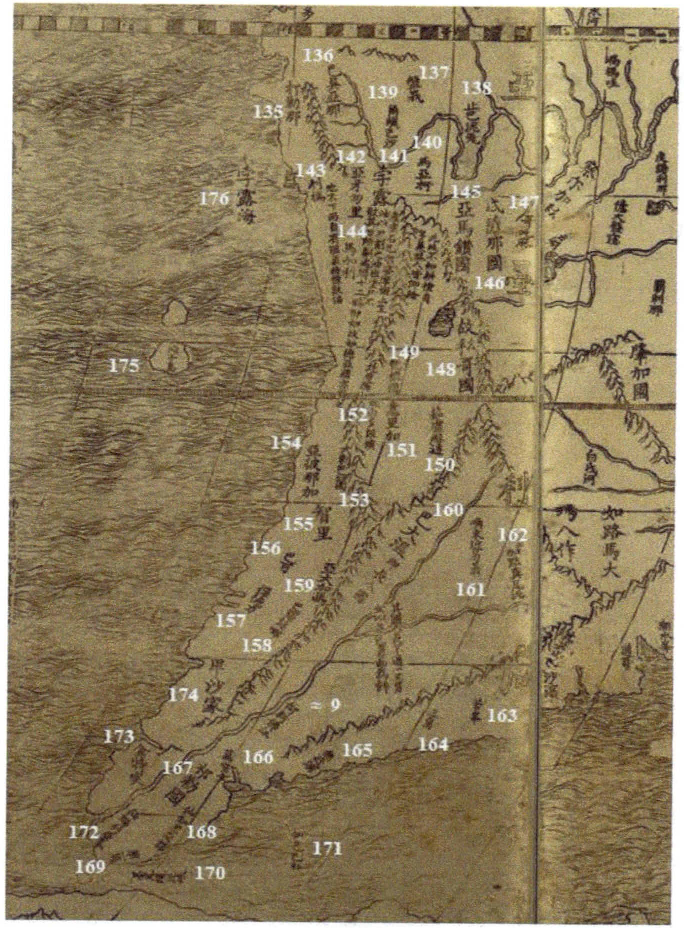

Fig. A5.34. Extracted from Fig. 5.3, bottom right panel, public domain.

• 135	打勒那	Da Le Na	Talena[424]	On the S. America-KWQ, 打勒那 (5°S, 87.5°W) may be in today's Ecuador (1.8°S, 78.2°W) or north Peru (1.8°S, 78.2°W). Talena is not depicted on any of the three European maps. There is no written history or a modern English name for the place.
136	巴吳亞那	Ba Wu Ya Na	Paguana[425]	On the S. America-KWQ, 巴吳亞那 (4°S, 84.5°W) is today's Paguana (9.6°S, 74.9°W), which is a locality in central Peru and has an elevation of 3,894 m.[426] Peru was home to the oldest civilization in the Americas and to the Inca Empire (1438–1533/1572), and the largest and most advanced state in pre-Columbian America. It was conquered by the Spanish Empire in the sixteenth century.[427] On the S. America-Mercator (1569), Paguana is depicted near the Amazon River basin at 6.0°S, 59.0°W, to the far east from its actual geographical location, and similarly on the S. America-Ortelius (1570) at 6.5°S, 65.5°W. The S. America-Plancius (1594) does not depict Paguana.
• 137	盤峩	Pan E	Pango	On the S. America-KWQ, 盤峩 (3.5°S, 79.5°W) is Pango (7.9°S, 78.5°W),[428] which is in Otuzco, La Libertad, Peru. There is no written history or a modern English name for the place.
• 138	止巴泥瓦	Zhi Ba Ni Wa	Cepaniua[429]	On the S. America-KWQ, 止巴泥瓦 (4.7°S, 75.5°W) is in today's Peru. There is no written history or a modern English name for the place.
• 139	角蠟巴沙	Jiao La Ba Sha	Chiola pascia[430]	On the S. America-KWQ, 角蠟巴沙 (5°S, 81°W) is in today's northern Peru or southern Ecuador. There is no written history or a modern English name for the place.
• 140	馬亞柯	Ma Ya Ke	Maiaco[431]	On the S. America-KWQ, 馬亞柯 (8.5°S, 78.5°W) is in today's northern Peru. There is no written history or a modern English name for the place.

(*Continued*)

(*Continued*)

Item #	Name	Pinyin	Etymology	Analysis of history, politics, geography; comparison or comment
141	孛露	Bei Lu	Peru: the name may be derived from *Birú*, the name of a local ruler in the early sixteenth century.[432]	On the S. America-KWQ, 孛露 (9.5°S, 81.5°W) was today's Peru (9.2°S, 75.0°W). The Peruvian territory was home to several ancient cultures including the Inca Empire (1438–1533/1572), the largest empire in the pre-Columbian Americas. Spanish conquistadors arrived in 1522 and established a viceroyalty that encompassed most of its South American territories.[433] The Incas's territory included a large portion of western South America, centred on the Andean Mountain ranges, from southern Colombia to northern Chile, between the Pacific Ocean in the west and the Amazon rainforest in the east. The S. America-Mercator (1569) does not show Peru; on the S. America-Ortelius (1570), Peru is depicted at 11°S, 78°W; and on the S. America-Plancius (1594), at 12.5°S, 89°W.
@ 9	[孛露] 此地不知耕種自多菓蓏人好仰給産香名巴尔婆摩樹上生油以刀劃之油出塗尸不敗其刀所劃處周十二時卽如故如德亞國 亦有之		[Bei Lu] Ci Di Bu Zhi Geng Zhong Zi Duo Guo Luo Ren Hao Yang Gei Chan Xiang Ming Ba Er Po Mo Shu Shang Sheng You Yi Dao Hua Zhi You Chu Tu Shi Bu Bai Qi Dao Suo Hua Chu Zhou Shi Er Shi Ji Ru Gu Ru De Ya Guo Yi You Zhi	My translation is as follows: Peru: In this land, people do not know farming, but fruits and vegetables grow naturally and abundantly. People like what they are provided with. There is a kind of incense that they call Ba Er Po Mo balsam oil which is produced by a kind of tree. With a knife, an incision can be made into a tree trunk, and the oil is extracted. When the oil is smeared over corpses, they do not decompose. The place where the knife was drawn heals after twelve hours. This balsam oil can also be found in the Chiefdom of Ru De Ya. The balsam of Peru comes from the trunk of a tree grown in Central and South America. In the early period of the Spanish invasion of the region, balsam was collected in Central America and shipped to Callao (the port of Lima) in Peru, then shipped onward to Europe. It acquired the name of "Peru" because it was shipped through Peru.[434]

142	亞牙勿里	Ya Ya Wu Li	Aiauiri[435]	On the S. America-KWQ, 亞牙勿里 (10°S, 82.5°W) is today's Aiauiri in Peru.[436] It is a settlement of the province and corregimiento of Lamoa in Peru.[437] In its vicinity are some forts, which were built by the Indians in the time of their gentilism and are now in a state of great dilapidation.[438] On all the three sixteenth century European maps, "Aiauiri" is depicted as "Aiauari". On the S. America-Mercator (1569), Aiauari is depicted at 2°S, 72.5°W; on the S. America-Ortelius (1570), at 3°S, 78°W; and on the S. America-Plancius (1594), at 1°S, 78°W.
143	利禡	Li Ma	Lima: the city was founded in 1535 under a different name. But this name quickly fell into disuse, and *Lima* became the city's name of choice.[439]	On the S. America-KWQ, 利禡 (10°S, 85°W) is today's Lima (12°S, 77°W). The longitude position is more westward than its actual position. But on the S. America-Ortelius (1570), Lima (11.5°S, 93.5°W) shows even a larger longitudinal error. Similar large longitudinal errors also show up on the S. America-Mercator (1569) and the S. America-Plancius (1594). Lima was named by indigenous people in the agricultural region known by local Peruvians as Limaq before the colonial days. The history of Lima, the capital of Peru, began with its foundation by Francisco Pizarro (c. 1478–1541) in 1535.[440]
@ 10	[利禡] 地不下雨自有濕氣稜種數倍		[Li Ma] Di Bu Xia Yu Zi You Shi Qi Leng Zhong Shu Bei	My translation is as follows: Lima: It does not rain in this land, but the air is naturally humid. When crops are planted, the harvest multiplies. Peru is one of the five largest producers of avocado, blueberry, artichoke and asparagus, and its agriculture is considerably diversified. Peru has good agriculture; Lima is in Peru.
144	馬加利	Ma Jia Li	Hacari[441]	On the S. America-KWQ, 馬加利 (15°S, 82°W) should be in today's Peru, but its longitude position is clearly more westward than what it should be. Hacari is not depicted on the S. America-Mercator (1569) and the N. America-Plancius (1594), but it is depicted on the S. America-Ortelius (1570) at 15°S, 92°W, more westward than on the S. America-KWQ. There is no written history or a modern English name for the place.

(*Continued*)

(Continued)

Item #	Name	Pinyin	Etymology	Analysis of history, politics, geography; comparison or comment
• 145	亞馬鑽國	Ya Ma Zuan Guo	Amazonas is now a state of Brazil	On the S. America-KWQ, 亞馬鑽國 (12.5°S, 75°W) may represent some of the lost kingdoms of the Amazon, which correlate with today's Amazonas (5°S, 63°W), which is a state of Brazil, located in the northwestern region of Brazil. In 1541, the name "Amazonas" was originally given to the Amazon River by the Spanish explorer and conquistador Francisco de Orellana (1511–1546)[442] when he managed to sail the length of the Amazon, claiming to have come across a warlike tribe of Indians, with whom he fought. He thought they were like the Amazons of Greek mythology, hence giving the river the name *Río de las Amazonas*.[443] 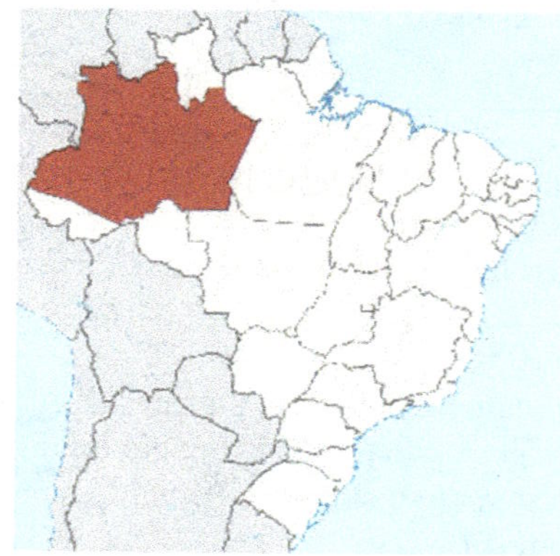 Fig. A5.35. Map of Amazonas (Brazilian State; see endnote 444, public domain). The three European maps all place *Amazonas* (they meant to denote the Amazon Basin) in the eastern part of today's Brazil, close to the estuary of the Amazon River, and none of them depicts Amazonas, the kingdoms in the northwestern region of today's Brazil. See Amazones (the Amazon Basin) in Item 204 for more explanations.

146	氐薩那國	Di Sa Na Guo	Tisnada[445]	On the S. America-KWQ, 氐薩那國 (12.4°S, 73.5°W) is depicted in the proximity of the Amazonas in the northwestern region in today's Brazil. On the S. America-Mercator (1569), Tisnada is depicted at 5°S, 51.5°W; on the S. America-Ortelius (1570), at 1°S, 60.5°W; and on the S. America-Plancius (1594), at 3°S, 55.5°W. All these locations are near the Equator and closer to the Atlantic coast in today's Brazil, but are different from that on the S. America-KWQ. There is no written history or a modern English name for the place.
• 147	哥吒麻	Ge Zha Ma	Cociama[446]	On the S. America-KWQ, 哥吒麻 (12.4°S, 72°W) is depicted in the proximity and to the east of 氐薩那國. It should be in today's Brazil. There is no written history or a modern English name for the place.
• 148	故私哥國	Gu Si Ge Guo	City State of Cusco/Cuzco	On the S. America-KWQ, 故私哥國 (20°S, 74°W) is the City State of Cusco or Cuzco (1197–1438; 13.5°S, 72.0°W), which was a small kingdom based in the city of Cusco in southeastern Peru near the Urubamba Valley of the Andes mountain range.[447] It was founded by the Incas and, through warfare or peaceful assimilation, it began to grow and was succeeded by the Inca Empire (1438–1533). Neither the Kingdom of Cusco nor the Inca Empire appears on any of the three European maps, because both were no longer in existence when the maps were drawn by the three European cartographers (although the city was still there). But 故私哥國/City State of Cusco or Cuzco is depicted on the S. America-KWQ. There is no reason for Matteo Ricci to draw in 1602 an extinct city state on the S. America-KWQ. The fact that 故私哥國/City State of Cusco or Cuzco is depicted on this map proves that the map he used to draw the final *S. America-KWQ is of an era before 1438.*

(*Continued*)

(*Continued*)

Item #	Name	Pinyin	Etymology	Analysis of history, politics, geography; comparison or comment
				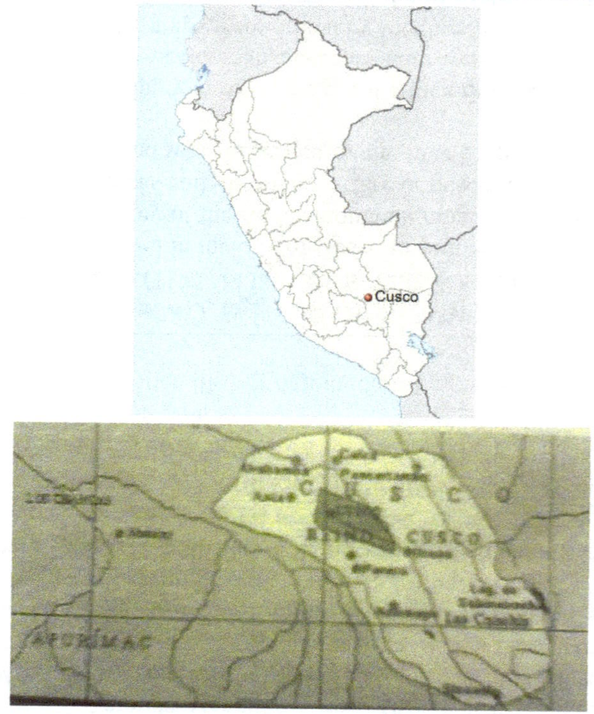 Fig. A5.36. The top panel shows the location of Cusco within Peru (see endnote 448, public domain); the bottom panel shows the map of the City State of Cusco (see endnote 449, public domain).

• 149	乾樤國	Gan Ze Guo	*Chience, regno di*[450]	On the S. America-KWQ, 乾樤國 (22°S, 78°W) is the Kingdom of Chience. It is depicted to the southwest of the City State of Cusco. It should be in today's Peru. There is no written history or a modern English names for the place.
150	茹盧麻達	Ru Lu Ma Da	Giurumada[451]	On the S. America-KWQ, 茹盧麻達 (25.5°S, 74.5°W) is Giurumada in Peru. The actual latitudinal reading of Giurumada can be deduced as follows: 意里加 (Arica; Item 151) is in northern Chile and very close to the Peru border; its actual latitude is 18.5°S; since on the S. America-KWQ, 茹盧麻達 (Giurumada) and 意里加 (Arica) have similar latitudes, we can deduce that the actual latitude of Giurumada should only be a little south of 18.5°S. Hence, we know that on the S. America-KWQ, the latitude of 茹盧麻達/Giurumada at 25.5°S is depicted southward than its actual position. On the S. America-Mercator (1569), *Giurumada* is depicted at 25°S, 72.5°W; on the S. America-Ortelius (1570), at 25°S, 75.5°W; and on the S. America-Plancius (1594), at 25.5°S, 71.0°W. There is no written history or a modern English name for the place.
151	意里加	Yi Li Jia	Arica[452]	On the S. America-KWQ, 意里加 (24.5°S, 77.5°W) is today's Arica (18.5°S, 70.3°W), which is a commune and a port city in northern Chile. Arica is located only 18 km/11 mi south of the border with Peru. Archaeological findings indicate that Arica was inhabited by different indigenous groups dating back 10,000 years. Spaniards settled the land in 1541.[453] By 1545, Arica was the main export entrepot for Bolivian silver and held a crucial role as one of the leading ports of the Spanish Empire. On the S. America-Mercator (1569), Arica is depicted as a port city at 18.5°S, 83.5°W; and on the S. America-Ortelius (1570), at 18.5°S, 88.0°W; but it is not depicted on the S. America-Plancius (1594).

(*Continued*)

(*Continued*)

Item #	Name	Pinyin	Etymology	Analysis of history, politics, geography; comparison or comment
				On the S. America-KWQ, Arica is depicted incorrectly not only with its latitude reading, but also depicted as an inland city. This indicates that the Chinese had only limited knowledge about the city and the nearby silver mines in Potosi (Item 152), but were not involved in the silver mining and did not know Arica was a port city.
152	北度西山	Bei Du Xi Shan	Potosi[454]	Located in the Bolivian Tin Belt, Potosí (20.7°S, 66.7°W), founded in 1545 as a mining town, has the world's largest silver deposit and has been mined since the sixteenth century.[455] The S. America-KWQ depicts 北度西山 (Potosi; 22°S, 80°W) to the northwest of 意里加 (24.5°S, 77.5°W; Arica). But on a modern map, Potosi (20.7°S, 66.7°W) is to the southeast of Arica (18.5°S, 70.3°W). These *rare* geographical missteps show that the Chinese were not familiar with these two places, another indication that they were not involved in the silver mining. Potosí is not depicted on the S. America-Mercator (1569) and the S. America-Plancius (1594). However, the S. America-Ortelius (1570) depicts Potosí at 20.5°S, 79°W, not surprisingly too far westward, while correctly depicting the relative locations of Arica and Potosi.
@ 11	此山多銀礦		Ci Shan Duo Yin Kuang	My translation is as follows: This mountain has many silver mines. The analyses in Items 151 and 152 explain that the Chinese knew of the silver mines but might not have been involved in silver mining.
153	大剌巴箇	Da La Ba Ge	*Tarapacá*	On the S. America-KWQ, 大剌巴箇 (27.5°S, 80.0°W) is *San Lorenzo de Tarapacá*, also known simply as *Tarapacá* (20.2°S, 70.2°W), which is a town in the region of the same name in today's Chile. When Spanish explorer Diego de Almagro reached the settlement in 1536, it was already inhabited by locals.[456]

				On the S. America-Mercator (1569), *Tarapacá* is depicted at 31°S, 83.5°W; and on the S. America-Ortelius (1570), at 21°S, 89°W. The longitudinal readings are too far westward. The S. America-Plancius (1594) does not depict *Tarapacá*.
• 154	亞彼那加	Ya Bi Na Jia	Iapeinachia[457]	On the S. America-KWQ, 亞彼那加 (28°S, 81.5°W) is in today's Chile. There is no written history or a modern English name for the place.
155	智里	Zhi Li	Chile or Chili: this name has various sources. It could stem from the Incas, or a town in Peru, or is an indigenous American word, or the warble of a bird locally known as trile.[458]	On the S. America-KWQ, 智里 (31°S, 80.5°W) is in today's Chile (35.7°S, 71.5°W). About 10,000 years ago, migrating indigenous peoples settled in fertile valleys and coastal areas of what is today's Chile. The Incas briefly extended their empire (Inca Empire; 1438–1533) into what is now northern Chile.[459] In 1520, Ferdinand Magellan (1480–1521)[460] was the first European to set foot on what is now Chile. "Chili" was in use in English until the early twentieth century. On the S. America-Mercator (1569), Chili is depicted at 34°S, 85°W; on the S. America-Ortelius (1570), the name "Chile" is not depicted; and on the S. America-Plancius (1594) it is at 31.5°S, 75.5°W.
156	巴可	Ba Ke	Paca	On the S. America-KWQ, 巴可 (34°S, 81°W) is a place in today's Chile. The S. America-Mercator (1569) depicts Paca at 32°S, 85.5°W; the S. America-Ortelius (1570), at 32°S, 89.0°W; but Paca is not depicted on the S. America-Plancius (1594). There is no written history or a modern English name for the place.
• 157	徑得力	Jing De Li	Chinteli[461]	On the S. America-KWQ, 徑得力 (36°S, 81.5°W) is a place in today's Chile. There is no written history or a modern English name for the place.

(*Continued*)

(Continued)

Item #	Name	Pinyin	Etymology	Analysis of history, politics, geography; comparison or comment
• 158	巴羅謎哪	Ba Luo Mi Na	Palomina[462]	On the S. America-KWQ, 巴羅謎哪 (37°S, 78.5°W) is a place in today's Chile. There is no written history or a modern English name for the place.
159	亞大加馬	Ya Da Jia Ma	Atacama Plateau	On the S. America-KWQ, 亞大加馬 (35.5°S, 78°W) is the Atacama Desert (23.9°S, 69.1°W), which is a desert plateau in South America covering a 1,600 km/990 mi strip of land on the Pacific coast and is to the west of the Andes Mountains.[463] The longitudinal reading of 亞大加馬 is too westward. 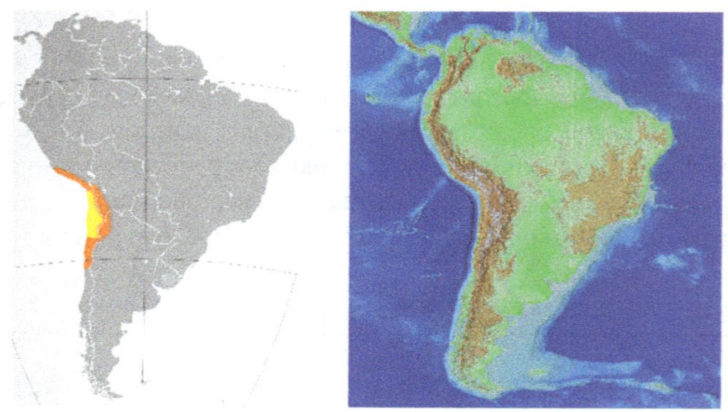 Fig. A5.37. The left panel shows a map of the Atacama Desert (see endnote 464, public domain); the area most defined as Atacama is in yellow-orange colour to the west of the Andes Mountains. In orange are the outlying arid areas. The right panel is a map (see endnote 465, public domain) of South America showing the Andes Mountains running along the entire western part (roughly parallel to the Pacific coast) of the continent.

				From about 1530 to 1700, the Europeans emphasised Atacama's remote but strategic political location rather than its desert climate.[466] On the S. America-Mercator (1569), Atacama is depicted at 34.0°S, 83.5°W; on the S. America-Ortelius (1570), at 33.0°S, 89.0°W. Their longitudinal readings are even more western than that on the S. America-KWQ. The S. America-Plantius (1594) depicts some mountains in Chile but does not depict "Atacama" explicitly.
160	巴大溫即長人國	Ba Da Wen Ji Chang Ren Guo	Patagonia/the Country of the Tall People: in the Patagonia region, people were taller than the Europeans.	On the S. America-KWQ, 巴大溫即長人國 (34°S, 73.5°W) is the Patagonia (41.8°S, 68.9°W). It is a geographical region that encompasses the southern end of South America in Argentina and Chile. The region comprises the southern section of the Andes Mountains, lakes, fjords and glaciers in the west, and deserts, tablelands and steppes to the east.[467] Patagonia is bounded by the Pacific Ocean on the west, the Atlantic Ocean to the east, and the Strait of Magellan, the Beagle Channel and the Drake Passage to the south. At the time of the Spanish arrival, Patagonia was inhabited by multiple indigenous tribes.[468] The name Patagonia comes from the word *patagón*. Magellan used this term in 1520 to describe the indigenous tribes of the region, whom his expedition thought to be giants. The people he called the Patagons are now believed to have been the Tehuelche, who tended to be taller than Europeans of the time.[469]

(*Continued*)

(*Continued*)

Item #	Name	Pinyin	Etymology	Analysis of history, politics, geography; comparison or comment
				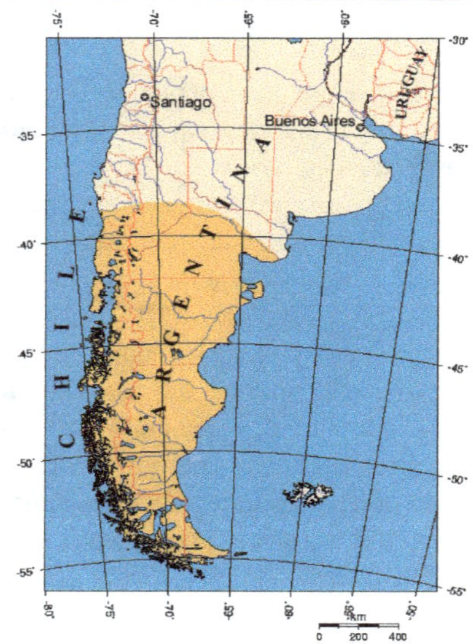 Fig. A5.38. Location of Patagonia (see endnote 470, public domain); Chile is at left and Argentina is at right. Patagonia is not depicted on the S. America-Plancius (1594), but it is depicted as *Patagonvm Regio vbi incole Sunt gigantes* on the S. America-Mercator (1569) at 36°S, 72.5°W and on the S. America-Ortelius (1570) at 35°S, 75°W, respectively. These coordinates are comparable to those on the S. America-KWQ.

@ 12	其国人长不过丈男女以各色画面为饰	Qi Guo Ren Chang Bu Guo Zhang Nan Nu Yi Ge Se Hua Mian Wei Shi	My translation and comment (between square brackets) are as follows: The people of this country are not taller than one *Zhang* [c. 10 feet], and the men and women coloured their faces as decoration. In the Chinese record, there is a Country of the Tall People mentioned in the chapter titled "*Da Huang Dong Jing*《大荒東經》" or "Classic of the Great Wilderness: East" in the Chinese classic *Shan Hai Jing*《山海經》or *Classic of Mountains and Seas/Shan Hao Jing*.[471] The book is a Chinese classic text and a compilation of mythic geography and beasts, and may have existed since as early as the fourth century B.C.[472] The era it describes can be traced back to the Xia Dynasty (c. 2070 B.C.–1600 B.C.). The content involves detailed descriptions of locations in the cardinal directions of the Mountains, Regions Beyond Seas, Regions Within Seas and Wilderness. For the Country of the Tall People, it has the following: 東海之外, 大荒之中, 有山名曰大言, 日月所出。有波谷山者, 有大人之國 My translation and comment (between square brackets) are as follows: Beyond the East China Sea, in the great wilderness, there is a mountain named Dayan. It is where the sun and the moon rise from. There is also a Bo Gu Mountain [here 波 "Bo" means "hilly like waves" and 谷 "Gu" means "valley"] where the Country of the Tall People is located. In 1520, Magellan reached the Patagonia/the Country of the Tall People and *did* discover that the Patagons (Tehuelche) tended to be taller than Europeans of the time.[473]

(*Continued*)

(*Continued*)

Item #	Name	Pinyin	Etymology	Analysis of history, politics, geography; comparison or comment
				Earlier in Item 71, I have mentioned that Da Han (it is the pinyin of 大汉, meaning "Great Guy" or "Big Guy", in today's Alaska) on an old Asian world map is associated with an ancient Chinese geography: the *Classic of Mountains and Seas,* also known as *Shan Hai Jing.*[474] Here is another case showing that Chinese knowledge and exploration of the Americas may be traced to a few thousand years ago, long before Christopher Columbus "discovery" of the New World.
• 161	唃末近曷聶	Gu Mo Jin Ge Nie	Chiomochin-honié[475]	On the S. America-KWQ, 唃末近曷聶 (32.5°S, 69.5°W) is in today's Argentina. There is no written historical record or a modern English name for this place.
• 162	加默真瓦泥	Jia Mo Zhen Wa Ni	Chiamécenuani[476]	On the S. America-KWQ, 加默真瓦泥 (34.5°S, 66°W) is in today's Argentina. There is no written historical record or a modern English name for this place.
• 163	苦峰	Ku Feng	*Cu, promontorio*[477]	On the S. America-KWQ, 苦峰 (43°S, 64°W) is of Chinese origin and in today's Argentina. A promontory is a raised mass of land (hill or headland) that projects into a lowland or a body of water (in which case it is a peninsula). 苦 means "bitter, or tough, or hardship" and here 峰 means "a raised mass of land, or hill". 苦峰 is the name of this hill or headland. This headland should be in today's Argentinian Patagonia (Eastern Patagonia), which is for the most part a region of steppe-like plains, rising in a succession of thirteen abrupt terraces about 100 m/330 ft at a time. This may explain that a long range of "mountains" is depicted in Eastern Patagonia.[478] Hence, to climb these mountains must be a tough job to accomplish. There is no written historical record or a modern English name for this headland.

164	三角	San Jiao	*C. de 3 puntas*	On the S. America-KWQ, 三角 (43.5°S, 66.5°W) is in today's Argentina. On the S. America-Mercator (1569), *C. de 3 puntas* is depicted at 42.2°S, 55.0°W; and on the S. America-Ortelius (1570), at 42.5°S, 59.0°W; it is not depicted on the S. America-Plancius (1594). There is no written historical record or a modern English name for this cape.
165	無氏澳	Wu Di Ao	*Bahia sin fondo* or the Bottomless Bay, San Matías Gulf, in Patagonia Argentina	*Bahia sin fondo* or the Bottomless Bay can be literally translated as 無氏澳. 無 means "no or none", 氏 is pronounced similarly as "底" meaning "bottom", and 澳 means "natural bay". On the S. America-KWQ, 無氏澳 (46°S, 72.5°W) is today's *Bahia sin fondo* or the Bottomless Bay (46°S, 67.6°W),[479] San Matías Gulf, in Patagonia Argentina. At the time of the Spanish arrival, Patagonia was inhabited by multiple indigenous tribes. On the S. America-Mercator (1569), *Bahia sin fondo* is depicted at 42.4°S, 55.2°W; and on the S. America-Ortelius (1570), at 42.5°S, 59.5°W; it is not depicted on the S. America-Plancius (1594).
≈ 9	河剌佛江	He La Fo Jiang	*Orufes p.*[480]	This river cannot be found on modern maps of South America in which major rivers are drawn.[481] But the S. America-KWQ and the three European maps all depict this long river which flows in the north–south direction through almost the entire length of Patagonia. This river seems to exist only in the cartographers' imagination.
166	萬室女	Wan Shi Nu	*Verginik le diecimila*[482]	On the S. America-KWQ, 萬室女 (46.5°S, 77.5°W) is in today's Argentina. On the America-Ortelius (1570), there is *C. 11000 Virginus* at 52°S, 74°W; and on the S. America-Plancius (1594), there is *C. de Virgin Maria al: das 11000 Virgines* at 51°S, 57°W. The S. America-Mercator does not depict such a cape. There is no written historical record or a modern English name for this place.

(*Continued*)

(*Continued*)

Item #	Name	Pinyin	Etymology	Analysis of history, politics, geography; comparison or comment
• 167	祁勒國	Qi Le Guo	Chico[483]	On the S. America-KWQ, 祁勒國 (48°S, 79°W) is the town of Chile Chico (Spanish for *Little Chile*; 46.6°S, 71.8°W) located on the south shore of General Carrera Lake in today's Chile. The town is about 6 km/4 mi west of the border with Argentina.
• 168	仙如里亞諾	Xian Ru Li Ya Ruo	*Giuliano, S.*[484]	On the S. America-KWQ, 仙如里亞諾 (51.5°S, 77.5°W) is in today's Chile. There is no written historical record or a modern English name for this place.
169	第一角	Di Yi Jiao	Cape Froward; Spanish: *Cabo Froward*	The name 第一角 is of Chinese origin. On the S. America-KWQ, 第一角 (53.5°S, 79.5°W) is today's Cape Froward (53.9°S, 71.3°W; see Fig. 5.4 in the main text) which is the southernmost point of the South American mainland. It is in the Magallanes region of Chile, along the north shore of the Magellan Strait, being the southern tip of the Brunswick Peninsula. In January 1587, the English corsair Thomas Cavendish (1560–1592) named the place Cape Froward after the region's rough climate (strong rains and winds).[485] On the S. America-Mercator (1569), there is *C. de Seado* (53°S, 78.5°W) along the southern shore of the Magellan Strait and at the western side of the Strait. It is obviously not the southernmost point of mainland South America. On the S. America-Ortelius (1570), there is *C. de Seado* (52°S, 85°W) along the northern shore of Magellan Strait and at the western side of the Strait. It is also not the southernmost point of the South American mainland. Finally, on the S. America-Plancius (1594), *C. de Seddo* (54°S, 57°W) is depicted on the southern shore and the eastern side of the Strait, hence, it cannot be the southernmost point of the South American mainland. The above three European maps show large differences in their longitudinal readings of *C. de Seddo*, depending on its location at the eastern or at the western side of the Magellan Strait. This shows that they did not know that *C. de Seddo*/Cape Froward is the southernmost tip of the South American mainland.

				On the 1520 first map of the Magellan Strait, *Cap de Seado* is on the north shore of the strait and is the southernmost point of the South American mainland.[486] Only the S. America-KWQ correctly depicts the geographical position of 第一角 as today's Cape Froward — the southernmost tip of the South American mainland.
170	墨瓦蠟泥峡	Mc Wa La Ni Xia	Magellan Strait	On the S. America-KWQ, 墨瓦蠟泥峡 (54.5°S, 75.5°W) is the Magellan Strait (53.5°S, 70.8°W). The Strait of Magellan is a navigable sea route in southern Chile separating the South American mainland to the north and *Tierra del Fuego* to the south, see Fig. 5.4 in the main text. The Strait of Magellan has been inhabited by indigenous Americans for thousands of years. Ferdinand Magellan (1480–1521) was the first European to navigate the strait in 1520.[487] On the S. America-Mercator (1569), the Strait of Magellan is depicted at 54°S, 72.5°W; on the S. America-Ortelius (1570), the Strait of Magellan (54°S, 78°W) is unnamed; and on the S. America-Plancius, the Strait of Magellan is depicted at 54°S, 61°W.
171	仙色巴松	Xian Se Ba Song	*San Sebastián Bay*	On the S. America-KWQ, 仙色巴松 (52°S, 70°W) is today's San Sebastián Bay (53.1°S, 68.2°W), which should partly be enclosed by the 17 km/10.6 mi-long El Páramo spit that protrudes from the north.[488] However, the S. America-KWQ depicts it as an ocean beyond the eastern coast of Argentina. San Sebastián, which shares the same name with the bay, is a village in Argentina, located on the southwestern shore of San Sebastián Bay, in the present Province of Tierra del Fuego, Antarctica and South Atlantic Islands. *Tierra del Fuego* was spotted by Ferdinand Magellan in 1520.[489] Readers are referred to Chapter 2 of this book, which has a full analysis of *Tierra del Fuego*. On the S. America-Mercator (1569), there is *Baia de bapass anegadas* at 37.5°S, 51.5°W. and *Golfo de S. Se* at 56°S, 45°W near *Terra Australis Incognita*. On the S. America-Ortelius (1570), neither *Baia de bapass anegadas* nor *Golfo de S. Sebastiano is depicted*. On the America-Plancius (1594), *I. de S. Sebastian* is depicted at 25°S, 43°W, but *Golfo de S. Sebastiano* is not depicted.

(Continued)

(*Continued*)

Item #	Name	Pinyin	Etymology	Analysis of history, politics, geography; comparison or comment
172	仙瑪里亞峯	Xian Ma Li Ya Feng	*S. Maria. C. de*[490]	On the S. America-KWQ, 仙瑪里亞峯 (51.5°S, 81.5°W; Zaoshajia) is in today's Southern Chile. On the S. America-Mercator (1569), *C. de S. Maria* (42.5°S, 96.5°W) is depicted at the westernmost location of the South American mainland. On the S. America-Ortelius (1570), apart from *C. de S. Maria* (42.5°S, 103°W) depicted at the westernmost location of the South American mainland, there is also *C. de Maria* (34°S, 50.5°W) on the eastern side of Argentina. Ortelius apparently did not know where he should place *C. de Maria*. Both the S. America-Mercator (1569) and the S. America-Ortelius (1570) depict the shape of the South American mainland poorly. The shape of South American mainland improves a lot on the S. America-Plancius (1594), but *C. de S. Maria* (35°S, 47°W) is depicted on the eastern side of Argentina. There is no written historical record or a modern English name for this cape.
173	金特突	Jin Te Tu	*Quintete*	On the S. America-KWQ, 金特突 (47°S, 83.5°W) is in today's Chile. There is no written historical record or a modern English name for this place.

• 174	早沙家	Zao Sha Jia	Zaoshajia	On the S. America-KWQ, 早沙家 (42°S, 81.5°W); Zaoshajia is in today's Chile. There is no written historical record or a modern English name for this place.
• 175	椰子島	Sha Dao	Cocos Island; "Isla del Coco" is also known as "Treasure Island".	On the S. America-KWQ, 椰子島 (21°S, 92°W) should be today's Cocos Island (5.5°N, 87°W), off the Pacific coast of Costa Rica. Here the S. America-KWQ gives very poor latitudinal value for this island. The first European to discover this island was Juan de Cabezas (also known as Juan de Grado) in 1526.[491] But the island name does not show up on any of the three European maps.
176	孛露海	Bei Lu Hai	Sea of Peru	On the S. America-KWQ, the sea to the west of Peru is named 孛露海 (Sea of Peru). To the west of 孛露海 (Sea of Peru) lies 東南海 — the ocean to the west of 孛露海 (Sea of Peru) and to the southeast of China as shown on the full map of KWQ — which is today's South Pacific Ocean. In 1513, Vasco Núñez de Balboa (c. 1475–1519) was the first European to sight the Pacific Ocean,[492] which he called *Mar del Sur* (South Sea). In 1521, a Spanish expedition led by the Portuguese navigator Ferdinand Magellan was the first recorded European crossing of the Pacific Ocean, Magellan then naming it the "peace-ful sea". On the S. America-Mercator (1569), the sea to the west of Peru is depicted as *El Mar Pacifico*; on the S. America-Ortelius (1570), as *Mar Pacifico*; and on the S. America-Plancius (1594), also as *El Mar Pacifico*. The Sea of Peru really means the "Pacific Sea of Peru".

(Continued)

(Continued)

Item #	Name	Pinyin	Etymology	Analysis of history, politics, geography; comparison or comment
Zone VI: South America (Eastern Region)				

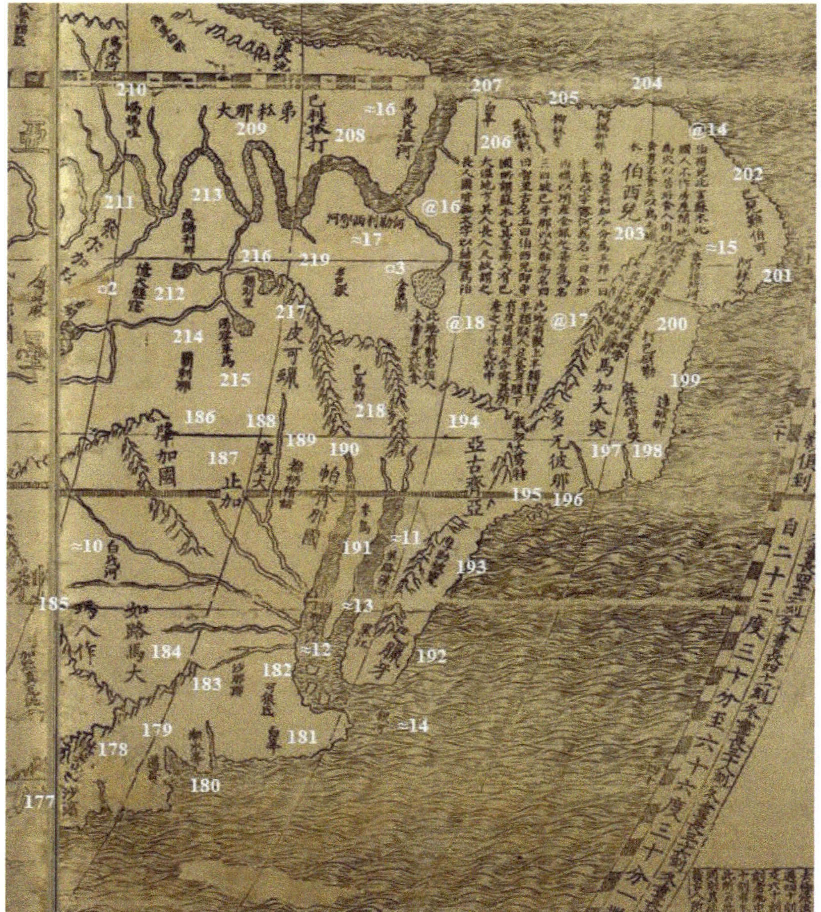

Fig. A5.39. Extracted from Fig. 5.3, bottom right panel, public domain.

177	沙濱	Sha Bin	Arenagarda[493]	沙 means "sand" and 濱, "coast". *Arena* in Spanish also means "sand". On the S. America-KWQ, 沙濱 (41°S, 60.5°W) may be today's Arenas Verdes (38.6°S, 58.6°W), which is a town near the South Atlantic coast in the province of Buenos Aires, Argentina. The Spanish navigators Juan Díaz de Solís (c. 1470–1516) and Sebastian Cabot (c. 1474–c. 1557) visited the territory that is now Argentina in 1516 and 1526, respectively.[494] *Arenagarda* is depicted on the S. America-Mercator (1569) at 37°S, 50°W and on the S. America-Ortelius (1570) at 37°S, 59.5°W, but not on the S. America-Plancius (1594).
• 178	其其山巴	Qi Qi Shan Ba	Chichichampa	On the S. America-KWQ, 其其山巴 (38°S, 61°W) is in today's Argentina. Now we start to see that the S. America-KWQ gives a longitudinal reading more eastward than the actual value of the place. There is no written historical record or a modern English name for this place.
• 179	邁哥	Mai Ge	Mayco[495]	On the S. America-KWQ, 邁哥 (39°S, 56.5°W) is in today's Argentina. There is no written historical record or a modern English name for this place.
• 180	潮水峯	Chai Shui Feng	Literally, it means "cape of tidal waters".	The name 潮水峯 (38°S, 54°W) is of Chinese origin. This headland should be in today's Argentina. This item should not be mixed up with 水潮峰 (Item 73; Cape Peirce) in Alaska. There is no written historical record or a modern English name for this cape.
181	白峯	Bai Feng	Spanish: *Peña Blanca*; English: White Rock; Italian: *Bianco, capo*	On the S. America-KWQ, 白峯 (38.1°S, 51°W) is in today's Argentina. On the S. America-Mercator (1569), *C. Blanco* is depicted at 36°S, 46.5°W, near *Rio de la Plata*; and on the S. America-Ortelius (1570), at 36°S, 52.5°W, also near *Rio de la Plata*. The S. America-Plancius (1594) does not depict this cape. There is no written historical record or a modern English name for this cape.

(Continued)

(*Continued*)

Item #	Name	Pinyin	Etymology	Analysis of history, politics, geography; comparison or comment
• 182	可狼氏	Ke Lang Di	Spanish: *Colante*; Italian: *Colamti*	On the S. America-KWQ, the headland 可狼氏 (35°S, 51.5°W) should be in today's Argentina. There is no written historical record or a modern English name for this place.
• 183	沙那耶	Sha Na Ye	Sciananes[496]	On the S. America-KWQ, 沙那耶 (36°S, 53.5°W) may be in today's Argentina or Uruguay. There is no written historical record or a modern English name for this place.
184	如路馬大	Ru Lu Ma Da	Giurumata[497]	On the S. America-KWQ, 如路馬大 (32°S, 60.5°W) should be in today's Argentina. On the S. America-Mercator (1569), Giurumata is depicted at 26°S, 72.5°W; on the S. America-Ortelius (1570), at 24°S, 75.5°W; and on the S. America-Plancius (1594), at 25°S, 66°W. There is no written historical record or a modern English name for this place.
185	瑪八作	Ma Ba Zuo	Mapazo[498]	On the S. America-KWQ, 瑪八作 (31.5°S, 64.5°W) should be in today's Argentina. On the S. Amarica-Mercator (1569), *Mapazo* is depicted at 7°S, 69.5°W; on the S. America-Ortelius (1570), at 8°S, 75°W; and on the S. America-Plancius (1594), at 6°S, 74.5°W. On these three European maps, Mapaza is depicted by the Amazon River in today's Brazil and close to the Equator, whereas 瑪八作 on the S. America-KWQ is in today's Argentina, quite a distance away. There is no written historical record or a modern English name for this place.

186	肇加國	Zao Jia Guo	Charcas	On the S. America-KWQ, 肇加國 (21°S, 62.5°W) was the ancient Indian Chiefdom of Charcas. The Andean area (a vast windy plateau of the central Andes in Peru and Bolivia) or today's Bolivia was known then as Charcas. By the fifteenth century, the region was controlled largely by independent nomadic groups of Aymara-speaking Indians. In c. 1430, the Inca emperor began conquests southward and Aymara territories ultimately formed a major part of the Inca Empire. The Inca Empire lasted from 1438 to 1533/1572.[499] The Spaniards conquered the Incas in 1532 and conquered Bolivia in 1538,[500] which became part of the Vice-royalty of Peru. When the son of the last monarch of the Neo-Inca State was executed by the Spanish in 1572, the Inca Empire ended.[501] Hence, we can be sure that by the time the three European cartographers drew their maps, Charcas had long gone from history and Inca was no longer an empire. The S. America-KWQ depicts only 肇加國/Charcas, not the Inca Empire, showing that the map reflects the political landscape prior to c. 1430 before Charcas was conquered by the Inca people. The S. America-Ortelius (1570) depicts Charcas as a huge country at 28°S, 85°W, further westward and southward in Peru or Bolivia than 肇加國 (21°S, 62.5°W) on the S. America-KWQ. The other two European maps do not depict Charcas. Why does Charcas appear on the S. America-Ortelius (1570) when it was long gone? This gives a strong implication that Ortelius, without knowing the true history of Charcas, copied an old Chinese map on which 肇加國/Charcas appears. Besides, his location for the kingdom is far from reality.

(Continued)

(*Continued*)

Item #	Name	Pinyin	Etymology	Analysis of history, politics, geography; comparison or comment
187	止加	Zhi Jia	Chinca or Chica[502]	On the S. America-KWQ, 止加 (23.5°S, 58°W) is almost right on the Tropic of Capricorn and in the inland area in today's Argentina. However, on the S. America-Ortelius (1570), Chinca is at 29.8°S, 93°W in today's Chile and by the Pacific Ocean coast. The other two European maps do not depict Chica/Chinca. On a modern map, there is Chinca (0.8°N, 79.6°W) in Ecuador.[503] But this location is very different from those on the S. America-KWQ and the S. America-Ortelius (1570). There is no written historical record or a modern English name for this place.
188	寧瓦大	Ning Wa Da	Ningatas[504]	On the S. America-KWQ, 寧瓦大 (21°S, 57.5°W) is in today's Argentina. On the S. America-Mercator (1569), Ningatas is depicted at 28°S, 55.5°W by *Rio Peti*; and similarly on the S. America-Ortelius (1570), at 28°S, 61°W by *Rio Peti*. The S. America-Palancius (1594) does not depict Ningatas. There is no written historical record or a modern English name for this place.
• 189	都柄稽詣	Du Bing Ji Yi	Tupimchii[505]	On the S. America-KWQ, 都柄稽詣 (23.1°S, 54.5°W) is in today's Argentina. There is no written historical record or a modern English name for this place.
• 190	帕齊那國	Pa Qi Na Guo	*Pésina, regno*[506]	On the S. America-KWQ, 帕齊那國 (24°S, 52.5°W) is in today's Argentina. There is no written historical record or a modern English name for this place.

<table>
<tr>
<td>• 191</td>
<td>半昰</td>
<td>Ban Dao</td>
<td>On the S. America-KWQ, the long strip of land separating the Paraná River and the Uruguay River beyond their confluence Punta Gorda is denoted as a peninsula (半島).</td>
<td>The Paraná River in South America flows north–south and becomes a flood plain between the Argentine provinces of Entre Ríos, Santa Fe and Buenos Aires before emptying into the Río de la Plata which starts at Punta Gorda (the confluence of the Uruguay and Paraná rivers). The Río de la Plata flows eastward into the South Atlantic Ocean, see Fig. A5.40. The Rio de la Plata forms part of the border between Argentina and Uruguay.

Fig. A5.40. Map showing the Río de la Plata drainage basin including major tributaries and cities (see endnote 507, public domain).
The long strip of land separating the Paraná River and the Uruguay River beyond their confluence Punta Gorda is denoted as a peninsula (半島) on the S. America-KWQ.</td>
</tr>
</table>

(*Continued*)

(*Continued*)

Item #	Name	Pinyin	Etymology	Analysis of history, politics, geography; comparison or comment
				Such a long strip of land also appears on all the three sixteenth century European maps discussed in this chapter, but none explicitly denotes it as a "peninsula". However, in today's map (Fig. A5.40), such a long strip of land or peninsula does not show up. Perhaps the four maps all exaggerated the length of that strip of land.
≈ 10	白氐河	Bai Di He	Peti River; Spanish: *Peti, Rio*[508]	On the S. America-KWQ, 白氐河 (*Rio Peti*) may be in today's Argentina. But the Technical University of Munich Database for Hydrological Time Series of Inland Waters (DAHITI) shows that the Peti River (8.8°S, 52.5°W)[509] is in the Amazon River Basin, not a tributary of the great river of Paraná in Argentina. The three European maps also depict *Rio Peti* as a tributary of the great river of Paraná. But on modern maps, there is no Peti River in Argentina.
≈ 11	吴路漢河	Wu Lu Han He	Portuguese: *Rio Uruguai.* The name "Uruguay" comes from the language Guaraní.	On the S. America-KWQ, 吴路漢河 is the Uruguay River (a major river in South America). It flows from north to south and forms parts of the boundaries of Brazil, Argentina and Uruguay. The river joins the Paraná River at Punta Gorda to form the *Río de la Plata,* which flows eastward into the South Atlantic Ocean. The Portuguese navigator Ferdinand Magellan (1480–1521)[510] reached the estuary in 1520 and explored it briefly before his expedition continued its circumnavigation of the globe. Between 1526 and 1529, the Italian explorer Sebastian Cabot (c. 1474–c. 1557)[511] made a detailed study of the estuary and explored the Uruguay and Paraná rivers. On the S. America-Mercator (1569), the name of the river is blurred and unrecognisable. On the S. America-Ortelius (1570), the river is depicted as *Rio de Vruan* (same as *Rio de Uruan*), but on the S. America-Plancius (1594) the river is unnamed.

≈ 12	伯剌那大江	Bo La Na Da Jiang	It is the great river of Paraná; Spanish: *El gran rio de Paraná*	The great river of Paraná[512] is the longest river in Argentina, and the third largest river in the American Continent, after the Mississippi in the United States and the Amazon in Brazil. The river flows from Brazil and through Paraguay and Argentina and its riverbed in the middle reaches is the natural border between Paraguay and Argentina. The river merges with the Paraguay River (the second major river of the *Rio de la Plata* Basin, after the Paraná River) and then farther downstream with the Uruguay River to form the *Río de la Plata*. The river is navigable and flows into the Atlantic. The first European to go up the Paraná River was the Venetian explorer Sebastian Cabot (c. 1474–c. 1557) in 1526.[513] The great river of Paraná is depicted on the three European maps, but on the S. America-Plancius (1594) it is unnamed.
≈ 13	黑江	Hei Jiang	*Río Negro* in Uruguay	黑江 (*Río Negro*) or Black River is the main tributary of the Uruguay River, see Fig. A5.40. Charrúas Amerindians were the first people to inhabit the Uruguay area. The arrival of Europeans in the region dates from 1516, when the territory was discovered by Spanish explorer Juan Díaz de Solís (c. 1470–1516), who sailed on *Río de la Plata*.[514] The Portuguese first explored the region of today's Uruguay in 1512–1513.[515] The first permanent settlement on the territory of today's Uruguay was founded by Spanish Jesuits in 1624 at Villa Soriano on the *Río Negro*.[516] The S. America-KWQ and the three European maps all depict 黑江/*Río Negro*, but on the S. America-Plancius (1594) the river is very short and unnamed.
≈ 14	銀河	Yin He	*Río de la Plata*	On the S. America-KWQ, 銀河 (35°S, 45.5°W) is the *Río de la Plata* (35°S, 55°W), which literally means "river of silver". 銀 means "silver" and 河, "river". The *Río de la Plata* begins at the confluence of the Uruguay and Paraná rivers at Punta Gorda and flows eastward into the South Atlantic Ocean. Though it is generally spoken of as a river, the *Río de la Plata* is considered by some geographers to be a large bay or marginal sea of the Atlantic Ocean.

(*Continued*)

(*Continued*)

Item #	Name	Pinyin	Etymology	Analysis of history, politics, geography; comparison or comment
				The Portuguese were the first Europeans to explore the *Río de la Plata* in 1512–1513, followed by the Spanish in 1516. The river received its name because the Spanish explorers acquired silver trinkets trading with the indigenous people in the region.[517]
192	巴臘牙	Ba La Ya	Plaia [518]	On the S. America-KWQ, 巴臘牙 (33°S, 46°W) is in today's Uruguay. Here the longitudinal reading is too far eastward, placing it in the South Atlantic Ocean. Plaia is also depicted on the S. America-Ortelius (1570) at 32°S, 50.5°W, which places it in Uruguay or southern Brazil. There is no written historical record or a modern English name for this place.
193	摩勒被甕	Mo Lei Bei Weng	Morpion[519]	On the S. America-KWQ, 摩勒被甕 (25°S, 45.5°W) is in today's South Brazil. Again, the longitudinal reading is eastward in the South Atlantic Ocean. Morpion is also depicted on the S. America-Ortelius (1570) at 23.5°S, 50°W in today's Brazil. There is no written historical record or a modern English name for this place.
194	亞古齊亞	Ya Gu Qi Ya	Acutia[520]	On the S. America-KWQ, 亞古齊亞 (22°S, 45°W) is in today's southern Brazil. On the S. America-Ortelius (1570), *Acvtia* (same as *Acutia*) is depicted as a vast area at 18°S, 55.5°W in western-central Brazil. In the book titled *History of Brazil*,[521] Robert Southey has a description of how the Spanish explorers reached Acutia.
• 195	峨勿大葛特	E Wu Da Ge Te	Ngoutacote[522]	On the S. America-KWQ, 峨勿大葛特 (20.5°S, 42.5°W) is in today's South Brazil. There is no written historical record or a modern English name for this place.

196	多兀彼那	Duo Wu Bi Na	Tuoupeina[523]	On the S. America-KWQ, 多兀彼那 (21°S, 40°W) is in today's central-southern Brazil. Tuoupeina is depicted as Tuouopina on the S. America-Plancius (1594) at 21°S, 39.5°W. There is no written historical record or a modern English name for this place.
• 197	馬加大突	Ma Jia Da Tu	Machiatatu[524]	On the S. America-KWQ, 馬加大突 (17°S, 38°W) is in today's eastern Brazil. There is no written historical record or a modern English name for this place.
@ 13	此處人臥無床褥得但結繩為網旁高中窪兩頭以木樁掛之偃臥其中行即為轎		Ci Chu Ren Wo Wu Chuang Ru De Dan Jie Sheng Wei Wang Pang Gao Zhong Wa Liang Tou Yi Mu Zhuang Gua Zhi Yan Wo Qi Zhong Xing Ji Wei Jiao	My translation and comment (between square brackets) are as follows: The indigenous people of this land do not sleep in beds with mattresses, but tie ropes to make nets [hammocks]. The sides of the net are higher than the centre and the net is hung from wooden poles at the two ends. They lie in these suspension beds and can be carried as in a litter. Christopher Columbus wrote that when he landed in the Caribbean, he found the indigenous people resting in hammocks, a bed made from cotton netting and suspended between two trees or poles.[525] Columbus's letter verifies that what he saw in the Caribbean was also true at 馬加大突 (Machiatatu) in South America. None of the three European maps contain similar descriptions about indigenous Americans' lifestyle and related inventions.
• 198	麻尔碍曷突	Ma Er Ai He Tu	Morghaehotu[526]	On the S. America-KWQ, 麻尔碍曷突 (18°S, 37°W) is in today's eastern Brazil. There is no written historical record or a modern English name for this place.
• 199	達坡那	Da Po Na	Tapuona[527]	On the S. America-KWQ, 達坡那 (18°S, 35°W) is in the eastern coast of Brazil. There is no written historical record or a modern English name for this place.

(Continued)

(*Continued*)

Item #	Name	Pinyin	Etymology	Analysis of history, politics, geography; comparison or comment
• 200	打巴砑勒	Da Ba Ya Le	Tapaialé[528]	On the S. America-KWQ, 打巴砑勒 (15°S, 36°W) is near the eastern coast of Brazil. There is no written historical record or a modern English name for this place.
• ≈ 15	革那非所河	Ge Na Fei Suo He	Italian: *Cona feisuo, fiume*[529]	On the S. America-KWQ, 革那非所河 (11°S, 34.5°W) may be the Salgado River (9.1°S, 35.2°W)[530] in today's Alagoas state of eastern Brazil. It flows into the Atlantic Ocean on the border between Japaratinga and Maragogi municipalities. The pronunciation of 革那非所河 is not close to that of Salgado River, but their coordinates and the flow direction of the river are similar.
201	阿林荅	A Lin Da	*Alinde, C. de*[531]	On the S. America-KWQ, 阿林荅 (10.5°S, 31.5°W) is in today's eastern Brazil. On the S. America-Ortelius (1570), *C. de Alinde* (2.5°S, 46°W) is depicted near the faraway outlet of *Maragnon fl.* There is no written historical record or a modern English name for this place.
202	巴兒難伯可	Ba Er Nan Bo Ke	*Pernambuco, R.*[532]: the origins of the name are debated, but must have been derived from the now-extinct Tupi language.[533]	On the S. America-KWQ, 巴兒難伯可 (7.5°S, 31°W) is Pernambuco (8.3°S, 37.8°W) in the eastern corner of today's Brazil. The S. America-KWQ depicts the place more eastward than its actual location. Prior to discovery and colonisation by Portugal, numerous Tupi-Guarani-speaking tribes of indigenous peoples lived in Pernambuco. Colonial Brazil comprises the period from 1500, with the arrival of the Portuguese, until 1815.[534] On the S. America-Mercator (1569), Pernambuco is depicted at 8.1°S, 29°W; on the S. America-Ortelius (1570), at 7.5°S, 37°W; and on the S. America-Plancius (1594), at 7°S, 34°W. There are no large differences between the coordinates on the S. America-KWQ and those on the three European maps.

203	伯西兒	Bc Xi Er	*Brazil*, *Brasilia* or *Bresilia*	On the S. America-KWQ, 伯西兒 (5.5°S, 38.0°W) is in today's eastern Brazil. As is explained in annotation @15, the Chinese name of Brazil was 蘇木 (Sumu), which was the name of Brazil's famous production of the special kind of hardwood (brazilwood). The name of the hardwood gave the territory its name.[535] Earlier European names for the land of what became Brazil started as *Ilha de Vera Cruz* ("Island of the True Cross"), upon the Portuguese discovery of the land in 1500. Later, the 1516 map of Martin Waldseemüller drops his earlier *America* designation and refers to it as *Brasilia sive Terra Papagalli* ("Brasilia, or the Land of Parrots"). The first "official" use of the term appears in 1516, when King Manuel I of Portugal invested the Portuguese captain Cristóvão Jacques as *governador das partes do Brasil* ("governor of the parts of Brazil").[536] On the S. America-Mercator (1569), *Bresilia* is depicted at 7°S, 37.5°W based on 1504 geographical information; on the S. America-Ortelius (1570), at 12.5°S, 49°W; and on the S. America-Plancius (1594) it is depicted as *Brasilia* at 14°S, 39.5°W. Mercator's coordinates based on 1504 geographical information are closest to those on the S. America-KWQ. These coordinates represent Portuguese earlier knowledge about this new land soon after its discovery in 1500.
@ 14	伯西兒此言蘇木此國人不作房屋開地爲穴以居好食人肉但食男不食女以鳥毛織衣		Bi Xi Er Ci Yan Su Mu Ci Guo Ren Bu Zuo Fang Wu Kai Di Wei Xue Yi Ju Hao Shi Ren Rou Dan Shi Nan Bu Shi Nu Yi Niao Mao Zhi Yi	My translation and comment (between square brackets) are as follows: Brazil is called Sumu [this is a kind of precious hardwood]. The indigenous people do not build houses but use caves as habitat. They like to eat human flesh, but only eat men's, not women's flesh. They use bird feathers to make clothing. In the book titled *True History: An Account of Cannibal Captivity in Brazil* (German: *Warhaftige Historia und beschreibung eyner Landtschafft der Wilden Nacketen, Grimmigen Menschfresser-Leuthen in der Newenwelt America gelegen*),[537] the German soldier Hans Staden in 1557 describes his two trips to the new world.

(*Continued*)

(*Continued*)

Item #	Name	Pinyin	Etymology	Analysis of history, politics, geography; comparison or comment
				While held captive by the Tupinambá near Curitiba, Brazil, he observed the Tupinambá nature, culture and lifestyle. Staden's experience was one of the first accounts of cannibalism by a European. The annotation on the S. America-KWQ reconfirms cannibalism practiced by the indigenous tribes in Brasilia. However, none of the three European maps annotates cannibalism in Brazil, showing their lack of real exploration experiences about this part of the world.
@ 15	南亞墨利加今分爲五邦一曰孛露以孛露河爲名二 曰金加西蠟以所産金銀之甚多爲名三曰坡巴牙那以大郡為名四曰智里古名五曰伯西兒即中國所謂蘇木也其至南又在巴大搵地方其人長八尺故謂之長人國皆無文字以結繩為治		Nan Ya Mo Li Jia Jin Fen Wei Wu Bang Yi Yue Bei Lu Yi Bei Lu He Wei Ming Er Yue Jin Jia Xi La Yi Suo Chan Jin Yin Zhi Shen Duo Wei Ming San Yue Po Ba Ya Na Yi Da Jun Wei Ming Si Yue Zhi Li Gu Ming Wu Yue Bo Xi Er Ji Zhong Guo Suo Wei Su Mu Ye Qi Zhi Nan You Zai Ba Da Wen Di Fang Qi Ren Chang Ba Chi Gu Wei Zhi Chang Ren Guo Jie Wu Wen Zi Yi Jie Sheng Wei Zhi	My translation and comment (between square brackets) are as follows: Currently, South America is divided into five states: the first is Peru, named after the rivers in Peru; the second is *Castilla de Oro*, named after its rich gold and silver resources; the third is Popayán, named after its grand county; the fourth is Chile, named after its ancient name; and the fifth is Brasilia, which is named by the Chinese as Sumu because the place produces sappanwood known by the Chinese as Sumu. Further to the south in Patagonia, people measure around eight *Chi* [one *chi* is about 1/3 m], hence the state is called "the Country of Tall People". They do not have writing and rule the country by tying knots. The above annotation explains the political landscape in South America prior to the arrival of the Europeans, because all these five states had existed before the Europeans arrived. Among the five state names, two have Chinese origins: Sumu is the ancient Chinese name for Brazil and the Country of Tall People is first recorded in the Chinese *Shan Hai Jing* around 400 B.C.; three are of indigenous origin: Peru, Popayán and Chile; only one is of European origin: *Castilla de Oro* or *del Oro,* which was the name given by the Spanish settlers at the beginning of the sixteenth century to the Central American territories from the Gulf of Urabá, near today's Colombian–Panamanian border, to the Belén River,

			as I have explained in Item 123. The entire continent is depicted as 南亞墨利加 (South America); this name, together with *Castilla de Oro,* was the name used by Matteo Ricci to edit his Chinese source map of the S. America-KWQ to make it look more like a European map of his era. It is particularly interesting and significant to know that this annotation is an important revealation: the Chinese called the then Brazil a different name, "Sumu", because the sappanwood grown there was named Sumu in China. How could Matteo Ricci know that the Chinese call Brazil a different name, Sumu, if they had not been there? After Zheng He's seventh voyage was over in the 1430s, the Ming emperor quickly imposed a Sea Ban, hence, no ocean-going ships could sail to the Americas. When the ban was lifted in 1567, there were no longer seafaring ships capable of carrying sailors and goods to more distant seas to reach the Americas. Hence, the annotation is telling us that before the 1430s, the Chinese had already sailed to Sumu!
@ 16	此地有獸上半類狸下半類猴人足梟耳腹下有皮可張可合容其所產之子休息於中	Ci Di You Shou Shang Ban Lei Li Xia Ban Lei Hou Ren Zu Xiao Er Fu Xia You Pi Ke Zhang Ke He Rong Qi Suo Chan Zhi Zi Xiu Xi Yu Zhong	My translation and comment (between square brackets) are as follows: In this land, there are wild animals. The upper half of their bodies is like a raccoon and the lower half is like a monkey. They have human-like feet and owl-like ears. Under their belly, they have a skin pocket [a belly pouch on the females of the marsupials] which can open and close to shelter the young. All marsupials including wallabies, kangaroos and opossums originated in South America, a genetic study suggests; but marsupials — a group of mammals known for toting their young in belly pouches on the females — are still common in South America, too.[538] None of the three European maps has annotations about marsupials in Brazil.

(*Continued*)

(*Continued*)

Item #	Name	Pinyin	Etymology	Analysis of history, politics, geography; comparison or comment
204	阿瑪加那	A Ma Jia Na	*Amazones*[539]; in Greek mythology, it is the Amazons (Ancient Greek: Ἀμαζόνες *Amazónes*). According to Greek mythology, *Amazons* were a warlike tribe of women.	On the S. America-KWQ, 阿瑪加那 (2°S, 39.5°W) is to the east of today's Amazon Basin,[540] which is the part of South America drained by the Amazon River and its tributaries. Many complex and large-scale indigenous societies, mainly chiefdoms, developed towns and cities around the region of the Amazon River. In 1500, Vicente Yáñez Pinzón (c. 1462–after 1514) was the first documented European to sail up the Amazon River.[541] In 1541, Francisco de Orellana (1511–1546) was the first European to travel from the origins of the upstream river basins, situated in the Andes, to the mouth of the river.[542] The name *Amazonas* is thought to be taken from the indigenous warriors that attacked this expedition, mostly women, that reminded him of the mythical Greek female Amazon warriors from the ancient Hellenic culture.[543] The name "Amazonas" was originally given to the Amazon River — *Río de las Amazonas* — the name *Amazones* was then used to denote the region where these woman warriors lived. On the S. America-Mercator (1569), *Amazones* is depicted at 8.5°S, 53.5°W; on the S. America-Ortelius (1570), at 8.5°S, 55°W; and on the S. America-Plancius (1594), at 10°S, 51.5°W. The three European maps depict *Amazones* as a region near the eastern side of Amazon River Basin, which is different from the 亞馬鑽國 (Item 145) in the northwestern side of the Basin depicted on the S. America-KWQ. The full basin is shown in Fig. A5.41.

				 Fig. A5.41. Map of the Amazon River Basin (the southern Guianas, not marked on this map, are a part of the basin; see endnote 544, public domain).
205	椰林峯	Ye Lin Feng	*Palmar, capo di*[545]	On the S. America-KWQ, 椰林峯 (2.5°S, 42°W) is in today's northeastern Brazil. 椰 means "coconut", 林, "forest" and 峯, "cape or peak". Both the palm tree and the coconut tree belong to the palm family, although there are differences in them. *C. de Palmar* is depicted on the S. America-Ortelius (1570) at 1°S, 40.5°W. There is no written historical record or a modern English name for this cape.

(*Continued*)

(*Continued*)

Item #	Name	Pinyin	Etymology	Analysis of history, politics, geography; comparison or comment
• 206	毗麻剌	Pi Ma La	*Pimala*[546]	On the S. America-KWQ, 毗麻剌 (2.6°S, 44.5°W) is in today's northeastern Brazil. There is no written historical record or a modern English name for this place.
207	白峯	Bai Feng	*Blanco, cabo*	On the S. America-KWQ, 白峯 (1°S, 46°W) correlates with today's *Morro Branco* (White Mountain; 9.5°S, 45°W)[547] in Oeiras, Piaui, northeast Brazil. 白 means "white" and 峯, "cape or peak". 白峯 means *Blanco, cabo* or "White Cape" or "White Mountain". On the S. America-Ortelius (1570), *Blanco, cabo* (1°N, 58°W) is north of the Equator and too far to the west. There is no written historical record or a modern English name for this cape or mountain.
≈ 16	馬良温河	Ma Liang Wen He	*Maragnon fl.* or *Río Maragnon*; English: Maragnon River	The Amazon River was initially known by Europeans as the *Marañón*, and the Peruvian part of the river is still known by that name today, see A41. It later became known as *Río de las Amazonas* in Spanish and Portuguese. For most of the eighteenth–nineteenth and into the twentieth century, the *Marañón* was generally considered the source of the Amazon. The Amazon basin[548] is the largest drainage basin in the world. The river system originates in the Andes Mountains of Peru and travels through Ecuador, Colombia, Venezuela, Bolivia and Brazil before emptying into the Atlantic Ocean. About 80 percent of the basin is covered with Amazon rainforest.[549] It is estimated that 400 or more tribal groups lived here. The origin of the river name has been explained in Items 145 and 204. On the S. America-Ortelius (1570), there are two rivers south of the Equator and they are in the northern part of South America, flowing into the Atlantic Ocean: *Maragnon fl.* and *Amazomun uel oregliana fl.* But these two rivers are connected near Picova.

				On the S. America-Plancius (1594), south of the Equator and in the northern part of South America, there is only the *Maragnon fl.*, flowing into the Atlantic Ocean, but it is joined by a major tributary named *Maragnõ fl.* at *Picora.* The situation is like that depicted on the S. America-Mercator (1569) and the S. America-KWQ. When comparing the Amazon River on a modern map shown in Fig. A5.41, we can see that the Amazon River has several major southern tributaries, but only Ortelius shows two rivers flowing into the Atlantic Ocean on his 1570 map.
• 208	巴利孤打	Ba Li Gu Da	Paricuta[550]	On the S. America-KWQ, 巴利孤打 (2.5°S, 56.5°W) seems to be erroneously placed in today's Brazil. Paricuta[551] was in the British colony of British Guiana, on the northern coast of South America and to the north of the Equator. Since 1966, British Guiana has been known as the independent nation of Guyana.
209	弟私那大	Di Si Na Da	Tisnada[552]	On the S. America-KWQ, 弟私那大 (2.5°S, 60°W) is close to the Amazon River and shculd be in today's Brazil. On the S. America-Mercator (1569), Tisnada is depicted at 5°S, 52.5°W; on the S. America-Ortelius (1570), at 1°S, 60.5°W; and on the S. America-Plancius (1594), at 2.5°S, 56°W. There is no written historical record or a modern English name for this place.
210	喝祃哇	Kuai Ma Wa	Aiaui[553]	On the S. America-KWQ, 喝祃哇 (2.6°S, 67°W) should be in today's Brazil. On the S. America-Ortelius (1570), there is *Aiauiri Sama* at 2.5°S, 67°W; and on the S. America-Placius (1594), there is *Aiauiri Sama* at 2°S, 68°W. Both have similar coordinates like that of 喝祃哇, hence, they may denote the same place with different names. There is no written historical record or a modern English name for this cape or mountain.

(*Continued*)

(*Continued*)

Item #	Name	Pinyin	Etymology	Analysis of history, politics, geography; comparison or comment
● 211	察尔加私	Cha Er Jia Si	Chaerjiasi	On the S. America-KWQ, the latitude of 察尔加私 (9.5°S, 69.5°W) is too far north than Charcas (18.3°S, 65.8°W), which has been identified as 肇加國 (Charcas; Item 186; 21°S, 62.5°W). Hence, 察尔加私 is not Charcas. There is no written historical record or a modern English name for this place.
● ¤ 2	麻剌加岳湖	Ma La Jia Yue Hu	*Maracayo, lago*[554]	On the S. America-KWQ, 麻剌加岳湖 (15°S, 69.5°W) may be in today's Brazil, because the map can produce a longitudinal error around 10° further west than a modern map. There is no written historical record or a modern English name for this lake.
● 212	億大盤窪	Yi Da Pan Wa	Itapanua[555]	On the S. America-KWQ, 億大盤窪 (11.5°S, 66°W) most likely was in today's Brazil. There is no written historical record or a modern English name for this place.
● 213	皮諾利那	Pi Nuo Li Na	Pinolina[556]	On the S. America-KWQ, 皮諾利那 (8°S, 63.5°W) should be in today's Brazil. There is no written historical record or a modern English name for this place.
214	霸剌那	Ba La Na	Paraná	On the S. America-KWQ, 霸剌那 (16°S, 63.4°W; Paraná) is an inland area. On modern maps, there is Paraná (31.7°S, 60.5°W) in Argentina,[557] which is an island city. But there is also a Paraná (24°S, 51°W) in Brazil,[558] which is bordered on the east by the Atlantic Ocean. The Paraná on the S. America-KWQ is inland, which should correspond to the Paraná in Argentina, but its latitude shows a large error; this is somewhat unusual on the S. America-KWQ.

				Such a latitudinal error is a little smaller for *Parana* (19.5°S, 65.5°W) depicted on the S. America-Ortelius (1570), but the longitudinal error is a little larger, and it is a huge area. Neither the S. America-KWQ nor the S. America-Ortelius (1570) gives satisfactory coordinates of Paraná. Before the discovery of this region by European explorers in the sixteenth century, indigenous populations inhabited the region for thousands of years. Colonisation of the state by settlers started in the sixteenth century.
• 215	渴察革馬	Ke Cha Ge Ma	Cociacoma[559]	On the S. America-KWQ, 渴察革馬 (15°S, 60.5°W) should be in today's Argentina. There is no written historical record or a modern English name for this place.
• 216	題別里	Tie Bie Li	Tipilli[560]	On the S. America-KWQ, 題別里 (12°S, 59.5°W) may be in today's Brazil. There is no written historical record or a modern English name for this place.
217	皮可蠟	Pi Ke La	Picora[561]	On the S. America-KWQ, 皮可蠟 (14°S, 56.5°W) should be in today's Brazil. On the S. America-Mercator (1569), Picora is depicted at 8.5°S, 59.5°W; on the S. America-Ortelius (1570), at 8.5°S, 64°W; and on the S. America-Plancius (1594), at 9°S, 65°W. There is no written historical record or a modern English name for this place.
• 218	巴烏的	Ba Wu De	Pauti[562]	On the S. America-KWQ, 巴烏的 (17°S, 52.5°W) may be in today's Brazil or Uruguay. There is no written historical record or a modern English name for this place.
• 219	多巴欲	Duo Ba Yu	Tuopaiü[563]	On the S. America-KWQ, 多巴欲 (12°S, 54°W) should be in today's Brazil. There is no written historical record or a modern English name for this place.

(Continued)

(*Continued*)

Item #	Name	Pinyin	Etymology	Analysis of history, politics, geography; comparison or comment
≈ 17	何勒利西那河	He Lei Li Xi Na He	*Oregliana fl.*	On the S. America-KWQ, 何勒利西那河 is a tributary flowing into the great 馬良温河 (Amazon River; Item ≈ 16) and emptying into the Atlantic Ocean. The first European to discover this river was Vicente Yáñez Pinzón (c. 1441–c. 1493), in 1500[564]; Francisco de Orellana (1511–1546) surveyed its length and course in 1542 and gave the river its name after reporting pitched battles with tribes of female warriors.[565] On the S. America-Mercator (1569), there are four named tributaries plus one major tributary (unnamed) of the unnamed Amazon River. But none of the four named tributaries is named *Oregliana fl.* However, on the S. America-Ortelius (1570), there are two major rivers to the south of the Equator and in the northern part of South America, flowing into the Atlantic Ocean: the *Maragnon fl.* and the *Amazomun uel oregliana fl.*; and these two rivers are connected at Picora. This river structure is *unrealistic* and completely different from that depicted on the S. America-KWQ. It is also inconsistent with the modern map. On the S. America-Plancius (1594), there is a *Maragnon flu*, and a long and vast southern tributary named *Maragnõ fl.* joined at *Picora.* But to the east of *Maragnon flu*, there are two unnamed tributaries. Their confluences with the Amazon River are at 8°S, 58°W and 6.5°S, 54°W, respectively. The former can be identified as 何勒利西那河 (Item ≈ 17; its confluence with 馬良温河, Item ≈ 16, is at 7.5°S, 50.5°W). 何勒利西那河 would correspond to Tapajósa river on the modern map shown in Fig. A5.41, which is a major tributary of the Amazon River. Its confluence with the Amazon River is at Santarém (2.4°S, 54.7°W). The river systems of the S. America-KWQ and the three European maps are different, especially on the S. America-Ortelius (1570).

● ¤ 3	金魚湖	Jin Yu Hu	Spanish: *Dorado, laguna del.*[566] English: Goldfish Lagoon	The name 金魚湖 is of Chinese origin. It means "Goldfish Lagoon/Lake". On the S. America-KWQ, it is depicted as the source of 何勒利西那河 (*Oregliana fl.*; Item ≈ 17). It is peculiar that the Amazon River and its tributaries flow into or out of six lakes on the S. America-KWQ. This is not seen on the other three European maps: on the S. America-Mercator (1569), no such lakes are depicted; on the S. America-Ortelius (1570), only three small lakes in the northwestern region of the South American continent are shown; and on the S. America-Plancius (1594), only one lake is depicted at the source of *Maragnon flu.* The Amazon basin has few large lakes surrounded by highlands. But along the Amazon River in South America, oxbow lakes (a U-shaped lake or pool forms when a wide meander of a river is cut off, creating a free-standing body of water) are common in river bends, and their still waters provide a unique habitat for many species of fish and aquatic animals. We can read (see @17) to find that indeed animals like apes live around the Goldfish Lagoon.
@ 17	此地有獸名猿人常見其飲食		Ci Di You Shou Ming Yuan Ren Chang Gian Qi Yin Shi	My translation is as follows: In this land, there is a kind of animal called ape which is frequently seen eating and drinking around. Apes were originally distributed in Africa and Southeast Asia, and eventually spread all over the world through long migrations. More than thirty million years ago, monkeys rafted across the Atlantic to South America.[567]
● 220	南亞墨利加	Nan Ya Mo Li Jia	South America	See explanation in Item 58.

Endnotes

[1]Library of Congress. "Kun yu wan guo quan tu." www.loc.gov, Library of Congress, 2022, loc.gov/item/2010585650.

[2]Ronnie Po-Chia Hsia. *Matteo Ricci and the Catholic Mission to China, 1583–1610: A Short History with Documents*. Indianapolis, IN, USA: Hackett Publishing Company, Inc., 2016.

[3]Eugenio Garin. "Ritratto Di Paolo dal Pozzo Toscanelli." *JSTOR*, vol. 12, no. 3, 1957, pp. 241–257. https://www.jstor.org/stable/26068390.

[4]The private collection of Roy Winkelman. "Atlantic Ocean, Toscanelli, 1474." https://etc.usf.edu/, Maps ETC, 2022, etc.usf.edu/maps/pages/10200/10258/10258.htm.

[5]"File:500 map by Juan de la Cosa rotated.jpg." *Wikipedia Commons: The Free Media Repository*, Wikimedia Foundation, 3 Jan. 2021, upload.wikimedia.org/wikipedia/commons/4/4a/1500_map_by_Juan_de_la_Cosa_rotated.jpg.

[6]National Geographic Society. "North America: Physical Geograph." https://education.nationalgeographic.org/, National Geographic Society, 2022, education.nationalgeographic.org/resource/north-america-physical-geography; Thomas F. McIlwraith, *et al. North America: The Historical Geography of a Changing Continent*. Lanham, MD, USA: Rowman & Littlefield Publishers, 2001.

[7]National Geographic Society Maps. *National Geographic: South America Classic Enlarged Wall Map*. Washington, DC, USA.: National Geographic Maps, 2016.

[8]Thomas T. Veblen, *et al. The Physical Geography of South America*. Oxford, UK: Oxford University Press, 2007.

[9]Felipe Fernández-Armesto. *Amerigo: The Man Who Gave His Name to America*. New York, NY, USA: Random House, 2008.

[10]"File:Waldseemuller map closeup with America.jpg." *Wikipedia Commons*, 3 Mar. 2014, upload.wikimedia.org/wikipedia/commons/d/db/Waldseemuller_map_closeup_with_America.jpg.

[11]Jennifer Raff. *Origin: A Genetic History of the Americas*. New York, NY, USA: Twelve, 2022.

[12]Jennifer Raff. "Journey into the Americas." *Scientific America*, vol. 324, no. 5, May 2021, pp. 24–29.

[13]Charlotte Harris Rees. "Chapter 11: Pacific Ocean currents." *Secret Maps of the Ancient World*, Bloomington, Ind., U.S.A.: AuthorHouse, 2008, pp. 89–91.

[14]Jeremy DeSilva. *Did the First Americans Arrive via Land Bridge?* New York, NY, USA: Twelve, 2022.

[15]Noel B. Gerson and Philip Vail. *The Magnificent Adventures of Henry Hudson*. UK: Sapere Books, 2021.

[16]Roberto Velasco Alonso, *et al. Aztec Empire*. New York, NY, USA: Guggenheim Museum, 2004.

[17]"Mercator 1569 world map." *Wikipedia Commons*, 11 Apr. 2022, upload.wikimedia.org/wikipedia/commons/4/4b/Mercator_1569_world_map_composite.jpg.

[18]Ortelius, Abraham, Aegid. Coppenius Diesth, and Humphrey Llwyd. *Theatrvm Orbis Terrarvm*. Antverpiae: Apud Aegid. Coppenium Diesth, 1570. [Map] Retrieved from the Library of Congress, www.loc.gov/item/98687183/; loc.gov/resource/g3200m.gct00003/?sp=18&st=single.

[19]"File:1594 double hemisphere world map by Petrus Plancius.jpg." *Wikipedia: The Free Encyclopedia*, Wikimedia Foundation, 21 Mar. 2020, upload.wikimedia.org/wikipedia/commons/0/0d/1594_double_hemisphere_world_map_by_Petrus_Plancius.jpg.

[20]"File:Kunyu Wanguo Quantu by Matteo Ricci Plate 4-6.jpg." *Wikipedia Commons*, 23 Oct. 2020, upload.wikimedia.org/wikipedia/commons/b/b3/Kunyu_Wanguo_Quantu_by_Matteo_Ricci_Plate_4-6.jpg.

[21]Andrew Burke and David Else. *The Gambia & Senegal*, Fort Mill, SC, USA: Lonely Planet, 2002, p. 236.

[22]"Patagonian Giants." https://lib-dbserver.princeton.edu, Princeton University, 2022, lib-dbserver.princeton.edu/visual_materials/maps/websites/pacific/magellan-strait/patgonian-giants.html.

[23]Anonymous. Anne Birrell, ed. and trans. *The Classic of Mountains and Seas*. London, UK: Penguin Classics, 1999.
[24]*Ibid.*
[25]"File:South America southern tip pol.png." *Wikipedia*, 30 Apr. 2016, upload.wikimedia.org/wikipedia/commons/1/17/South_America_southern_tip_pol.png.
[26]The Editors. "The Strait of Magellan: 250 Years of Maps (1520–1787)." http://lib-dbserver.princeton.edu, Princeton University, 2022, lib-dbserver.princeton.edu/visual_materials/maps/websites/pacific/magellan-strait/magellan-strait-maps.html.
[27]Richard Hakluyt. *The Principal Navigations, Voyages and Discoveries of the English Nation (Works. Extra Series/Hakluyt Society)*. Cambridge, UK: Cambridge University Press, 1965.
[28]Margot, *et al*. "Evidence for European presence in the Americas in AD 1021." *Nature*, vol. 601, 2022, pp. 388–391. https://doi.org/10.1038/s41586-021-03972-8.
[29]Peter Firstbrook. *The Voyage of the Matthew: John Cabot and the Discovery of North America*. Toronto, Ontario, Canada: McClelland & Stewart, 1997.
[30]Jacques Cartier, *et al*. *The Voyages of Jacques Cartier*. Toronto, Ontario, Canada: University of Toronto Press, 1993.
[31]"File:Lake Melville-Hamilton Inlet-Groswater Bay.png." *Wikipedia Commons*, 3 Oct. 2020, commons.wikimedia.org/wiki/File:Lake_Melville-Hamilton_Inlet-Groswater_Bay.png.
[32]Sheng-Wei Wang. *The Last Journey of the San Bao Eunuch, Admiral Zheng He*. Hong Kong, China: Proverse Hong Kong, 2019.
[33]"File:Mississippi watershed map 1.jpg." *Wikipedia Commons*, 18 Oct. 2020, commons.wikimedia.org/wiki/File:Mississippi_watershed_map_1.jpg.
[34]Laura M. Chmielewski. *Jacques Marquette and Louis Jolliet: Exploration, Encounter, and the French New World*. Oxfordshire, UK: Routledge, 2017; August William Derleth and H. Lawrence Hoffman. *Father Marquette and the Great Rivers*. San Francisco, CA, USA: Ignatius Press, 1998.
[35]Charles Hudson, *et al*. Lawrence A. Clayton, *et al*., eds. *The De Soto Chronicles: The Expedition of Hernando de Soto to North America in 1539–1543*. Tuscaloosa, AL, USA: University Alabama Press, 1995.
[36]Charles Hudson. *Knights of Spain, Warriors of the Sun: Hernando de Soto and the South's Ancient Chiefdoms*. Athens, GA, USA: University of Georgia Press, 1998; Marvin T. Smith (Valdosta State University). "Late Prehistoric/Early Historic Chiefdoms (ca. A.D. 1300–1850)." www.georgiaencyclopedia.org, Georgia Humanities, University of Georgia Press, 2022, georgiaencyclopedia.org/articles/history-archaeology/late-prehistoric-early-historic-chiefdoms-ca-a-d-1300-1850.
[37]*Ibid.*
[38]"Cusco: Location within Peru." *Wikipedia*, 2 July 2022, upload.wikimedia.org/wikipedia/commons/thumb/7/7b/Peru_location_map.svg/1682px-Peru_location_map.svg.png.
[39]"File:Kigdomofcuscomap.JPG." *Wikimedia Commons*, 11 Oct. 2020, commons.wikimedia.org/wiki/File:Kigdomofcuscomap.JPG.
[40]UNESCO World Heritage Centre. "City of Cuzco." https://whc.unesco.org/, UNESCO World Heritage Convention, 2022, whc.unesco.org/en/list/273.
[41]Thomas C. Patterson. *The Inca Empire: The Formation and Disintegration of a Pre-Capitalist State*. Oxford, UK: Berg Publishers, 1992.
[42]Roberto Velasco Alonso, *et al*. *Aztec Empire*. *Op. cit*.
[43]"File:Zhenghemap.jpg." *Wikimedia Commons*, 23 Oct. 2021, commons.wikimedia.org/wiki/File:Zhenghemap.jpg.
[44]Sheng-Wei Wang. *The Last Journey of the San Bao Eunuch, Admiral Zheng He*. *Op. cit*.

[45]Sheng-Wei Wang. "Chapter five: Travelling from the Gulf of St. Lawrence to the country of Fengdu." *The Last Journey of the San Bao Eunuch, Admiral Zheng He*, *op. cit*., pp. 170–176.
[46]William C. Massey. "Tribes and languages of Baja California." *Southwestern Journal of Anthropology*, vol. 5, no. 3, 1949, pp. 272–307. http://www.jstor.org/stable/3628506.
[47]Mike Walker, *et al*. "Formal definition and dating of the GSSP (Global Stratotype Section and Point) for the base of the Holocene using the Greenland NGRIP ice core, and selected auxiliary records." (PDF) *Journal of Quaternary Science*, vol. 24, no. 1, 2009, pp. 3–17. https://onlinelibrary.wiley.com/doi/abs/10.1002/jqs.1227.
[48]Junxuan Fan and Xudong Hou. "International Chronostratigraphic Chart." International Commission on Stratigraphy. Archived from the original on 13 Jan. 2017. Retrieved 18 June 2016.
[49]Randall Ray Loudamy. *Cortés after the Conquest of Mexico: Constructing Legacy in New Spain*. Submitted to the Faculty of the Graduate College of the Oklahoma State University in partial fulfilment of the requirements for the Degree of DOCTOR OF PHILOSOPHY, Dec. 2013. https://shareok.org/bitstream/handle/11244/14979/Loudamy_okstate_0664D_13123.pdf?sequence=1&isAllowed=y.
[50]Woodes Rogers. *A Cruising Voyage Round the World: The Adventures of an English Privateer*. Torrington, WY, USA: The Narrative Press, 2004.
[51]*Ibid*.
[52]Rolando González-José, *et al*. "Craniometric evidence for Palaeoamerican survival in Baja California." *Nature*, vol. 425, 2003, pp. 62–65; Paul Rivet. "Recherches anthropologiques sur la Basse-Californie." *Journal de la Société des Américanistes de Paris*, vol. 6, 1909, pp. 147–253.
[53]Sheng-Wei Wang. *The Last Journey of the San Bao Eunuch, Admiral Zheng He*. *Op. cit*.
[54]"File:Girolamo de Verrazzano's 1529 map of the East Coast of America.jpg." *Wikipedia Commons*, 10 Nov. 2020, commons.wikimedia.org/wiki/File:Girolamo_de_Verrazzano's_1529_map_of_the_East_Coast_of_America.jpg.
[55]Emerson W. Baker. *American Beginnings: Exploration, Culture, and Cartography in the Land of Norumbega*, Lincoln, NE, USA: University of Nebraska Press, 1994, p. 87; Arthur James Weise. *The Discoveries of America to the Year 1525*, New York, NY, USA: G. P. Putnam's Sons, 1884, pp. 357–361.
[56]Eben Norton Horsford. *The Discovery of the Ancient City of Norumbega: A Communication to the President and Council of the American Geographical Society (1890)*. Whitefish, MT, USA: Kessinger Publishing, LLC, 2010.
[57]William D. Williamson. *The History of the State of Maine; From Its First Discovery, A.D. 1602, to the Separation, A.D. 1820, Inclusive. With an Appendix and General Index; Volume 2*. Sydney, New South Wales, Australia: Wentworth Press, 2016, pp. 27, 266, 293.
[58]T. J. Brasser. "Early Indian-European Contacts." In Bruce G. Trigger, ed. *Handbook of North American Indians. Vol. 15: Northeast*, Washington, DC, USA: Smithsonian Institution Press, 1978, pp. 79–80.
[59]"File:Acadia 1754.png." *Wikipedia Commons*, 17 July 2021, upload.wikimedia.org/wikipedia/commons/9/93/Acadia_1754.png.
[60]Jack Anderson. "History of the St. Mary's River." www.stmarysriverassociation.com, St. Mary's River Association, 2022, stmarysriverassociation.com/uploads/6/7/6/5/67652351/history_of_the_st._marys_river__nxpowerlite_copy_.pdf.
[61]*Ibid*.
[62]Huang Shijian (黄时鉴) and Gong Yingyan (龚缨晏). *Li Ma Dou Shi Jie Di Tu Yan Jiu*《利玛窦世界地图研究》or *Research on Matteo Ricci's World Map*. Shanghai, China: Shanghai Chinese Classics Publishing House (上海古籍出版社), 2004, p. 201.

[63]Robert Leavitt. "Maliseet and Mi'kmaq Languages." www2.gnb.ca, Aboriginal Affairs of New Brunswick, Canada, 2022, www2.gnb.ca/content/gnb/en/departments/aboriginal_affairs/wolastoqiyik/languages.html.
[64]Ramsay Cook, ed. *The Voyages of Jacques Cartier*. Toronto, Ontario, Canada: University of Toronto Press (Scholarly Publishing Division), 2017.
[65]Maurice A. Mook. "Algonquian Ethnohistory of the Carolina Sound." *Journal of the Washington Academy of Sciences*, vol. 34, no. 6, 1944, pp. 181–228.
[66]Frederick Webb Hodge. *Handbook of American Indians North of Mexico*. Scituate, MA, USA: Digital Scanning Inc., 2003.
[67]Huang Shijian (黄时鉴) and Gong Yingyan (龚缨晏). *Op. cit.*, p. 202.
[68]U.S. Geological Survey. "National Hydrography Dataset high-resolution flowline data." https://apps.nationalmap.gov/, U.S. Geological Survey, 2022, apps.nationalmap.gov/viewer/.
[69]Christopher C. Meyers and David Williams. *Georgia: A Brief History*. Macon, GA, USA: Mercer University Press, 2012.
[70]"File:Gacolony.png." *Wikipedia Commons*, 30 Sept. 2022, commons.wikimedia.org/wiki/File:Gacolony.png.
[71]Huang Shijian (黄时鉴) and Gong Yingyan (龚缨晏). *Op. cit.*, p. 202.
[72]Robert H. Fuson. *Juan Ponce de Leon and the Spanish Discovery of Puerto Rico and Florida*. Lincoln, NE, USA: University of Nebraska Press, 1999.
[73]*Ibid.*
[74]John H. Hann and Bonnie G. Mcewan. *The Apalachee Indians and Mission San Luis*. Gainesville, FL, USA: University Press of Florida, 1998.
[75]Alvar Nunez Cabeza de Vaca. Harold Augenbraun, ed. Fanny Bandelier, trans. *Chronicle of the Narvaez Expedition*. London, UK: Penguin Classics, 2002.
[76]Paul Schneider. *Brutal Journey: The Epic Story of the First Crossing of North America*, New York, NY, USA: Henry Holt, 2006, pp. 102–103.
[77]Aaron Davis. J. W. Williamson, ed. "On the Naming of Appalachia." *An Appalachian Symposium: Essays Written in Honor of Cratis D. Williams*. Boone, NC, USA: Appalachian State University Press, 1977.
[78]David Ewing Duncan. *Hernando De Soto: A Savage Quest in the Americas*. New York, NY, USA: Crown, 1996; Charles Hudson, *et al.* Lawrence A. Clayton, *et al.*, eds. *The De Soto Chronicles: The Expedition of Hernando de Soto to North America in 1539–1543. Op. cit.*
[79]"File:MobileAlabamaCoosa3.png." *Wikipedia Commons*, 15 Sept. 2020, upload.wikimedia.org/wikipedia/commons/3/32/MobileAlabamaCoosa3.png.
[80]Marcius Willson. *American History: Comprising Historical Sketches of the Indian Tribes; A Description of American Antiquities (1855)*. Whitefish, MT, USA: Kessinger Publishing, LLC, 2007.
[81]Huang Shijian (黄时鉴) and Gong Yingyan (龚缨晏). *Op. cit.*, p. 195.
[82]Huang Shijian (黄时鉴) and Gong Yingyan (龚缨晏). *Op. cit.*, p. 200.
[83]*Ibid.*
[84]"File:1562 Americæ Gutiérrez.JPG." *Wikipedia Commons*, 28 Apr. 2020, commons.wikimedia.org/wiki/File:1562_Americæ_Gutiérrez.JPG.
[85]Marvin T. Smith (Valdosta State University). "Late Prehistoric/Early Historic Chiefdoms (ca. A.D. 1300–1850)." *Op. cit.*
[86]Robbie Franklyn Ethridge. *From Chicaza to Chickasaw: The European Invasion and the Transformation of the Mississippian World, 1540–1715*, Chapel Hill, NC, USA: University of North Carolina Press, 2010, p. 62.

[87]"File:DeSoto Map Leg 2 HRoe 2008.jpg." *Wikipedia Commons*, 29 Sept. 2020, commons.wikimedia.org/wiki/File:DeSoto_Map_Leg_2_HRoe_2008.jpg.
[88]"File:OrteliusWorldMap1570.jpg." *Wikipedia Commons*, 26 May 2009, en.wikipedia.org/wiki/File:OrteliusWorldMap1570.jpg.
[89]Huang Shijian (黄时鉴) and Gong Yingyan (龚缨晏). *Op. cit*., p. 207.
[90]Trails of History. "Discover the Monongahela Culture Archaeology of Southwestern, Pennsylvania." http://twipa.blogspot.com/, The State Museum of Pennsylvania, 2022, twipa.blogspot.com/2019/10/discover-monongahela-culture.html; James B. Richardson III, *et al*. "The disappearance of the Monongahela: Solved?" *Archaeology of Eastern North America*, vol. 30, 2002, pp. 81–96. https://www.jstor.org/stable/40914458; Varna G. Boyd and Kathleen A. Furgerson. Katherine Marie Benitez and Amy Kristina Benitez, eds. "The Mystery of the Monongahela Indians." http://www.phmc.state.pa.us, The United States Department of Transportation, The Federal Highway Administration and The Pennsylvania Department of Transportation, 2022, www.phmc.state.pa.us/portal/communities/archaeology/files/mysteryofmonindians.pdf.
[91]"File:Fort Ancient Monongahela cultures HRoe 2010.jpg." *Wikipedia Commons*, 21 Sept. 2020, commons.wikimedia.org/wiki/File:Fort_Ancient_Monongahela_cultures_HRoe_2010.jpg.
[92]Huang Shijian (黄时鉴) and Gong Yingyan (龚缨晏). *Op. cit*., p. 193.
[93]John A. Dickinson and Brian Young. *A Short History of Quebec*. Montreal, Quebec, Canada: McGill-Queen's University Press, 2008.
[94]Jacques Cartier, *et al*. *The Voyages of Jacques Cartier*. *Op. cit*.
[95]"File:Saint Lawrence map-fi.png." *Wikipedia Commons*, 23 Oct. 2021, upload.wikimedia.org/wikipedia/commons/e/ee/Saint_Lawrence_map-fi.png.
[96]Anonymous. *The Definitive Treaty of Peace and Friendship between His Britannick Majesty, the Most Christian King, and the King of Spain [microform]: Concluded at Paris the 10th day of February, 1763; to Which the King of Portugal Acceded on the Same Day*. Auckland, New Zealand: Legare Street Press, 2021.
[97]R. Douglas Francis, *et al*. *Journeys: A History of Canada*, Boston, MA, USA: Cengage Learning, 2009, p. 51.
[98]Canadian Museum of History. "The Explorers Étienne Brûlé 1615–1621." www.historymuseum.ca, Virtual Museum of New France, 2022, historymuseum.ca/virtual-museum-of-new-france/the-explorers/etienne-brule-1615-1621.
[99]Willem & Jan Blaeu. "Willem & Jan Blaeu: Extrema Americae Versus Boream ubi Terra Nova Nova Francia." www.themaphouse.com, The Map House, 2022, themaphouse.com/search_getamap.aspx?id=111921&ref=CAN2763.
[100]Clements R. Markham. *The Journal of Christopher Columbus during His First Voyage 1492–1493: Documents Relating the Voyages of John Cabot and Gaspar Corte Real 1893*. London, UK: Facsimile Publisher, 2015.
[101]*Ibid*.
[102]Peter Firstbrook. *The Voyage of the Matthew: John Cabot and the Discovery of North America*. *Op. cit*.
[103]Kevin Major. *As Near to Heaven by Sea: A History of Newfoundland and Labrador*. Toronto, Ontario, Canada: Penguin Canada, 2002.
[104]"File:Gerard de Jode 1593 Map Northern hemisphere.jpg." *Wikipedia Commons*, 23 Oct. 2021, commons.wikimedia.org/wiki/File:Gerard_de_Jode_1593_Map_Northern_hemisphere.jpg.
[105]"File:Le Canada ou Nouvelle France.jpg." *Wikipedia Commons*, 24 Oct. 2021, commons.wikimedia.org/wiki/File:Le_Canada_ou_Nouvelle_France.jpg.
[106]"File:Map by nicolo zeno 1558.jpg." *Wikipedia Commons*, 4 Dec. 2022, upload.wikimedia.org/wikipedia/commons/7/78/Map_by_nicolo_zeno_1558.jpg.

[107]"File:Canada Newfoundland and Labrador location map 2.svg." *Wikipedia Commons*, 7 Nov. 2021, upload.wikimedia.org/wikipedia/commons/thumb/d/d0/Canada_Newfoundland_and_Labrador_location_map_2.svg/1512px-Canada_Newfoundland_and_Labrador_location_map_2.svg.png.

[108]Dale Mackenzie Brown. "The Fate of Greenland's Vikings." https://archive.archaeology.org/, Archaeological Institute of America, 2022, archive.archaeology.org/online/features/greenland/.

[109]Museum of thc North. "Thule Culture." www.uaf.edu, Museum of the North at University of Alaska, 2022, uaf.edu/museum/collections/archaeo/online-exhibits/paleo-eskimo-cultures/thule/.

[110]Andrew Evans. "Is Iceland Really Green and Greenland Really Icy?" www.nationalgeographic.com, National Geographic Society, 2022, nationalgeographic.com/science/article/iceland-greenland-name-swap.

[111]Fernando Garcés de los Fayos. "Greenland: The challenge of managing a key geostrategic territory." www.europarl.europa.eu, Policy Department (Directorate-General for External Policies, European Union), 2022, europarl.europa.eu/RegData/etudes/briefing_note/join/2014/522332/EXPO-AFET_SP(2014)522332_EN.pdf.

[112]Soaial Sciemces and Humanities Research Council of Canada. "The Portuguese Explorers." https://web.archive.org/, Newfoundland and Labrador Heritage, 2022, web.archive.org/web/20160508193835/, http://www.heritage.nf.ca/articles/exploration/portuguese.php.

[113]Huang Shijian (黄时鉴) and Gong Yingyan (龚缨晏). *Op. cit*., p. 188.

[114]Huang Shijian (黄时鉴) and Gong Yingyan (龚缨晏). *Op. cit*., p. 199.

[115]Huang Shijian (黄时鉴) and Gong Yingyan (龚缨晏). *Op. cit*., p. 196.

[116]"File:Ruysch map.jpg." *Wikipedia Commons*, 15 May 2018, commons.wikimedia.org/wiki/File:Ruysch_map.jpg.

[117]"File:Mercator north pole 1595.jpg." *Wikipedia Commons*, 14 Mar. 2008, en.wikipedia.org/wiki/File:Mercator_north_pole_1595.jpg.

[118]"File:Waldseemuller map 2.jpg." *Wikipedia Commons*, 21 June 2009, en.wikipedia.org/wiki/File:Waldseemuller_map_2.jpg.

[119]"File:Arctic Ocean - en.png." *Wikipedia Commons*, 23 Oct. 2021, commons.wikimedia.org/wiki/File:Arctic_Ocean_-_en.png.

[120]"File:Kunyu Wanguo Quantu by Matteo Ricci Plate 1-3.jpg." *Wikipedia Commons*, 23 Oct. 2020, upload.wikimedia.org/wikipedia/commons/b/b8/Kunyu_Wanguo_Quantu_by_Matteo_Ricci_Plate_1-3.jpg.

[121]A. Mountain. *Tale of Semyon Dezhnev with His Comrades*. Russia: Россия: издание Обь-Иртышского союза кооперативов (publication of the Ob-Irtysh Union of Cooperatives), 1919.

[122]Matthew A. Henson. *A Journey for the Ages: Matthew Henson and Robert Peary's Historic North Pole Expedition*. New York, NY, USA: Skyhorse, 2016.

[123]Paul Chiasson. "Chapter four: The mark of an ancient canal." *Written in the Ruins: Cape Breton Island's Second Pre-Columbian Chinese Settlement*, Dundurn, Saskatchewan, Canada: Dundurn, 2016, p. 72.

[124]Natural Resources Canada. "The Atlas of Canada: Groswater Bay, Newfoundland and Labrador." https://web.archive.org/, Natural Resources Cabada, 2022, web.archive.org/web/20121002160500/, http://atlas.nrcan.gc.ca/site/english/search/search_results?expression=Groswater.

[125]Library and Archives Canada. "John Davis, the Master Navigator." https://epe.lac-bac.gc.ca/, Library and Archives Canada, 2022, epe.lac-bac.gc.ca/100/206/301/lac-bac/explorers/www.collectionscanada.gc.ca/explorers/h24-1350-e.html.

[126]William Baillie Hamilton. *Place Names of Atlantic Canada*. Toronto, Ontario, Canada: University of Toronto Press, 1996.

[127]"File:1593 Americae Pars Borealis Florida, Baccalaos, Canada, Corterealis.jpg." *Wikimedia Commons*, 14 June 2021, commons.wikimedia.org/wiki/File:1593_Americae_Pars_Borealis_Florida,_Baccalaos,_Canada,_Corterealis.jpg.
[128]"File:Lake Melville-Hamilton Inlet-Groswater Bay.png." *Wikipedia Commons*, 3 Oct. 2020, upload.wikimedia.org/wikipedia/commons/0/05/Lake_Melville-Hamilton_Inlet-Groswater_Bay.png.
[129]World Ocean Review. "The Arctic and Antarctic — Natural realms at the poles." https://worldoceanreview.com/, World Ocean Review, 2022, https://oceanrep.geomar.de/id/eprint/54472/1/WOR6_en.pdf; Chris Burn. "The Polar Night." *The Aurora Research Institute Scientist Report*, no. 4, Mar. 1996, https://nwtresearch.com/sites/default/files/the-polar-night.pdf.
[130]*Ibid.*
[131]Swinburne University of Technology. "Constellation." https://astronomy.swin.edu.au/, Cosmos, 2022, astronomy.swin.edu.au/cosmos/c/Constellation.
[132]Robert A. Holland. *The Mississippi River in Maps & Views: From Lake Itasca to the Gulf of Mexico.* New York, NY, USA: Rizzoli, 2008.
[133]William Iseminger. *Cahokia Mounds: America's First City*. Gloucestershire, UK: The History Press, 2010.
[134]James L. Shaffer and John T. Tigges. *The Mississippi River: Father of Waters*. Chicago, IL, USA: Arcadia Publishing, 2000.
[135]Charles Hudson, *et al.* Lawrence A. Clayton, *et al*., eds. *The De Soto Chronicles: The Expedition of Hernando de Soto to North America in 1539–1543. Op. cit.*
[136]Laura M. Chmielewski. *Jacques Marquette and Louis Jolliet: Exploration, Encounter, and the French New World. Op. cit.*; August William Derleth and H. Lawrence Hoffman. *Father Marquette and the Great Rivers. Op. cit.*
[137]Geronimo de Chaves. "*La Florida*." https://encyclopediavirginia.org/, Virginia Humanities, 2020, encyclopediavirginia.org/3446-919b9bf8531cf97/.
[138]Huang Shijian (黄时鉴) and Gong Yingyan (龚缨晏). *Op. cit*., p. 195.
[139]Bernal Diaz Del Castillo. John M. Cohen, trans. *The Conquest of New Spain.* London, UK: Penguin Books, 1963.
[140]Alvar Nunez Cabeza De Vaca. Martin A. Favata and Jose B. Fernandez, trans. *The Account: Alvar Nunez Cabeza de Vaca's Relacion.* Houston, USA: Arte Publico Press, 1963.
[141]Marcel and Deborah van den Broecke. "Cartographica Neerlandica Topographical names for Ortelius Map No. 11." https://orteliusmaps.com, orteliusmaps, 2022, orteliusmaps.com/topnames/ort11.html.
[142]Huang Shijian (黄时鉴) and Gong Yingyan (龚缨晏). *Op. cit*., p. 191.
[143]Oxford University Press. "Timeline: Spanish empire." www.oxfordreference.com, Oxford Reference, 2022, oxfordreference.com/view/10.1093/acref/9780191737671.timeline.0001.
[144]Robert H. Fuson. *Juan Ponce de Leon and the Spanish Discovery of Puerto Rico and Florida. Op. cit.*
[145]"File:Gastaldi Nueva Hispania Tabula Nova 1548 UTA.jpg." *Wikipedia Commons*, 14 Jan. 2022, upload.wikimedia.org/wikipedia/commons/0/01/Gastaldi_Nueva_Hispania_Tabula_Nova_1548_UTA.jpg.
[146]Buddy Levy. *Conquistador: Hernán Cortés, King Montezuma, and the Last Stand of the Aztecs.* New York, NY, USA: Bantam, 2009.
[147]Bernal Diaz del Castillo. John M. Cohen, trans. *The Conquest of New Spain.* London, UK: Penguin Books, 1963.
[148]*Ibid.*
[149]Huang Shijian (黄时鉴) and Gong Yingyan (龚缨晏). *Op. cit*., p. 188.
[150]De Agostini G. Dagli Orti (photographer). "Tenochtitlan." https://education.nationalgeographic.org/, National Geographic Society, 2022, education.nationalgeographic.org/resource/tenochtitlan.

[151]Smithsonian National Zoological Park. "Hummingbirds." https://archive.ph/, Migratory Bird Centre, 2022, archive.ph/20120716064758/http://nationalzoo.si.edu/scbi/migratorybirds/webcam/hummingbirds.cfm.
[152]American Ornithologists' Union. *Check-List of North American Birds: The Species of Birds of North America from the Arctic through Panama, Including the West Indies and Hawaiian Islands*. Washington, DC, USA: American Ornithologists' Union, 1998.
[153]Huang Shijian (黄时鉴) and Gong Yingyan (龚缨晏). *Op. cit*., p. 195.
[154]Huang Shijian (黄时鉴) and Gong Yingyan (龚缨晏). *Op. cit*., p. 193.
[155]Arnoldo C. Vento. *Mestizo: The History, Culture, and Politics of the Mexican and the Chicano*, Lanham, MD, USA: University Press of America, 1997, p. 65.
[156]Buddy Levy. *Conquistador: Hernán Cortés, King Montezuma, and the Last Stand of the Aztecs. Op. cit.*
[157]Huang Shijian (黄时鉴) and Gong Yingyan (龚缨晏). *Op. cit*., p. 195.
[158]Josefina Mateos Madrigal. *El capitán Lázaro de Cebreros: Cebreros en los siglos XVI-XVII: cebrereños en el Nuevo Mundo (Spanish edition)*. Ávila, Spain: Ayuntamiento de Cebreros, 2006.
[159]Donald E. Chipman. *Nuno de Guzman and the Province of Panuco in New Spain, 1518–1533*. Glendale, CA, USA: Arthur H. Clark Co., 1967.
[160]"File:Michoacan in Mexico (location map scheme).svg." *Wikipedia Commons*, 18 Sept. 2021, commons.wikimedia.org/wiki/File:Michoacan_in_Mexico_(location_map_scheme).svg.
[161]Valle Rafael Heliodoro. *Semblanza de Cristóbal de Olid.* Ciudad de México, México: Fondo de Cultura Económica, 1998.
[162]Colin M. MacLachlan and Jaime E. Rodriguez O. *The Forging of the Cosmic Race: A Reinterpretation of Colonial Mexico*, Berkeley, CA, USA: University of California Press, 1980, p. 107.
[163]Robert Himmerich Y Valencia. *The Encomenderos of New Spain, 1521–1555*. Austin, TX, USA: University of Texas Press, 1991; Donald E. Chipman. *Nuno de Guzman and the Province of Panuco in New Spain, 1518–1533. Op. cit.*
[164]Huang Shijian (黄时鉴) and Gong Yingyan (龚缨晏). *Op. cit*., p. 195.
[165]Huang Shijian (黄时鉴) and Gong Yingyan (龚缨晏). *Op. cit*., p. 187.
[166]Lansing B. Bloom and Paul A. F. Walter, eds. *The New Mexico History Review, Vol. 1*. Santa Fe, NM, USA: The Museum Press of the Historical Society of New Mexico, 1826, p. 212; as found in the electronic version on the internet at https://www.google.com/books/edition/New_Mexico_Historical_Review/K3IRAAAAIAAJ?hl=en&gbpv=1&dq=%22%22&pg=PA212&printsec=frontcover.
[167]Robert S. Weddle. "Álvarez de Pineda, Alonso (unknown-1520)." *Op. cit.*
[168]Huang Shijian (黄时鉴) and Gong Yingyan (龚缨晏). *Op. cit*., p. 190.
[169]J. H. Simpson. *Coronado's March in Search of the Seven Cities of Cibola, and Discussion of Their Probable Location*. London, UK: Forgotten Books, 2018; Willie Drye. "Seven Cities of Cibola." www.nationalgeographic.com, National Geographic Society, 2022, nationalgeographic.com/history/article/seven-cities-of-cibola.
[170]Paul Chiasson. *The Island of Seven Cities: Where the Chinese Settled When They Discovered America*. Toronto, Ontario, Canada: Random House Canada, 2006.
[171]Paul Chiasson. *The Island of Seven Cities: Where the Chinese Settled When They Discovered America. Op. cit*., pp. 290–297.
[172]Sheng-Wei Wang. "Chapter five: Travelling from the Gulf of St. Lawrence to the country of Fengdu." *The Last Journey of the San Bao Eunuch, Admiral Zheng He*, *op. cit*., p. 162.

[173]Paul Chiasson. "Chapter 18: Gavin and Gold." *The Island of Seven Cities: Where the Chinese Settled When They Discovered America, op. cit.*, pp. 290–297; T.C. Bell's booklet, 'Cape Breton Island'; courtesy of T.C. Bell.
[174]Huang Shijian (黄时鉴) and Gong Yingyan (龚缨晏). *Op. cit.*, p. 191.
[175]Kenan Heise. "Is Long-Lost Chilaga the Camelot of Checagou?" www.chicagotribune.com, Chicago Tribune, 2022, chicagotribune.com/news/ct-xpm-1987-03-04-8701170765-story.html.
[176]Bridgette Russell. "Who was Jean Baptiste Pointe DuSable?" www.fieldmuseum.org, Field Museum, 2022, fieldmuseum.org/blog/who-was-jean-baptiste-pointe-dusable.
[177]Denys Delâge. "Aboriginal influence on the Canadians and French at the time of New France." In Gordon Christie, ed. *Aboriginality and Governance: A Multidisciplinary Approach*, Penticton Indian Reserve, British Columbia, Canada: Theytus Books, 2006, p. 28.
[178]Carolyn Harris. "The Queen's land." *Canada's History*, vol. 97, no. 4, 2017, pp. 34–43.
[179]Huang Shijian (黄时鉴) and Gong Yingyan (龚缨晏). *Op. cit.*, p. 208.
[180]Editors. "American Journeys." https://americanjourneys.org/, Wisconsin Historical Society, 2021, americanjourneys.org/aj-072/summary/.
[181]da Nizza Marco. Adolph Francis Alphonse Bandelier, ed. Fanny Bandelier, trans. "A Relation of the Reverend Father Frier Marco de Nica, Touching His Discovery of the Kingdom of Ceuola or Cibola, Situate about 30 Degrees of Latitude, to the North of Nueva Espanna." https://content.wisconsinhistory.org/, Wisconsin Historical Society, https://americanjourneys.org/aj-072/index.asp.
[182]Huang Shijian (黄时鉴) and Gong Yingyan (龚缨晏). *Op. cit.*, p. 194.
[183]Huang Shijian (黄时鉴) and Gong Yingyan (龚缨晏). *Op. cit.*, p. 191.
[184]Joseph Kinsey Howard. *Montana: High, Wide, and Handsome*. Lincoln, NE, USA: Bison Books, 2003.
[185]Juliet Morrow and Stuart Fiedel. *New Radiocarbon Dates for the Clovis Component of the Anzick Site, Park County, Montana in Paleoindian Archaeology — A Hemispheric Perspective*. Gainesville, FL, USA: University Press of Florida, 2006; Douglas Owsley and David Hunt. "Clovis and Early Archaic Crania from the Anzick Site (24PA506), Park County, Montana." *Plains Anthropologist*, vol. 46. no. 176, 2001, pp. 115–124.
[186]"File:1768 Vaugondy Map of California and Alaska — Geographicus — CartedeLaCalifornie-vaugondy-1768.jpg." *Wikipedia Commons*, 12 Jan. 2022, commons.wikimedia.org/wiki/File:1768_Vaugondy_Map_of_California_and_Alaska_-_Geographicus_-_CartedeLaCalifornie-vaugondy-1768.jpg.
[187]Meriwether Lewis and William Clark. Bernard DeVoto, ed. *The Journals of Lewis and Clark*. New York, NY, USA: Harper Perennial, 1997.
[188]Huang Shijian (黄时鉴) and Gong Yingyan (龚缨晏). *Op. cit.*, p. 207.
[189]Huang Shijian (黄时鉴) and Gong Yingyan (龚缨晏). *Op. cit.*, p. 191.
[190]Editors. "The confluence where waters meet and marine life emerges." https://parcmarin.qc.ca/, Saguenay-St. Lawrence Marine Park, 2022, parcmarin.qc.ca/the-confluence.
[191]Ramsay Cook, ed. *The Voyages of Jacques Cartier*. *Op. cit.*
[192]*Ibid.*
[193]Louise Phelps d. Kellogg. *The French Regime in Wisconsin and the Northwest*. Generic, 2021, p. 42; as found in the electronic version on the internet at http://books.google.com.hk/books?id=8_aDQwNeY98C&lpg=PA41-IA1&ots=_gLZT7ZjW_&pg=PA41-IA1&redir_esc=y#v=onepage&q&f=false.
[194]André Thevet. "La cosmographie universelle d'Andre Thevet cosmographe dv roy. Illvstree de diverses figvres des choses plvs remarqvables veves par l'auteur, & incogneues de noz anciens & modernes." https://nla.gov.au/, National Library of Australia, 2022, nla.gov.au/nla.obj-236886032/view.

[195]"1597 Wytfiet Map of Northern Canada (first specific map of central Canada)." www.geographicus.com, Geographicus Rare Antique Maps, 2022, geographicus.com/P/AntiqueMap/conibasregio-wytfliet-1597; Barry Lawrence Ruderman Map Collection. "Conibas Regio cum Vicinis Gentibus." https://exhibits.stanford.edu/, Stanford Libraries, 2022, exhibits.stanford.edu/ruderman/catalog/pp681zd2975.
[196]G. Mercator and J. Hondius. "1606 Mercator Hondius Map of the Arctic (First Map of the North Pole)." www.geographicus.com, Geographicus Rare Antique Maps, 2022, www.geographicus.com/P/AntiqueMap/NorthPole-mercator-1606.
[197]Derek Hayes. *First Crossing: Alexander Mackenzie, His Expedition Across North America, and the Opening of the Continent*. Seattle, WA, USA: Sasquatch Books, 2002.
[198]Stephen J. Déry. "Map of the Hudson Bay Basin." www.researchgate.net, ResearchGate GmbH, 2022, researchgate.net/figure/Map-of-the-Hudson-Bay-Basin-showing-the-location-of-rivers-with-outlets-into-Hudson-Bay_fig1_253141706.
[199]Ransom A. Myers, *et al*. "The influence of Hudson Bay runoff and ice-melt on the salinity of the inner Newfoundland Shelf." *Atmosphere-Ocean*, vol. 28. no. 2, 1990-06-01, pp. 241–256. doi:10.1080/07055900.1990.9649377.
[200]William A. Gough (updated). "Map of the Hudson Bay region." www.researchgate.net, ResearchGate, 2022, researchgate.net/figure/Map-of-the-Hudson-Bay-region_fig1_232913301; Stephen J. Déry (updated). "Map of the Hudson Bay Basin." www.researchgate.net, ResearchGate, 2022, researchgate.net/figure/Map-of-the-Hudson-Bay-Basin-showing-the-location-of-rivers-with-outlets-into-Hudson-Bay_fig1_253141706.
[201]*Ibid.*
[202]Huang Shijian (黄时鉴) and Gong Yingyan (龚缨晏). *Op. cit*., p. 200.
[203]"File:Map from Magin 1611 BHoU-p6.png." *Wikipedia Commons*, 3 Nov. 2021, commons.wikimedia.org/wiki/File:Map_from_Magin_1611_BHoU-p6.png.
[204]Kat Eschner. "How One Quest for the Northwest Passage Ended at the Icy Mouth of Disappointment River." www.smithsonianmag.com, Smithsonian Magazine, 2022, smithsonianmag.com/smart-news/how-one-quest-northwest-passage-ended-mouth-disappointment-river-180964005/.
[205]"File:Mercator north pole 1595.jpg." *Wikipedia Commons*, 17 Aug. 2016, commons.wikimedia.org/wiki/File:Mercator_north_pole_1595.jpg.
[206]Douglas Hunter. *Half Moon: Henry Hudson and the Voyage That Redrew the Map of the New World*. London, UK: Bloomsbury Press, 2009.
[207]Luke Foxe and Thomas James. Christy Miller, ed. *The Voyages of Captain Luke Foxe of Hull and Captain Thomas James V2: To Hudson's Bay in 1632*. Whitefish, MT, USA: Kessinger Publishing, LLC, 2010.
[208]*Ibid.*
[209]Joseph Burr Tyrrell. *David Thompson, Explorer*. London, UK: Forgotten Books, 2019.
[210]"File:Liard river map with borders.png." *Wikipedia Commons*, 2 Mar. 2022, commons.wikimedia.org/wiki/File:Liard_river_map_with_borders.png.
[211]"File:Mackenzie River basin map.png." *Wikipedia Commons*, 22 Sept. 2022, upload.wikimedia.org/wikipedia/commons/4/45/Mackenzie_River_basin_map.png.
[212]Huang Shijian (黄时鉴) and Gong Yingyan (龚缨晏). *Op. cit*., p. 194.
[213]Felipe Fernández-Armesto. *Amerigo: The Man Who Gave His Name to America. Op. cit.*
[214]Erin Allen. "How Did America Get Its Name?" https://blogs.loc.gov/, Library of Congress Blog, 2022, blogs.loc.gov/loc/2016/07/how-did-america-get-its-name/.
[215]Huang Shijian (黄时鉴) and Gong Yingyan (龚缨晏). *Op. cit*., p. 190.

[216]James Stuart Olson and Robert Shadle. *Historical Dictionary of European Imperialism*, Santa Barbara, CA, USA: Greenwood Publishing Group, 1991, p. 109.
[217]Alan Rayburn. *Naming Canada: Stories about Canadian Place Names*, Toronro, Ontario, Canada: University of Toronto Press, 2001, pp. 14–22.
[218]John A. Dickinson and Brian Young. *A Short History of Quebec. Op. cit.*
[219]Jacques Cartier, *et al. The Voyages of Jacques Cartier. Op. cit.*
[220]Huang Shijian (黄时鉴) and Gong Yingyan (龚缨晏). *Op. cit.*, p. 194.
[221]Huang Shijian (黄时鉴) and Gong Yingyan (龚缨晏). *Op. cit.*, p. 193.
[222]Huang Shijian (黄时鉴) and Gong Yingyan (龚缨晏). *Op. cit.*, p. 191.
[223]Huang Shijian (黄时鉴) and Gong Yingyan (龚缨晏). *Op. cit.*, p. 195.
[224]USDA. "PLANTS Database." https://plants.usda.gov/home, Natural Resources Conservation Service of the United States Department of Agriculture, 2022, plants.usda.gov/home.
[225]Jim Pojar and Andy MacKinnon. *Plants of the Pacific Northwest Coast*. Sisters, OR, USA: Partners Publishing, 2004.
[226]Staff Writers. "About Oregon's Forests." www.oregon.gov, Oregon Department of Forestry, 2022, oregon.gov/ODF/ForestBenefits/Pages/AboutForests.aspx#:~:text=Oregon%20is%20home%20to%20some%20of%20the%20world.
[227]USDA. "PLANTS Database." *Op. cit.*
[228]"File:Ortelius – Maris Pacifici 1589.jpg." *Wikipedia*, 8 July 2012, en.wikipedia.org/wiki/File:Ortelius_-_Maris_Pacifici_1589.jpg.
[229]Jon Bell. *On Mount Hood: A Biography of Oregon's Perilous Peak*. Seattle, WA, USA: Sasquatch Books, 2013.
[230]National Geographic Maps. *Mount Rainier National Park*. Washington, DC, USA: National Geographic Society Maps, 2018.
[231]NGS. "Mount Hood Highest Point." https://www.ngs.noaa.gov/, U.S. National Geodetic Survey data sheet, 2022, www.ngs.noaa.gov/cgi-bin/ds_mark.prl?PidBox=RC2244.
[232]Mount Rainier National Park. "Mount Reiner." www.nps.gov, National Park Service, 2022, nps.gov/mora/index.htm.
[233]"File:Mount Hood reflected in Mirror Lake, Oregon.jpg." *Wikipedia*, 25 May 2012, upload.wikimedia.org/wikipedia/commons/e/ef/Mount_Hood_reflected_in_Mirror_Lake%2C_Oregon.jpg.
[234]"File:Mount Rainier from west.jpg." *Wikimedia Commons*, 10 Aug. 2020, upload.wikimedia.org/wikipedia/commons/e/eb/Mount_Rainier_from_west.jpg.
[235]George Vancouver. *The Exploration of the Columbia River by Lieutenant W.R. Broughton, October, 1792: An Extract from the Journal of Captain George Vancouver*. Longview, TX, USA: Longview, 1926; Department of History at Portland State University. "Document | Lieutenant William Broughton's Encounter at Cathlapotle — Excerpt." http://publichistorypdx.org/, Public History PDX, 2022, publichistorypdx.org/projects/chinook/document-lieutenant-william-broughtons-encounter-at-cathlapotle-excerpt/.
[236]George Vancouver. W. Kaye Lamb, ed. *The Voyage of George Vancouver 1791–1795: Four Volume Set*. Oxfordshire, UK: Routledge, 1984.
[237]*Ibid.*
[238]Hendon Harris. *The Asiatic Fathers of America*, Taipei, Taiwan: Wen Ho Printing Co., 1975, p. 54.
[239]David Malakoff. "Searching For Etzanoa." *American Archaeology*, vol. 20, no. 1, 14 Mar. 2016. https://www.archaeologicalconservancy.org/searching-for-etzanoa/.
[240]Pierluigi Portinaro and Franco Knirsch. *The Cartography of North America: 1500–1800*. New York, NY, USA: Chartwell Books, 1999; Frank Wilson Blackmar. *Kansas, Vol. 2: A Cyclopedia of State History,*

Embracing Events, Institutions, Industries, Counties, Cities, Towns, Prominent Persons, Etc. London, UK: Forgotten Books, 2018; as found in the electronic version on the internet at https://web.archive.org/web/201404ing/07073154/http://skyways.lib.ks.us/genweb/archives/1912/q/quivira.html.

[241]Peter Lauridsen. Julius E. Olson and Frederick Schwatka, ctrs. *Russian Explorations, 1725–1743; Vitus Bering: The Discoverer of Bering Strait.* Chicago, IL, USA: Leopold Classic Library, 2016.

[242]"File:Ortelius Maris Pacifici 1589.jpg." *Op. cit.*

[243]Arturo Giraldez. *The Age of Trade: The Manila Galleons and the Dawn of the Global Economy*. Lanham, MD, USA: Rowman & Littlefield Publishers, 2015.

[244]"File:Gulfofalaskamap.png." *Wikipedia Commons*, 31 July 2022, upload.wikimedia.org/wikipedia/commons/d/d3/Gulfofalaskamap.png.

[245]Ray Hilborn. "Map of Alaska Peninsula and Bristol Bay." www.researchgate.net, ResearchGate, 2022, researchgate.net/figure/Map-of-Alaska-Peninsula-and-Bristol-Bay_fig2_33515373.

[246]Huang Shijian (黄时鉴) and Gong Yingyan (龚缨晏). *Op. cit*., p. 199.

[247]William Frame and Laura Walker. *James Cook: The Voyages*. Montreal, Quebec, Canada: McGill-Queen's University Press, 2018.

[248]Huang Shijian (黄时鉴) and Gong Yingyan (龚缨晏). *Op. cit*., p. 189.

[249]Marleen Felius. *Cattle Breeds: An Encyclopedia*. Chicago, IL, USA: Trafalgar Square Publishing, 2007.

[250]Pekka Hämäläinen. "The rise and fall of Plains Indian horse cultures." *Journal of American History*, vol. 90, no. 3, 2003, pp. 833–862. doi:10.2307/3660878.

[251]Hendon Harris. Edited and Abridged by Charlotte Harris Rees. *The Asiatic Fathers of America: Chinese Discovery & Colonization of Ancient America*, Lynchburg, VA, USA: Warwick House Publishing, 2006, p. 78.

[252]*Ibid.*

[253]Hendon Harris. *The Asiatic Fathers of America*, *op. cit*., p. 36.

[254]Huang Shijian (黄时鉴) and Gong Yingyan (龚缨晏). *Op. cit*., p. 195.

[255]Hendon Harris. *The Asiatic Fathers of America*, *op. cit*., p. 48.

[256]Anonymous. Anne Birrell, ed. and trans. *The Classic of Mountains and Seas. Op. cit.*

[257]"File:Relief map of USA Alaska.png." *Wikipedia Commons*, 3 Apr. 2022, commons.wikimedia.org/wiki/File:Relief_map_of_USA_Alaska.png.

[258]Michael Balter. "Ancient infants buried together in Alaska suggest long journey to the Americas." www.science.org, Science, 26 Oct. 2015, science.org/content/article/ancient-infants-buried-together-alaska-suggest-long-journey-americas.

[259]*Ibid.*

[260]Jeremy DeSilva. *Did the First Americans Arrive via Land Bridge? Op. cit.*; Lizzie Wade. "New map of Alaska's ancient coast supports theory that America's first people arrived by boat." www.science.org, Science, 30 May 2018, www.science.org/content/article/new-map-alaska-s-ancient-coast-supports-theory-america-s-first-people-arrived-boat.

[261]Huang Shijian (黄时鉴) and Gong Yingyan (龚缨晏). *Op. cit*., p. 188.

[262]Editors. "Kenai Fjords." www.nps.gov, National Park Alaska, 2022, nps.gov/kefj/index.htm.

[263]Cape Peirce can be found on Google Maps at https://www.google.com/maps/place/Cape+Peirce/@58.5527882,-161.7773658,15z/data=!3m1!4b1!4m5!3m4!1s0x56e1f7dafd550c7f:0x8da9e9ae9473c4fa!8m2!3d58.5527778!4d-161.768611.

[264]*Ibid.*

[265]J. C. Beaglehole. *The Journals of Captain James Cook on His Voyages of Discovery: Volume III, Part I and II: The Voyage of the Resolution and Discovery 1776–1780*. London, UK: Hakluyt Society, 2015.

[266]Siu-Leung Lee. "Chinese Mapped America Before 1430." *Proceedings of the ICA*, vol. 1, 16 May 2018. https://www.researchgate.net/publication/325189032_Chinese_Mapped_America_Before_1430.
[267]"File:Ortelius — Maris Pacifici 1589.jpg." *Op. cit.*
[268]James H. Marsh, ed. "Arctic Archipelago." *The Canadian Encyclopedia*. Toronto, Ontario, Canada: Hurtig Publishers, 1988.
[269]"File:Chukchi Sea map.png." *Wikipedia Commons*, 20 Oct. 2020, upload.wikimedia.org/wikipedia/commons/5/5a/Chukchi_Sea_map.png.
[270]Dederal State Budget Institution. "Wrangel Island." http://eng.ostrovwrangelya.org/, State Nature Reserve, 2022, eng.ostrovwrangelya.org/historyotkr.html.
[271]Sergey V. Gusev, *et al.* "Sea mammal hunters of Chukotka, Bering Strait: Recent archaeological results and problems." *World Archaeology*, vol. 30, no. 3, Feb. 1999, pp. 354–369. doi:10.1080/00438243.1999.9980417.
[272]Huang Shijian (黄时鉴) and Gong Yingyan (龚缨晏). *Op. cit.*, p. 201.
[273]N. N. Dikov. "The earliest sea mammal hunters of Wrangell Island." *Arctic Anthropology*, vol. 25, no. 1, 1988, pp. 80–93. JSTOR 40316156.
[274]Joel Berger. "Studying a Polar Menagerie on an Island in Arctic Russia." e360.yale.edu, Yale School of the Environment, 2022, e360.yale.edu/features/studying_a_polar_menagerie_on_an_island_in_arctic_russia.
[275]A. E. Nordenskiöld. *Facsimile-Atlas to the Early History of Cartography with Reproductions of the Most Important Maps Printed in the XV and XVI Centuries*, New York, USA: Dover Publications, 1973, p. 40.
[276]Anonymous author, *et al. Geographical Explorations. Early Expeditions to the Region of Bering Sea and Strait, Etc. (*Notes *on an original manuscript chart of Bering's ... manuscript chart of his second expedition.)* London, UK: British Library, 2011.
[277]Robert Murphy. *The Haunted Journey: The Heroic Story of Vitus Bering's Discovery of Alaska*. New York, NY, USA: Doublesday, 1961.
[278]Huang Shijian (黄时鉴) and Gong Yingyan (龚缨晏). *Op. cit.*, p. 192.
[279]Huang Shijian (黄时鉴) and Gong Yingyan (龚缨晏). *Op. cit.*, p. 195.
[280]Huang Shijian (黄时鉴) and Gong Yingyan (龚缨晏). *Op. cit.*, p. 202.
[281]Gordon Whittaker. "The Study of North Mesoamerican Place-Signs." www.iai.spk-berlin.de, iai.spk-berlin, 2022, www.iai.spk-berlin.de/fileadmin/dokumentenbibliothek/Indiana/Indiana_13/IND_13_Whittaker.pdf.
[282]"*File:Isthmus of Tehuantepec*-aeac.jpg." *Wikipedia Commons*, 13 May 2022, commons.wikimedia.org/wiki/File:Isthmus_of_Tehuantepec-aeac.jpg.
[283]Max L. Moorhead. "Hernán Cortés and the Tehuantepec Passage." *The Hispanic American Historical Review*, vol. 29, no. 3, 1949, pp. 370–379. https://doi.org/10.2307/2508457.
[284]Huang Shijian (黄时鉴) and Gong Yingyan (龚缨晏). *Op. cit.*, p. 186.
[285]Huang Shijian (黄时鉴) and Gong Yingyan (龚缨晏). *Op. cit.*, p. 202.
[286]Dennis Herrick. *Winter of the Metal People: The Untold Story of America's First Indian War.* Mechanicsburg, PA, USA: Sunbury Press, Inc., 2013.
[287]Huang Shijian (黄时鉴) and Gong Yingyan (龚缨晏). *Op. cit.*, p. 187.
[288]Huang Shijian (黄时鉴) and Gong Yingyan (龚缨晏). *Op. cit.*, p. 194.
[289]Ron Stewart. "Looking Back 500 Years: The Tiguex War." www.nps.gov, National Park Service, 2022, nps.gov/coro/learn/historyculture/the-tiguex-war.htm.
[290]Albert B. Elsasser. "Explorations of Hernando Alarcón in the Lower Colorado River Region, 1540." *Journal of California and Great Basin Anthropology*, vol. 1, no. 1, Summer 1979, pp. 8–37.

[291]Gulf of California is shown on Google Maps at https://www.google.com/maps/place/%E5%8A%A0%E5%88%A9%E7%A6%8F%E5%B0%BC%E4%BA%9E%E7%81%A3/@27.2640547,-115.3476342,6z/data=!3m1!4b1!4m5!3m4!1s0x86cbcc2e363e4c59:0xd366a88891d18a70!8m2!3d26.7312935!4d-110.7122465.
[292]"File:Colorado River basin map.png." *Wikipedia Commons*, 6 Apr. 2022, upload.wikimedia.org/wikipedia/commons/d/d1/Colorado_River_basin_map.png.
[293]Jim Turner. *The Mighty Colorado River: From the Glaciers to the Gulf*. Tucson, AZ, USA: Rio Nuevo Publishers, 2016.
[294]Huang Shijian (黄时鉴) and Gong Yingyan (龚缨晏). *Op. cit*., p. 191.
[295]Lansing B. Bloom and Paul A. F. Walter, eds. *The New Mexico History Review, Vol. 1. Op. cit.*
[296]*Ibid.*
[297]Huang Shijian (黄时鉴) and Gong Yingyan (龚缨晏). *Op. cit*., p. 199.
[298]E. Charles Adams, ed. "Marata: A New Mexico Kingdom (With a Note on Totonteac)." *Op. cit.*
[299]Huang Shijian (黄时鉴) and Gong Yingyan (龚缨晏). *Op. cit*., p. 191.
[300]Francisco Vázquez de Coronado (1540). "Coronado's Report to Viceroy Mendoza, Part IV." www.digitalhistory.uh.edu, Digital History, 2021, www.digitalhistory.uh.edu/disp_textbook.cfm?smtID=3&psid=1253.
[301]Huang Shijian (黄时鉴) and Gong Yingyan (龚缨晏). *Op. cit*., p. 208.
[302]Baja California is shown on Google Maps at https://www.google.com/maps/place/%E5%A2%A8%E8%A5%BF%E5%93%A5%E4%B8%8B%E5%8A%A0%E5%88%A9%E7%A6%8F%E5%B0%BC%E4%BA%9E/@30.3448948,-117.7757724,7z/data=!3m1!4b1!4m5!3m4!1s0x80d7700ca877ddd3:0xfca4fd9f0318de8e!8m2!3d30.8406338!4d-115.2837585; Joseph Wood Krutch. *The Forgotten Peninsula: A Naturalist in Baja California*. Tucson, AZ, USA: University of Arizona Press, 1986.
[303]Charles E. Chapman. *A History of California: The Spanish Period*. New York, NY, USA: The MacMillan Co., 1921.
[304]Editors. "The Explorers, 1492–1774." https://sandiegohistory.org/, San Diago History Centre, 2022, sandiegohistory.org/archives/books/explorers/ch2/.
[305]Dick Schwenkmeyer. "Sierra de San Pedro Mártir." www.sdnhm.org, San Diego Natural History Museum, 2022, sdnhm.org/oceanoasis/fieldguide/sanpedromartir.html.
[306]Editors. "Eusebio Francisco Kino." www.nps.gov, National Park Service, 2022, nps.gov/tuma/learn/historyculture/eusebio-francisco-kino.htm.
[307]Eleanor Dart O'Bryon. *Coming Home from Devil Mountain*. Open Library; Harbinger House, 1989.
[308]Ruben Castro Valdez. "Map of Baja California peninsula and mountain ranges." www.researchgate.net, ResearchGate, 2022, researchgate.net/figure/Map-of-Baja-California-peninsula-and-mountain-ranges-The-positions-of-meteorological_fig4_262705222.
[309]Henry R. Wagner. *Juan Rodriguez Cabrillo: Discoverer of the Coast of California*. Whitefish, MT, USA: Kessinger Publishing, LLC, 2010.
[310]As found in the electronic version on the internet at http://www.csun.edu/~pjk77863/Exploits%20of%20Esplandian.pdf.
[311]John Walton Caughey. *California, 2nd ed.* Prentice Hall, 1955.
[312]Nathan Masters. "Cortés' California misadventure." https://news.usc.edu, USC News, 2022, news.usc.edu/46890/cortes-california-misadventure/; Buddy Levy. *Conquistador: Hernán Cortés, King Montezuma, and the Last Stand of the Aztecs*. *Op. cit.*
[313]Eyewitness Accounts of Early American Exploration and Settlement. "The Voyage of Francisco de Ulloa, 1539." www.americanjourneys.org, Wisconsin Historical Society, 2022, americanjourneys.org/aj-113/summary/index.asp; Nathan Masters. "Cortés' California misadventure." *Op. cit.*
[314]Henry R. Wagner. *Juan Rodriguez Cabrillo: Discoverer of the Coast of California*. *Op. cit.*

[315]Hubert Howe Bancroft. *History of California*, Fort Oglethorpe, GA, USA: The History Company, 1884, p. 68.
[316]SDHC Research Services. "Sebastian Vizcaíno." https://sandiegohistory.org/, San Diego History Centre, 2022, sandiegohistory.org/archives/biographysubject/vizcaino/.
[317]Gaspar de Portolá. *Diary of Gaspar de Portolá During the California Expedition of 1769–1770*. Open Library, 2017.
[318]Theodore E. Treutlein. *San Francisco Bay: Discovery and Colonization*, 1769–1776. San Francisco, CA, USA: California Historical Society, 1968.
[319]"File:Rocas Alijos-map.png." *Wikipedia Commons*, 13 June 2022, commons.wikimedia.org/wiki/File:Rocas_Alijos-map.png.
[320]Abraham Ortelius. "America." www.vims.edu, Virginia Institute of Marine Science, 2022, www.vims.edu/about/_archive/cbh_maps/america.php.
[321]N. M. Fenneman and D. W. Johnson. "Physiographic divisions of the conterminous U. S." https://water.usgs.gov/, U.S. Geological Survey, 2022, water.usgs.gov/GIS/metadata/usgswrd/XML/physio.xml.
[322]"File:Mexico Baja California Sur location map.svg." *Wikipedia Commons*, 13 Sept. 2020, upload.wikimedia.org/wikipedia/commons/thumb/b/bd/Mexico_Baja_California_Sur_location_map.svg/2002px-Mexico_Baja_California_Sur_location_map.svg.png.
[323]"File:CaboSanLucas ISS012-E-7151 annotated.jpg." *Wikimedia Commons*, 25 Apr. 2022, commons.wikimedia.org/wiki/File:CaboSanLucas_ISS012-E-7151_annotated.jpg.
[324]Robert Ryal Miller. "Cortés and the first attempt to colonize California." *Calif History QJ Calif Hist Soc*, vol. 53, no. 1, 1974, pp. 4–16; John Leddy Phelan. "Chapter three: Hernán Cortés, the Moses of the New World." *The Millennial Kingdom of the Franciscans in the New World*. Berkeley, CA, USA: University of California Press, 1971, pp. 33–34; R. C. Stein. *The World's Greatest Explorers: Hernando Cortés*. Chicago, IL, USA: Chicago Press Inc., 1991.
[325]Rebekah K. Nix. "The Gulf of California: A Physical, Geological, and Biological Study." http://activetectonics.asu.edu, University of Texas at Dallas, 2022, activetectonics.asu.edu/BAJA/gulf_cal.pdf.
[326]Michael D. Coe and Stephen D. Houston. *The Maya*. London, UK: Thames & Hudson, 2015.
[327]Simon Martin and Nikolai Grube. *Chronicle of the Maya Kings and Queens: Deciphering the Dynasties of the Ancient Maya*. London, UK. and New York, USA.: Thames & Hudson, 2000.
[328]"File:Map of Central America.png." *Wikimedia Commons*, 6 Sept. 2020, upload.wikimedia.org/wikipedia/commons/d/d8/Map_of_Central_America.png.
[329]"File:Mayamap.png." *Wikimedia Commons*, 27 Sept. 2020, upload.wikimedia.org/wikipedia/commons/c/c5/Mayamap.png.
[330]Lyle Campbell. *American Indian Languages: The Historical Linguistics of Native America*. Oxford, UK: Oxford University Press, 1997.
[331]W. George Lovell, *et al. Strike Fear in the Land: Pedro de Alvarado and the Conquest of Guatemala, 1520–1541*. Norman, OK, USA: University of Oklahoma Press, 2020.
[332]Huang Shijian (黄时鉴) and Gong Yingyan (龚缨晏). *Op. cit*., p. 185.
[333]Bernal Díaz del Castillo. Janet Burke and Ted Humphrey, trans. *The True History of the Conquest of New Spain*. Indianapolis, IN, USA: Hackett Publishing Company, Inc., 2012.
[334]Marshall Howard Saville. *The Discovery of Yucatan in 1517 by Francisco Hernández de Córdoba*. Generic, 2020.
[335]Laurence Bergreen. *Columbus: The Four Voyages, 1492–1504*. London, UK: Penguin Books, 2012.
[336]Philip Ainsworth Means. *History of the Spanish Conquest of Yucatan and of the Itzas*. Charleston, SC, USA: BiblioBazaar, 2009.

[337]Bernal Diaz del Castillo. John M. Cohen, trans. *The Conquest of New Spain. Op. cit.*
[338]Felipe Fernández-Armesto. *Amerigo: The Man Who Gave His Name to America. Op. cit.*
[339]Julian Granberry and Gary Vescelius. *Languages of the Pre-Columbian Antilles*. Tuscaloosa, AL, USA: University of Alabama Press, 2004.
[340]*Ibid.*
[341]Sandra Riley. *The Lucayan Taino: First People of the Bahamas*. Southampton, UK: Crystal Parrot House, 2013.
[342]Clements R. Markham. *The Journal of Christopher Columbus during His First Voyage 1492–93 Documents Relating the Voyages of John Cabot and Gaspar Corte Real 1893. Op. cit.*
[343]"File:CIA map of the Bahamas.png." *Wikipedia Commons*, 12 Sept. 2017, commons.wikimedia.org/wiki/File:CIA_map_of_the_Bahamas.png.
[344]Julian Granberry and Gary Vescelius. *Languages of the Pre-Columbian Antilles. Op. cit.*
[345]*Ibid.*
[346]Huang Shijian (黄时鉴) and Gong Yingyan (龚缨晏). *Op. cit.*, p. 186.
[347]Julian Granberry and Gary Vescelius. *Languages of the Pre-Columbian Antilles. Op. cit.*
[348]*Ibid.*
[349]*Ibid.*
[350]Huang Shijian (黄时鉴) and Gong Yingyan (龚缨晏). *Op. cit.*, p. 199.
[351]"File:Inagua in Bahamas (zoom).svg." *Wikipedia Commons*, 5 Oct. 2020, commons.wikimedia.org/wiki/File:Inagua_in_Bahamas_(zoom).svg.
[352]Alfred Caldecott. *The Church in the West Indies*, London, UK: Frank Cass and Co., 1898, p. 11.
[353]Na Na. *General History of the Caribbean UNESCO Volume 6: Methodology and Historiography of the Caribbean, 1st ed,* CA, USA: Palgrave Macmillan, 2009, p. 635.
[354]Lisa L. Owens. *A Journey with Hernán Cortés*. Minneapolis, MN, USA: Lerner Publications, 2017.
[355]Editors. "The History of Jamaica." https://jis.gov.jm/information/jamaican-history/, Jamaica Information Service, 2022, jis.gov.jm/information/jamaican-history.
[356]Central Intelligence Agency. *The CIA World Factbook 2021–2022*. New York, NY, USA: Skyhorse, 2021; as found in the electronic version on the internet at https://www.cia.gov/the-world-factbook/countries/bermuda/.
[357]Editors. "The History of Jamaica." *Op. cit.*
[358]*Ibid.*
[359]Gregory C. McIntosh. *The Piri Reis Map of 1513*, Athens, GA, USA: University of Georgia Press, 2000, p. 88.
[360]Patricia A. Schechter. "Adelante Hermanas de la Raza', Josefina Silva de Cintron and Puerto Rican Women's Feminismo. — The New York's World Fair: 19391940." *Exploring the Decolonial Imaginary: Four Transnational Lives*. New York, NY, USA: MacMillan, 2012.
[361]"File:Caribbean general map.png." *Wikipedia Commons*, 25 Sept. 2022, upload.wikimedia.org/wikipedia/commons/9/98/Caribbean_general_map.png.
[362]James Lazell. *Island: Fact and Theory in Nature*, Berkeley, CA, USA. University of California Press, 2005, p. 382.
[363]"File:SVG Map of Virgin Islands.svg." *Wikipedia Commons*, 26 July 2022, upload.wikimedia.org/wikipedia/commons/thumb/f/fd/SVG_Map_of_Virgin_Islands.svg/2560px-SVG_Map_of_Virgin_Islands.svg.png.
[364]Huang Shijian (黄时鉴) and Gong Yingyan (龚缨晏). *Op. cit.*, p. 208.
[365]Susana Goulart Costa. *Azores: Nine Islands, One History = Açores: Nove Ilhas, Uma Historia*. Berkeley, CA, USA: Institute of Governmental Studies Press, 2008.

[366]Duane W. Roller. *Through the Pillars of Herakles: Greco-Roman Exploration of the Atlantic*. Oxfordshire, UK: Routledge, 2005.
[367]"File:Medici-Laurentian Atlas 1351.jpg." *Wikiedia Commons*, 22 Sept. 2020, commons.wikimedia.org/wiki/File:Medici-Laurentian_Atlas_1351.jpg.
[368]Abraham Cresques. *Catalan Atlas of 1375*. Norwalk, CN, USA: Abaris Books, 1978.
[369]Carlos Melo Bento. *História dos Açores: Da descoberta a 1934*, Ponta Delgada (Azores), Portugal: Câmara Municipal de Ponta Delgada, 2008, p. 27.
[370]Central Intelligence Agency. *The CIA World Factbook 2021–2022. Op. cit.*
[371]The Smithsonian Institution. "Bermuda — History and Heritage." www.smithsonianmag.com, Smithsonian Magazine, 2022, smithsonianmag.com/travel/bermuda-history-and-heritage-14340790/.
[372]Apollodorus. Robin Hard, trans. *The Library of Greek Mythology (Oxford World's Classics)*. Oxford, UK: Oxford University Press, 2008.
[373]*Ibid.*
[374]"File:OrteliusWorldMap1570.jpg." *Op. cit.*
[375]ICON Group International, Inc. *Honduras: Webster's Timeline History, 507–2001*. Las Vegas, NV, USA: ICON Group International, 2010.
[376]U.S. Department of State. "Honduras (02/08)." https://search.usa.gov/, U.S. Department of State, 2022, search.usa.gov/search?query=Honduras&commit=Search&utf8=✓&affiliate=dos_stategov.
[377]"File:CIA map of Central America.png." Wikipedia Commons, 23 June 2021, upload.wikimedia.org/wikipedia/commons/9/94/CIA_map_of_Central_America.png.
[378]Huang Shijian (黄时鉴) and Gong Yingyan (龚缨晏). *Op. cit*., p. 186.
[379]Huang Shijian (黄时鉴) and Gong Yingyan (龚缨晏). *Op. cit*., p. 204.
[380]John S. Henderson. "The Valley de Naco: Ethnohistory and Archaeology in Northwestern Honduras." *Ethnohistory*, vol. 24, no. 4, 1977, pp. 363–377. https://doi.org/10.2307/481388. Accessed 11 Apr. 2022.
[381]John Eoghan Kelly. *Pedro de Alvarado, conquistador*. Port Washington, NY, USA: Kennikat Press, 1971.
[382]Huang Shijian (黄时鉴) and Gong Yingyan (龚缨晏). *Op. cit*., p. 192.
[383]BBC editors. "Nicaragua profile — Timeline." www.bbc.com, BBC News, 2022, bbc.com/news/world-latin-america-19909695.
[384]*Ibid.*
[385]"File:Map of South America.png." *Wikipedia Commons*, 19 Oct. 2021, commons.wikimedia.org/wiki/File:Map_of_South_America.png.
[386]Pascual de Andagoya. Clements R. Markham, trans. *Narrative of the Proceedings of Pedrarias Davila in the Provinces of Tierra Firme, or Catilla del Oro: And of the Discovery of the South Sea and the Coast of Peru and Nicaragua*. Cambridge, UK: Cambridge University Press, 2010.
[387]*Ibid.*
[388]Charles Loftus Anderson (Creator). *Old Panama and Castilla del Oro*. New South Wales, Australia: Wentworth Press, 2016.
[389]*Ibid.*
[390]"File:Castilla del Oro.jpg." *Wikipedia Commons*, 24 May 2022, commons.wikimedia.org/wiki/File:Castilla_del_Oro.jpg.
[391]Huang Shijian (黄时鉴) and Gong Yingyan (龚缨晏). *Op. cit*., p. 197.
[392]William H. Prescott. *The History of the Conquest of Peru*. New York, NY, USA: 2005.
[393]Huang Shijian (黄时鉴) and Gong Yingyan (龚缨晏). *Op. cit*., p. 190.
[394]Huang Shijian (黄时鉴) and Gong Yingyan (龚缨晏). *Op. cit*., p. 198.

[395]Runk, J. Velásquez. "Creating Wild Darién: Centuries of Darién's imaginative geography and its lasting effects." *Journal of Latin American Geography*, vol. 14, no. 3, 2015, pp. 127–156. doi:10.1353/lag.2015.0032; Andrew Niall Egan. *Crossing the Darien Gap: A Daring Journey through a Forbidding and Enchanting and Roadless Jungle That Is the Only Link by Land between North America and South America*. London, UK: Adventura Publishing, 2008.
[396]*Ibid*.
[397]Franz Lidz. "Tracking Balboa." *Smithsonian. Washington DC: Smithsonian Institution*, vol. 44, no. 5, 2013, pp. 32–36; Angel De Altolaguirre Y Duvale. *Vasco Nunez De Balboa (1914)*. Whitefish, MT, USA: Kessinger Publishing, LLC, 2010.
[398]A. Jimenez-Landi. *Alonso De Ojeda*. Madrid, Spain: Aguila, 1966.
[399]Franz Lidz. "Tracking Balboa." *Op. cit*.; Angel De Altolaguirre Y Duvale. *Vasco Nunez De Balboa (1914)*. *Op. cit*.
[400]"File:Map of Darién Gap-en.svg." *Wikimedia Commons*, 6 May 2022, upload.wikimedia.org/wikipedia/commons/thumb/5/5d/Map_of_Darién_Gap-en.svg/2025px-Map_of_Darién_Gap-en.svg.png.
[401]"File:New Caledonia in Darien2.jpg." *Wikimedia Commons*, 11 Sept. 2022, commons.wikimedia.org/wiki/File:New_Caledonia_in_Darien2.jpg.
[402]"File:Balboa Voyage 1513.PNG." *Wikimedia Commons*, 10 Nov. 2021, commons.wikimedia.org/wiki/File:Balboa_Voyage_1513.PNG.
[403]Huang Shijian (黄时鉴) and Gong Yingyan (龚缨晏). *Op. cit*., p. 186.
[404]H. Micheal Tarver. *The History of Venezuela*. New York, NY, USA: St. Martin's Griffin, 2006.
[405]Hugh Thomas. *Rivers of Gold: The Rise of the Spanish Empire, from Columbus to Magellan*. New York, NY, USA: Random House, 2005.
[406]H. Micheal Tarver. *The History of Venezuela*. *Op. cit*.
[407]Huang Shijian (黄时鉴) and Gong Yingyan (龚缨晏). *Op. cit*., p. 205.
[408]"File:1562 Diego Gutierrez Amazonas.jpg." *Wikimedia Commons*, 12 Oct. 2020, commons.wikimedia.org/wiki/File:1562_Diego_Gutierrez_Amazonas.jpg.
[409]*Ibid*.
[410]Oxford Dictionary. "Guyana." www.lexico.com, Oxford Dictionary, 2022, https://www.dictionary.com/browse/guyana.
[411]Odeen Ishmael. *The Guyana Story: From Earliest Times to Independence*. Bloomington, IN, USA: Xlibris, 2015.
[412]Huang Shijian (黄时鉴) and Gong Yingyan (龚缨晏). *Op. cit*., p. 201.
[413]Alfred Russel Wallace. *A Narrative of Travels on the Amazon and Rio Negro*. New York, NY, USA: Dover, 1972.
[414]Buddy Levy. *River of Darkness: Francisco Orellana and the Deadly First Voyage through the Amazon*. New York, NY, USA: Diversion Books, 2022.
[415]Ann Byers. *Pedro Alvares Cabral: First European Explorer of Brazil*. New York, NY, USA: Rosen Young Adult, 2016.
[416]"File:Negroamazonrivermap.png." *Wikimedia Commons*, 20 Feb. 2023, upload.wikimedia.org/wikipedia/commons/e/e5/Negroamazonrivermap.png.
[417]Huang Shijian (黄时鉴) and Gong Yingyan (龚缨晏). *Op. cit*., p. 204.
[418]Huang Shijian (黄时鉴) and Gong Yingyan (龚缨晏). *Op. cit*., p. 206.
[419]Antonio Carlos Diegues, *et al*. "Artisanal fisheries in Brazil." https://nupaub.fflch.usp.br/, University of São Paulo, 2022, nupaub.fflch.usp.br/sites/nupaub.fflch.usp.br/files/artisanal2.pdf.
[420]C. & J. Tundisi Teixeira. "Primary production and phytoplnakton in equatorial waters." *Bulletin of Marine Science*, vol. 17, no. 4, 1967, pp. 884–891.

[421]Daniel J. Hopkins, ed. "West Indies." *Merriam-Webster's Geographical Dictionary, 3rd ed.*, Springfield, MA, USA: Merriam-Webster Inc., 2001, p. 1298.
[422]Alfred Caldecott. *The Church in the West Indies*. *Op. cit*.
[423]"File:1562 Americæ Gutiérrez.JPG." *Wikimedia Commons*, 28 Apr. 2020, commons.wikimedia.org/wiki/File:1562_Americæ_Gutiérrez.JPG.
[424]Huang Shijian (黄时鉴) and Gong Yingyan (龚缨晏). *Op. cit*., p. 188.
[425]Huang Shijian (黄时鉴) and Gong Yingyan (龚缨晏). *Op. cit*., p. 187.
[426]Editors. "Paguana." mapcarta.com, Mapcarta, 2022, mapcarta.com/20212868.
[427]Cecilia Pardo and Jago Cooper. *Peru: A Journey through Time*. London, UK: British Museum, 2021.
[428]Pango on the map, Peru can be located on Google maps at https://www.google.com/maps/place/Pango,+%E7%A7%98%E9%AD%AF/@-7.8979478,-78.55672,17z/data=!3m1!4b1!4m5!3m4!1s0x91ad995aa1cddb55:0xcfc3496c095ff1d6!8m2!3d-7.8985987!4d-78.5554381.
[429]Huang Shijian (黄时鉴) and Gong Yingyan (龚缨晏). *Op. cit*., p. 187.
[430]Huang Shijian (黄时鉴) and Gong Yingyan (龚缨晏). *Op. cit*., p. 193.
[431]Huang Shijian (黄时鉴) and Gong Yingyan (龚缨晏). *Op. cit*., p. 197.
[432]Raúl Porras Barrenechea. *El nombre del Perú*. Lima, Peru: Talleres Gráficos P. L. Villanueva, 1968, p. 8.
[433]Kim MacQuarrie. *The Last Days of the Incas*. New York, NY, USA: Simon & Schuster, 2008.
[434]As found in the electronic version on the internet at https://books.google.com.hk/books?id=QiY4AQAAMAAJ&q=%22balsam+of+peru%22+callao&pg=PA247&redir_esc=y#v=snippet&q=%22balsam%20of%20peru%22%20callao&f=false.
[435]Huang Shijian (黄时鉴) and Gong Yingyan (龚缨晏). *Op. cit*., p. 195.
[436]Anonymous. *La Prima Parte Dell'Istorie Del Peru Dove Si Tratta L'Ordine Delle Provincie Delle Citta Nuove In Quel Paese Edificate I Riti & Amp 1557 [Leather Bound]*. New South Wales, Australia: Generic Publishing, 2016.
[437]LLILAS Benson. "The geographical and historical dictionary of America and the West Indies [volume 1]." https://fromthepage.lib.utexas.edu, FromThePage, 2022, fromthepage.lib.utexas.edu/display/read_all_works?article_id=1778.
[438]*Ibid*.
[439]Carlos Aguirre and Charles F. Walker, eds. *The Lima Reader: History, Culture, Politics*. Durham, NC, USA: Duke University Press Books, 2017.
[440]John Paul Zronik. *Francisco Pizarro: Journeys Through Peru and South America*. New York, NY, USA: Crabtree Publishing Company, 2005.
[441]Huang Shijian (黄时鉴) and Gong Yingyan (龚缨晏). *Op. cit*., p. 197.
[442]Buddy Levy. *River of Darkness: Francisco Orellana and the Deadly First Voyage through the Amazon*. *Op. cit*.
[443]*Ibid*.
[444]"File:Amazonas in Brazil.svg." *Wikimedia Commons*, 12 Sept. 2022, upload.wikimedia.org/wikipedia/commons/thumb/3/30/Amazonas_in_Brazil.svg/2073px-Amazonas_in_Brazil.svg.png; author TUBS, date 18 Oct. 2011, titled "Location of Amazonas in Brazil."
[445]Huang Shijian (黄时鉴) and Gong Yingyan (龚缨晏). *Op. cit*., p. 189.
[446]Huang Shijian (黄时鉴) and Gong Yingyan (龚缨晏). *Op. cit*., p. 200.
[447]Editors. "City of Cuzco." https://whc.unesco.org/, UNESCO World Heritage Centre, 2022, whc.unesco.org/en/list/273/.
[448]"File:Peru location map.svg." *Wikipedia Commons*, 2 July 2022, upload.wikimedia.org/wikipedia/commons/thumb/7/7b/Peru_location_map.svg/1682px-Peru_location_map.svg.png.

[449]"File:Kigdomofcuscomap.JPG." *Wikimedia Commons*, 11 Oct. 2020, commons.wikimedia.org/wiki/File:Kigdomofcuscomap.JPG.
[450]Huang Shijian (黄时鉴) and Gong Yingyan (龚缨晏). *Op. cit*., p. 201.
[451]Huang Shijian (黄时鉴) and Gong Yingyan (龚缨晏). *Op. cit*., p. 198.
[452]Huang Shijian (黄时鉴) and Gong Yingyan (龚缨晏). *Op. cit*., p. 205.
[453]John L. Rector. *The History of Chile.* New York, NY, USA: St. Martin's Griffin, 2005.
[454]Huang Shijian (黄时鉴) and Gong Yingyan (龚缨晏). *Op. cit*., p. 188.
[455]Kris Lane. *Potosi: The Silver City That Changed the World.* Berkeley, CA, USA: University of California Press, 2021.
[456]Editora. "Pueblo de Tarapacá." www.monumentos.gob.d, Consejo de Monumentos Nacionales de Chile, 2022, monumentos.gob.cl/monumentos/zonas-tipicas/pueblo-tarapaca.
[457]Huang Shijian (黄时鉴) and Gong Yingyan (龚缨晏). *Op. cit*., p. 195.
[458]Anonymous. *Chile, A Country Study*. Ann Arbor, MI, USA: University of Michigan Library, 1982.
[459]Insight Guides. *Insight Guides Chile & Easter Islands.* London, UK: Insight Guides, 2019.
[460]Laurence Bergreen. *Over the Edge of the World Updated Edition: Magellan's Terrifying Circumnavigation of the Globe*. Boston, MA, USA: Mariner Books, 2019.
[461]Huang Shijian (黄时鉴) and Gong Yingyan (龚缨晏). *Op. cit*., p. 200.
[462]Huang Shijian (黄时鉴) and Gong Yingyan (龚缨晏). *Op. cit*., p. 187.
[463]Molly Aloian. *The Andes*. New York, NY, USA: Crabtree Publishing Company, 2011.
[464]"File:Atacama map.svg." *Wikimedia Commons*, 3 June 2022, upload.wikimedia.org/wikipedia/commons/thumb/5/51/Atacama_map.svg/1462px-Atacama_map.svg.png.
[465]"File:Andes.png." *Wikimedia Commons*, 1 Oct. 2020, upload.wikimedia.org/wikipedia/commons/9/9b/Andes.png.
[466]Richard Francaviglia. *Imagining the Atacama Desert: A Five-Hundred-Year Journey of Discovery*. Salt Lake City, UT, USA: University of Utah Press, 2018.
[467]Jimmy Langman, ed. *Patagonia in Photos*. Pucón, Chile: Patagon Journal, 2018.
[468]Chris Moss. *Patagonia: A Cultural History*. Oxford, UK.: Signal Books, 2008.
[469]Antonio Pigafetta. R. A. Skelton, trans. *Magellan's Voyage: A Narrative Account of the First Circumnavigation.* New Haven, CN, USA: Yale University Press, 1969; Raven Garvey. *Patagonian Prehistory: Human Ecology and Cultural Evolution in the Land of Giants*. Salt Lake City, UT, USA: University of Utah Press, 2021; as found in the electronic version on the internet at https://lib-dbserver.princeton.edu/visual_materials/maps/websites/pacific/magellan-strait/patgonian-giants.html.
[470]"File:Pat map.PNG." *Wikimedia Commons*, 12 Dec. 2021, upload.wikimedia.org/wikipedia/commons/a/a7/Pat_map.PNG.
[471]Anonymous. Anne Birrell, ed. and trans. *The Classic of Mountains and Seas*." *Op. cit.*
[472]*Ibid.*
[473]Antonio Pigafetta. R. A. Skelton, trans. *Magellan's Voyage: A Narrative Account of the First Circumnavigation. Op. cit*.; Raven Garvey. *Patagonian Prehistory: Human Ecology and Cultural Evolution in the Land of Giants*. *Op. cit.*
[474]Hendon Harris. *The Asiatic Fathers of America*, *op. cit*., p. 48.
[475]Huang Shijian (黄时鉴) and Gong Yingyan (龚缨晏). *Op. cit*., p. 200.
[476]Huang Shijian (黄时鉴) and Gong Yingyan (龚缨晏). *Op. cit*., p. 190.
[477]Huang Shijian (黄时鉴) and Gong Yingyan (龚缨晏). *Op. cit*., p. 194.
[478]Colin McEwan, *et al*., eds. *Patagonia: Natural History, Prehistory and Ethnography at the Uttermost End of the Earth.* Princeton, NJ, USA: Princeton University Press, 1997.

[479]Richard Hakluyt. *The Principal Navigations Voyages Traffiques and Discoveries of the English Nation (Cambridge Library Collection - Maritime Exploration) (Volume 12) Illustrated Edition*, Cambridge, UK: Cambridge University Press, 17 Apr. 2014, p. 86; as found in the electronic version on the internet at <https://books.google.com.hk/books?id=mV4HAwAAQBAJ&pg=PA86&lpg=PA86&dq=Baia+sin+fondo&source=bl&ots=fZATWonEzS&sig=ACfU3U08nn2X8znUjoR2Zl1VfLej4yfskQ&hl=zh-TW&sa=X&ved=2ahUKEwi5k5bI-9b1AhVzTWwGHYiPCO8Q6AF6BAg3EAM#v=onepage&q=Baia%20sin%20fondo&f=false>.
[480]Huang Shijian (黄时鉴) and Gong Yingyan (龚缨晏). *Op. cit.*, p. 193.
[481]"Category:Maps of rivers of South America." *Wikipedia Commons*, 17 Jan. 2018, commons.wikimedia.org/wiki/Category:Maps_of_rivers_of_South_America.
[482]Huang Shijian (黄时鉴) and Gong Yingyan (龚缨晏). *Op. cit.*, p. 201.
[483]Huang Shijian (黄时鉴) and Gong Yingyan (龚缨晏). *Op. cit.*, p. 191.
[484]Huang Shijian (黄时鉴) and Gong Yingyan (龚缨晏). *Op. cit.*, p. 189.
[485]Richard Hakluyt. *The Principal Navigations, Voyages and Discoveries of the English Nation (Works. Extra Series/Hakluyt Society)*. *Op. cit.*
[486]The Editors. "The Strait of Magellan: 250 Years of Maps (1520–1787)." *Op. cit.*
[487]*Ibid.*
[488]F. I. Isla, *et al*. "Gravel drift and wind effects on the macrotidal San Sebastian Bay, Tierra del Fuego, Argentina." *Marine Geology*, vol. 97, 1991, pp. 1–2. doi:10.1016/0025-3227(91)90027-2.
[489]The Editors. "The Strait of Magellan: 250 Years of Maps (1520–1787)." *Op. cit.*
[490]Huang Shijian (黄时鉴) and Gong Yingyan (龚缨晏). *Op. cit.*, p. 189.
[491]J. Lines. *Diario de Costa Rica (Spanish edition)*. May 12, 1940.
[492]Franz Lidz. "Tracking Balboa." *Op. cit.*; Angel De Altolaguirre Y Duvale. *Vasco Nunez De Balboa (1914)*. *Op. cit.*
[493]Huang Shijian (黄时鉴) and Gong Yingyan (龚缨晏). *Op. cit.*, p. 193.
[494]John A. Crow. *The Epic of Latin America*. Berkeley, CA, USA: University of California Press, 1992.
[495]Huang Shijian (黄时鉴) and Gong Yingyan (龚缨晏). *Op. cit.*, p. 206.
[496]Huang Shijian (黄时鉴) and Gong Yingyan (龚缨晏). *Op. cit.*, p. 193.
[497]Huang Shijian (黄时鉴) and Gong Yingyan (龚缨晏). *Op. cit.*, p. 192.
[498]Huang Shijian (黄时鉴) and Gong Yingyan (龚缨晏). *Op. cit.*, p. 204.
[499]Thomas C. Patterson. *The Inca Empire: The Formation and Disintegration of a Pre-Capitalist State*. *Op. cit.*
[500]BCC staff writer. "Bolivia profile — Timeline." www.bbc.com, BBC News Services, 2022, bbc.com/news/world-latin-america-18727510.
[501]Peter O. Koch. *The Spanish Conquest of the Inca Empire*. Jefferson, NC, USA: McFarland & Company, 2007.
[502]Huang Shijian (黄时鉴) and Gong Yingyan (龚缨晏). *Op. cit.*, p. 187.
[503]Chinca is shown on Google maps at https://www.google.com/maps/place/Chinca,+%E5%8E%84%E7%93%9C%E5%A4%9A/@0.7666884,-79.5920878,15z/data=!3m1!4b1!4m5!3m4!1s0x8fd4c512d64a50df:0x805e4ae8f97c65db!8m2!3d0.766667!4d-79.583333.
[504]Huang Shijian (黄时鉴) and Gong Yingyan (龚缨晏). *Op. cit.*, p. 206.
[505]Huang Shijian (黄时鉴) and Gong Yingyan (龚缨晏). *Op. cit.*, p. 201.
[506]Huang Shijian (黄时鉴) and Gong Yingyan (龚缨晏). *Op. cit.*, p. 195.
[507]"File:Riodelaplatabasinmap.png." *Wikimedia Commons*, 7 Aug. 2022, upload.wikimedia.org/wikipedia/commons/1/13/Riodelaplatabasinmap.png.
[508]Huang Shijian (黄时鉴) and Gong Yingyan (龚缨晏). *Op. cit.*, p. 189.

509"Peti, River." www.tum.de, Technical University of Munich, 2022, dahiti.dgfi.tum.de/en/4710/.
510Antonio Pigafetta and R. A. Skelton, trans. *Magellan's Voyage: A Narrative Account of the First Circumnavigation. Op. cit.*
511J. A. Williamson. *The Voyages of John and Sebastian Cabot*. London, UK: G. Bell & Sons, 1937.
512Chris D. Metcalfe, *et al. The Paraná River Basin: Managing Water Resources to Sustain Ecosystem Services*. Oxfordshire, UK: Routledge, 2020; as found in the electronic version on the internet at https://books.google.com.hk/books?id=OPDqDwAAQBAJ&pg=PT155&lpg=PT155&dq=Paran%C3%A1&source=bl&ots=syfRj17KPL&sig=ACfU3U1zefsaMPuqsQ7rfgXXgFhwg5CMbQ&hl=zh-TW&sa=X&ved=2ahUKEwiNpMPTzpb5AhXND94KHdfOAnI4HhDoAXoECAIQAw#v=onepage&q=Paran%C3%A1&f=false.
513Fabián Bognanni. "Some controversies regarding the arrival place of sebastian cabot in itati, corrientes (Argentina): A historical archaeological approach." www.researchgate.net, ResearchGate, 2022, researchgate.net/publication/321157629_Some_controversies_regarding_the_arrival_place_of_sebastian_cabot_in_itati_corrientes_Argentina_A_historical_archaeological_approach.
514As found in the electronic version on the internet at https://ia800502.us.archive.org/10/items/juandazdesol00medi/juandazdesol00medi.pdf.
515Leslie Bethell. *The Cambridge History of Latin America, Volume 1, Colonial Latin America*, Cambridge, UK: Cambridge University Press, 1984, p. 257.
516Rex A. Hudson and Sandra W. Meditz, eds. "Uruguay a country study." http://memory.loc.gov/, Library of Congress, 2022, memory.loc.gov/master/frd/frdcstdy/ur/uruguaycountryst00huds_0/uruguaycountryst00huds_0.pdf.
517*Ibid.*
518Huang Shijian (黄时鉴) and Gong Yingyan (龚缨晏). *Op. cit.*, p. 187.
519Huang Shijian (黄时鉴) and Gong Yingyan (龚缨晏). *Op. cit.*, p. 207.
520Huang Shijian (黄时鉴) and Gong Yingyan (龚缨晏). *Op. cit.*, p. 195.
521Robert Southey. "Chapter VII." *History of Brazil, Vol. 1*, Charleston, SC, USA: Nabu Press, 2020, p. 177.
522Huang Shijian (黄时鉴) and Gong Yingyan (龚缨晏). *Op. cit.*, p. 200.
523Huang Shijian (黄时鉴) and Gong Yingyan (龚缨晏). *Op. cit.*, p. 191.
524Huang Shijian (黄时鉴) and Gong Yingyan (龚缨晏). *Op. cit.*, p. 197.
525R. H. Major, ed. *Select Letters of Christopher Columbus, with Other Original Documents, Relating to His Four Voyages to the New World*. London, UK: Hakluyt Society, 2010; as found in the electronic version on the internet at https://books.google.com.hk/books?id=lAgkDwAAQBAJ&pg=PT361&dq=hammock++christopher+columbus&hl=en&newbks=1&newbks_redir=0&sa=X&redir_esc=y#v=onepage&q=hammock%20&f=false.
526Huang Shijian (黄时鉴) and Gong Yingyan (龚缨晏). *Op. cit.*, p. 202.
527Huang Shijian (黄时鉴) and Gong Yingyan (龚缨晏). *Op. cit.*, p. 201.
528Huang Shijian (黄时鉴) and Gong Yingyan (龚缨晏). *Op. cit.*, p. 188.
529Huang Shijian (黄时鉴) and Gong Yingyan (龚缨晏). *Op. cit.*, p. 198.
530As provided in the electronic version by the Brazilian Ministry Transport) on the internet at https://www.gifex.com/imapa/americas/md_alagoas_brazil.pdf.
531Huang Shijian (黄时鉴) and Gong Yingyan (龚缨晏). *Op. cit.*, p. 197.
532Huang Shijian (黄时鉴) and Gong Yingyan (龚缨晏). *Op. cit.*, p. 188.
533Leslie Bethell, ed. *Colonial Brazil*, Cambridge, UK: Cambridge University Press, 1987.
534*Ibid.*
535Boris Fausto. *A Concise History of Brazil*. Cambridge, UK: Cambridge University Press, 1999, p. 9.

536Laura de Mello e Souza. "O nome do Brasil." *Revista de Historia da USP*, n. 145, 2001.
537Hans Staden and Neil Whitehead. Michael Harbsmeier, trans. *Hans Staden's True History: An Account of Cannibal Captivity in Brazil*. Durham, NC, USA: Duke University Press Books, 2008.
538Maria A. Nilsson, *et al*. David Penny, ed. "Tracking Marsupial Evolution Using Archaic Genomic Retroposon Insertions." www.ncbi.nlm.nih.gov, National Library of Medicine, 2022, ncbi.nlm.nih.gov/pmc/articles/PMC2910653/.
539Huang Shijian (黄时鉴) and Gong Yingyan (龚缨晏). *Op. cit*., p. 197.
540M. Goulding, *et al. The Smithsonian Atlas of the Amazon*. Washington, DC, USA: Smithsonian Books, 2003.
541Samuel Eliot Morison. *The European Discovery of America, the Southern Voyages 1492–1616 (1974), the Northern Voyages A.D. 500–1600 (1971) (The European Discovery of America, Vol. 1 (1971) and Vol. 2 (1974))*. Oxford, UK: Oxford University Press, 1993.
542Buddy Levy. *River of Darkness: Francisco Orellana and the Deadly First Voyage through the Amazon. Op. cit.*
543Apollodorus. Robin Hard, trans. *The Library of Greek Mythology. Op. cit.*
544"File:Amazonriverbasin basemap.png." *Wikimedia Commons*, 3 Aug. 2022, upload.wikimedia.org/wikipedia/commons/0/02/Amazonriverbasin_basemap.png.
545Huang Shijian (黄时鉴) and Gong Yingyan (龚缨晏). *Op. cit*., p. 203.
546Huang Shijian (黄时鉴) and Gong Yingyan (龚缨晏). *Op. cit*., p. 199.
547White Mountain is shown on Google Maps at https://www.google.com/maps/search/White+Mountain,+Brazil/@43.9338511,-71.814866,10z/data=!3m1!4b1.
548Lilian del Castillo-Laborde. "Chapter 15: The Amazon River Basin: Features, history and legal framework." https://brill.com/, Water Law and Cooperation in the Euphrates-Tigris Region, 2022, brill.com/view/book/edcoll/9789004258358/B9789004258358_016.xml.
549*Ibid.*
550Huang Shijian (黄时鉴) and Gong Yingyan (龚缨晏). *Op. cit*., p. 187.
551Julian H. Steward. *Handbook of South American Indians, Vol. 5: The Comparative Ethnology of South American Indians*, London, UK: Forgotten Books, 2017, p. 275.
552Huang Shijian (黄时鉴) and Gong Yingyan (龚缨晏). *Op. cit*., p. 194.
553Huang Shijian (黄时鉴) and Gong Yingyan (龚缨晏). *Op. cit*., p. 202.
554Huang Shijian (黄时鉴) and Gong Yingyan (龚缨晏). *Op. cit*., p. 202.
555Huang Shijian (黄时鉴) and Gong Yingyan (龚缨晏). *Op. cit*., p. 206.
556Huang Shijian (黄时鉴) and Gong Yingyan (龚缨晏). *Op. cit*., p. 190.
557"Location of Paraná in Argentina." *Wikipedia*, 22 Mar. 2022, en.wikipedia.org/wiki/Paraná,_Entre_Ríos.
558The Paraná in Brazil is shown on Google Maps at https://www.google.com/maps/place/%E5%B7%B4%E8%A5%BF%E5%B7%B4%E6%8B%89%E6%8B%BF/@-24.595051,-53.5663727,7z/data=!3m1!4b1!4m5!3m4!1s0x94db0b9430b8629d:0xe893fd5063cef061!8m2!3d-25.2520888!4d-52.0215415.
559Huang Shijian (黄时鉴) and Gong Yingyan (龚缨晏). *Op. cit*., p. 204.
560Huang Shijian (黄时鉴) and Gong Yingyan (龚缨晏). *Op. cit*., p. 208.
561Huang Shijian (黄时鉴) and Gong Yingyan (龚缨晏). *Op. cit*., p. 189.
562Huang Shijian (黄时鉴) and Gong Yingyan (龚缨晏). *Op. cit*., p. 187.
563Huang Shijian (黄时鉴) and Gong Yingyan (龚缨晏). *Op. cit*., p. 191.
564Albert F. Kunze. "The Amazon: Has it been discovered?" http://www.smithsonianeducation.org, Smithsonian Education, 2022, smithsonianeducation.org/scitech/impacto/graphic/amazon/article_3.htm.

[565]*Ibid.*
[566]Huang Shijian (黄时鉴) and Gong Yingyan (龚缨晏). *Op. cit.*, p. 196.
[567]Riley Black. "More Than 30 Million Years Ago, Monkeys Rafted Across the Atlantic to South America." www.smithsonianmag.com, Smithsonian Museum, 2022, smithsonianmag.com/science-nature/monkeys-raft-across-atlantic-twice-180974637/.

Index

B

Printed in the USA
CPSIA information can be obtained
at www.ICGtesting.com
JSHW060018021123
51142JS00004B/7

9 789811 271083